Untangling General Relativity

The Intuitive Self-Study Guide

Simon Sherwood

WILEY

Registered Offices
John Wiley & Sons, Inc., 111 River Street, Hoboken, NJ 07030, USA
John Wiley & Sons Ltd, New Era House, 8 Oldlands Way, Bognor Regis, West Sussex, PO22 9NQ, UK

For details of our global editorial offices, customer services, and more information about Wiley products visit us at www.wiley.com.

The manufacturer's authorized representative according to the EU General Product Safety Regulation is Wiley-VCH GmbH, Boschstr. 12, 69469 Weinheim, Germany, e-mail: Product_Safety@wiley.com.

Library of Congress Cataloging-in-Publication Data:

Name: Sherwood, Simon. author
Title: Untangling general relativity : the intuitive self-study guide /
  Simon Sherwood.
Description: Hoboken, NJ : Wiley, [2026] | Includes bibliographical
  references and index.
Identifiers: LCCN 2025032176 | ISBN 9781394355853 (paperback) | ISBN
  9781394355877 (epdf) | ISBN 9781394355860 (epub)
Subjects: LCSH: General relativity (Physics)
Classification: LCC QC173.6 .S545 2026
LC record available at https://lccn.loc.gov/2025032176

Cover Design: Wiley
Cover Image: © MirageC/Getty Images

Printed and bound by CPI Group (UK) Ltd, Croydon, CR0 4YY

C9781394355853_280925

For Rachel, Saskia and Gabriella

Contents

Acknowledgements

This book wouldn't exist without the kind assistance of several experts in the field. Dr Matthew von Hippel of the Niels Bohr International Academy helped with the initial idea and overall structure. The contribution of Dr Andy Mummery of Oxford University was invaluable. He took time out from his study of black holes to edit and review the entire book. Dr Joachim Harnois-Deraps of Newcastle University also deserves special thanks. He applied his expertise in cosmology to edit in depth the third module, while Dr Andreea Font of Liverpool John Moores University provided an initial critique. Any errors that remain are mine and mine alone.

I want to thank Diane Pengelly for highlighting points in the draft text that might have been unclear for a lay reader. At Wiley, I am particularly grateful to my commissioning editor, Martin Preuss, for his faith and encouragement with both my first book and this one.

Finally, I must thank my family. My wife Rachel and my daughters Saskia and Gabriella have now survived five years of exposure to physics lectures, text-books and the arrival of various physicists at our house. In spite of Rachel's sparkling intellect, I still haven't managed to entice her into the world of physics, but her support and that of my children have underpinned all my efforts.

Introduction

Why Write This Book?

I remember my introduction to general relativity. It was an online course that bombarded my brain with a blizzard of mathematics: parallel transport, covariant derivatives, Riemann tensors, Ricci tensors, Christoffel symbols (known affectionately as Cripes-they're-awful symbols), the energy-momentum tensor, the Schwarzschild metric, the cosmological constant... I could go on and on. My gut reaction was to recoil in horror and wonder how any normal human being can understand the subject. I suspect this is the reaction of many students, which is why I have written this book.

The truth is that, contrary to common opinion, the basics of general relativity are not difficult to master. Why? Because if it is taught steadily step by step, it is very intuitive. Ponder the following. In 1907, Einstein reasoned that gravity might be described in terms of the curvature of spacetime. It took him over eight years to fully develop the theory of general relativity. Eight years during which he was one of the most admired physicists of his time. Imagine spending that much time on one theory. Why was Einstein absolutely convinced that he was right? Because the intuitive logic that underpins general relativity is so strong... as I hope to show you.

Why Read This Book?

Any of you who are familiar with my book on quantum mechanics, *Quantum Untangling*, will have a good sense of the maths required to make the most of this one. Knowledge equivalent to A-level maths (UK) or an AP calculus course (US) should be sufficient. For the most part you will require no more than basic calculus. The calculations in general relativity can be laborious and convoluted, but the steps involved are simple and repetitive.

The first module, *The Essentials*, gives an overview of the subject. The aim is to build a broad understanding of general relativity and the Einstein field equations. The second module, *Vacuum Curvature*, discusses the curvature of spacetime in a vacuum and how this leads to both the gravitational effects we see day to day and the bizarre mathematics of black holes and gravitational waves. The third module, *Cosmology*, examines large-scale spacetime curvature in the presence of energy and mass. This allows physicists to model the shape and development of the universe.

Some topics may appeal to you more than others. You may be itching to study the basic theory, a fanatic about black holes, or fascinated by the future of the universe. Whichever modules appeal, I will be a happy man if you learn something new that interests you or inspires you. Please leave a review, even if you don't read the whole book. It will encourage others... and I must confess that each positive review lifts my spirits.

Good luck with your studies! Let me end with a couple of thoughts that inspired me in writing both this and my earlier book, *Quantum Untangling*:

> *If you can't explain it simply, you don't understand it well enough.*
>
> –Attributed to Albert Einstein

> *The real problem in speech is not precise language. The problem is clear language.*
>
> –Richard Feynman

Module I

The Essentials

Chapter 1

Overview

Box 1.1 Einstein's apology to Newton

Newton, forgive me; you found the only way that, in your times, was just about possible for a man with the highest powers of thought and creativity. The concepts you created still guide us today in our thinking in physics, although we now know that they have to be replaced by others, more remote from the realm of immediate experience. Einstein in his autobiography

---

Let me start with a warning. This overview is a brief summary of what is to come. If you follow even a fraction of it, well done. And, if you don't, *do not be concerned*. You will have a much better grip on things when I step through them slowly. Think of this overview as a road map to check occasionally as you progress step by step towards a full understanding of the subject. Hopefully you will return and say: Ah yes... I get it now.

Einstein's theory of general relativity is radically different to Newton's (see Einstein's apology in Box 1.1). It has often been summarised with a statement along the lines of *matter tells space how to curve, space tells matter how to move*. As we dig into the subject, you will learn that it is not just *matter*, but every source of energy and momentum (together called energy-momentum in this book) that creates the curvature, and it is not just *space*, but the combination of space and time (together called spacetime) that is curved. In spite of these inaccuracies, there is merit in using the simple description above as a starting point.

Einstein's theory, which has copious evidence to back it, is that the presence of a source of energy-momentum distorts the surrounding spacetime, slowing clocks and altering lengths, relative to measurements made by an observer further from the source. For example, distant observers would see a clock near the source tick more slowly than their own. As a result, the straight-line paths, along which undisturbed objects might normally travel, are distorted and we observe the objects accelerate towards the source of energy-momentum. The energy-momentum distorts the spacetime (*matter tells space how to curve*) and the distorted spacetime changes the natural path of travel that objects follow (*space tells matter how to move*).

By the end of this book, I hope that readers will have a slightly more sophisticated view of the relationship between energy-momentum and the curvature of spacetime. Rather than thinking of energy-momentum *causing* a particular curvature of spacetime, I encourage you to think of energy-momentum and its related curvature of spacetime as being two sides of the same coin. Mess with the energy-momentum and the spacetime curvature changes. Mess with the spacetime curvature and the energy-momentum changes. Think less of the relationship as $A \rightarrow B$ and more of it as $A = B$. As you will see, this is what Einstein's equations say.

1.1 Einstein Field Equations

Let's jump straight to the punchline. Equation 1.1 shows the Einstein field equations.[1] We will be discussing them a lot, so I will abbreviate them as *the EFEs*. If you understand what they mean and why they make sense, then you

---

1 Technical point: I am ignoring the cosmological constant for now. It will be addressed later.

Untangling General Relativity: The Intuitive Self-Study Guide, First Edition. Simon Sherwood.
© 2026 John Wiley & Sons Ltd. Published 2026 by John Wiley & Sons Ltd.

can stop reading and throw this book away. On the other hand, these equations will look weird and unfamiliar to any readers who are new to the subject.

$$G_{\mu\nu} \;=\; \frac{8\pi G}{c^4}\, T_{\mu\nu}, \qquad \textit{Einstein field equations,}$$

$$\textit{Spacetime curvature} \Longleftrightarrow \textit{Energy-momentum.} \tag{1.1}$$

The EFEs show the relationship between the way that spacetime is curved ($G_{\mu\nu}$) and the amount of energy and momentum in that spacetime ($T_{\mu\nu}$). The two little symbols (μ and ν) in $G_{\mu\nu}$ and $T_{\mu\nu}$ indicate that they are components of *tensors*, which are sort of spiced-up vectors. The factor $\frac{8\pi G}{c^4}$ is just a constant, where c is the speed of light and G is the gravitational constant. It is the same G that appears in Isaac Newton's equation for the gravitational force of attraction between two objects of mass m and M at distance r apart (as shown in Equation 1.2, which shows the inverse square relationship between the force F and the separation r).

$$F = \frac{GMm}{r^2}, \qquad \textit{Newton's inverse square law of gravitational attraction.} \tag{1.2}$$

There is an important difference between tensors and vectors. I suspect (and hope) you know that a vector in space can be expressed in terms of three coordinates (x, y, z). For example, we might want to identify a particular point p in space. We set an origin at $(0, 0, 0)$ and can express p in terms of a vector that starts at the origin and ends at p. If our point p is a distance of 2 metres in the x direction, 4 metres in the y direction and 3 metres in the z direction, we can label p as $(x, y, z) = (2, 4, 3)$. In flat space, the distance from the origin to p is $\sqrt{x^2 + y^2 + z^2}$.

When you start dealing with curvature, you need a more complicated mathematical beast. You have to allow for the fact that the distance from the origin to p may not always be $\sqrt{x^2 + y^2 + z^2}$. It may vary depending on where you are. Indeed, the directions may not even be *orthogonal*. By this, I mean that the x, y and z axes may become bent and no longer be at right angles to each other. Suppose that the z axis tilts slightly in the x direction. A step away from the origin in the z direction also involves moving away from the origin in the x direction. This makes things more complicated. Simple vector notation of (x, y, z) isn't enough to fully describe the space or allow us to calculate how far p is from the origin. We need a (3×3) matrix to determine distances. This has three components describing a step in the x direction: (xx) to cover the basic movement along x, plus two further components (xy) and (xz) to describe what the step in the x direction means in terms of a change in y and z. Similarly, there are three components to describe a step in the y direction (yx, yy, yz) and three for a step in the z direction (zx, zy, zz). This (3×3) matrix is an example of a tensor.

The $G_{\mu\nu}$ in the EFEs is any component of the tensor $[G_{\mu\nu}]$, which is a (4×4) matrix. $T_{\mu\nu}$ is the corresponding component from the tensor $[T_{\mu\nu}]$, which also is a (4×4) matrix (I will use square brackets when referring to the whole matrix to distinguish from its individual components). They are (4×4) matrices because they contain information about spacetime, so they include time as a fourth dimension. Another important tensor $[g_{\mu\nu}]$ (with a small g) is called the *metric* that describes the actual curvature of spacetime itself, i.e. how a step along one axis (such as time t) relates to distance moved in each direction, potentially including movement in other directions (such as x). Equation 1.3 compares the structure of a spacetime vector with this sort of spacetime tensor. Each tensor contains 16 components compared with the 4 components of a spacetime vector.

$$\textit{Spacetime tensor:} \qquad \begin{bmatrix} tt & tx & ty & tz \\ xt & xx & xy & xz \\ yt & yx & yy & yz \\ zt & zx & zy & zz \end{bmatrix},$$

$$\textit{Spacetime vector:} \qquad (t, x, y, z). \tag{1.3}$$

We will be discussing tensors in detail later, so do not worry if they puzzle you. I just want you to understand that spacetime tensors have 16 components. This means that the EFEs as shown in Equation 1.1 are actually 16 equations, one for each component. For clarity, I have listed in Equation 1.4 three randomly chosen examples of these equations. Note that throughout this book I will use capitals for the indices of tensors simply because it is easier to read when written in the text. It turns out that 6 of the 16 equations are repeats because, in the case of the EFEs, components with reversed indices are equal e.g. $T_{XZ} = T_{ZX}$, leaving only 10 independent equations.

$$G_{TT} = \frac{8\pi G}{c^4} T_{TT}, \qquad G_{TX} = \frac{8\pi G}{c^4} T_{TX}, \qquad G_{YZ} = \frac{8\pi G}{c^4} T_{YZ}, \quad and\ so\ on... \qquad (1.4)$$

Aarrgghh! This sounds awful. 16 related equations, 10 of which are independent. Do not panic, it is simpler than it sounds. For much of this book we will be studying the implications of the formula for spacetime, which is a vacuum (whether or not it is near to a source of energy-momentum). In this case $[G_{\mu\nu}] = 0$, indicating that both sides of all 16 equations, including those in Equation 1.4, equal 0. An example might be the gravitational curvature of the spacetime in the vacuum surrounding a star or near a black hole.

I should add one technical point for clarity. The word *tensor* is also used for more complicated mathematical objects. For example, when discussing curvature, we will touch on the Riemann tensor, which for spacetime is a mammoth ($4 \times 4 \times 4 \times 4$) monster. In this book, when I use the term *tensor*, I will always be referring to (4×4) tensors such as $[G_{\mu\nu}]$, $[T_{\mu\nu}]$ and $[g_{\mu\nu}]$ (see structure in Equation 1.3), unless I specify otherwise.

The major advantage of using tensors is that the resulting equations, such as the EFEs, work in all coordinate systems. For example, in addition to working with Cartesian coordinates (t, x, y, z), they work in spherical coordinates (t, r, θ, ϕ) (don't worry if you are unfamiliar with spherical coordinates, which will be explained in detail later). Some randomly chosen examples from the EFEs in spherical coordinates are shown in Equation 1.5.

$$G_{TR} = \frac{8\pi G}{c^4} T_{TR}, \qquad G_{\theta\theta} = \frac{8\pi G}{c^4} T_{\theta\theta}, \qquad G_{R\phi} = \frac{8\pi G}{c^4} T_{R\phi}, \quad and\ so\ on... \qquad (1.5)$$

1.2 Gravity as Curved Spacetime

For a moment, I want to step back from the abyss of complicated equations and talk about Einstein's insight. In his theory of *special relativity*, he showed that time is different for an object when it is observed moving. The clock of a moving object ticks more slowly than that of a stationary object (*time dilation*). The spatial dimension also changes, contracting in the direction of motion (*length contraction*). The result is that time and space are no longer independent entities and combine into what is called *Minkowski spacetime*. We will look at the maths behind this in Chapter 2. For now, please just accept it as a proven truth.

The theory of special relativity doesn't address the presence of mass or gravitational acceleration, but Einstein would have wondered about the time dilation for an object accelerating in a gravitational field. Putting gravity to one side for a moment, let's consider the time dilation associated with any object as its velocity increases. To keep things simple, let's consider an accelerating clock. Special relativity tells us that the clock ticks progressively slower as it speeds up. The acceleration of the clock leads to a slowing in the passage of its time relative to the clock of a stationary observer whose velocity is not changing.

Einstein's genius was to turn the whole thing on its head. According to special relativity, the change in velocity associated with a particular acceleration profile leads to a change in the profile of the passage of time. What about the reverse? Might a change in the profile of the background passage of time lead to acceleration? Instead of the acceleration leading to a change in time... a change in time leads to acceleration. Hats off to Einstein. What a cool idea!

Why might Einstein think such an effect could be linked to gravitational acceleration? In addition to being a super-nifty concept, it helps explain something that was nagging in Einstein's mind. It explains why all objects experience the same gravitational acceleration. Starting from Newton's law in Equation 1.2 and using $F = ma$, gravitational acceleration a of an object of mass m can be expressed in terms of the attracting mass M and distance r. It does *not* depend on the mass of the object m. The acceleration a towards the attracting mass M is the same for all objects attracted by the mass. This is shown in Equation 1.6:

$$a = \frac{F}{m} \quad \Longrightarrow \quad a = \frac{GM}{r^2}, \qquad Gravitational\ acceleration\ (as\ in\ Equation\ 1.2). \qquad (1.6)$$

Initially, this may not strike you as odd, but it is. Compare it with the electromagnetic force. The acceleration of an electrically charged particle in an electromagnetic field depends on its mass and charge. It is not the same for all particles. Why is gravity so different? Well, if gravity is caused by a change in the background dimensions of spacetime around the attracting mass, that will have the same accelerating effect on everything. Mystery explained.

Before moving on, it is worth taking a moment to admire Einstein's brainwave. A change in the background passage of time can be described as introducing some curvature into spacetime. Why might such a change affect how particles move? This is difficult to visualise. The best I can do is the analogy of two bugs living on a two-dimensional surface. Let's say they set off walking in a parallel direction alongside each other. If the surface is flat, like a tabletop, they will walk on, always remaining the same distance apart. However, suppose the surface is curved like the surface of the earth (this is called *positive curvature*). Imagine the bugs start at the equator and take parallel paths north. The paths start parallel but will meet at the north pole. If the bugs did not know the earth is a sphere, they might mistake this for a force of attraction between them. Oooh! That sounds a bit like gravity.

As noted earlier, the relationship between coordinate units in any spacetime is described by its *metric tensor* [$g_{\mu\nu}$] (one of those tensor things). If you mess with the metric tensor it changes the *geodesics*, which are best described as the shortest possible line between two points and are the paths taken through spacetime by an undisturbed particle. In Newton's physics, the geodesics are straight lines that the particle follows at constant velocity when unaffected by gravity or any other accelerating force. Messing with the background passage of time changes the geodesics and can cause particles to accelerate away from the straight-line trajectory. We will use a crude model to show that time dilation in the metric can lead to a pattern of acceleration that matches that of gravity. Einstein's idea works.

1.3 The Equivalence Principle

Einstein had another brilliant brainwave called the *equivalence principle*, which he described as his *happiest thought* in developing the theory of general relativity. Einstein considered what you experience if you freefall in a gravitational field. The answer is *nothing*. A modern example is that an astronaut on the space station in orbit is weightless and feels no gravitational force. I should note that it is possible to detect gravitational acceleration over a larger scale because of what are called tidal forces that come from the difference in acceleration at different locations. However, Einstein concluded that theoretically, at a microscopic level, there is no way in a vacuum to tell the difference between freefall acceleration in a gravitational field and no acceleration at all. Wow! In Einstein's words: *Now it came to me... in a gravitational field (of small spatial extension) things behave as they do in a space free of gravitation.*

In Chapter 6, I will use the scenario of a field of dust particles to illustrate the implications. The dust particles start off stationary. The old Newtonian view is that they feel an attractive gravitational force that pulls them towards each other. In contrast, Einstein's equivalence principle says that the particles feel no force. Rather, the dust particles follow geodesic paths, and it is the geodesics that draw together in the presence of the dust field. In a vacuum their geodesic paths would remain stationary relative to each other (i.e. stay still while moving through time). However, in the dust field the *volume of space* separating the geodesics reduces over time. As a result, the dust particles gradually get closer to each other.

Einstein was looking for the relationship between the presence of energy-momentum and the change in separation of geodesics. From his theory of special relativity, he knew space and time cannot be treated separately. The solution needs to be based on spacetime. His intuitive leap was to conclude that the presence of energy-momentum reduces the *volume of spacetime* separating geodesics.

This leads us to what is called the *Ricci tensor*, a (4 × 4) tensor labelled [$R_{\mu\nu}$]. It is the curvature measure that quantifies any change to the overall spacetime separation between infinitely close adjacent parallel spacetime geodesics, i.e. any change in the volume of spacetime separating them. In the presence of energy-momentum, the Ricci tensor is non-zero, and the spacetime volume separating parallel geodesics changes. If the region of spacetime contains no energy-momentum, its Ricci tensor is zero, which I label as *Ricci-0*.

In addition to helping us understand how energy-momentum distorts spacetime, this places major limitations on the way that spacetime can curve in a vacuum. We know such curvature occurs as we see the gravitational effect of a massive star on objects in the vacuum surrounding it. But Ricci-0 curvature means there is no change in the overall spacetime volume separation between infinitely close adjacent parallel geodesics; the curvature does not change the volume of spacetime separating them. The way this works is far from obvious. In fact, you will learn that this sort of spacetime curvature is only possible in a universe with four or more dimensions.

1.4   Working Out the Details

Einstein's biggest headache, which cost him several years' work, was in quantifying the exact relationship between the energy-momentum in a region and the curvature it induces. Let me introduce you to the (4 × 4) energy-momentum tensor $[T_{\mu\nu}]$, whose components appear on the right of the EFEs (see Equation 1.1).

Why a (4 × 4) tensor? As an example, consider the rest mass energy of this book. It cannot *be in* a small fixed volume of spacetime. It must flow through it. If stationary in space, it will flow through the time dimension t into the same volume of space a moment later. If it is also moving through space, the energy can also flow out through the x, y or z dimensions. To track the flow of energy we need four entries, one for the flow through each dimension. Plus four more for the flow each of x-momentum, y-momentum and z-momentum. That is 16 entries. Don't worry if this puzzles you. All will be revealed in Chapter 8.

Einstein figured out the structure of the EFEs as shown in Equation 1.7. The Einstein tensor $[G_{\mu\nu}]$ is a somewhat complicated mix involving the Ricci tensor $[R_{\mu\nu}]$, which measures the change in spacetime separation of parallel geodesics, and the Ricci scalar R, which is a number calculated from the components of the Ricci tensor. Working out the values of $[R_{\mu\nu}]$ and R is not easy. Each involves an intricate calculation using derivatives of the components of the metric tensor $[g_{\mu\nu}]$. This mathematical headache is one of the challenges of general relativity.

Returning to the topic of Equation 1.7, you may wonder why the relationship is so complicated. It turns out that many of the 16 components of the (4 × 4) Ricci tensor $[R_{\mu\nu}]$ are interrelated, so change one and it changes others. In Chapters 9 and 10, I will explain all this and how to derive the value of the constant $\frac{8\pi G}{c^4}$, which appears in the EFEs and typically is labelled k.

$$G_{\mu\nu} = R_{\mu\nu} - \frac{1}{2} R g_{\mu\nu} = k T_{\mu\nu}, \qquad\qquad k = \frac{8\pi G}{c^4}. \qquad\qquad (1.7)$$

Stepping back from the equations, gravitational curvature can be explained as follows. The presence of energy-momentum causes Ricci curvature, which changes the separation between parallel geodesics. This distorts the spacetime metric of the region containing the energy-momentum, dilating time (clocks run slower) and contracting spatial lengths. It also distorts adjacent regions of spacetime but, in the case of a vacuum, it is in a subtly different way that gradually diminishes at greater distance from the source of energy-momentum.

One obvious question is *why* energy-momentum distorts the spacetime metric. This I cannot answer. Perhaps such questions are best left to philosophers. However, I will show you that the spacetime metric $[g_{\mu\nu}]$ and energy-momentum are not independent entities. In fact, the relationship between mass, energy and momentum can be derived directly from the spacetime metric (this appears in Section 2.8). Later in Chapter 14 of the book, I will show you that the conservation laws defining energy and momentum come directly from the spacetime metric $[g_{\mu\nu}]$. So, while I cannot answer the fundamental question, at least I can show you that energy-momentum and the structure of the spacetime metric are inextricably intertwined.

1.5   Gimme, Gimme, Gimme... Some Hard Evidence

Before you get too excited about Einstein's work, remember that he was trying to find field equations to extend Newton's laws, so a decent match is not enough to start the champagne flowing. However, the good news is that the match is not perfect. In stronger gravitational fields, Einstein's results begin to differ from Newton's inverse square law. One example is that Einstein's theory explains a small anomaly in the orbit of Mercury. It also predicts light will bend in a gravitational field twice as much as you would calculate based on Newton's theory. And what do observations tell us? It is the EFEs that give us the right results. So, yes, you can pop those champagne corks.

Nowadays, we have such accurate clocks that there is abundant proof of the time dilation experienced in a gravitational field. In fact, when designing GPS satellites, the difference in the passage of time at different altitudes has to be adjusted for. Without that, clock errors would quickly render our GPS satellite systems useless.

1.6   The Cosmological Constant

By the end of the first module of the book, we will have covered the basics of Einstein's theory and I will introduce you to his final tweak: the cosmological constant, generally labelled Λ. It appears as an additional term on the left side of the EFEs as shown in Equation 1.8.

$$G_{\mu\nu} - \Lambda g_{\mu\nu} \;=\; \frac{8\pi G}{c^4}\, T_{\mu\nu}, \qquad \textit{The EFEs with the cosmological constant,}$$
$$\textit{Spacetime curvature} \iff \textit{Energy-momentum.}$$

(1.8)

Einstein added the cosmological constant because without it, his EFEs predicted an expanding or collapsing universe, whereas he believed at the time that the universe was static. When the astronomer Edwin Hubble showed from astronomical observations that the universe is expanding, Einstein abandoned Λ, reportedly describing it as: *my biggest blunder.* However, there is now strong evidence that a positive Λ term really does exist.

Don't try to figure out the maths at this stage. I will step through that later. Focus on what a positive Λ term means. Adding it to the EFEs means that the spacetime curvature $[G_{\mu\nu}]$ is not zero when $[T_{\mu\nu}]$ is zero (a vacuum). You could say that spacetime *always* has some background Ricci curvature. This indicates the presence of some sort of background energy that physicists choose to describe as *dark energy* or *vacuum energy*, even though they have no real idea what it is. Personally, I prefer the term vacuum energy and will tend to refer to it as such, but dark energy is more compelling as a brand name. Newspapers love it. The name *dark energy* has stuck. Hard as I may try, this is not a battle I am going to win!

1.7   Vacuum Curvature

In the second module, we will examine the vacuum solutions to the EFEs. They describe the gravitational curvature of spacetime for a vacuum that is close to a mass. The spacetime is influenced by the nearby mass, so it has curvature, but the vacuum itself has no energy-momentum present (any effect from the cosmological constant is small enough to ignore). The vacuum solutions are important for two reasons. First, they are a good model for the gravity around us, both on earth and in the solar system. Second, we actually have found some solutions! The second point may seem trite, but the EFEs are complicated non-linear equations that, except in the simplest scenarios, are impossible to solve with pen and paper.

As discussed in Section 1.3, for any curvature of spacetime that is a vacuum, the Ricci tensor is zero: $[R_{\mu\nu}] = 0$, i.e. every $R_{\mu\nu}$ component in it is zero. I am guessing that some readers are unimpressed. The curvature of spacetime in a vacuum is Ricci-0. So what? So... a lot. This is a big deal. We will calculate what it means for the gravitational curvature of a vacuum near a simple mass (by simple, I mean it is spherically symmetrical, not spinning and not electrically charged). It turns out there is only *one* possible form of curvature that gives $[R_{\mu\nu}] = 0$. This is called the *Schwarzschild metric*, which we will derive and discuss in Chapter 12.

Drop in some values of constants to make Schwarzschild's solution match the strength of gravity as we experience it and the result is, to a very good degree of approximation, Newton's equations for gravity, including his inverse square law of attraction. But the result comes with mathematical consequences. There is little flexibility in how the curvature in the metric can work in a vacuum. We have to take the Schwarzschild metric (and related solutions for rotating and charged masses) seriously, even where the gravitational force is extreme. We will spend several chapters discussing this metric and the Kerr metric (its equivalent for a rotating mass), especially their weird implications for black holes with *event horizons*, where a distant observer would see approaching clocks slow down. The closer the clock to the event horizon, the slower it ticks. That is right. From the perspective of a distant observer, if the clock ever were to reach the event horizon of a black hole, it would stop!

General relativity's prediction of black holes was initially controversial, but modern observatories allow us to observe them. For example, about 26,000 light years away at the heart of our Milky Way Galaxy sits Sagittarius A^*, a black hole with a mass some four million times that of the sun (see Box 1.2 if you need a reminder that a light year is a measure of distance, not time).

We will also discuss gravitational waves. These are small perturbations in the curvature of spacetime that can propagate through the vacuum. Being massless, they travel at the speed of light. We can use the maths behind

Ricci-0 spacetime to show that they are sinusoidal waves with two possible polarisations. You may have heard that gravitational waves were first directly detected in 2015 by the combined efforts of the Laser Interferometer Gravitational-Wave Observatory (LIGO) and the Virgo Interferometer. These ultrasensitive observatories picked up the aftermath of a collision between two black holes. It released such an intense burst of gravitational waves that it could be detected here on earth, over a billion light years away.

Box 1.2   What is a light year?

A light year is a measure of *distance*, not time. It is the distance that light travels in a year and is equivalent to about 9,500 billion kilometres. For context it is over 60,000 times further than the distance to the sun, which is about 8 light minutes away.

1.8   Cosmology

We shall not ignore what the EFEs tell us about spacetime when it contains energy-momentum. His equations are an essential tool in studying the origin and development of the universe as a whole—the study of *cosmology*.

1.8.1   The Expanding Universe

Let's return to Edwin Hubble's discovery. The universe is expanding in the sense that distant galaxies are moving away from us: on average, the further the galaxy is away, the faster its receding velocity. The relationship is summarised in Equation 1.9. H generally is called the Hubble constant, which is a bit misleading because while H is a constant in terms of its current value, nobody is suggesting it has always had the same value in the past or will have in the future. I prefer to call it the Hubble *parameter* $H(t)$ to reflect that it can vary over time. Anyway, tracking back in time we conclude (with good supporting evidence) that the universe expanded from a much denser hotter state about 14 billion years ago: the *Big Bang*. Although there is much similarity in the maths, the theories of Newton and Einstein provide very different descriptions of the subsequent expansion.

$$H(t) = \frac{v}{r}, \qquad \textit{H is the Hubble parameter, v = velocity and r = distance away.} \qquad (1.9)$$

Newton realised that his law of gravity doesn't naturally accommodate a static universe. It must either be expanding or contracting. This is clear from his Equation 1.2 that shows all objects accelerate towards each other. Let me explain. Imagine throwing a ball up in the air. With a normal throw, it rises to a high point and then falls back down. Alternatively, if you throw it fast enough (very fast indeed), its speed will be higher than *escape velocity* giving it enough energy to overcome the earth's gravitational pull and fly off into the distance. Thus, the ball will either move up and then down or, if above escape velocity, move up and up. But, however you throw the ball, it cannot end up sitting static at a fixed distance above you. There is no stable static scenario. However, Newton was religious and chose to ignore the possibility that the universe is expanding or contracting (he concluded it is held static by a *voluntary agent*, God).

How might Newton describe the expansion that Hubble observed? Perhaps he would hypothesise that the galaxies were accelerated by the Big Bang in a massive explosion that flung them out through space, while constantly feeling a dragging force of gravity pulling them back together.

Einstein's explanation is very different from Newton's. There is no initial accelerating explosion. Nor do galaxies or other objects feel an ongoing decelerating force. It is the very fabric of time and space that is developing. Each galaxy can consider itself stationary (or moving steadily without acceleration). The motion of galaxies relative to each other is driven by the curvature of spacetime that alters the geodesic paths followed by undisturbed objects. In layman's language the expansion, which we call the Big Bang, is not galaxies being flung out *through* space. It is the expansion *of* space carrying galaxies along for the ride.

In the third module of the book, we will model the development of the universe. In the study of cosmology, physicists make the broad assumption that the universe on an extremely large scale is the same viewed from any location (*homogeneous*) and in any direction (*isotropic*). It is treated mathematically as having a constant energy-density throughout. It may help you to think of all the galaxies, on the grandest of scales, being like tiny flecks of dust in a cloud scattered smoothly across the universe. These assumptions are described as the *cosmological principle*. We will use this principle to derive the *Friedmann equations*. These equations quantify how the rate of expansion of the universe changes with the presence of various types of energy.

1.8.2 An Accelerating Expansion

The next step is to examine observational data. This brings some big surprises. The way that galaxies rotate reveals the gravitational pull of a large amount of what is called *dark matter*. Measurements show this represents about 85% of the total matter in the universe. We know it is not electrically charged and, except for its gravitational effect, interacts only weakly with other matter. Otherwise, we would have detected it long ago. But we still don't know what it is.

Another surprise is the importance of *dark (vacuum) energy*. In order to reconcile the Friedmann equations with observational evidence, you have to incorporate a positive cosmological constant: $\Lambda > 0$. The presence of what a layman might call normal mass/energy (matter and radiation) has a braking effect on the expansion of the universe. This is exactly what you would expect from classical Newtonian gravity pulling objects together. However, recent research indicates that the rate of expansion of the universe is *accelerating*. This requires the presence of dark (vacuum) energy, which appears to be the dominant component of the energy mix of the universe today.

How can this vacuum energy be so important without us noticing it day to day? If we quantify it in terms of mass (using $E = mc^2$), it would be the equivalent of only a few hydrogen atoms per cubic metre of space. To put things in context, the air around you is about $1 \, \text{kg m}^{-3}$, so the air contains over $10^{26} \times$ more energy per cubic metre. However, there is a huge amount of vacuum relative to the presence of galaxies and stars, so, even though negligible except on cosmological scales, vacuum energy is believed to represent well over half of the total energy in the universe. Based on current estimates it is about 70% of it. Wow!

Let me offer comfort to those who fear the spatial expansion of the universe may explode their heads. The electromagnetic interaction between electrons and protons dictates the stability and size of the objects around us. It (along with the strong force) also dictates the structure and size of atoms. These forces are way way stronger than the effect of this miniscule spatial expansion. The same is true of the gravitational force that balances the position of planets in our solar system and the movement of stars in a galaxy. Expansion of the universe has no effect on the size of an atom, the solar system, a galaxy—or your head—because it is immediately offset by these much stronger balancing forces. But it will affect the relative position of galaxies, which typically are millions of light years apart and will grow evermore distant from each other as time passes.

What does this mean for the universe? It appears that it will continue to expand faster and faster until all the mass/radiation energy is so widespread that its *density* becomes negligible (this is known as *heat death of the universe*). The galaxies we see in the night sky will gradually recede out of view, leaving us surrounded by a growing sea of darkness.

1.8.3 The Big Picture

The final topic discussed in cosmology will be *inflation*. The current consensus is that there was a period of very rapid expansion of the universe early in its development, prior to it being flooded with radiation and particles. This completes the cosmological story of the Big Bang as currently understood: inflation, then the universe as a hot dense plasma of radiation and particles, then expansion and cooling, then the formation of galaxies; and, finally, in the distant future, heat death of the universe.

However, there are many unanswered questions in cosmology. For example, it is a bit embarrassing to know so little about dark matter and dark (vacuum) energy, which between them represent about 95% of the total energy in the universe.

I also must mention the wider challenge of tying together general relativity with quantum mechanics. General relativity has proved itself in test after test as an accurate description of gravity. The same is true of quantum mechanics for particle interactions involving the electromagnetic, weak and strong forces. Yet the two have little in common. In the final chapter of the book I will touch on a few of the better-known efforts to unite them, including *string theory* and *loop quantum gravity*. But the frustrating truth is that after many decades of work, the solution still eludes us.

1.9 The Field Equations in Full Form

So there we have it. I have assaulted your brains with an abundance of concepts and maths. Box 1.3 summarises the EFEs including the rather simple-looking expression for curvature in a vacuum. I have included details on the conventions I use. This will be meaningless to you at the moment but will be useful for reference in the future. And let me repeat my warning from the start of this chapter. This overview is a map of what is to come. You are *not* expected to understand the details at this stage. All will become clear. Let's get started.

Box 1.3   Einstein Field Equations (EFEs)

$$\text{\textit{General Form:}} \quad G_{\mu\nu} = \frac{8\pi G}{c^4} T_{\mu\nu}, \qquad \text{\textit{where}} \quad G_{\mu\nu} = R_{\mu\nu} - \frac{1}{2} R g_{\mu\nu}, \tag{1.10}$$

$$\text{\textit{Vacuum Form:}} \quad R_{\mu\nu} = 0, \qquad \text{\textit{because}} \quad T_{\mu\nu} = 0, \tag{1.11}$$

$$\text{\textit{Full Form with } } \Lambda: \quad G_{\mu\nu} - \Lambda g_{\mu\nu} = \frac{8\pi G}{c^4} T_{\mu\nu}. \tag{1.12}$$

$G_{\mu\nu}$ is any component of the (4×4) Einstein tensor $[G_{\mu\nu}]$. Similarly, $T_{\mu\nu}$, $R_{\mu\nu}$ and $g_{\mu\nu}$ are the corresponding components of the energy-momentum tensor $[T_{\mu\nu}]$, the Ricci tensor $[R_{\mu\nu}]$ and the spacetime metric tensor $[g_{\mu\nu}]$. R is the Ricci scalar (a number).
   Λ is the cosmological constant. G is Newton's gravitational constant. c is the speed of light.

Conventions: Metric: $< + - - - >$ Ricci tensor: $R_{AB} = R^C_{\ ACB}$

Chapter 2

Special Relativity

Any readers who know my book on quantum mechanics will find much of this chapter familiar. It is an abridged version of the first three chapters of *Quantum Untangling*. This is because the same maths of special relativity sits behind both general relativity and quantum mechanics... making it all the more frustrating that nobody has yet developed a convincing theory of quantum gravity that combines the two.

If you find this chapter taxing, you may want to take a look at the more detailed material in *Quantum Untangling* if a copy is available, or work through the excellent online lecture series by Dr Larry Lagerstrom at Stanford University.[1]

2.1 Relativity

Einstein's *theory of special relativity* rests on two postulates that apply for all observers, however fast they are moving relative to each other:

- The speed of light is constant for all observers.
- The laws of physics are the same for all observers.

In 1865, some 40 years before Einstein's breakthroughs, the physicist James Maxwell showed that light is an electromagnetic wave. A fluctuating electric field generates a fluctuating magnetic field that in turn generates a fluctuating electric field... resulting in a self-sustaining wave. Maxwell's equations showed that light must move at a specific speed ($c \approx 300{,}000$ km s^{-1}). But this specific speed is a speed relative to what? Most scientists assumed that light waves move through some underlying substance (that they called *ether*). Einstein had an altogether different view.

His *theory of special relativity* requires no special substance or ether. The speed of light is constant for all observers. The laws of physics are the same for all observers. This may not sound revolutionary, but it is. This indeed is how the universe works and, as we will show, a direct consequence is that for an object moving relative to you, you will see:

1. Its time change, ageing more slowly: *time dilation*.
2. Its length contract in the direction of motion: *length contraction*.
3. Its clocks fall out of sync. Imagine two clocks are synchronised on a stationary rocket, one at the front and one at the back. An observer seeing the rocket pass at high speed sees them out of sync. The front clock shows a later time than the back clock: *leading clocks lag*.

Special relativity throws up many apparent paradoxes. For example, if I observe a rocket passing me at high speed, I will see the rocket's clock running more slowly than mine. But from the perspective of an observer on the rocket, it is me that is moving and he/she sees my clock running more slowly. This apparent paradox focuses on time dilation. Things don't appear to add up when you think about any one of the three distortions on a stand-alone basis. The solution to this (and other apparent paradoxes) depends on combining all three distortions.

---

1 https://online.stanford.edu/courses/som-y0009-understanding-einstein-special-theory-relativity (as of February 2025).

Untangling General Relativity: The Intuitive Self-Study Guide, First Edition. Simon Sherwood.
© 2026 John Wiley & Sons Ltd. Published 2026 by John Wiley & Sons Ltd.

Before digging further into the maths of special relativity, I want to reflect on Einstein's overarching philosophy of *relativity*. Much of his work was driven by his philosophical stance. In his own words: *I believe in intuition and inspiration... at times I feel certain I am right while not knowing the reason.* He built special relativity on the basis that, in terms of the basic laws of physics, nothing changes to an observer when moving. In physics, an unaccelerated object is said to be in an *inertial frame*. In layman's language, the observer feels nothing in an inertial frame in a vacuum, whatever speed we might see them moving at. Unaccelerated motion is best described as motion relative to another object.

It was a natural step for Einstein to extend this philosophy to include acceleration in a gravitational field. Again, in his own words: *I was sitting in a chair in the patent office at Bern when all of sudden a thought occurred to me: if a person falls freely he will not feel his own weight. I was startled. This simple thought made a deep impression on me. It impelled me toward a theory of gravitation.* In *special* relativity, there is no *local* way to detect unaccelerated motion in a vacuum. In *general* relativity, there is no *local* way to detect accelerated motion due to a gravitational field in a vacuum (freefall). Einstein sensed a similarity. But I am getting ahead of things. Let's forget all the talk of gravity and return to the topic of special relativity.

2.2 The Speed of Light Is Constant: So What?

So what? So a lot! To get to grips with special relativity, it is essential that you understand why the constant speed of light messes with the classical laws of physics. Let's start with a simple example. Imagine that you are in the park throwing a ball for your dog Pooch to chase. You want to know how fast Pooch and the ball move, so you mark out evenly-spaced stripes on the ground. You throw the ball, which Pooch chases. One second later you take a photo, which allows you to see exactly how far the ball and Pooch have moved.

Suppose your photo shows that after 1 second, the distance between you and Pooch is 5 metres and the distance between Pooch and the ball is 5 metres. This is illustrated at the top of Figure 2.1. From your perspective, you conclude that the speed of Pooch relative to you is $5 \, \mathrm{m \, s^{-1}}$ and that the speed of the ball relative to Pooch is also $5 \, \mathrm{m \, s^{-1}}$. Classically you would conclude that Pooch also sees the ball moving away from him (for he is a male dog) at $5 \, \mathrm{m \, s^{-1}}$. I am sorry to tell you that you would be wrong.

The good news is that you would only be very very slightly wrong at these speeds. However, at higher speeds the discrepancy becomes significant. Let me illustrate by introducing you to Turbo-Pooch, an imaginary dog that can travel at half the speed of light. You repeat your experiment, but this time you flash a pulse of light for Turbo-Pooch to chase. Turbo-Pooch runs at half the speed of light (0.5*c*). Again, you take a photo after 1 second. At that moment, there are 150,000 kilometres between you and Turbo-Pooch and 150,000 kilometres between Turbo-Pooch and the light pulse. This is illustrated at the bottom of Figure 2.1.

You conclude that Turbo-Pooch moved 150,000 kilometres in 1 second, which is $\frac{c}{2}$, and that the light pulse moved 300,000 kilometres in 1 second, which is *c*. That all makes sense, but what about the speed of the light pulse relative to Turbo-Pooch? The distance between them is 150,000 kilometres so it must be $\frac{c}{2}$ *from your perspective*. All of this is true, but the final three words in the last sentence are crucial, so I will repeat them: *from your perspective*.

Let's switch now and examine what happens from Turbo-Pooch's point of view. In physics, this is called *changing reference frame*. If Turbo-Pooch measures how fast the light moves away from him, it *must always* be *c*. That is Einstein's law. Therefore, Turbo-Pooch, making the same measurement as you, sees the light pulse 300,000 kilometres away from him after 1 second. This is very different from the 150,000 kilometres after 1 second that you measure.

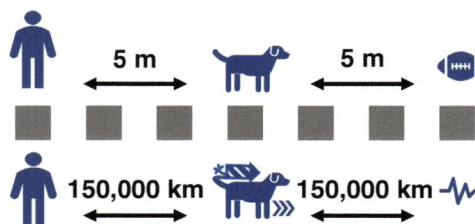

Figure 2.1 The problem with a constant speed of light.

I want to be clear that this conundrum is unavoidable if the speed of light is constant. Let me repeat it. If, from your perspective, Turbo-Pooch is moving in the direction of the light pulse at $\frac{c}{2}$, you will measure the distance between dog and light pulse as 150,000 kilometres after 1 second on your clock, but Turbo-Pooch must measure it as 300,000 kilometres after 1 second on his clock, because that is the speed of light.

What can possibly explain these two very different views of the same thing? The reason is that your and Turbo-Pooch's measurements of time and space differ. In terms of time, your two clocks run at different speeds. The two of you disagree on how long a second is. In terms of space, you will measure distances differently. The two of you will disagree on the distance between the stripes on the ground. Let's take a look at the mathematics of these distortions.

2.3 The Invariant Interval Equation

The speed of light is constant for all observers. As a result, time and space are distorted when viewed by a moving observer. We can derive the underlying mathematics of this distortion by considering how a model clock looks if viewed stationary versus in motion. Our clock is going to be a light clock. This simplifies the mathematics because we know the speed of light is the same for all observers. So the light in a stationary clock travels at c and, if you observe that clock moving relative to you at velocity v, you will still see the light in it travel at c. Simple!

So, let's consider clocks that work by emitting a flash of light up and down. The clocks tick every time the light flash arrives. An observer measures the time between each tick for a stationary clock. In general relativity, the time measured by an observer who is stationary relative to the clock is called *proper time* and typically is labelled τ (pronounced *tau*). So, the stationary clock registers the observer's proper time. A short moment on the stationary clock is $\Delta\tau$. This is shown as (a) in Figure 2.2.

If the observer sees an identical clock fly by at high speed, then the clock physically moves along between each tick. Tracking the path of the *light* between ticks, it is slightly longer than for a stationary clock. This is shown as (b) in Figure 2.2. As the light travels a longer path in the moving clock and the speed of light is constant, the time between ticks for a moving clock must be longer for the observer. Comparing the time between ticks on each clock, the moving clock ticks more slowly. We will call the time between ticks of the moving clock Δt.

The relationship between the tick of a stationary clock ($\Delta\tau$) and a moving clock (Δt) can be determined from the triangle in the bottom right-hand corner of Figure 2.2. The distance travelled by the light in the stationary clock (a) is the time taken multiplied by the velocity of light: $c\Delta\tau$. By a similar calculation, the light in the moving clock (b) travels $c\Delta t$. The clock itself moves Δx distance over the same time. Simple Pythagoras gives us: $(c\Delta\tau)^2 + (\Delta x)^2 = (c\Delta t)^2$. This is called the invariant interval equation and, now that we have derived it, we have the maths of special relativity at our mercy. It is more typically written as in Equation 2.1 below (we use notation such as dx to indicate an infinitesimally small movement through time or space).

$$c^2\,d\tau^2 = c^2\,dt^2 - dx^2. \tag{2.1}$$

This represents an enormous shift in the maths of space and time. In Newtonian physics, time was assumed to be *absolute*. We learn now that the tick of a clock (i.e. the passage of time) changes for an object that is moving

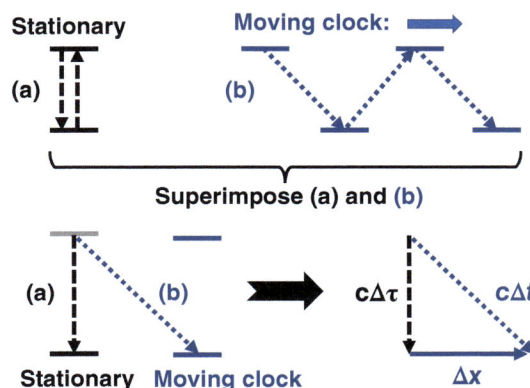

Figure 2.2 Determining the distortion of time and space.

relative to another. We have a new understanding of space and time. Physicists refer to this as *Minkoswki spacetime.* Equation 2.1 is shown using only one spatial dimension. The motion of the clock can be extended simply to three spatial dimensions. The important thing is the total distance that the clock travels during the interval, so the full invariant interval equation becomes:

$$c^2\,d\tau^2 = c^2\,dt^2 - dx^2 - dy^2 - dz^2. \tag{2.2}$$

The proper time $d\tau$ of any object, clock or person, is invariant in the sense that everyone can agree on it. Consider a clock that is assessed by multiple observers travelling past it at different speeds. Each will *measure* it to have a different length of tick dt. But they can all *calculate* the invariant interval $d\tau$, which is the length of the clock tick as measured by an observer who is stationary relative to the clock. This sort of quantity is described as *Lorentz invariant* because, in special relativity, the mathematical calculation used to switch between reference frames of different observers is called a Lorentz transform. Don't worry about the details. Just remember that all observers will agree on a *Lorentz invariant* quantity. Examples are the speed of light c, rest mass m and the proper time of any clock or observer $d\tau$.

2.4 Time Dilation Quantified

We have seen that time dilates for a moving object in the sense that a second of time that we observe on a moving clock will be longer than a second observed on the same clock when it is stationary. If we observe a moving clock, or man, or woman, or anything, we will see each second of its/his/her life last a little longer than a second of our life. Moving things age more slowly.

How does the speed of motion change the observed time, dt versus our proper time, $d\tau$? We can derive this from the invariant interval Equation 2.1, shown again below for simplicity in one spatial dimension (i.e. the clock is moving in the x direction). With some simple maths, we can derive the relationship between dt and $d\tau$ as shown below in Equation 2.4. Remember that $\frac{dx}{dt}$ is velocity v.

$$c^2\,d\tau^2 = c^2\,dt^2 - dx^2, \tag{2.3}$$

$$\frac{c^2\,d\tau^2}{c^2\,dt^2} = 1 - \frac{dx^2}{c^2\,dt^2},$$

$$\left(\frac{d\tau}{dt}\right)^2 = 1 - \frac{v^2}{c^2},$$

$$\left(\frac{dt}{d\tau}\right)^2 = \frac{1}{1 - \frac{v^2}{c^2}},$$

$$dt = \sqrt{\frac{1}{1 - \frac{v^2}{c^2}}}\,d\tau,$$

$$dt_{(moving)} = \gamma d\tau_{(static)}, \qquad \textit{where time dilation is}: \ \ \gamma = \sqrt{\frac{1}{1 - \frac{v^2}{c^2}}}. \tag{2.4}$$

From now on, we will use the generally accepted physics notation γ (pronounced gamma) for the time dilation effect, which is usually called the *Lorentz factor*. It varies from 1 when $v = 0$ up to infinity when $v = c$ (i.e. if the clock moved at light speed). It is helpful to remember that γ is always greater than 1. The second/hour/day we observe on a moving clock is always longer than the second/hour/day observed on a stationary clock. It is worth reminding ourselves why, by quickly referring back to Figure 2.2 and remembering that we see the flash of light in our moving light clock travel further than that of an identical stationary clock.

How does this affect elapsed time? It is important to get your head around this. It means that the elapsed time shown on a moving clock is *shorter* than the elapsed time on the stationary clock. For a moving clock, we see time dilate. Each second/day/year is longer, so if something is moving, it ages more slowly. I find it helps to think of time dilation as the moving clock *slowing down*. Consider a scenario where $\gamma = 1.5$ (about 0.75c or 220,000 km s^{-1}). We see each second on the moving clock take 1.5 seconds on our stationary clock. By the time 4 years have passed for the moving clock, our stationary clock reads 6 years. Elapsed time is shorter for something moving. This can be confusing. Try to remember that *proper time* has the shortest seconds (i.e. the clock ticks fastest),

Table 2.1 The effect of time dilation.

Speed	Time discrepancy	Scenario
0	None	Stationary
360 km h^{-1}	1 second per 600,000 years	Formula 1 racing car
1,800 km h^{-1}	1 second per 20,000 years	Supersonic jet
40,000 km h^{-1}	1 second per 50 years	Spacecraft (Apollo 11)
30,000 km s^{-1} (0.1c)	1 second per 3 days	Almost around the world in 1 second
0.5c	1 second per 6 seconds	Half the speed of light
0.998c	Time x 15	Muon created by cosmic ray
0.99999999c	Time x 7,000	Proton in the Large Hadron Collider

so elapsed *proper time* is always longest. How big is this distortion? We can quantify the effect at different velocities using γ to calculate the discrepancy in elapsed time at various speeds. This is shown in Table 2.1.

The effect is small even for objects that we consider to be moving fast, so it is not surprising that the distortion was missed pre-Einstein. However, it becomes very significant at high velocities. The muon example in Table 2.1 is interesting. Muons are created when cosmic rays hit the earth's upper atmosphere. They are very short-lived (about 2 microseconds) and can be detected at the earth's surface only because their high speed increases their lifespan 15-fold from our perspective.

For non-relativistic velocities we can use the approximation for time dilation shown in Equation 2.5, which is $\gamma \approx 1 + \frac{1}{2}\frac{v^2}{c^2}$. This is based on a Taylor expansion (see Box 2.1). This approximation is accurate to within 1 in 100 (1% error) up to 30,000 km s^{-1} (0.1c). At slower speeds it is even more accurate. Up to speeds of 2,000–3,000 km s^{-1}, the approximate speed of electrons in a hydrogen nucleus, it is good to better than one part in 10,000 (0.01% error).

$$\frac{dt}{d\tau} = \gamma = \sqrt{\frac{1}{1-\frac{v^2}{c^2}}} \approx 1 + \frac{1}{2}\frac{v^2}{c^2}. \tag{2.5}$$

Box 2.1 Taylor expansions are useful for making approximations

We can use a Taylor expansion to approximate γ when v is small compared to c. If you have not come across Taylor expansions, it is worth reading about them. They express a function in terms of an infinite series that can be used to create useful approximations. For γ, we use the following Taylor series:

$$\sqrt{\frac{1}{1-x}} = 1 + \frac{1}{2}x + \frac{3}{8}x^2 + \frac{5}{16}x^3\ldots \quad \Rightarrow \quad \sqrt{\frac{1}{1-\frac{v^2}{c^2}}} = 1 + \frac{1}{2}\frac{v^2}{c^2} + \frac{3}{8}\frac{v^4}{c^4} + \frac{5}{16}\frac{v^6}{c^6}\ldots$$

At non-relativistic speeds, terms of $\frac{v^4}{c^4}$ and higher can be ignored so: $\gamma \approx 1 + \frac{1}{2}\frac{v^2}{c^2}$.

2.5 Length Contraction

In addition to the distortion in time, there is a distortion in the length of objects and distances. As an object moves faster, a stationary observer will see its length shorten in the direction of motion. Imagine somebody flying by you at very high speed holding a metre ruler upwards (i.e. not in the direction of motion). The metre would appear the same length to both of you. However, if our high-speed friend shifts the metre ruler so it is pointing forwards (i.e. in the direction of motion), then you will see it shrink in length. This is called length contraction. Length or distance in the direction of motion, shown below as *dL*, *contracts* in the same proportion γ that time *dilates* (i.e. it is inversely proportional):

$$Time: \ dt_{(moving)} = \gamma d\tau_{(static)} \qquad Distance: \ dL_{(moving)} = \frac{dL_{(static)}}{\gamma} \qquad\qquad (2.6)$$

The distortion of length is small for everyday objects. The supersonic jet listed in Table 2.1 shrinks in length about one part in a billion billionth, but it is significant at higher speeds. For example, a 100-metre-long rocket passing at $0.5c$ will be shrunk to about 87 metres due to this length contraction. Similarly, if you are stationary relative to a planet 10 light years away and then suddenly speed up towards it at $0.5c$ (or you can consider it moving towards you), the distance from you to the planet shrinks to 8.7 light years.

Something emerges that is surprising to many unless you have the imagination of an Einstein (see Box 2.2). Suppose you want to travel in your lifetime to a planet 500 light years away. Is this allowed by special relativity? The answer is *yes*. For example, if you could accelerate up to $0.998c$ (similar to the muon in Table 2.1) then, for you, the distance to the planet would contract from 500 light years to about 30 light years. At $0.998c$, you in your rocket would arrive in just over 30 years on your clock. But for me, watching your progress from here on earth, there is no shortcut, so I would see your journey take just over 500 years, and be worm-food by the time you arrive. A lovely idea but, in practical terms, the energy involved is beyond fantastic and the acceleration would squash you to a pulp. Doh!

Box 2.2 Einstein's imagination

How could Einstein possibly take this all seriously? For it to work, you need to accept time dilation, length contraction and clocks going out of sync. It is a tribute to him that he could ignore all acquired wisdom and leap in a new direction. In his own words: *I am enough of the artist to draw freely upon my imagination. Imagination is more important than knowledge. Knowledge is limited. Imagination encircles the world.*

2.6 Leading Clocks Lag

Another effect of special relativity is that clocks placed apart on a moving object appear out of sync. This is not going to feature prominently in our study of general relativity, but is worth a mention because it plays an important part in resolving some of the more perplexing apparent paradoxes in special relativity. Consider the following scenario (see Figure 2.3). Two clocks (A and B) are synchronised using a flash from a light exactly between them. When the flash of light hits the clocks, they are programmed to show the same time, say 12.00. Somebody stationary relative to the clocks (as in the top of Figure 2.3) sees them trigger at the same time... perfectly in sync.

How does this appear to an observer who sees the clocks moving at high speed to the right (as in the lower part of Figure 2.3)? Things look very different. Clock B moves to the right as the flash of light travels towards it. The flash of light arrives at clock B before the flash of light arrives at clock A. To that observer, clocks A and B are *not* in sync... clock A is slightly behind clock B. The time discrepancy depends on the velocity v and the distance

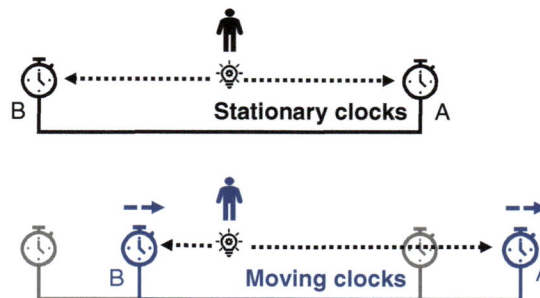

Figure 2.3 Leading clocks lag: if the clocks are moving, the synchronizing light flash must travel further to reach the leading clock, so the time on that clock lags.

d between them in the direction of motion (d is measured on a stationary basis to avoid the distortion of length contraction):

$$\text{Leading clock lags by } \quad \frac{dv}{c^2}. \tag{2.7}$$

2.7 Adding Things Up: An Apparent Paradox

Let me pull things together with an example of how *time dilation*, *length contraction* and *leading clocks lag* combine to produce a coherent mathematical picture. If you cannot follow the resolution of this apparent paradox, do not worry. You will not need it to understand the material on general relativity. I include it for a bit of fun and because many students are perplexed by what appear to be paradoxes in special relativity. I will use an Earth man compared with a Rocket woman passing at high speed. No sexism is intended. It allows me to distinguish using *he* and *she*.

At some stage, most students wonder at the symmetry of special relativity. If Rocket woman passes at great speed, Earth man sees her clock run slowly. But from the perspective of Rocket woman, it is the earth that is rushing past, so she will see his clock run slowly. How can we reconcile this?

Figure 2.4 shows our scenario. Earth man is stationary relative to a planet 8 light years away. He has synchronised his clock with the distant planet clock. In theory (and that is enough), this could be done using radio transmissions, although I grant that it would take a long time! Rocket woman flies by at 0.8c heading to the distant planet.

Let's start with Earth man's view of what happens (see the clock times in the left box of Figure 2.4). As the rocket passes earth, both Earth man and Rocket woman mark the time on their clocks and set it as zero. So, upon departure, the Earth man's view is that the earth, rocket and planet clocks all read zero. He sees the rocket travel a distance of 8 light years at 0.8c so it takes 10 years on his clock. However, he sees time dilate on Rocket woman's clock. The time dilation factor γ is 1.67, so the elapsed journey time on the rocket clock upon arrival is only 6 years.

Let's turn to Rocket woman (see the clock times in the right box of Figure 2.4). She sees the distant planet rushing towards her at 0.8c. This creates length contraction, so the distance to the planet from her perspective is not 8 light years but 4.8 light years ($\gamma = 1.67$). Her clock shows 6 years on arrival, so she agrees that she has travelled at 0.8c (4.8 ÷ 0.8 = 6 years). You can see that length contraction is essential for special relativity to be consistent. Without it, Rocket woman would have travelled 8 light years distance in 6 years, which would be faster than light speed.

Now for the trickier piece of the paradox: from Rocket woman's perspective, the earth and planet are moving at 0.8c, so she sees the earth and planet clocks run slowly compared to hers. During the journey, she sees only 3.6 years pass on the earth and planet clocks compared to the 6 years that pass on her clock ($\gamma = 1.67$). How can Earth man see 10 years elapse on the planet clock between departure and arrival, while Rocket woman sees only 3.6 years elapse on the same clock?

For Rocket woman, the earth and planet are moving. It may help you to imagine a huge solid ruler connecting the earth and the planet. For Rocket woman, this ruler moves past her with the earth leading at the front and

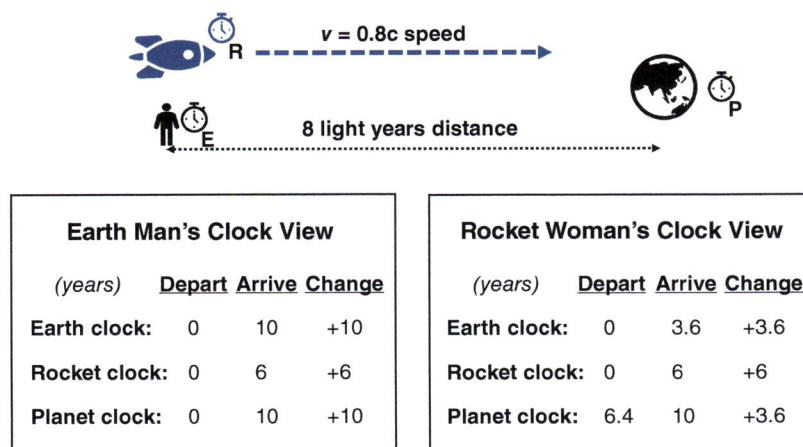

Earth Man's Clock View			
(years)	Depart	Arrive	Change
Earth clock:	0	10	+10
Rocket clock:	0	6	+6
Planet clock:	0	10	+10

Rocket Woman's Clock View			
(years)	Depart	Arrive	Change
Earth clock:	0	3.6	+3.6
Rocket clock:	0	6	+6
Planet clock:	6.4	10	+3.6

Figure 2.4 Different observers, different views. To make sense of the scenario, remember that for the Rocket woman at departure, the planet clock is ahead of the earth clock (leading clocks lag).

the planet at the rear, so she sees the earth and planet clocks out of sync (leading clocks lag). On departure, she sees the earth clock set at zero, but for her, the planet clock already reads 6.4 years: $\frac{dv}{c^2} = 8 \times 0.8$. For our Rocket woman, the planet clock reads 6.4 years on departure plus 3.6 years that elapse during the journey, so the planet clock does indeed read 10 years when she arrives.

Upon arrival at the planet, from the Rocket woman's perspective, the earth clock reads 3.6 years while the planet clock reads 10 years. Why? Well, she is still moving at 0.8c and leading clocks lag so, to her, the earth clock is still 6.4 years behind the planet clock. Everything is consistent when you account for all three distortions: time dilation, length contraction and that leading clocks lag.

If you really want to fry your mind, consider what happens if Rocket woman turns around and returns to earth at 0.8c. In turning around, the earth clock goes from being her leading clock to the rear clock. Her turn involves a massive deceleration followed by a massive re-acceleration back home. During this, the earth clock goes from lagging 6.4 years behind the planet clock to being 6.4 years ahead. From her perspective, 12.8 years more time passes on the earth clock than on the planet clock during the turn. Upon her return to earth, 12 years will have elapsed on her clock (6 years out, 6 years back) versus 20 years on the earth clock (from her perspective, 3.6 years out, 3.6 years back, plus 12.8 years synchronisation change). It is this that drives the twin paradox. I imagine you know the paradox. Two twins separate. One heads off on a distant journey and comes back younger than the other. Yes... that is a true consequence of special relativity.

2.8 Energy and Momentum

Einstein's $E = mc^2$ is probably the most famous equation in the world. As I hope you know, it shows the energy E of a mass m at rest. With a few symbols, it tells us that mass and energy are interchangeable and even gives us the exchange rate. How on earth did Einstein figure this out from special relativity? Well, actually it is fairly obvious... so let's get to it.

We can use the invariant interval equation to show what special relativity tells us about the momentum of an object (mv in classical physics). This is Equation 2.1 shown again below as 2.8 including only one spatial dimension, so motion is in the x direction. For the first step, we divide by invariant *proper time $d\tau$*. Then, if we want a classical momentum term (mv), we need to develop a velocity term $\frac{dx}{dt}$ and introduce mass m defining it as *rest mass* to be sure it is invariant for all observers. This can all be done in short order and there is no way Einstein would have missed it:

$$c^2\,d\tau^2 = c^2 dt^2 - dx^2, \tag{2.8}$$

$$c^2\,\frac{d\tau^2}{d\tau^2} = c^2\,\frac{dt^2}{d\tau^2} - \frac{dx^2}{d\tau^2}, \tag{2.9}$$

$$c^2 = c^2\frac{dt^2}{d\tau^2} - \frac{dx^2}{dt^2}\frac{dt^2}{d\tau^2}, \qquad Remember\ \frac{dt}{d\tau} = \gamma,$$

$$c^2 = c^2\gamma^2 - v^2\gamma^2,$$

$$m^2 c^2 = (mc\gamma)^2 - (mv\gamma)^2. \qquad where\ mv\ is\ classical\ momentum\ (mv). \tag{2.11}$$

It should be obvious to you that the left side of Equation 2.11 is Lorentz invariant, meaning that all observers will agree on the value for an object however fast they are moving relative to it. It must be, because the rest mass m of the object and the speed of light c are constants. The far-right term of Equation 2.11 is the square of $mv\gamma$. The difference between mv and $mv\gamma$ only is significant at relativistic speeds when the value of γ diverges noticeably from 1 (see Table 2.1 for a reminder). It is not a major leap to define $mv\gamma$ as relativistic momentum, which I will give the usual label p. Clearly, this momentum term is *not* the same for all observers. The faster an observer is moving relative to an object, the higher the momentum they will measure.

Any increase in momentum is offset according to Equation 2.11 by an increase in the other term that involves $mc\gamma$, so that the equation balances to the constant $m^2 c^2$. An observer passing at speed sees momentum increase *and* something else increase. Put yourself inside Einstein's head. Can you imagine the excitement mounting? We can expand $mc\gamma$ using the approximation for γ from Equation 2.5 to give Equation 2.12.

$$mc\gamma \approx mc\left(1 + \frac{1}{2}\frac{v^2}{c^2}\right) = mc + \frac{1}{2}m\frac{v^2}{c} = \frac{1}{c}\left(mc^2 + \frac{1}{2}mv^2\right). \tag{2.12}$$

As velocity increases, the last term ($mc^2 + \frac{1}{2}mv^2$) increases with $\frac{1}{2}mv^2$, so exactly as classical energy increases due to kinetic energy. This makes sense. An observer seeing the object pass at higher speed will see it have higher momentum and higher energy. Equation 2.11 tells us how these two increases balance out. Let's neaten things up and multiply both sides of Equation 2.12 by the speed of light and call it relativistic energy E (this label is different from the usage in my other book *Quantum Untangling*).[2]

$$E = mc^2\gamma \quad \approx \quad mc^2 + \frac{1}{2}mv^2. \tag{2.13}$$

We can substitute this definition of E into Equation 2.11 to give Equation 2.14 that shows the relationship between relativistic energy E and relativistic momentum p in the Minkowski spacetime of special relativity. If you are not familiar with this, then look long and hard. It is one of the key equations in physics. Einstein's famous equation for the energy of a stationary mass, $E = mc^2$, is easily derived from Equation 2.14 because at zero velocity, $p = 0$. It can also be derived from Equation 2.13 setting v to zero.

$$m^2c^2 = (mc\gamma)^2 - (mv\gamma)^2,$$
$$= \left(\frac{E}{c}\right)^2 - p^2,$$
$$m^2c^4 = E^2 - p^2c^2. \tag{2.14}$$

Remember that all we have done is look at the rate of change of the invariant interval equation. Nothing more. This means that the relationship between energy and momentum (Equation 2.14) and the relationship between time and space (Equation 2.8) share a common root. The two are shown below for comparison, both showing all spatial dimensions. The underlying relationship is called the *metric*, which we will discuss in the next chapter.

$$c^2\,d\tau^2 = c^2\,dt^2 - dx^2 - dy^2 - dz^2,$$
$$m^2c^4 = E^2 - p_x^2c^2 - p_y^2c^2 - p_z^2c^2. \tag{2.15}$$

I suspect that Einstein figured this out quickly. He was very intuitive and, despite urban legend, he was also an able mathematician (see Box 2.4). As is usual in physics, the final proof of $E = mc^2$ came in the form of experimental results, and both it and the other tenets of special relativity have been confirmed by countless particle experiments.

2.9 Energy, Momentum, Time and Space

It is worth taking a moment to consider what this means for the relationship between energy, momentum, time and space. The formula for relativistic momentum should come as no surprise. Classical momentum is $m\frac{dx}{dt}$. In contrast, relativistic momentum p is $m\frac{dx}{d\tau}$. The difference is that relativistic momentum uses the velocity of the object over invariant proper time τ, rather than over observer time t (see Equation 2.16).

$$\text{Relativistic Momentum}: \quad p = mv\gamma = m\frac{dx}{dt}\gamma = m\frac{dx}{dt}\frac{dt}{d\tau} = m\frac{dx}{d\tau}. \tag{2.16}$$

The equivalent relationship for relativistic energy may come as more of a surprise. Relativistic energy E is $mc^2\frac{dt}{d\tau}$ (see Equation 2.17). The c^2 factor appears because of the units we humans use for time versus space. Ignoring this, you see that what we call energy is mass multiplied by time dilation.

$$\text{Relativistic Energy}: \quad E = mc^2\gamma = mc^2\frac{dt}{d\tau}. \tag{2.17}$$

Putting this another way, relativistic momentum is mass multiplied by the speed of the object through space, while relativistic energy is mass multiplied by its speed through time.

---

2 In *Quantum Untangling*, I use bold labels such as **E** and **p** for *relativistic* energy and momentum. I haven't used the bold labels in this book, because there is rarely a need to refer to classical quantities.

Box 2.3 Hermann Minkowski (1864–1909)

Minkowski, one of Einstein's professors, famously introduced his lecture on spacetime with: *Henceforth space by itself, and time by itself, are doomed to fade away into mere shadows, and only a kind of union of the two will preserve an independent reality.*

He died suddenly of appendicitis at only 44 years old.

2.10 Summary

This chapter started with two simple and sensible-sounding postulates: *the speed of light is constant for all observers* and *the laws of physics are the same for all observers*. But Einstein's two simple postulates wreak havoc on the old common-sense view that space and time are independent. Newtonian space is replaced by Minkoswki spacetime, a new, more intricate, picture of the fabric of the universe (see Box 2.3 for Minkowski's famous quote).

Our analysis used scenarios with clocks that work on flashes of light. This simplifies things because the speed of light is constant for all observers, so the law of addition of velocities is simple: whatever the velocity relative to an observer, c is still c. Comparing stationary and moving light clocks gives us the invariant interval equation and leads us to three distortions that occur to objects that are moving: time dilation (increasing with γ), length contraction (decreasing with γ) and moving clocks appearing out of sync (leading clocks lag). These distortions are forced on us if we want a consistent framework of space and time with the speed of light constant. Are you joking, Mr Einstein? But it is no joke. This is how things work.

We also examined what the invariant interval equation means for energy and momentum. This leads to Einstein's famous equation for rest mass energy $E = mc^2$ and gives a different perspective on the relationship between energy-momentum and spacetime. Relativistic momentum is mass multiplied by an object's velocity through space. Relativistic energy is mass multiplied by its velocity through time.

In Chapter 3, we will take a closer look at the structure of spacetime. The invariant interval equation will play a key role and don't forget how this drives the relationship between energy and momentum (both these relationships are shown in Equation 2.15).

Box 2.4 Einstein: mathematical dunce and lazy dog?

Legend has it that Einstein was poor at maths. While he may not have been a mathematical prodigy, this seems at odds with his many papers and fabulous breakthroughs. On closer examination, the story does not hold up. When he was 7 years old, his mother noted: *He was again number one, his report card was brilliant.*

So where did the story come from? His last report card exists from Aargau school when he was 17 years old. Of a possible 1–6, he scored 6 for physics, 6 for algebra and 6 for geometry. This was the highest possible score, *but* the school had changed its marking system that very year from 6 being the lowest grade to 6 being the highest. Is it confusion over this that besmirched his academic reputation?

Or maybe it was his attitude at university. Einstein studied under Hermann Minkowski in Zurich. He had a reputation for not paying attention and skipping lectures. Minkowski described him as a *lazy dog* and commented, upon the publication of the theory of general relativity:

I really would not have believed him capable of it.

Chapter 3

The Metric

The spacetime metric $[g_{\mu\nu}]$ is just another way of writing the *invariant interval equation*. My aim in this chapter is to give you a decent understanding of what the metric is, illustrate how it contains information on curvature and then use a simple model (with the help of some dung beetles) to show curvature can create an effect like gravity.

3.1 The Minkowski Metric

The easiest way to explain the metric is with an example. In Chapter 2, we saw what the constant speed of light means for spacetime. We analysed how a clock runs at different rates for different observers. This gave us the invariant interval equation shown again as Equation 3.1. We call this Minkowski spacetime. Let me remind you what the equation says. The left side shows the perspective of an observer stationary relative to the clock. The time between ticks is the clock's proper time and is shown as $d\tau$. The right side shows the perspective of somebody who sees the clock moving. Measured against this observer's clock, the time between ticks, dt is longer than $d\tau$ because of time dilation. This observer also sees the clock move between ticks and, if that movement is taken into account in the way shown in the equation, it always gives the same value.

$$c^2\, d\tau^2 = c^2\, dt^2 - dx^2 - dy^2 - dz^2, \tag{3.1}$$
$$= c^2\, (dt)(dt) - 1\,(dx)(dx) - 1\,(dy)(dy) - 1\,(dz)(dz), \quad using\ (t,x,y,z), \tag{3.2}$$
$$(cd\tau)^2 = 1\,(c\,dt)(c\,dt) - 1\,(dx)(dx) - 1\,(dy)(dy) - 1\,(dz)(dz), \quad using\ (ct,x,y,z). \tag{3.3}$$

Equations 3.2 and 3.3 show the invariant interval written using two slightly different coordinates. In Equation 3.3, the coordinates are (ct,x,y,z). This puts the coordinate units of time and space into a 1:1 relationship. In layman's language, it uses the right exchange rate between them. I will tend to use the (ct,x,y,z) coordinate system for more advanced calculations. When I do so, I will signal it with (ct,x) instead of (t,x). It simplifies the maths and will not take long to get used to.

Now step back and look at this from a broader point of view. If something is moving through Minkowski spacetime, the left side of Equation 3.1 is a common measure that everybody can agree on and therefore, so is the *total* on the right side (as the right side equals the left). Think of the right side as the rough equivalent of a metre rule for distances through spacetime. This is called the *metric* $[g_{\mu\nu}]$ of the spacetime and is written as a (4×4) matrix as shown in Equation 3.4. The indices of the columns and rows of the matrix are both ordered (ct,x,y,z). To be sure you see how this works, have a look at the matrix and compare it with the Minkoswki invariant interval equation as expanded in Equation 3.3. Minkowski spacetime is so frequently used in calculations that it has its own symbol $[\eta_{\mu\nu}]$, which you will see in some texts (I will not spring this symbol on you without clearly identifying it). Also, beware of different conventions (see Box 3.1).

Untangling General Relativity: The Intuitive Self-Study Guide, First Edition. Simon Sherwood.
© 2026 John Wiley & Sons Ltd. Published 2026 by John Wiley & Sons Ltd.

$$\text{Minkowski metric } (ct, x, y, z): \quad [g_{\mu\nu}] = \begin{bmatrix} 1 & 0 & 0 & 0 \\ 0 & -1 & 0 & 0 \\ 0 & 0 & -1 & 0 \\ 0 & 0 & 0 & -1 \end{bmatrix}, \text{ often labelled } [\eta_{\mu\nu}].$$ (3.4)

The neat thing is that any version of spacetime can be quantified this way. If you know $[g_{\mu\nu}]$ and the coordinate basis, which is (ct, x, y, z) in the case of Equation 3.4, you have a full picture of the spacetime. If that spacetime becomes curved or distorted, this will show up as a change to the entries in its $[g_{\mu\nu}]$ matrix. And don't be intimidated by the little μ and ν symbols. They simply show it is a matrix with multiple components. Convention is to use Greek symbols when working in four-dimensional (4-D) spacetime, so each index covers four variables: $\mu = (t, x, y, z)$ and $\nu = (t, x, y, z)$. To make the script as legible as possible, I will use capitals for the indices of each tensor component. In the case of Equation 3.4, $g_{TT} = 1$, $g_{XX} = g_{YY} = g_{ZZ} = -1$ and all the other entries such as g_{TX} are zero. Let me remind you that I use brackets when referring to the whole matrix, such as $[g_{\mu\nu}]$. The absence of these brackets indicates I am referring to a single component, such as g_{XX}.

Box 3.1 Beware of different conventions for the metric

The invariant interval equation can be written in terms of time (1) or space (2):

1. $c^2 d\tau^2 = c^2 dt^2 - dx^2 - dy^2 - dz^2,$ *Note coordinates here are* (t, x, y, z)
2. $ds^2 = -c^2 dt^2 + dx^2 + dy^2 + dz^2.$

The crucial thing is that the left-hand term in the equation is invariant for all observers and that time and space offset each other in the formula. In this book, I use (1) with time positive. This is described as *signature* $< + - - - >$. The alternative is $< - + + + >$ with space positive, in which case all the entries in any $[g_{\mu\nu}]$ matrix, including that of Equation 3.4, change sign i.e. positive terms become negative and vice versa. While the two signatures are equally legitimate, you must be careful to remember which you are using and to be consistent in order to avoid $+/-$ errors appearing in your work.

It is important to note that the entries in the $[g_{\mu\nu}]$ matrix *do depend* on the coordinate system used. An example is the switch to spherical coordinates, which we will need later in the book. Equation 3.5 shows the invariant interval equation in spherical coordinates, and Equation 3.6 shows how this appears in the Minkowski metric. Don't worry about spherical coordinates for now. I just want to highlight that the metric changes.

$$(c\,d\tau)^2 = (c\,dt)^2 - dr^2 - r^2\,d\theta^2 - r^2 \sin^2\theta\,d\phi^2,$$ (3.5)

$$\text{Minkowski metric } (ct, r, \theta, \phi): \quad [g_{\mu\nu}] = \begin{bmatrix} 1 & 0 & 0 & 0 \\ 0 & -1 & 0 & 0 \\ 0 & 0 & -r^2 & 0 \\ 0 & 0 & 0 & -r^2 \sin^2\theta \end{bmatrix}.$$ (3.6)

The underlying spacetime is not changed by any switch in coordinates (of course). Different coordinate versions of the same underlying spacetime are called *diffeomorphisms*. How nice of physicists to choose a simple word! And this is where the wonderful power of *tensors* comes in handy. The Einstein field equations are built with components of tensors: the metric tensor $[g_{\mu\nu}]$, the Einstein tensor $[G_{\mu\nu}]$ and the energy-momentum tensor $[T_{\mu\nu}]$. And... drum roll please... *being tensors, their components all transform between coordinate systems in exactly the same way.*

3.2 Einstein's Tensor and the Metric

Now you know what the metric is, I want briefly to bring the Einstein field equations, which I will label EFEs, into the story. The EFEs, as in Box 1.3 in Chapter 1, are shown again as Equation 3.7 (I will ignore the cosmological constant until later).

$$G_{\mu\nu} = \frac{8\pi G}{c^4} T_{\mu\nu}, \qquad \text{where } G_{\mu\nu} = R_{\mu\nu} - \frac{1}{2} R g_{\mu\nu}.$$

(3.7)

Spacetime curvature $\Longleftrightarrow$ *Energy-momentum*

Expressed in the simplest way possible, each EFE consists of components from two tensors and a constant k giving: $G_{\mu\nu} = k T_{\mu\nu}$. Don't forget the EFEs are actually 16 equations, one for each entry in the tensors' (4×4) matrices: $G_{TT} = k T_{TT}$, $G_{TX} = k T_{TX}$... 10 are independent (6 are repeats). The component on the right side is from the energy-momentum tensor $[T_{\mu\nu}]$, which has information on energy-momentum (no surprise). For now, we will focus on the component from the Einstein tensor $[G_{\mu\nu}]$ on the left. This contains information on spacetime curvature.

The component $G_{\mu\nu}$ is shown in more detail on the far right of Equation 3.7. It can be broken down into corresponding components of the Ricci tensor $[R_{\mu\nu}]$, and metric tensor $[g_{\mu\nu}]$ (multiplied by the Ricci scalar, a number). You don't know what the Ricci tensor and scalar are yet, but here is the link to the spacetime metric: *every component of $[G_{\mu\nu}]$ is a function of, and only of, the metric tensor $[g_{\mu\nu}]$ and its derivatives*. The components of $[R_{\mu\nu}]$ and the scalar R are complicated mixes of various derivatives of the components of $[g_{\mu\nu}]$. Complicated they are, but they and therefore $[G_{\mu\nu}]$ are pure functions of the spacetime metric (there are no other inputs in the calculation).

I hope this helps you see the fundamental importance of the metric $[g_{\mu\nu}]$ (which you should remember is just the invariant interval equation written in matrix form). Before moving on, I want to highlight again that the EFEs are built of tensor components, which all transform between coordinate systems in the same way. Consider, for example, changing from a Cartesian basis (ct, x, y, z) to spherical coordinates (ct, r, θ, ϕ). This will have an effect on the entries in the $[T_{\mu\nu}]$ matrix but will have the same effect on all the other tensors, so $[G_{\mu\nu}]$ will transform in the same way. The EFEs still work albeit in different coordinates. Instead of $G_{TT} = k T_{TT}$, $G_{XX} = k T_{XX}$, $G_{YY} = k T_{YY}$... we have $G_{TT} = k T_{TT}$, $G_{RR} = k T_{RR}$, $G_{\theta\theta} = k T_{\theta\theta}$...

If the EFEs work in one coordinate system, they work for all. This property is called *covariance* and was very important to Einstein. If you think about it, it makes sense. Physics should not depend on which coordinates you choose. At one stage, Einstein became disheartened and gave up on finding a covariant version of his equations... but thankfully he got there in the end.

Some readers may wonder what the value of $[G_{\mu\nu}]$ is for the metric of Minkowski spacetime. It is zero. $[G_{\mu\nu}]$ is a measure of the *curvature* in $[g_{\mu\nu}]$ and Minkowski spacetime is what we call *flat*.

3.3 Distortion in the Metric

Let's get back to our main topic: the metric. The metric is important because without its information you may get a very distorted view of things. Think of it as being like the scale of a map. If you don't have the scale, then you have no real idea what the coordinates mean. Let me start with some simple examples in flat space of how the distortion of coordinates is reflected in the metric.

Let's start with what is called *Euclidean space*. This has no curvature. It is the boring old space you are used to. We will start in two-dimensions (2-D) using (x, y) Cartesian coordinates (see Box 3.2). The separation between any two points is (from Pythagoras) $\sqrt{\Delta x^2 + \Delta y^2}$. This gives us the invariant interval equation shown as Equation 3.8, which is expressed in metric form below it. Note that as this is not Minkowski spacetime, I show spatial separation as ds^2 (this is convention) and positive for simplicity. Also note that another convention is to use the symbol $[g_{ij}]$ when referring purely to the spatial parts of a metric. The use of i and j instead of the Greek letters μ and ν indicates it is only space, rather than 4-D spacetime.

$$ds^2 = dx^2 + dy^2, \tag{3.8}$$

$$\text{Cartesian 2-D spatial metric } (x,y): [g_{ij}] = \begin{bmatrix} 1 & 0 \\ 0 & 1 \end{bmatrix}. \tag{3.9}$$

Not only is this flat space, it's coordinate system is also simple and clean in the sense that the effect of moving a step in either direction (*dx* or *dy*) is the same distance (value of *ds*) everywhere on the *manifold* (a fancy word for *surface* that includes possible higher-dimensional surfaces). When you did maths at school, you will have used this. While you may have found it hard to handle some calculations involving *x* and *y* in flat space, believe me, it is a doddle compared to more convoluted metrics.

Box 3.2 Cartesian coordinates and René Descartes

Cartesian coordinates were introduced by René Descartes (1596–1650), whose name in Latin was Renatus Cartesius, in his book *Geometry* in 1637. What we now label as (*x*, *y*) were originally called the *abscissa* and the *ordinate*. This was a huge leap forwards because it allowed mathematicians to express geometry in numerical terms.

Descartes was unimpressed by the education he received at school, writing: *I came to think I had gained nothing from my attempts to become educated but increasing recognition of my ignorance.* On 10 November 1619, Descartes had three life-changing dreams. In the third, a stranger presented him a poem starting with the words: *What is and is not.* He believed it to be a message from God telling him to question everything rather than to depend on past wisdom. Searching for something that he could rely on and build upon, his famous starting point was *cogito ergo sum*, or in English, *I think therefore I am.*

Let me give you a different example. For simplicity, I will stick with flat space, but switch to 2-D polar coordinates. Figure 3.1a gives a reminder of how 2-D polar coordinates work. Rather than *x* and *y*, we indicate the position of a point (shown as a black dot) using the coordinates (r, θ), where *r* is the distance from the origin and θ is the angle subtended at the origin. If you increase the value of the coordinate *r* by a small amount *dr*, the black point moves (as shown in red) to a point distance *dr* away. The distance moved is the same whatever the value of θ. However, if you make a small change to the angle θ, increasing it by $d\theta$, the black point moves to that shown in green. The distance moved is $r\, d\theta$ if the angle is measured in radians, as is the case throughout this book (for a refresher on radians, see Box 3.3). The distance moved depends on the value of coordinate *r* at your starting point. Thus, the actual distance moved when you increase *r* by *dr* is simply *dr*, but the actual distance moved when you increase θ by $d\theta$ is $r\, d\theta$.

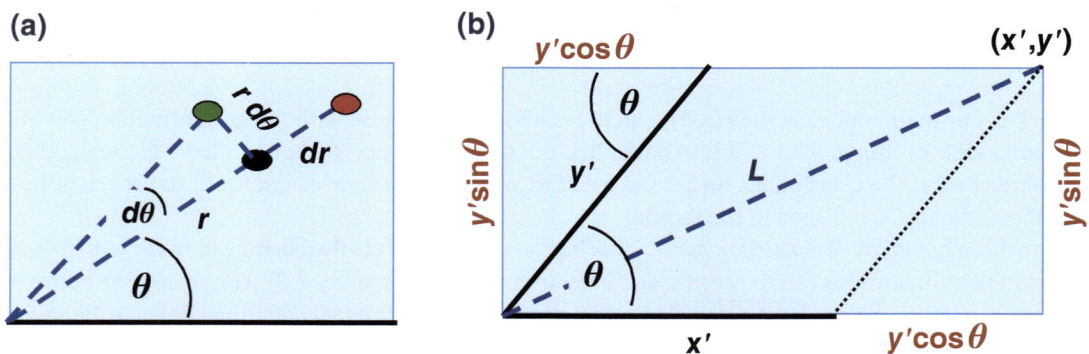

Figure 3.1 Examples of different coordinate systems.

If you combine a small change to r of dr with a small change to θ of $d\theta$, Pythagoras reveals that the square of the actual distance moved is the sum of the squares. We show this with the invariant interval equation in Equation 3.10 and the metric in Equation 3.11.

$$ds^2 = dr^2 + r^2\, d\theta^2,$$ (3.10)

$$2\text{-}D\ polar\ metric\ (r,\theta)\colon\ [g_{ij}] = \begin{bmatrix} 1 & 0 \\ 0 & r^2 \end{bmatrix}.$$ (3.11)

Box 3.3 Radians

2π radians is equivalent to $360°$, π radians is $180°$ and so forth. As the circumference of a circle is $2\pi r$, the small change $d\theta$ in Figure 3.1a at distance r out from the origin, is a move of $r\,d\theta$ around the circumference of a circle of radius r.

The problem comes when you start taking derivatives. Suppose I give you some velocity data without saying that it uses coordinates (r,θ) and not (x,y). You want to work out if the track is a *geodesic* (the name for the natural path followed by an undisturbed object). On a flat surface, this is a straight line travelled at constant speed. In terms of (x,y) coordinates, the velocity v in both the x and y directions, which I will label v^x and v^y, are constant: $\frac{\partial v^x}{\partial t} = 0$ and $\frac{\partial v^y}{\partial t} = 0$ (if you need a reminder on derivatives, check out Box 3.4 plus Box 3.11 at the end of this chapter gives a general refresher on the rules of differentiation).

Box 3.4 Total derivative or partial derivative?

This is an important distinction when something is affected by two or more variables. As a simple example, consider the function: $k = 2t + 4x$. The *partial* derivative is shown by ∂ and is the change with respect to *one* variable (say, t) while keeping the other (x) constant. In this case, increasing t by 1 increases k by 2, so $\frac{\partial k}{\partial t} = 2$. Imagine that we track k from (t,x) of $(0,0)$ to $(4,3)$. In this simple example, the partial derivatives are constant values, so we can calculate the change in k as:

$$\Delta k = \frac{\partial k}{\partial t}\Delta t + \frac{\partial k}{\partial x}\Delta x = (2)4 + (4)3 = 20.$$

The *total* derivative uses the symbol d and is calculated when all variables are allowed to vary. On the track above, k has changed by 20 over $t = 4$ seconds, so on average over the track: $\frac{dk}{dt} = 5$.

It would be natural for you to look at the data and search for the telltale zero-value velocity derivatives. However, this is completely wrong in (r,θ) coordinates. Zero-value would mean a constant value of $\frac{\partial \theta}{\partial t}$ and $\frac{\partial r}{\partial t}$, which is a circle if $\frac{\partial r}{\partial t} = 0$ or a spiral if not. It certainly is *not* a straight-line geodesic. This sort of derivative, if taken without reference to the metric, is like a map without any hint of how the scale works... no use at all.

My next example is another coordinate shift in flat space as shown in Figure 3.1b. Rather than using x and y, which are at right angles, I have switched to x' and y' which are not. They are angled to each other at θ and so are not *orthogonal*. Working out the metric is a little trickier, but hopefully the figure will allow you to understand the calculation. If we label the distance to point (x',y') as L then Pythagoras gives Equation 3.12 and for Equation 3.13 you need to remember that: $\sin^2\theta + \cos^2\theta = 1$.

$$L^2 = (x' + y'\cos\theta)^2 + (y'\sin\theta)^2,$$ (3.12)

$$= x'^2 + 2x'y'\cos\theta + (y'\cos\theta)^2 + (y'\sin\theta)^2,$$ (3.13)

$$= x'^2 + y'^2 + 2x'y'\cos\theta.$$

$$Invariant\ interval\ equation\colon\ ds^2 = dx'^2 + dy'^2 + 2\cos\theta\, dx'dy',$$ (3.14)

$$2\text{-}D\ non\text{-}orthogonal\ metric\ (x',y')\colon\ [g_{ij}] = \begin{bmatrix} 1 & \cos\theta \\ \cos\theta & 1 \end{bmatrix}.$$ (3.15)

This gives us the invariant interval in Equation 3.14, which is expressed in metric form in Equation 3.15. You can see that there is a $dx'dy'$ term in the invariant interval equation. This indicates that the coordinate axes (x' and y') are *not* orthogonal. Moving a bit in the x' direction affects the y' value and vice versa. This shows up in the metric in the $dx'\,dy'$ entry (top right in the matrix) and the $dy'\,dx'$ entry (bottom left). These entries are always equal, so they get half the total each ($\cos\theta + \cos\theta = 2\cos\theta$).

The idea of coordinate axes not being orthogonal may strike you as strange, and you will be pleased to hear that virtually all the metrics discussed in this book have orthogonal coordinates, which I refer to as *diagonal metrics* (i.e. no off-diagonal elements). However, later in the book there are a couple of weirder metrics when off-diagonal elements appear (e.g. the Kerr metric and gravitational waves). I will be sure to warn you when this is the case.

3.4 Curvature, Dung Balls and a First Hint of Gravity

All the examples in Section 3.3 are different coordinate systems for the same flat surface. To remind you of the technical term, the metrics are diffeomorphisms of the flat 2-D space. They are relevant because curved spacetime poses the same challenge. Generally, the metric must be taken into account when taking derivatives.

We will delve into that problem in a moment. First, let's step away from the maths and think about a more intuitive example. Again, it is 2-D, the advantage being that you can visualise a curved 2-D surface by imagining it in 3-D (mathematicians describe this as *embedding* the 2-D surface in 3-D space). I want to show you why curvature interested Einstein. The question we need to ask is whether the curvature of a surface can lead to something that might be mistaken for gravitational attraction... and the answer is... *yes*.

Consider how life would look living on a sphere. Well, now that is convenient because we *do* live on the surface of a sphere called planet Earth. What happens if two people set off on parallel paths? Well, if they set off from the equator heading north, their paths converge at the north pole. On a sphere, all parallel paths converge (note that lines of latitude are *not* parallel for somebody on the surface). If you have doubts, grab a football and try it. Draw two little marks side by side pointing in the same direction, then follow them along and watch them converge. This is illustrated in Figure 3.2. For some strange reason, physicists prefer bugs to humans, so I will follow that trend.

I want you to imagine that you are a tiny dung beetle travelling on the surface of the sphere in Figure 3.2. You and your friend start on two parallel paths, but the paths gradually draw together. Having no idea that you are on a curved surface, you might conclude there is some strange force accelerating you towards your smelly dung-pal. But it is curious because you cannot feel the force. You do some experiments and discover that this weird acceleration is the same even if you push a huge ball of dung in front of you. The mysterious acceleration appears to be identical for all objects, whatever their mass. Does this sound familiar? It should—it is just like gravity!

As I pointed out in the overview, one of the surprising features of gravitational acceleration is that it is the same, whatever the mass of the test object you use. Drop an object on earth in a vacuum (to avoid air resistance) and it will accelerate at $9.8\ \mathrm{m\,s^{-2}}$ whether it is a lead weight, an apple or a feather. You can see this using $F = ma$ with Newton's equation for gravitational force. The gravitational acceleration is independent of the mass m of the attracted object, as shown in Equation 3.16 (a is the acceleration, M is the attracting mass, G is Newton's gravitational constant and r is the distance apart).

$$\text{Gravitational force}\qquad F = \frac{GMm}{r^2}\qquad \Longrightarrow \qquad a = \frac{GM}{r^2}\qquad \text{Gravitational acceleration.}\qquad (3.16)$$

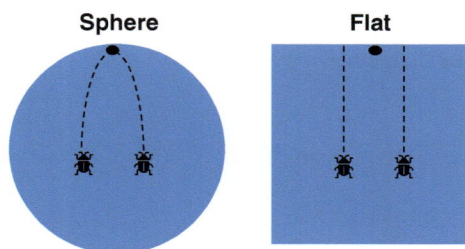

Figure 3.2 The two bugs on the sphere start parallel but appear to be drawn together.

Let's take our dungy adventure a step further. You and your friend, not liking the stinky smell of each other, decide to resist getting closer together. You find a long stick and hold it between you. While you walk, you push against it to stay apart. But if you roll one of your dung balls in front of you, it veers towards your pal. Do you see the analogy? It is us on the surface of the earth. Our natural path is downwards, but is resisted by the electromagnetic repulsion of the atoms in the ground. As a result, we feel a force against the ground just as the dung beetles would against their stick. And when we drop something, it follows its natural path just like the dung ball in our story.

This comparison may be laboured, but it is important to understand why Einstein was so convinced that space-time curvature might be a good explanation of gravity. He was particularly struck by two things: that gravitational acceleration is the same for all objects and that we feel no acceleration in gravitational freefall (as astronauts know well from weightlessness). The fit with curvature was too good for him to ignore.

3.5 A Mathematical Challenge

In order to tie gravity to curvature, we are going to need to understand the mathematics of what is happening to the little dung beetles. What is it about curvature that changes the geodesics and draws them together? The first step is to figure out the metric of the spherical surface they are crawling along. The left of Figure 3.3 gives you the mathematics behind the metric. The bugs are living on the surface of the sphere.

We define where they are on the surface using the coordinates θ and ϕ, both measured in radians. As shown, θ is the angle down from the north pole, then ϕ is the angle around the sphere at that level (i.e. at that latitude if it were the earth). Note that there is no r radial coordinate because the whole surface has the same value for r. The bugs are constrained to live on the surface, so r is a constant not a coordinate. For simplicity, we set the sphere radius as $r = 1$ (called a *unit* sphere).

We need to figure out what a small step in each coordinate direction means in terms of spatial separation on the surface of the sphere. As shown on the far left of the figure, if a bug takes a step $d\theta$ radians in the θ direction, the distance moved on the surface is $r\,d\theta$, which is simply $d\theta$ because $r = 1$. However, the answer is more complicated for a step $d\phi$ in the ϕ direction. The $d\phi$ step is around a circle, but the radius of the circle depends on the value of θ. The radius of the circle at this latitude is $\sin\theta$, so the distance moved on the surface by a step of $d\phi$ radians is $(\sin\theta\,d\phi)$. This leads to the invariant interval equation and metric shown as Equation 3.17.

$$ds^2 = d\theta^2 + \sin^2\theta\,d\phi^2,$$

$$\text{Surface of a unit sphere}, (\theta, \phi): \quad [g_{ij}] = \begin{bmatrix} 1 & 0 \\ 0 & \sin^2\theta \end{bmatrix}. \tag{3.17}$$

Why do the parallel paths of the dung beetles draw closer together? Let's position the coordinates so the beetles are travelling in the θ direction (we are free to line things up as we wish because of the spherical symmetry), and let's start them off $d\phi$ apart. As they progress, they continue to crawl in the θ direction and they continue to stay $d\phi$ apart *but* the value of $d\phi$ in terms of spatial separation ds reduces. This is shown on the right of Figure 3.3.

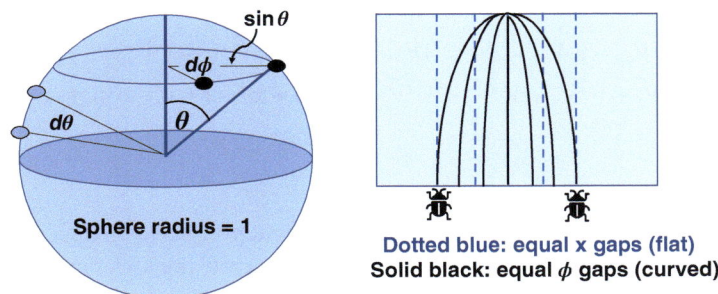

Dotted blue: equal x gaps (flat)
Solid black: equal ϕ gaps (curved)

Figure 3.3 Finding the metric of the surface of a unit sphere.

This makes them veer away from the straight lines that they would have followed if the space were flat. The recipe for this is in the metric. The $g_{\phi\phi}$ entry of $\sin^2\theta$ tells us that the invariant length ds of a step in the $d\phi$ direction depends on θ.

In the case of the sphere, the surface *is* curved and, as a result, the geodesics draw together. However, it can be hard to tell the difference between a surface that really is curved versus the distortion caused by the choice of coordinates. Compare the metric for the sphere's surface in Equation 3.17 with Equation 3.11, which is the metric for flat space expressed in polar coordinates. The former is curved. The latter is flat. Do you see the problem? There is no easy way to tell just by eye-balling the metric. Fortunately, a bunch of super-smart mathematicians have laboured over the last couple of centuries to understand curvature, as you will learn in the next few chapters.

3.6 Upper and Lower Indices

This seems as good a moment as any to address a complication that arises when you are not using the classic Euclidean/Cartesian spatial dx, dy, dz metric. I apologise for the next two sections. It isn't the most stimulating topic, but it is important. A vector in Euclidean/Cartesian space can be expressed in terms of three orthogonal components (x, y, z), with a step in the dx, dy or dz directions all being an equivalent distance. However, with more complicated metrics, you have to consider upper and lower indices. You will see two distinct notations relating to components of vectors such as V^x and V_x. These are typically described as contravariant (the V^x type) versus covariant (the V_x type). I advise you to *ignore* this nomenclature unless you are a mathematician. It will just confuse you. It is better to think of the upper index as the regular *vector coordinates* while the lower index is its *dual coordinates*.

Why do we need two indices? Let me show you the problem using Minkowski spacetime and the coordinates (t, x, y, z). In this case, the invariant interval is the sum of $c^2 dt^2 - dx^2 - dy^2 - dz^2$ and the metric $[g_{\mu\nu}]$ is as shown in Equation 3.18 (compare back with Equation 3.4 if needed, noting that in this case the time coordinate is in units of t rather than ct).

$$\text{Minkowski metric } (t, x, y, z): \quad [g_{\mu\nu}] = \begin{bmatrix} c^2 & 0 & 0 & 0 \\ 0 & -1 & 0 & 0 \\ 0 & 0 & -1 & 0 \\ 0 & 0 & 0 & -1 \end{bmatrix}. \tag{3.18}$$

Suppose we have a vector V that has components $V^t = t$, $V^x = x$, $V^y = y$ and $V^z = z$. We need to know its magnitude, labelled $|V|$, in units of the invariant interval because this is the only measure that makes any sense. If we were dealing with Euclidean/Cartesian space (x, y, z), we could simply say: $|V|^2 = V^x V^x + V^y V^y + V^z V^z = x^2 + y^2 + z^2$. However, this doesn't work with Minkowski spacetime (or any other complicated metric). For example, in Minkowski spacetime we know that $|V|^2 \neq V^t V^t + V^x V^x + V^y V^y + V^z V^z$. To match the invariant interval, the $V^t V^t$ needs to be multiplied by c^2 and each of the space terms such as $V^x V^x$ needs to be multiplied by -1.

The solution is to create matching components measured in dual coordinates, which we label with lower indices such as V_t. We design these so that: $|V|^2 = V^t V_t + V^x V_x + V^y V_y + V^z V_z$. For this to work for Minkowski spacetime, we set $V_t = c^2 V^t$, $V_x = -V^x$, $V_y = -V^y$ and $V_z = -V^z$ as shown in Equation 3.19.

$$\text{Vector: } V^t = t, V^x = x, V^y = y, V^z = z, \quad \text{Dual : } V_t = c^2 t, V_x = -x, V_y = -y, V_z = -z.$$
$$|V|^2 = V^t V_t + V^x V_x + V^y V_y + V^z V_z = t(c^2 t) + x(-x) + y(-y) + z(-z),$$
$$= c^2 t^2 - x^2 - y^2 - z^2,$$
$$= (c\tau)^2. \tag{3.19}$$

In the example above, the metric is diagonal, so the process of switching between upper and lower indices is fairly simple (I will address metrics with off-diagonal terms later). To generate the lower index version, you multiply by the relevant component of the metric $[g_{\mu\nu}]$. Thus, for example, $V_t = g_{TT} V^t$. You can see from the metric that g_{TT} is c^2, which gives: $V_t = g_{TT} V^t = c^2 t$. A similar calculation based on $g_{XX} = -1$ gives $V_x = g_{XX} V^x = -x$. This is called *index lowering* because the index on V^t is lowered to V_t.

Rather than write the long expression $V^tV_t + V^xV_x...$, physicists abbreviate it as $V^\mu V_\mu$ using the Einstein summation convention. The repeated symbol μ indicates that you must sum over all coordinates as shown in Equation 3.20. If you are unfamiliar with this, check out Box 3.5.

$$|V|^2 = V^\mu V_\mu = V^tV_t + V^xV_x + V^yV_y + V^zV_z. \tag{3.20}$$

Box 3.5 Einstein summation convention

The two key rules are:

1) *If a suffix appears twice, you sum over it*
2) *If a suffix appears only once, it can take any value*

In this book, I use capital letters for the indices of tensors to keep things legible, and I use bold face for dummy variables. However, the convention in most textbooks is:

- Greek indices (μ, ν...) indicate spacetime and go from 0–3 ($t = 0$, $x = 1$...)
- Roman indices (i, j...) indicate space only and go from 1–3 ($x = 1$, $y = 2$...)

You can use the lower index of a vector quantity for calculations instead of the upper index, and we often will in this book. It is just as valid an expression of coordinates to work with. Neither upper nor lower measure any proper interval until combined with the other. If this strikes you as odd, bear in mind that the upper index vector coordinates familiar to you, such as seconds for time and metres for distance, cannot be relied on in special or general relativity. One person's second or metre is not the same as another person's. An example of this in Minkowski spacetime is the time dilation and length contraction associated with a moving object (Sections 2.4 and 2.5).

If you work with the lower index form such as V_t, then in order to calculate V^t, you use *index raising*, which is the reverse of index lowering. You multiply with the relevant components of the *inverse metric tensor*, which is labelled $[g^{\mu\nu}]$. If, as in our example, the metric is diagonal (no off-diagonal components), then $g^{TT} = \frac{1}{g_{TT}}$, $g^{XX} = \frac{1}{g_{XX}}$ and so forth. As an example, Equation 3.21 shows index raising of V_t in Minkowski spacetime.

$$V^t = g^{TT}V_t = \frac{1}{g_{TT}}V_t = \frac{1}{c^2}V_t \quad \textit{Index raising, Minkowski metric, } (t,x,y,z). \tag{3.21}$$

The definition of the inverse metric tensor is that the matrix $[g^{\mu\nu}]$ multiplied with the matrix $[g_{\mu\nu}]$ gives the identity matrix showing that the action of the former undoes that of the latter. At the risk of showing the obvious, Equation 3.22 shows how this works for the Minkowski metric using (t,x,y,z) coordinates. Note again that I put brackets around the metric and inverse metric when handling the full matrix rather than a component. If you simply write $g^{\mu\nu}g_{\mu\nu}$, then under the Einstein summation convention it means the sum of all the components $(g^{TT}g_{TT} + g^{TX}g_{TX} + g^{TY}g_{TY} + g^{TZ}g_{TZ} + g^{XX}g_{XX}...)$ that in the case below would total four for the non-zero values $(g^{TT}g_{TT} + g^{XX}g_{XX} + g^{YY}g_{YY} + g^{ZZ}g_{ZZ} = 1 + 1 + 1 + 1 = 4)$. Unfortunately, in many texts you will see $g_{\mu\nu}$ used for both the matrix and its components, so you have to be a bit careful.

$$[g^{\mu\nu}][g_{\mu\nu}] = \begin{bmatrix} \frac{1}{c^2} & 0 & 0 & 0 \\ 0 & -1 & 0 & 0 \\ 0 & 0 & -1 & 0 \\ 0 & 0 & 0 & -1 \end{bmatrix} \begin{bmatrix} c^2 & 0 & 0 & 0 \\ 0 & -1 & 0 & 0 \\ 0 & 0 & -1 & 0 \\ 0 & 0 & 0 & -1 \end{bmatrix} = \begin{bmatrix} 1 & 0 & 0 & 0 \\ 0 & 1 & 0 & 0 \\ 0 & 0 & 1 & 0 \\ 0 & 0 & 0 & 1 \end{bmatrix}. \tag{3.22}$$

We tend to use the upper form of index for coefficients and the lower (dual) form for basis vectors as shown in Box 3.6. These indices are tiresome but essential in order to scale things. Just be grateful that at school you got to use Euclidean/Cartesian coordinates ($ds^2 = dx^2 + dy^2 + dz^2$) with the simple signature $<+++>$, where the spatial dimensions are positive (versus the $<+--->$ I am using for Minkowski spacetime). Using simple Euclidean/Cartesian $<+++>$ coordinates, there is no difference between upper and lower indices ($V^x = V_x$, $V^y = V_y$, $V^z = V_z$), so your teachers did not have to mention them.

Box 3.6 Dual-basis vectors

I suspect you will have come across basis vectors. As a simple example, consider in Euclidean/Cartesian space the vector $(x, y, z) = (2, 5, 3)$. This can be written out in full using the basis vectors $\vec{i}, \vec{j}$ and $\vec{k}$ as: $2\vec{i} + 5\vec{j} + 3\vec{k}$.

When working in curved space with vector quantities (such as velocity), we use the *dual* form for the basis vectors. These are labelled e_x, e_y and so forth (e is just a symbol for the basis vector and not to be confused with the exponential function). Thus, the vector in the example above becomes: $V^x\vec{e}_x + V^y\vec{e}_y + V^z\vec{e}_z = 2\vec{e}_x + 5\vec{e}_y + 3\vec{e}_z$.

The advantage of this is that it correctly scales each part of the vector to reflect the coordinates. But note that it means the length of each basis vector such as $\vec{e}_x$ may change as you move around the surface/space. For example, with the metric of the surface of a sphere as in Equation 3.17, the basis vector $\vec{e}_\phi$ varies across the surface because the basis vector's magnitude is $\sin\theta$, which changes as the value of θ changes.

In this way, the vector remains the same in any coordinate system. For example, a vector $\vec{V}$ in 2-D space can be written in terms of its Cartesian or polar components. Combining upper and lower indices scales the components, so the vector is unchanged: $\vec{V} = V^x\vec{e}_x + V^y\vec{e}_y = V^r\vec{e}_r + V^\theta\vec{e}_\theta$.

And what about a process such as differentiation, which measures rate of change? I hope it is obvious to you that, in this case, the coordinates of the numerator and denominator must *be the same*. Imagine that we are assessing the change in the value of the x component of vector V at two different places a and b. We want to know their magnitude relative to each other. This must be V^x at a divided by V^x at b, or it can be V_x at a divided by V_x at b. The two must both be upper index or both be lower index. Either is fine (indeed they are equivalent), but you cannot meaningfully compare V^x with V_x. The coordinates would not match. That would be, as they say in England, comparing apples with oranges.

In the study of general relativity you will typically see a partial derivative written as something like $\partial_x V^x$. This is the derivative of upper index V^x with respect to *upper* index coordinate x. Convention is to show the upper index denominator as ∂_x (i.e. the upper index version is labelled as lower index when below the line). Thus, we can still check that the number of upper and lower indices match in order to be sure the derivative is balanced. This means there are two equivalent ways of writing a derivative when the upper and lower coordinates match: $\partial_x V^x$ and $\partial^x V_x$. The first has both numerator and denominator upper index x and the second both lower index x. They are the same and are used interchangeably. An example using the Einstein summation convention is shown as Equation 3.23.

$$\partial^\mu V_\mu = \partial^t V_t + \partial^x V_x + \partial^y V_y + \partial^z V_z = \partial_\mu V^\mu = \partial_t V^t + \partial_x V^x + \partial_y V^y + \partial_z V^z. \tag{3.23}$$

3.7 Raising/Lowering Indices with Wonky Metrics (Off-Diagonal Terms)

If the metric is *diagonal*, it has no off-diagonal terms, so the axes are orthogonal. In this case, we can raise or lower indices using a single component of $[g_{\mu\nu}]$ as I have just shown you. For example: $V_x = g_{xx}V^x$. However, this does *not* work if you are using a metric with off-diagonal terms, which I laughingly call a *wonky* metric. In this book, we focus almost exclusively on scenarios with diagonal metrics, but let me introduce you to the somewhat more complicated general method of index raising/lowering that works even for wonky metrics, if only for you to refer back to later.

As an example, let's use the wonky non-orthogonal metric illustrated earlier in Figure 3.1b. The invariant interval and the metric tensor are shown again as Equation 3.24 (the same as Equation 3.15). For simplicity, I will take the liberty of continuing with the $g_{\mu\nu}$ spacetime notation rather than switching to g_{ij} spatial notation. If the upper index components are $V^{x'}$ and $V^{y'}$, then what are the lower index components $V_{x'}$ and $V_{y'}$? If you try to lower the index of the $V^{x'}$ component using $g_{x'x'}$, as you would for a diagonal metric, you multiply by one. That misses the $\cos\theta$ term in Equation 3.24, so it does not match the invariant interval. You cannot ignore the off-diagonal terms.

$$\text{Invariant interval equation: } ds^2 = dx'^2 + dy'^2 + 2\cos\theta\, dx'dy',$$

$$\text{2-D non-orthogonal metric } (x', y'): [g_{\mu\nu}] = \begin{bmatrix} 1 & \cos\theta \\ \cos\theta & 1 \end{bmatrix}. \tag{3.24}$$

The correct method of lowering the index, and this is important, is: $g_{\mu\nu}V^{\mu} = V_{\nu}$ with the repeated μ index being a dummy index under Einstein summation rules. This means that in order to lower the index of $V^{x'}$, you must use $g_{\mu x'}V^{\mu} = V_{x'}$. This looks fairly weird, so let me show you why it is the right answer. The key aim is that $V^{\mu}V_{\mu}$, which is $V^{x'}V_{x'} + V^{y'}V_{y'}$, must match the relevant invariant interval in Equation 3.24.

$$g_{\mu x'}V^{\mu} = g_{x'x'}V^{x'} + g_{y'x'}V^{y'} = (1)x' + (\cos\theta)y' = x' + \cos\theta\, y' = V_{x'}, \tag{3.25}$$

$$g_{\mu y'}V^{\mu} = g_{x'y'}V^{x'} + g_{y'y'}V^{y'} = (\cos\theta)x' + (1)y' = y' + \cos\theta\, x' = V_{y'}, \tag{3.26}$$

$$\implies \quad V^{\mu}V_{\mu} = V^{x'}V_{x'} + V^{y'}V_{y'} = x'(x' + \cos\theta\, y') + y'(y' + \cos\theta\, x'),$$

$$= x'^2 + y'^2 + 2\cos\theta\, x'y'. \tag{3.27}$$

Equation 3.25 shows how multiplying the vector V^{μ} by $g_{\mu x'}$ and summing over μ gives $V_{x'}$. Equation 3.26 shows V^{μ}, multiplied by $g_{\mu y'}$ to give $V_{y'}$. Equation 3.27 pulls things together by showing how the resulting value for $V^{x'}V_{x'} + V^{y'}V_{y'}$ matches the invariant interval equation.

Sadly, there is another complication when you come to index raising with a wonky metric, for which the calculation is: $g^{\mu\nu}V_{\mu} = V^{\nu}$. Working out the inverse matrix $[g^{\mu\nu}]$ becomes more difficult. Box 3.7 shows you how to calculate the inverse of a (2×2) matrix. Indeed, it becomes torturous to try to work out the inverse of a wonky spacetime (4×4) metric. You can see now why we try to restrict most of our scenarios in general relativity to those with diagonal metrics!

Box 3.7 Finding the inverse of a (2 × 2) matrix

$$[g_{ij}] = \begin{bmatrix} a & b \\ c & d \end{bmatrix}, \quad [g^{ij}] = \frac{1}{ad-bc}\begin{bmatrix} d & -b \\ -c & a \end{bmatrix}, \quad [g_{ij}][g^{ij}] = \begin{bmatrix} 1 & 0 \\ 0 & 1 \end{bmatrix}.$$

However if $b = c = 0$:
$$[g_{ij}] = \begin{bmatrix} a & 0 \\ 0 & d \end{bmatrix}, \quad [g^{ij}] = \begin{bmatrix} \frac{1}{a} & 0 \\ 0 & \frac{1}{d} \end{bmatrix}, \quad [g_{ij}][g^{ij}] = \begin{bmatrix} 1 & 0 \\ 0 & 1 \end{bmatrix}.$$

Let me end this gruelling section by expanding the general expression for index raising and lowering to work for any form of 4-D spacetime (with or without off-diagonal terms). An example is shown as Equation 3.28 below using Cartesian coordinates and lowering the x coordinate: $V_x = g_{\mu x}V^{\mu}$. It works with any coordinate system.

Note that if the axes are orthogonal (i.e. in the simpler case of a diagonal metric), then all the off-diagonal components of the metric and inverse metric tensors are zero, and this reduces to index lowering with $V_x = g_{XX}V^x$ and index raising with $V^x = g^{XX}V_x = \frac{1}{g_{XX}}V_x$.

$$V_x = g_{\mu x}V^{\mu} = g_{TX}V^t + g_{XX}V^x + g_{YX}V^y + g_{ZX}V^z \implies g_{XX}V^x \quad \text{if diagonal,}$$

$$V^x = g^{\mu x}V_{\mu} = g^{TX}V_t + g^{XX}V_x + g^{YX}V_y + g^{ZX}V_z \implies \frac{1}{g_{XX}}V_x \quad \text{if diagonal.} \tag{3.28}$$

I may have frightened you with all this. Be not afraid! We will almost exclusively use diagonal (not wonky) metrics. I show you this only to demonstrate why the general formula for raising and lowering indices takes the form it does, shown again for future reference as Equation 3.29. Box 3.8 contains a few test questions if you wish to check your knowledge.

$$\textit{Index lowering: } g_{\mu\nu}V^{\mu} = V_{\nu}, \qquad \textit{Index raising: } g^{\mu\nu}V_{\mu} = V^{\nu}. \tag{3.29}$$

Box 3.8 Understanding upper and lower indices: some questions

A few questions for you to mull over. For answers, see Box 3.10 at the end of this chapter.

1. If the metric is diagonal, how do you calculate V_x from V^x and vice versa?
2. For the surface of a unit sphere (Equation 3.17), if $V^\phi = \pi$, what is V_ϕ?
3. Non-orthogonal (wonky) metric (Equation 3.24): if $V^{y'} = 5$ and $V^{x'} = 2$, what is $V_{y'}$?
4. Using the Minkowski metric with spherical coordinates and the $< + - - - >$ signature (Equation 3.6): if $V_\theta = \frac{\pi}{r}$, what is V^θ?
5. What is the difference in value between $\partial^\mu V_\mu$ and $\partial_\mu V^\mu$?

3.8 Summary

Let's recap what we have covered so far. In Chapter 2, we looked at the mathematical implications of the constant speed of light. The outcome is that time and space are linked in a way that nobody could have imagined before Einstein. In this chapter, I showed you how the invariant interval equation of the Minkowski spacetime of special relativity can be expressed as a tensor called the *spacetime metric*, labelled $[g_{\mu\nu}]$, and written in the form of a (4×4) matrix.

Einstein did not know much about curvature when he started work on general relativity. However, he knew from special relativity that time and space are distorted when measured by a moving observer. He wondered if there might be background distortion in the spacetime metric around zones that contain mass/energy; some sort of curvature that we perceive as gravity.

To simplify discussion of this, I switched to 2-D spatial surfaces. In this case, being 2-D, the metrics are (2×2) matrices. We looked at how changing the coordinates alters the metric for the same flat surface, using first polar coordinates and then an example of axes that are not orthogonal to each other (i.e. not perpendicular). We also studied a metric with curvature, using as our example the 2-D surface of a unit sphere. These 2-D spatial metrics are shown again as Equation 3.30, all using the $< + + + >$ signature for simplicity.

$$
\begin{array}{cccc}
Flat\ (x,y) & Polar\ (r,\theta) & Angled\ (x',y') & Unit\ sphere\ (\theta,\phi) \\[4pt]
\begin{bmatrix} 1 & 0 \\ 0 & 1 \end{bmatrix} &
\begin{bmatrix} 1 & 0 \\ 0 & r^2 \end{bmatrix} &
\begin{bmatrix} 1 & \cos\theta \\ \cos\theta & 1 \end{bmatrix} &
\begin{bmatrix} 1 & 0 \\ 0 & \sin^2\theta \end{bmatrix}
\end{array}
\qquad (3.30)
$$

One lesson from these examples is that you cannot always tell if the surface is curved just by glancing at the metric. If the metric of the entire surface is free of coordinate variables and has no off-diagonal terms (such as the flat metric in 3.30), then you can be sure it is flat. However, the opposite is not true. You can see this by comparing the polar and unit sphere example metrics. They look similar, but the polar example is flat space while the sphere example is curved.

The presence of one of the coordinate variables in the metric (such as the r^2 in the polar example or the $sin^2\theta$ for the surface of the sphere) creates challenges because it means that the value of the polar $d\theta$ step and the sphere $d\phi$ step vary in terms of distance ds across the surface. The value depends on where you are on the surface. The presence of off-diagonal terms indicates the coordinate axes are not orthogonal and also creates problems. Fortunately, virtually all our work will be on diagonal metrics.

I introduced you to upper and lower indices. These can be a bit confusing but are essential for calculations to give answers based on the invariant interval of the metric, and those are the only answers that are meaningful. You can do calculations with the upper index variety (which I call vector coordinates) or with the lower ones (dual coordinates). And you can toggle between the two by multiplying with the appropriate components of the metric $[g_{\mu\nu}]$ (lowering) or its inverse $[g^{\mu\nu}]$ (raising). Things get complicated if the coordinates are not orthogonal to each other, because off-diagonal components appear in the metric. This is why the general formula for raising/ lowering indices in Equation 3.29 takes the form it does. Fortunately, there is no difference between upper and lower coordinates in flat Euclidean space using the $< + + + >$ signature and Cartesian coordinates (x, y, z), so you probably avoided all this complexity in your school exams (note that Einstein was no fan of exams, see Box 3.9).

Box 3.9 Einstein's nightmare

Einstein expressed his opinion of final school exams in a treatise titled *Nightmare*. I suspect many students will empathise with this theme:

Many have been plagued, into their later years, by nightmares whose origins trace back to the final school exams... which are furthermore harmful because they lower the level of teaching in the last school years. Instead of an exclusively substance-oriented occupation with the individual subjects, one too often finds a lapse into a shallow drilling of the students for the exams. Instead of in-depth teaching one gets a more or less showmanship exercise designed to give the class a certain lustre in front of the examiners.

More than anything else, there is one topic in this chapter that I want you to remember. Put the maths to one side, and think about the two dung beetles in Section 3.4. When you are next having a relaxed shower or soaking in the bathtub, please ponder these smelly friends crawling along the surface of their sphere. Their initially parallel paths inexorably draw together. From their perspective, they have no knowledge of curvature. As one beetle moves along, it sees itself (and other objects with it) accelerate towards the other beetle, always at the same rate whatever the mass. The acceleration is the same for all objects... just like gravity.

If the dung beetles allow their paths to draw together, they feel no force... just like freefall in a gravitational field. They feel a force only if they resist the natural motion (in my example, I propose they push against a stick held between them), just like our sense of the force between our feet and the ground. Is it any wonder that Einstein suspected a link between curvature and gravity? In his words: *the gravitational field exhibits a most remarkable property... bodies that are moving under the sole influence of a gravitational field receive an acceleration that does not in the least depend either on the material or on the physical state of the body.*

But here is a question for you contemplate. What happens when the dung beetles stop crawling? Their paths halt, so they no longer draw together. The effect disappears. Is there a parallel with gravity? Does this mean that gravity should disappear if we stop moving? Yes and no. Yes in theory, if we could stop. But no in reality, because our path is through spacetime, not just space. Unlike the dung beetles, we can *never stop*. Standing still as a stone, we continue to travel through time. That is why gravity is always there.

In Chapter 4, we are going to have to kiss some mathematical frogs. You will learn that we need a special covariant form of differentiation to handle curvature. And that leads to some weird functions called Christoffel symbols. Their structure is a bit convoluted and serpentine. However, the prize will be to go one step further and show mathematically that time dilation can look like gravity. So pour yourself a nice cup of tea and prepare for some slightly trickier maths. Don't worry. We will take it step by step.

Box 3.10 Understanding upper and lower indices: answers

1. If diagonal, the metric has no off-diagonal components, $V_x = g_{xx} V^x$ and $V^x = g^{xx} V_x$
2. $V_\phi = g_{\phi\phi} V^\phi = \sin^2 \theta \pi$
3. $V_{y'} = g_{\mu y'} V^\mu = g_{x'y'} V^{x'} + g_{y'y'} V^{y'} = 2\cos\theta + 5$
4. $V_\theta = g_{\theta\theta} V^\theta = (-r^2)\frac{\pi}{r} = -\pi r$. It is negative because of the $< + - - - >$ signature.
5. There is no difference. $\partial^\mu V_\mu$ uses lower indices for enumerators and denominators; $\partial_\mu V^\mu$ uses upper indices for both. As enumerators and denominators have the same index, any difference due to using upper and lower coordinates cancels out.

I hope readers are familiar with basic calculus, but some may not have had to use it for many years. The calculations in this book frequently involve differentiation, so I offer Box 3.11 as a simple refresher for those who need it.

Box 3.11 Differentiation: a refresher

Differentiation is used to calculate the rate at which something changes. For example, if you want the velocity of an object moving in the x direction, you calculate the change in x over time t, which is written as $\frac{dx}{dt}$. This is described as differentiating x *with respect to t*.

   The example below shows the basic rule. To differentiate with respect to t, multiply each term by the power of its t factor, and then reduce that power by one. Any term without a t disappears because it doesn't affect the rate of change of x (it only affects the x starting point at $t = 0$).

$$\text{If}: x = 3t^2 + 8t + 5, \implies \frac{dx}{dt} = 6t + 8.$$

   To differentiate with respect to x in the complicated example below, you multiply the first term by -1 because $\frac{1}{x} = x^{-1}$ and reduce the power by one to x^{-2}, which is $\frac{1}{x^2}$. You multiply the second by $\frac{1}{2}$ because $\sqrt{x} = x^{\frac{1}{2}}$ and reduce the power by one to $x^{-\frac{1}{2}}$, which is $\frac{1}{\sqrt{x}}$.

$$\text{If}: y = \frac{3}{x} - 4\sqrt{x} + 5x^3 - 7, \implies \frac{dy}{dx} = -\frac{3}{x^2} - \frac{2}{\sqrt{x}} + 15x^2.$$

   When dealing with trigonometric functions, you just have to learn the rules. A couple of important ones are that $\frac{d\sin\theta}{d\theta} = \cos\theta$ and $\frac{d\cos\theta}{d\theta} = -\sin\theta$.

   You can differentiate by parts using the rule: $\frac{d(ab)}{dx} = a\frac{db}{dx} + b\frac{da}{dx}$. For example (and this is used later in the book):

$$\frac{d\sin^2\theta}{d\theta} = \frac{d(\sin\theta\sin\theta)}{d\theta} = \sin\theta\frac{d\sin\theta}{d\theta} + \sin\theta\frac{d\sin\theta}{d\theta} = 2\sin\theta\cos\theta.$$

   Differentiating trigonometric functions can get tricky, but many sources on the Web give the basic rules. In this book, I always offer tips and pointers for any complicated calculations.

Chapter 4

Covariant Derivatives and Christoffel Symbols

In Chapter 1 *Overview*, I introduced Einstein's brainwave. Accelerating an object increases its velocity and, as a result, its time dilates relative to a stationary observer. If a certain profile of acceleration leads to a certain profile of time dilation, might the reverse be true? Might a change in the background passage of time lead to acceleration? Put another way, might a change to the spacetime metric explain gravitational acceleration? This idea sits at the core of general relativity. In this chapter, I need to introduce a few mathematical gizmos. I apologise that it is a bit heavy on the maths, but your reward will be a simple model of how spacetime curvature leads to what we call gravity.

4.1 Covariant Derivatives

With curvature, we cannot use the normal form of differentiation but must use a distinct approach called *covariant* differentiation to account for the variation in the value of the basis vectors as you move across the surface. To calculate covariant derivatives we use strange mathematical tools called *Christoffel symbols*. Read on.

Imagine a bug living in a one-dimensional (1-D) world. It has a rather boring life being only able to crawl in one direction (labelled as x). Each stride the bug takes is a small movement forwards in the x direction, so let's label each stride as dx. From the bug's perspective, each stride is the same length. However, distortion in the metric means that the actual distance covered by each stride dx varies depending on where the bug is. This may be a real distortion in the metric (we have discussed the concepts of time dilation and length contraction) or due to the choice of coordinates (a 2-D example is the $d\theta$ polar coordinate of the flat space metric in Equation 3.11). Whatever the reason, the actual distance covered by each stride dx varies as the bug moves along in the x direction.

This is shown in Figure 4.1. The top of the figure illustrates the change in dx by using an additional dimension to mimic distortion in the metric as hills and troughs. You can think of each stride being the same distance along the curvy line (the red dots), but with varying progress in the x direction. Don't forget that the extra dimension is just a visual aid and is not real (see Box 4.1).

Or you may find it easier (I do) simply to think of length contraction and expansion happening at different points along the x-axis. Either way, the result is that the length of each stride dx changes in terms of real distance ds along the x-axis. In the metric, the relationship between dx and ds is not constant. It changes with the value of x.

The fun starts when you work with vector quantities. A vector has a magnitude and a direction. You have probably seen before that a two-dimensional (2-D) vector might be labelled something like: $3\vec{i} + 2\vec{j}$ (this would be three in the x-direction plus two in the y-direction). For general relativity, instead of using $\vec{i}$ and $\vec{j}$ as labels for the basis vectors in the x and y directions, we label them $\vec{e}_x$, $\vec{e}_y$... and so forth to cover all coordinate choices (there is more detail back in Box 3.6).

Returning to the example of the bug on the line, let's label each stride with the basis vector label $\vec{e}_x$ as shown at the bottom of Figure 4.1. The distortion in the metric means that the length of the basis vector $\vec{e}_x$ changes in terms of proper distance ds at different points along the x-axis.

Untangling General Relativity: The Intuitive Self-Study Guide, First Edition. Simon Sherwood.
© 2026 John Wiley & Sons Ltd. Published 2026 by John Wiley & Sons Ltd.

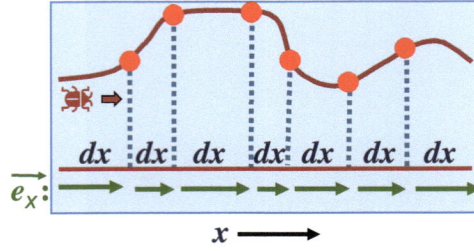

Figure 4.1 The need to use covariant derivatives of vectors.

Suppose that the bug walks steadily at a stride pace of 2 strides per second: $\vec{v} = 2\vec{e}_x$ and we want to calculate how its progress in the x direction changes as it moves along the line. If this was flat space with Cartesian coordinates (x, y, z) the basis vector $\vec{e}_x$ would not change. We would calculate with simple differentiation as shown in Equation 4.1. With a constant stride rate and a constant length of stride there would be no variation in the rate of progress.

$$\vec{v} = 2\vec{e}_x \implies \frac{\partial \vec{v}}{\partial x} = \frac{\partial(2)}{\partial x}\vec{e}_x = 0\vec{e}_x = 0, \qquad \textit{in flat space.} \tag{4.1}$$

The picture is trickier with distortions in the metric. The length of the basis vector $\vec{e}_x$ changes with the value of x, so we must use the product rule of differentiation to include this variation (if needed, refer back to Box 3.11 for the general rules of differentiation). This is called the *covariant derivative* and is labelled ∇ to distinguish it from the simpler process. Equation 4.2 shows the result for our example. Equation 4.3 shows it more generally for a vector $\vec{V}$ with V^x being the vector component in the x direction.

$$\vec{v} = 2\vec{e}_x \implies \nabla_x\vec{v} = \frac{\partial(2)}{\partial x}\vec{e}_x + 2\frac{\partial\vec{e}_x}{\partial x}, \qquad \textit{in curved space,} \tag{4.2}$$

$$\vec{V} = V^x\vec{e}_x \implies \nabla_x\vec{V} = \frac{\partial V^x}{\partial x}\vec{e}_x + V^x\frac{\partial\vec{e}_x}{\partial x}, \qquad \textit{more generally.} \tag{4.3}$$

As you can see, there is a new term in the derivative based on how the basis vector changes. And where do you find information on how the measure of the basis vector changes? Yes... in the metric $[g_{\mu\nu}]$.

At the risk of boring you, let me give a more complete example of a *vector field*. An example might be the movement of water on the surface of the ocean. At each location (x, y), the water is moving with a particular velocity. It has a speed and a direction. We can represent this with a velocity vector $\vec{v}$ at each point (its length is the speed and it points in the direction of motion).

Each velocity vector $\vec{v}$ can be decomposed into components in the x and y directions multiplied by unit basis vectors in those directions: $\vec{v} = v^x\vec{e}_x + v^y\vec{e}_y$. Again, we decide to calculate the partial derivative with respect to x taking into account possible changes to the basis vectors. This is shown as Equation 4.4. We need to take into account how v^x and v^y (the x and y components) *and* how the $\vec{e}_x$ and $\vec{e}_y$ basis vectors vary with the value of x. Tricky stuff!

$$\vec{v} = v^x\vec{e}_x + v^y\vec{e}_y,$$
$$\nabla_x\vec{v} = \frac{\partial v^x}{\partial x}\vec{e}_x + v^x\frac{\partial\vec{e}_x}{\partial x} + \frac{\partial v^y}{\partial x}\vec{e}_y + v^y\frac{\partial\vec{e}_y}{\partial x}, \qquad \textit{covariant derivative.} \tag{4.4}$$

Box 4.1 Adding dimensions is for illustration only

When discussing curvature, physicists frequently use an extra dimension to embed a surface. This is to help you visualise the distortion of the spacetime. Take the example of the bug crawling along the 1-D line in Figure 4.1. It may help you to think of the bug's progress slowing because of hills and troughs (adding an extra dimension), but be warned this is just for illustration. Einstein's theory of general relativity reveals that the presence of energy-momentum distorts spacetime, dilating time and contracting lengths. However, there is no suggestion that these distortions to 4-D spacetime involve some hidden fifth dimension.

In order to differentiate a vector (or tensor) or anything involving basis vectors,[1] we must use the covariant derivative if there is some distortion in the metric from curvature or the coordinates. Let me be more specific. If we are taking a derivative with respect to x, we must use the covariant derivative if the metric depends on x (i.e. if one of the components of $[g_{\mu\nu}]$ has a term involving x).

What is a covariant derivative? It is a combination of two things. The *normal derivative* measures the rate of change of something relative to the coordinates used. The *basis vector derivatives* measure the rate of change of those coordinates relative to the underlying spacetime (as encapsulated in the metric). Taken together, these two combine to give the covariant derivative, which is the true rate of change of something based on the underlying spacetime.

Let me challenge you with a question. Suppose we take the covariant derivative with respect to one of the coordinates (which I will label α) of the spacetime metric itself. What will the result be? What is $\nabla^{\alpha} g_{\mu\nu}$? The answer is in Box 4.7 at the end of the chapter, but think a bit before peeking. The underlying physics is not obvious, but perhaps you can take an intuitive guess.

In order to study the relationship between curvature and gravity, we are going to need to embrace covariant derivatives. Looking at Equation 4.3, it doesn't look too bad. Sorry, but it is, and unravelling it will be our next challenge.

4.2 Christoffel Symbols

For covariant differentiation, you need *Christoffel symbols*. If you have never heard of them, I doubt you are going to like them. They are known as *Cripes-they're-awful* symbols for a reason. Most definitions are not helpful. I cannot resist sharing a couple of quotes: *The Christoffel symbols provide a concrete representation of the connection of pseudo-Riemannian geometry in terms of coordinates on the manifold.* And the following (which, to be fair, is not such a bad effort): *The Christoffel symbol for a metric is the unique torsion-free connection such that the associated covariant derivative operator ∇ satisfies $\nabla g = 0$.*

I am hoping you want the hand-crafted version of Simon (your author), so here it is: *Christoffel symbols are nifty tools that correct your derivatives when working with weird metrics.* It is as simple as that. They are just mathematical tools. I will share some intuitive logic about them and show you how and when to use them, but if you are an obsessive mathematician who is desperate for a full derivation, I suggest you check out the videos of *Eigenchris*.[2] He does a great job of explaining what is a fairly dull topic. Or like me, you may be somebody who is happy to drive a car without opening up the bonnet and looking inside. Now it is time to introduce you properly to these Cripes-they're-awful devices.

Before hammering you with the maths, I should show you how Christoffel symbols fit into the Einstein Field Equations (EFEs). Equation 4.5 shows the EFEs in summary form again. In Chapter 3, I told you that the left side is purely a function of the metric.

$$G_{\mu\nu} \; = \; \frac{8\pi G}{c^4}\, T_{\mu\nu}, \qquad \text{where} \;\; G_{\mu\nu} = R_{\mu\nu} - \frac{1}{2}\, R\, g_{\mu\nu}, \tag{4.5}$$

Spacetime curvature $\Longleftrightarrow$ Energy-momentum.

Specifically, the crucial curvature variables in the EFEs (the Ricci tensor component $R_{\mu\nu}$ and Ricci scalar R) are combinations of the *Christoffel symbol* values for the metric. This makes sense because if there is no distortion of the basis vectors, there is no curvature in the spacetime. Therefore, to work with the EFEs, you need to know how to calculate Christoffel symbols.

---

1 Technical note: this does not apply to a scalar field because it is just a set of numbers with no direction and so does not involve any directional basis vectors. However, it does apply if you want the *second* derivative of a scalar field, because the first derivative of a scalar field is a vector field.

2 *Eigenchris, Tensor Calculus 15, Geodesics and Christoffel Symbols.* There is a series of videos covering, in detail, the maths of general relativity: https://www.youtube.com/watch?v=1CuTNveXJRc (as of February 2025).

4.2.1 What Are Christoffel Symbols?

Christoffel symbols are labelled Γ. They carry three indices, such as Γ_{XY}^{T}. I shall explain the indices in a moment, but note that I will use capitals for their indices (as I do for tensors), because it is easier to read than Γ_{xy}^{t}. We use Christoffel symbols to calculate the value of the derivatives of basis vectors. As a simple starting point, consider our bug-on-a-line 1-D example from Section 4.1 (x dimension only). We derived the formula of the covariant derivative in Equation 4.3, shown again as Equation 4.6. The difference between the usual derivatives you studied at school and covariant derivatives is that we must include a term to account for the variation in the value of the *basis vector* at different points on the line. In this 1-D example, we express the derivative of the basis vector using a Christoffel symbol, which is labelled Γ_{XX}^{X} as shown in Equation 4.7. All three indices of the symbol are the one (and only) dimension x.

$$\nabla_x \vec{V} = \frac{\partial V^x}{\partial x}\, \vec{e}_x + V^x \frac{\partial \vec{e}_x}{\partial x}, \qquad \textit{covariant derivative of vector } \vec{V} \textit{ with respect to } x, \qquad (4.6)$$

$$\frac{\partial \vec{e}_x}{\partial x} = \Gamma_{XX}^{X}\, \vec{e}_x, \qquad \textit{Christoffel symbol for the 1-D basis vector derivative.} \qquad (4.7)$$

The derivatives of the basis vectors are essential to understanding the curvature of a surface and its geodesics. Why? Using the analogy of a map, the derivatives of the basis vectors tell us how the scale changes over the surface. When we start discussing the curvature of general relativity, the derivatives of the basis vectors tell us how the length of a ruler and the ticks on a clock change as we move through time and space. The Christoffel symbols give us their values. Before we dig into the maths, you might want to check out Box 4.2 for a bit of history on how Einstein coped with all this.

Box 4.2 Christoffel, Levi-Civita and Einstein

Elwin Bruno Christoffel (1829–1900) invented covariant differentiation and Christoffel symbols in 1869, but it was Tullio Levi-Civita (along with Gregorio Ricci-Curbastro) who developed the connection with curvature more fully some 40 years later.

Levi-Civita helped Einstein to refine the use of curvature tensors in his theory. Einstein showed his admiration, writing: *I admire the elegance of your method of calculation. It must be nice to ride through these fields upon the horse of true mathematics, while the likes of us have to make our way laboriously on foot.* Asked in an interview what he most liked about Italy, Einstein is reputed to have answered: *Spaghetti and Levi-Civita.*

Why are there three indices on each Christoffel symbol? It is because, unfortunately, things get complicated when we work with more than one dimension. Let's move up to two dimensions, x and y. This was the case with the vector field we studied earlier, arriving at the ∇_x covariant derivative in Equation 4.4, shown again for ease of reference as Equation 4.8.

$$\vec{V} = V^x\, \vec{e}_x + V^y\, \vec{e}_y, \qquad \nabla_x \vec{V} = \frac{\partial V^x}{\partial x}\, \vec{e}_x + V^x \frac{\partial \vec{e}_x}{\partial x} + \frac{\partial V^y}{\partial x}\, \vec{e}_y + V^y \frac{\partial \vec{e}_y}{\partial x}. \qquad (4.8)$$

The right of Equation 4.8 includes $\frac{\partial \vec{e}_x}{\partial x}$ and $\frac{\partial \vec{e}_y}{\partial x}$. We also need $\frac{\partial \vec{e}_x}{\partial y}$ and $\frac{\partial \vec{e}_y}{\partial y}$ if we want the covariant derivative with respect to y, which is $\nabla_y \vec{V}$. Thus, there are four derivatives of basis vectors in 2-D. The expression for these in terms of Christoffel symbols is shown in Equation 4.9.

$$\frac{\partial \vec{e}_x}{\partial x} = \Gamma_{XX}^{X}\, \vec{e}_x + \Gamma_{XX}^{Y}\, \vec{e}_y, \qquad \frac{\partial \vec{e}_y}{\partial x} = \Gamma_{YX}^{X}\, \vec{e}_x + \Gamma_{YX}^{Y}\, \vec{e}_y,$$

$$(4.9)$$

$$\frac{\partial \vec{e}_x}{\partial y} = \Gamma_{XY}^{X}\, \vec{e}_x + \Gamma_{XY}^{Y}\, \vec{e}_y, \qquad \frac{\partial \vec{e}_y}{\partial y} = \Gamma_{YY}^{X}\, \vec{e}_x + \Gamma_{YY}^{Y}\, \vec{e}_y, \qquad \textit{2-D basis vector derivatives.}$$

Things are getting complicated. Note, for example, that in 2-D, the derivative of the x basis vector (top left of Equation 4.9) can point in the x direction (the $\vec{e}_x$ part) and all or partly in the y direction (the $\vec{e}_y$ part). The value of the Christoffel symbol Γ_{XX}^{X} is the magnitude of the component in the x direction and Γ_{XX}^{Y} is the magnitude in the y direction.

$$\boldsymbol{\Gamma}^{X}_{YZ} = \frac{2}{y} \qquad \qquad \frac{\partial \overrightarrow{e_y}}{\partial z} = \frac{2}{y} \overrightarrow{e_x}$$

Output component

Example value

Input basis vector

Derivative with respect to this

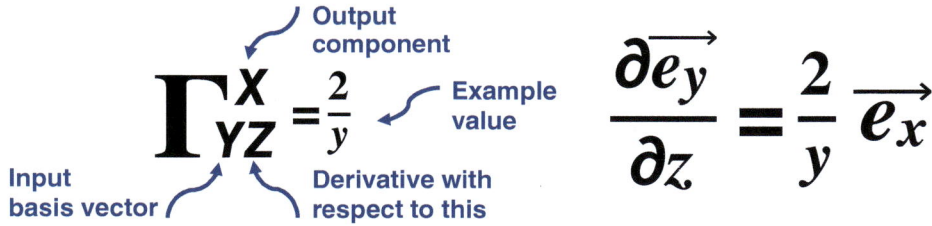

Figure 4.2 Christoffel symbols: explanation of indices and an example.

The bad news is that spacetime has four dimensions. This means that each derivative involves four Christoffel symbols, because it can have components in all four directions (one of which is time). The expression for two example derivatives is shown as Equation 4.10. You can probably see from the examples how the Christoffel symbol indices work, but there is also a summary in Figure 4.2.

$$\frac{\partial \vec{e}_x}{\partial x} = \Gamma^T_{XX}\,\vec{e}_t + \Gamma^X_{XX}\,\vec{e}_x + \Gamma^Y_{XX}\,\vec{e}_y + \Gamma^Z_{XX}\,\vec{e}_z,$$

$$\frac{\partial \vec{e}_t}{\partial y} = \Gamma^T_{TY}\,\vec{e}_t + \Gamma^X_{TY}\,\vec{e}_x + \Gamma^Y_{TY}\,\vec{e}_y + \Gamma^Z_{TY}\,\vec{e}_z, \qquad \textit{4-D basis vector derivatives.}$$

(4.10)

In total, there are 16 possible derivatives of the basis vectors in spacetime: four basis vectors varying as we move through spacetime in four possible directions. There are four Christoffel symbols for each one (each derivative can have components in four directions) giving a grand total of 64 Christoffel symbols for 4-D spacetime. Things are somewhat improved because, although not obvious from Figure 4.2, it turns out that the value is the same if you switch the two lower indices, so $\Gamma^A_{BC} = \Gamma^A_{CB}$, which means 24 are repeats, but that still leaves 40 distinct Christoffel symbols (yes - 40). Your reaction may be the same as mine when I first learned it: *Aarrgghh!*

4.2.2 Calculating the Value of Christoffel Symbols

The value of each Christoffel symbol is calculated using derivatives of the metric. Although each mathematical step is fairly straightforward, there are many of them, so it is easy to make mistakes. The formula is shown below in Box 4.3. Let's step through it slowly. Γ^A_{BC} is the Christoffel symbol we are calculating. Convention is to use Greek symbols for the indices, but I find that confusing, so I will avoid it. Index **D** in the formula is a dummy index, which means you have to work out the formula for every possible value of **D** and add them all together (I will put dummy indices in bold to make things as clear as possible). The g^{AD} term is that from the *inverse metric tensor $g^{\mu\nu}$* (check back to Section 3.6 for a reminder).

The first term inside the brackets is the derivative of one of the entries in the metric. For example, if the dummy index **D** was x and the indices B and C (which both come from the Christoffel symbol) were y and z respectively, then the derivative would be $\frac{\partial g_{XZ}}{\partial y}$, i.e. you differentiate the g_{XZ} entry from the metric with respect to y.

Box 4.3 Christoffel symbols: formula   **where D is a dummy index**

$$\Gamma^A_{BC} = \frac{1}{2}\, g^{AD} \left(\frac{\partial g_{DC}}{\partial B} + \frac{\partial g_{DB}}{\partial C} - \frac{\partial g_{BC}}{\partial D} \right).$$

Before I simplify things, let me reveal the total horror of what this calculation can look like. We sum over all possible values of the dummy index **D** so, in Cartesian coordinates, that means we need to add the results for **D** being t, x, y and z. For example, to calculate the Christoffel symbol Γ^T_{XX} in spacetime (which is 4-D), we need the steps in Equation 4.11, which is one long equation running to two lines.

$$\Gamma^T_{XX} = \frac{1}{2}g^{TT}\left(\frac{\partial g_{TX}}{\partial x} + \frac{\partial g_{TX}}{\partial x} - \frac{\partial g_{XX}}{\partial t}\right) + \frac{1}{2}g^{TX}\left(\frac{\partial g_{XX}}{\partial x} + \frac{\partial g_{XX}}{\partial x} - \frac{\partial g_{XX}}{\partial x}\right)\cdots$$

$$\cdots + \frac{1}{2}g^{TY}\left(\frac{\partial g_{YX}}{\partial x} + \frac{\partial g_{YX}}{\partial x} - \frac{\partial g_{XX}}{\partial y}\right) + \frac{1}{2}g^{TZ}\left(\frac{\partial g_{ZX}}{\partial x} + \frac{\partial g_{ZX}}{\partial x} - \frac{\partial g_{XX}}{\partial z}\right). \tag{4.11}$$

Hahaha! I have toyed with you enough. We are *not* going to need this 12-derivative nightmare, although it is best that you know it exists. Almost all of the complexity comes from potential *off-diagonal* entries in the metric. These entries exist only if the coordinates used in the metric are not orthogonal... if it is what I have been calling a wonky metric. If there are cross terms, some distortions can leak across from one coordinate to others because a small step in one direction involves a step in another. If you need to, check back to the metric of the non-orthogonal angled axes of Equation 3.15 in Section 3.3.

All of our detailed analysis will be with diagonal metrics, i.e. ones with no off-diagonal terms. This means the only non-zero entries in the metric are terms such as g_{TT}, g_{XX}, g_{YY} and g_{ZZ}. This dramatically (very dramatically) simplifies things. If the metric $[g_{\mu\nu}]$ is diagonal, then the inverse metric $[g^{\mu\nu}]$ also is diagonal. Therefore, with no off-diagonal terms in the metric $g^{TX} = g^{TY} = g^{TZ} = 0$, which wipes out 9 of the 12 terms in Equation 4.11. The first two derivative terms are also wiped out because $g_{TX} = 0$. This leaves only *one* term.

There is yet another simplification. For a diagonal metric, the entries in its inverse metric are, well, simply their inverse so, $g^{TT} = \frac{1}{g_{TT}}$. Taken together, this reduces Equations 4.11 to 4.12.

$$\Gamma^T_{XX} = -\frac{1}{2}\,g^{TT}\left(\frac{\partial g_{XX}}{\partial t}\right) = -\frac{1}{2g_{TT}}\left(\frac{\partial g_{XX}}{\partial t}\right), \qquad \textit{for diagonal metrics.} \tag{4.12}$$

Rejoice! The great news is that every Christoffel symbol for a *diagonal* metric reduces to only *one* derivative term (with no dummy index needed). It gets even better. For these diagonal metrics, any Christoffel symbol Γ^A_{BC} where the indices are all different ($A \neq B \neq C$) is zero. You can see this from the formula in Box 4.3. For a diagonal metric, dummy index **D** can only be A because any other value such as g^{AB} is zero. And, if **D** $= A$, and $B \neq C$, then all three terms in the bracket are zero.

We can use all of this to create a *Christoffel cheat sheet* for *diagonal* metrics (no off-diagonal terms). Personally, I would not undertake any Christoffel calculation without it. The formulas are shown in Box 4.4. The proof of the formulas, if anybody wants it, is at the end of this section in Box 4.5.

Box 4.4   Christoffel cheat sheet for diagonal metrics: A ≠ B ≠ C

$$\Gamma^A_{AA} = \frac{1}{2g_{AA}}\left(\frac{\partial g_{AA}}{\partial A}\right), \qquad \Gamma^A_{BB} = -\frac{1}{2g_{AA}}\left(\frac{\partial g_{BB}}{\partial A}\right),$$

$$\Gamma^A_{AB} = \Gamma^A_{BA} = \frac{1}{2g_{AA}}\left(\frac{\partial g_{AA}}{\partial B}\right), \qquad \Gamma^A_{BC} = 0.$$

The key to it is that $A \neq B \neq C$. Let me explain with an example. The cheat sheet also works for 2-D and 3-D diagonal metrics, so let's use the metric for flat space in polar coordinates as in Equation 3.30, shown again below as Equation 4.13.

$$\begin{bmatrix} 1 & 0 \\ 0 & r^2 \end{bmatrix}, \qquad \textit{2-D, flat, polar coordinates } (r, \theta) \textit{ so } g_{RR} = 1 \;\; g_{\theta\theta} = r^2. \tag{4.13}$$

This is a diagonal metric (no off-diagonal terms) so we can use the cheat sheet. Imagine you need the value of Γ^R_{RR} (remember I am using capital indices for legibility). You can see at once that it is of the Γ^A_{AA} type and read off the formula from the cheat sheet. The symbol $\Gamma^R_{\theta\theta}$ is of the Γ^A_{BB} type and $\Gamma^R_{R\theta}$ is of the Γ^A_{AB} type. The calculations are shown as Equation 4.14 (general rules of differentiation are in Box 3.11 if needed).

$$\Gamma^R_{RR} = \frac{1}{2g_{RR}} \left(\frac{\partial g_{RR}}{\partial r} \right) = \frac{1}{2(1)} \left(\frac{\partial (1)}{\partial r} \right) = 0,$$

$$\Gamma^R_{\theta\theta} = -\frac{1}{2g_{RR}} \left(\frac{\partial g_{\theta\theta}}{\partial r} \right) = -\frac{1}{2(1)} \left(\frac{\partial (r^2)}{\partial r} \right) = -r, \qquad (4.14)$$

$$\Gamma^R_{R\theta} = \Gamma^R_{\theta R} = \frac{1}{2g_{RR}} \left(\frac{\partial g_{RR}}{\partial \theta} \right) = \frac{1}{2(1)} \left(\frac{\partial (1)}{\partial \theta} \right) = 0.$$

For a bit of practice, calculate the values of $\Gamma^\theta_{\theta\theta}$, Γ^θ_{RR} and $\Gamma^\theta_{\theta R}$. The answers respectively are 0, 0 and $\frac{1}{r}$. With the cheat sheet, the Christoffel symbols become relatively simple to calculate. By the end of this book, you will be very grateful for this shortcut. One final piece of good news. I am using signature $< + - - - >$ with time positive in the metric and space negative. I suspect some of your professors and textbooks use the opposite signature $< - + + + >$. This does *not* change the value of the Christoffel symbols because the change of sign cancels out (try it and see). The cheat sheet works for both signatures.

Box 4.5 Christoffel cheat sheet detail for diagonal metrics: A ≠ B ≠ C

Christoffel Symbols: $\Gamma^A_{BC} = \frac{1}{2} g^{AD} \left(\frac{\partial g_{DC}}{\partial B} + \frac{\partial g_{DB}}{\partial C} - \frac{\partial g_{BC}}{\partial D} \right).$

If the metric has only diagonal terms, then the dummy variable $\mathbf{D} = A$ because for all other indices $g^{AD} = 0$. In diagonal metrics, only entries with the same index, such as g_{AA} are non-zero. For a diagonal metric, the inverse metric values can be calculated using $g^{AA} = \frac{1}{g_{AA}}$:

$$\Gamma^A_{AA} = \frac{1}{2} g^{AA} \left(\frac{\partial g_{AA}}{\partial A} + \frac{\partial g_{AA}}{\partial A} - \frac{\partial g_{AA}}{\partial A} \right) = \frac{1}{2g_{AA}} \left(\frac{\partial g_{AA}}{\partial A} \right),$$

$$\Gamma^A_{BB} = \frac{1}{2} g^{AA} \left(\frac{\partial g_{AB}}{\partial B} + \frac{\partial g_{AB}}{\partial B} - \frac{\partial g_{BB}}{\partial A} \right) = -\frac{1}{2g_{AA}} \left(\frac{\partial g_{BB}}{\partial A} \right),$$

$$\Gamma^A_{AB} = \Gamma^A_{BA} = \frac{1}{2} g^{AA} \left(\frac{\partial g_{AA}}{\partial B} + \frac{\partial g_{AB}}{\partial A} - \frac{\partial g_{BA}}{\partial A} \right) = \frac{1}{2g_{AA}} \left(\frac{\partial g_{AA}}{\partial B} \right),$$

$$\Gamma^A_{BC} = \frac{1}{2} g^{AA} \left(\frac{\partial g_{AC}}{\partial B} + \frac{\partial g_{AB}}{\partial C} - \frac{\partial g_{BC}}{\partial A} \right) = 0.$$

4.3 Summary

This chapter was all about covariant differentiation and Christoffel symbols. Christoffel symbols can be thought of simply as nifty tools that correct your derivatives when working with weird metrics. For curved spaces and/or those with coordinates that are not plain vanilla (x, y, z) Euclidean/Cartesian, or (t, x, y, z) flat Minkowski spacetime, you have to calculate covariant derivatives when differentiating a vector quantity. This adds extra terms because the basis vectors vary, so the derivatives of the basis vectors are non-zero. These additional terms are what the Christoffel symbols represent.

The value of a Christoffel symbol depends on a complicated combination of derivatives of the components in the metric (the formula is in Box 4.3). For diagonal metrics (i.e. with no off-diagonal components), I gave you a cheat sheet to make calculating these Christoffel symbols easier (Box 4.4 with a more detailed explanation in Box 4.5). In fact, for diagonal metrics, it is relatively easy.

We will be using Christoffel symbols in calculations throughout this book. They are essential in the process of covariant differentiation, which we use to quantify the difference between curved and flat spacetime. In Chapter 5, we will use Christoffel symbols to determine how geodesics change when spacetime is curved and show that Einstein's crazy idea of a link between time dilation and Newton's gravity was not so crazy after all. But check out Box 4.6 for an idea of his that really was crazy.

Box 4.6 Another of Einstein's crazy ideas

In 1914, Einstein embarked on an adulterous affair with his cousin. When Mileva (his wife at that time) found out, she left the family home. Later, she told Einstein that she had decided to move back in with him. In response he sent her a demand note, which was described by his biographer Hans Ohanian as *without doubt the oddest and most obnoxious item in the large collection of Einstein's papers*. After some vacillation (surprisingly), she left (unsurprisingly). Einstein had laid out the following conditions in writing:

(A). You will see to it (1) that my clothes and my wash are kept in order, (2) that three meals are regularly served to me in my room, (3) that my bedroom and my workroom are always kept in good order, and particularly that my desk is exclusively at my disposal.

(B). You renounce any personal relationship with me, except when absolutely required for social reasons. In particular, you renounce (1) that I sit with you at home, (2) that I go out with you or travel with you.

(C). You agree explicitly to observe the following points in your relationship with me. (1) You are not to expect tenderness from me nor are you to make any accusations, (2) when you direct any talk at me, you are to stop immediately if I demand it, (3) you are to leave my bedroom or workroom immediately without contradiction if I demand it.

(D). You agree not to denigrate me by word or by deed in the eyes of the children.

Box 4.7 The covariant derivative of the metric

Any covariant derivative of the spacetime metric is *zero*. A full proof requires delving into some more complicated physics, so let me try to give you an intuitive argument.

The covariant derivative with respect to a certain coordinate is the sum of two components (see, for example, Equation 4.6). It is the sum of (1) the rate of change of something with respect to the coordinate, plus (2) the rate of change of the coordinate with respect to the metric.

Consider any covariant derivative of the *metric* such as $\nabla^\alpha g_{\mu\nu}$, where α, μ and ν each can be any of the coordinates. (1) measures how the metric changes relative to the α coordinate. (2) measures how the α coordinate changes relative to the metric. So (2) is equal but opposite to (1). They cancel out and the answer is zero.

Therefore, covariant derivatives such as $\nabla^\alpha g_{\mu\nu}$ or $\nabla_\alpha g^{\mu\nu}$ are zero. This will become significant later in the book when we derive the EFEs and discuss the cosmological constant.

Chapter 5

The Geodesic Equation and Gravity

<table>
<tr><td>Box 5.1 Don't do it, Albert!</td></tr>
<tr><td>Max Planck followed Einstein's tortuous struggle. The following was Planck's advice to Einstein in regard to his search for the theory of general relativity:
As an older friend I must advise against it… In the first place you won't succeed; and even if you succeed, no one will believe you.</td></tr>
</table>

Now that you know how to calculate the value of Christoffel symbols, let's use them! In this chapter, we are going to work on the important question. Might what we perceive as gravitational acceleration be linked with time dilation? If we release an object (along with a clock to measure its time) in a gravitational field, it will accelerate towards an attracting mass. The theory of special relativity shows that, as the clock speeds up, we see its time tick more slowly. We can calculate how its time changes as it accelerates in the gravitational field.

Before general relativity, one would say that the clock's velocity from gravitational acceleration leads to its time dilation. Einstein's brilliant wheeze was to turn this on its head. Rather than the gravitational acceleration causing time dilation, he proposed that background time dilation causes the acceleration. Can curvature of spacetime by time dilation change the geodesics (the path of unaccelerated objects) in a way that resembles gravitational acceleration? Does this work? Spoiler alert: *yes*, out pops gravitational acceleration.

Once I have shown you that, the next challenge will be to find measurable differences between Einstein's theory and that of Newton (see Box 5.1 for Max Planck's doubts about whether this would be possible). A closer examination of the geodesics will reveal that there is a divergence between the theories for objects moving at high speed in strong gravitational fields.

5.1 A 2-D Model of Time Dilation and Gravitational Acceleration

Our cunning plan is to build into the spacetime metric $[g_{\mu\nu}]$, time dilation that matches the profile of a clock accelerating in a gravitational field. We will calculate (with the help of our super-duper Christoffel symbols) the geodesics of the spacetime in order to know how undisturbed objects would behave.

We will use the simplest possible model in 2-D: time t plus one spatial dimension x. The model is illustrated in Figure 5.1. The clock starts out at an infinite distance from the mass. As it picks up speed accelerating towards the mass, the time dilation factor means that the clock tick dt changes from the perspective of a stationary observer. Out on the (only) x dimension,[1] there is a lone stationary dung beetle minding its own business. If we build this profile of time dilation into the background metric, do we see the dung beetle accelerate towards the mass?

---

1 I am using x rather than r to avoid any suggestion that these are 3-D spherical coordinates.

Untangling General Relativity: The Intuitive Self-Study Guide, First Edition. Simon Sherwood.
© 2026 John Wiley & Sons Ltd. Published 2026 by John Wiley & Sons Ltd.

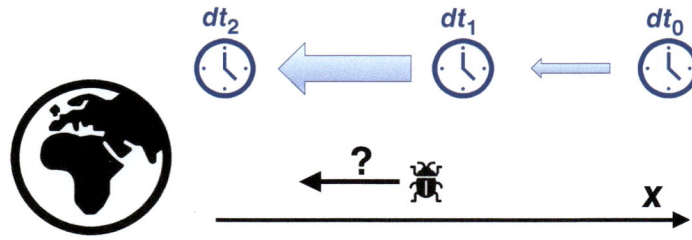

Figure 5.1 2-D model: Einstein's time dilation or Newton's gravity?

The first job is to calculate time dilation based on Newton's laws (i.e. Newton's law of gravity rather than Einstein's theory of general relativity). If we start the clock stationary far away[2] from the mass, its kinetic energy comes entirely from the change in gravitational potential energy, which is $\frac{GMm}{x}$, where G is Newton's gravitational constant, M is the attracting mass, m is the mass of the clock and x is their distance apart.[3] This means $v^2 = \frac{2GM}{x}$, where v is the clock's velocity at distance x from the attracting mass (see Equation 5.1).

$$\frac{GMm}{x} = \frac{1}{2}mv^2, \quad \Longrightarrow \quad v^2 = \frac{2GM}{x}. \tag{5.1}$$

If you cast your mind back to Chapter 2 on special relativity, you may remember that $\frac{dt}{d\tau} = \gamma$, which gives $d\tau = \frac{dt}{\gamma}$. Using the expression for γ that we derived long ago in Equation 2.4, we get the result in Equation 5.2. As a reminder, due to time dilation, $d\tau$ (stationary clock tick) is shorter than dt (clock seen moving).

$$d\tau = \frac{dt}{\gamma} = \sqrt{1 - \frac{v^2}{c^2}}\, dt, \quad \Longrightarrow \quad d\tau^2 = \left(1 - \frac{2GM}{xc^2}\right) dt^2. \tag{5.2}$$

The next step is to build this into our 2-D metric. Newton's gravitational acceleration leads to the relationship between time dilation (dt) and position (x) in Equation 5.2. Our question is the following: if that same relationship between dt and x exists because of the presence of energy-momentum, might it lead to an accelerating effect? Therefore, we adjust the 2-D Minkowski metric to reflect this new relationship as shown in Equation 5.3.

$$c^2\, d\tau^2 = c^2\, dt^2 - dx^2, \quad \Longrightarrow \quad c^2\, d\tau^2 = \left(1 - \frac{2GM}{xc^2}\right) c^2\, dt^2 - dx^2, \quad \textit{Adjusted 2-D model.} \tag{5.3}$$

Let's get back to our scenario. Now that we have the model metric, the next challenge is to find its geodesics, so we can see what happens to our dear old friend, the dung beetle.

5.2 The Geodesic Equation

The formula for finding geodesics is well known (see Box 5.2). If you are already comfortable with it, or you find the maths too much, you may want to skip to Section 5.3. In the following few paragraphs, I am going to derive the geodesic equation (as shown in Box 5.2) for a 2-D diagonal metric composed of time t and one spatial dimension x. This will help you see how and why the geodesic equation works. It also will improve your understanding of Christoffel symbols and why they play such a central role in general relativity.

---

2 Technical note: theoretically it must be an infinite distance away to be outside the gravitational field.
3 If you are unfamiliar with this formula, you can calculate it using Newton's inverse square law in Equation 1.2 and integrating with respect to distance.

Box 5.2 Geodesic equation where A and B are dummy indices

$$\frac{d^2x}{d\tau^2} = \frac{dv^x}{d\tau} = -\Gamma^X_{AB}\frac{d\mathbf{A}}{d\tau}\frac{d\mathbf{B}}{d\tau} = -\Gamma^X_{AB}\,v^{\mathbf{A}}\,v^{\mathbf{B}}.$$

Dummy indices **A** and **B** (in bold) can be any coordinate including x. Using the Einstein summation convention, we sum over all options. For example, for a 2-D space with only t and x dimensions, the full geodesic equation is (shifting the minus sign to the left):

$$-\frac{dv^x}{d\tau} = \Gamma^X_{TT}v^t v^t + \Gamma^X_{TX}v^t v^x + \Gamma^X_{XT}v^x v^t + \Gamma^X_{XX}v^x v^x,$$

$$= \Gamma^X_{TT}v^t v^t + 2\Gamma^X_{TX}v^t v^x + \Gamma^X_{XX}v^x v^x, \qquad \text{2-D geodesic equation.}$$

In physics lingo, an object in an inertial frame of reference follows a geodesic path. Newton would have described geodesics as the paths followed by objects that are not being accelerated. In good old flat Cartesian space (before Einstein messed things up with curvature), geodesics are straight lines travelled at constant velocity. The *derivative* of movement through space with respect to time t is zero. For the metric with just one spatial dimension (t,x), this is: $\frac{dv^x}{dt} = 0$.

Geodesics in curved spacetime are different. In this case, it is the *covariant derivative* of velocity with respect to *proper time* τ that must be zero: $\nabla_\tau \vec{v} = 0$.

The formula of this for the (t,x) metric is shown in Equation 5.4. We cannot ignore that time t and proper time τ may be different. Note also how the derivatives of the basis vectors add extra terms to the covariant derivative. Note that v^t is velocity through *time*. It is $\frac{dt}{d\tau}$. At anything but relativistic speeds, this is almost exactly 1.

$$\nabla_\tau \vec{v} = \frac{dv^t}{d\tau}\vec{e}_t + \frac{dv^x}{d\tau}\vec{e}_x + v^t\frac{d\vec{e}_t}{d\tau} + v^x\frac{d\vec{e}_x}{d\tau}, \qquad \text{for geodesics: } \nabla_\tau\vec{v}=0. \tag{5.4}$$

Equation 5.5 shows the formula from Equation 5.4 set to zero for geodesics. We can simplify things. In Equation 5.6, I use the vector dot product with $\vec{e}_x$. We know that the t and x dimensions are orthogonal because it is a diagonal metric. Those of you familiar with vector arithmetic know that $\vec{e}_x \cdot \vec{e}_x = 1$ and $\vec{e}_t \cdot \vec{e}_x = 0$, so the $\vec{e}_t \cdot \vec{e}_x$ term is zero. We only are interested in motion in the x direction, but must account for the derivative terms of both basis vectors, because they both may have components in the x direction.

$$0 = \frac{dv^t}{d\tau}\vec{e}_t + \frac{dv^x}{d\tau}\vec{e}_x + v^t\frac{d\vec{e}_t}{d\tau} + v^x\frac{d\vec{e}_x}{d\tau}, \qquad \text{for geodesics: } \nabla_\tau\vec{v}=0, \tag{5.5}$$

$$= \frac{dv^t}{d\tau}\vec{e}_t\cdot\vec{e}_x + \frac{dv^x}{d\tau}\vec{e}_x\cdot\vec{e}_x + v^t\frac{d\vec{e}_t}{d\tau}\cdot\vec{e}_x + v^x\frac{d\vec{e}_x}{d\tau}\cdot\vec{e}_x, \tag{5.6}$$

$$= \frac{dv^x}{d\tau} + v^t\frac{d\vec{e}_t}{d\tau}\cdot\vec{e}_x + v^x\frac{d\vec{e}_x}{d\tau}\cdot\vec{e}_x,$$

$$-\frac{dv^x}{d\tau} = v^t\frac{d\vec{e}_t}{d\tau}\cdot\vec{e}_x + v^x\frac{d\vec{e}_x}{d\tau}\cdot\vec{e}_x. \tag{5.7}$$

In Equation 5.8, we address the basis vector terms in Equation 5.7 starting with the derivative of the $\vec{e}_t$ basis vector. We break down the derivative using the product rule (if needed, refer to Box 3.11 for the rules of differentiation). In Equation 5.9 we use the appropriate Christoffel symbols to show the components of the partial derivatives of the basis vectors. In Equation 5.10 we cancel out the terms in the $\vec{e}_t$ direction because we are calculating the result in the $\vec{e}_x$ direction. Working through exactly the same steps for the derivative of the $\vec{e}_x$ basis vector gives Equation 5.11.

$$\frac{d\vec{e}_t}{d\tau} \cdot \vec{e}_x = \left(\frac{dt}{d\tau} \frac{\partial \vec{e}_t}{\partial t} + \frac{dx}{d\tau} \frac{\partial \vec{e}_t}{\partial x} \right) \cdot \vec{e}_x, \tag{5.8}$$

$$= \left(v^t \frac{\partial \vec{e}_t}{\partial t} + v^x \frac{\partial \vec{e}_t}{\partial x} \right) \cdot \vec{e}_x,$$

$$= \left(v^t \Gamma^T_{TT} \vec{e}_t + v^t \Gamma^X_{TT} \vec{e}_x + v^x \Gamma^T_{TX} \vec{e}_t + v^x \Gamma^X_{TX} \vec{e}_x \right) \cdot \vec{e}_x, \tag{5.9}$$

$$= v^t \Gamma^X_{TT} + v^x \Gamma^X_{TX}, \tag{5.10}$$

$$\frac{d\vec{e}_x}{d\tau} \cdot \vec{e}_x = v^t \Gamma^X_{XT} + v^x \Gamma^X_{XX}. \tag{5.11}$$

The hard work is done. We now substitute in the Christoffel symbol expressions from Equations 5.10 and 5.11 into Equation 5.7. This gives Equation 5.12, which is the geodesic equation for a 2-D (t, x) metric and matches that in Box 5.2. The more general formula is shown in Equation 5.13 for comparison.

$$-\frac{dv^x}{d\tau} = \Gamma^X_{TT} v^t v^t + \Gamma^X_{TX} v^t v^x + \Gamma^X_{XT} v^x v^t + \Gamma^X_{XX} v^x v^x, \quad \textit{for 2-D metric,} \tag{5.12}$$

$$= \Gamma^X_{TT} v^t v^t + 2\Gamma^X_{TX} v^t v^x + \Gamma^X_{XX} v^x v^x, \qquad \textit{Note: } \Gamma^X_{XT} = \Gamma^X_{TX},$$

$$-\frac{dv^x}{d\tau} = \Gamma^X_{AB} v^A v^B, \qquad \textit{Geodesic equation more generally.} \tag{5.13}$$

Many of you will be exhausted by the slog, but it should give you a good grasp of things if you battle through it and see how it works. If you have not absorbed it, don't worry. Many courses simply adopt the general geodesic equation that is in Box 5.2. It is a well-known formula. Let's look at what it tells us.

5.3 What Happens to the Dung Beetle?

Welcome back to those readers who skipped or skimmed the geodesic calculation. We are going to jump in and use the geodesic equation, so you have not missed too much. To know what happens to the dung beetle in Figure 5.1, we need to calculate the Christoffel symbols in the geodesic equation. Equation 5.14 shows the adjusted 2-D metric that we built to include time dilation equivalent to that which occurs in gravitational acceleration (for now, we are ignoring any length contraction). Below this metric is the relevant 2-D geodesic equation.

$$(c\,d\tau)^2 = \left(c^2 - \frac{2GM}{x} \right) dt^2 - dx^2, \qquad \textit{Time-dilated 2-D model,} \tag{5.14}$$

$$[g_{\mu\nu}] = \begin{bmatrix} c^2 - \dfrac{2GM}{x} & 0 \\ 0 & -1 \end{bmatrix}, \qquad \textit{2-D model metric } (t, x), $$

$$-\frac{dv^x}{d\tau} = \Gamma^X_{TT} v^t v^t + 2\Gamma^X_{TX} v^t v^x + \Gamma^X_{XX} v^x v^x, \quad \textit{2-D geodesic equation.} \tag{5.15}$$

We can simplify the geodesic equation. Using the cheat sheet (see earlier in Box 4.4), gives the expression for Γ^X_{XT} in Equation 5.16. The metric does not have any time dependency (i.e. none of the entries in the metric include variable t), so $\Gamma^X_{XT} = \Gamma^X_{TX} = 0$.

$$\Gamma^X_{XT} = \frac{1}{2g_{XX}} \left(\frac{\partial g_{XX}}{\partial t} \right) = \frac{1}{2(-1)} \left(\frac{\partial(-1)}{\partial t} \right) = 0. \tag{5.16}$$

We can simplify even further. The value of the dung beetle's speed through space v^x is the rate of change in the invariant interval equation of the dx term compared with the $c\,d\tau$ term, which is negligible because of the c factor so: $v^x \approx 0$. In layman's language, the exchange rate between time and space means that travelling 1 second through time is the equivalent of moving 300,000 kilometres through space.

On the other hand, the value of the dung beetle's speed through time v^t is $c\,dt$ compared to $c\,d\tau$, which is $\frac{dt}{d\tau}$. At non-relativistic speeds, t is almost exactly τ, so: $v^t = \frac{dt}{d\tau} \approx 1$. As far as I know, there are no relativistic high-speed dung beetles, so what I am calling approximations are correct to a fabulous degree of accuracy. Removing the zero-value Γ_{XT}^X term and substituting t for τ in Equation 5.15 gives Equation 5.17. The next steps are to ignore v^x as negligible compared with v^t and substitute with $v^t = 1$ to arrive at Equation 5.18.

$$-\frac{dv^x}{dt} = \Gamma_{TT}^X v^t v^t + \Gamma_{XX}^X v^x v^x, \qquad \text{substitute } t \text{ for } \tau, \tag{5.17}$$

$$\approx \Gamma_{TT}^X v^t v^t, \qquad \text{The value of } v^x \text{ is tiny versus } v^t,$$

$$\approx \Gamma_{TT}^X, \qquad \text{The value of } v^t \text{ is almost exactly 1.} \tag{5.18}$$

The last step (phew!) is to calculate the Christoffel symbol Γ_{TT}^X for the metric, using the cheat sheet. This gives the result on the left of Equation 5.19 (I have shifted the minus sign across). This matches gravitational acceleration (Equation 3.16 earlier).

$$\frac{dv^x}{dt} \approx -\Gamma_{TT}^X = \frac{1}{2g_{XX}}\left(\frac{\partial g_{TT}}{\partial x}\right) = \frac{1}{2(-1)}\frac{\partial}{\partial x}\left(c^2 - \frac{2GM}{x}\right),$$

$$= -\frac{GM}{x^2} \implies \frac{dv^x}{dt} = -\frac{GM}{x^2}. \tag{5.19}$$

What happens to the little dung beetle? It will accelerate towards the mass (that is what the negative sign in the acceleration formula means). Our model shows that, if a mass causes time dilation in the surrounding spacetime, then we will see surrounding objects attracted towards it. Einstein's crazy idea works! Special relativity shows that a particular acceleration profile leads to a particular change in the profile of the passage of time. We have now shown that a change in the profile of the background passage of time changes the geodesics and leads to acceleration. Instead of acceleration leading to a change in time... a change in time leads to acceleration. Cool!

This is all very exciting, but a smart student might ask Einstein: *Listen Albert, this curvature effect does look the same as the gravitational force. Your idea is really cute, but my old buddy Isaac had a cute idea too. Something about an apple falling on his head. How can I know which of you two is right?*

Einstein's and Newton's explanations of gravity are not slightly different. They are *completely* different. However, that wouldn't necessarily mean that one could distinguish experimentally between the two theories. Can we know if the poor dung beetle in our crude 2-D model is accelerating because of Einstein's spacetime curvature or because of Newton's gravity?

You will be pleased to know that the answer is *yes*. There are indeed measurable differences between the theories. There is experimental evidence. So, as it hurtles towards the mass, the dung beetle in our model can reflect that it has not all been for nothing.

5.4 Albert Versus Isaac: Differences Emerge

If we look closely, a significant theoretical difference emerges between the theories of Einstein and Newton, even at the level of our primitive 2-D model. The geodesic equation that we developed (Equation 5.17) has two contributing terms. We dismissed the second term that includes Γ_{XX}^X because, at dung-beetle speeds, it is insignificant. However, the term is there. The structure of the Christoffel symbol is shown in Equation 5.20. Its value is affected by any distortion in the g_{XX} component of the metric. This brings us to what Einstein called *the geometrical modification of space*, which in layman's language is the impact of length contraction on the metric.

$$\Gamma_{XX}^X v^x v^x = \frac{1}{2g_{XX}}\left(\frac{\partial g_{XX}}{\partial x}\right)v^x v^x. \tag{5.20}$$

If you need a refresher on length contraction, check back to Section 2.5. In special relativity, time dilation and length contraction always go hand in hand. It is a requirement for Minkowski spacetime to be mathematically consistent. For example, in the apparent paradox of Section 2.7, without length contraction, Rocket woman would travel at faster than light speed.

While time dilation is the factor γ, length contraction is $\frac{1}{\gamma}$. In terms of our crude model, if we accelerated a ruler alongside the clock in Figure 5.1, the length contraction would be inversely proportional to the time dilation. The temptation is to chuck this into the dx term of the invariant interval equation. This would add a term that is the inverse of the dt time dilation of Equation 5.3, resulting in Equation 5.21. Note that, in cases where $\frac{2GM}{xc^2}$ is small (it is only about 10^{-9} at the earth's surface), we can use the Taylor approximation shown.

$$(c\,d\tau)^2 = \left(1 - \frac{2GM}{xc^2}\right)(c\,dt)^2 - \left(1 - \frac{2GM}{xc^2}\right)^{-1} dx^2, \quad \textit{for small } x: \quad \frac{1}{(1-x)} \approx (1+x),$$

$$\approx \left(c^2 - \frac{2GM}{x}\right) dt^2 - \left(1 + \frac{2GM}{xc^2}\right) dx^2, \tag{5.21}$$

$$\begin{bmatrix} c^2 - \dfrac{2GM}{x} & 0 \\ 0 & -1 - \dfrac{2GM}{xc^2} \end{bmatrix}, \qquad \textit{2-D adjusted model metric } (t,x).$$

Things are not so simple. Ideally, we need to measure length distortion (contraction) at one instant in time, so $dt = 0$, but we cannot. What is simultaneous for the moving ruler is not simultaneous for a stationary observer (leading clocks lag). Also, from the perspective of the moving ruler, length contraction will change the distance x to the mass. If we build this into the metric for an equivalent stationary ruler, then whose value of x is correct? And if you include three spatial dimensions, you end up in spherical coordinates with the equivalent of the x term (radius r) appearing as a variable elsewhere in the metric. Complications, complications...

And yet, when we later derive the Schwarzschild metric (the first exact solution found for the Einstein Field Equations [EFEs]), you will see that the metric in Equation 5.21 proves to be a decent illustrative model. I should also note that, before the Schwarzschild metric was found, Einstein used a similar approximation to Equation 5.21, albeit with all three spatial dimensions (later, we will study this too).

So I am cheating a bit. I am using information that we will derive only later in the book. But it is my book, so with a reminder that you cannot simply inject things like time dilation and length contraction into the metric, let's investigate what this spatial distortion does to the geodesics. Starting again from Equation 5.17, we can calculate the value of the Γ^X_{XX} term. This substitutes the value of g_{XX} from the metric in Equation 5.21 into the formula in Equation 5.20. The result is Equation 5.23.

To make the velocities v^t and v^x comparable, we need to show v^x as a proportion of the speed of light, which I label u^x. Thus: $u^x = \frac{v^x}{c}$. The two are equivalent if we set the speed of light to ($c = 1$) as shown in Equation 5.24. The most important point, which may not be obvious at first glance, is that the $u^t u^t$ and $u^x u^x$ terms in Equation 5.24 are *additive*. The term $u^x u^x$ is always a *negative* value because it is $\frac{(dx)^2}{(d\tau)^2}$, which is negative. As you can see clearly in the invariant interval equation, dx^2 and $d\tau^2$ have opposite signs.

$$\frac{dv^x}{d\tau} = -\Gamma^X_{TT}\, v^t v^t - \Gamma^X_{XX}\, v^x v^x, \tag{5.22}$$

$$= -\frac{GM}{x^2} v^t v^t - \frac{1}{2}\left(-1 + \frac{2GM}{xc^2}\right)\frac{\partial}{\partial x}\left(-1 - \frac{2GM}{xc^2}\right)v^x v^x,$$

$$\approx -\frac{GM}{x^2}\left(v^t v^t - \frac{v^x v^x}{c^2}\right), \tag{5.23}$$

$$= -\frac{GM}{x^2}\left(u^t u^t - u^x u^x\right), \qquad (t,x). \tag{5.24}$$

Leaving the numbers to one side, what is this telling us? In general relativity, you will sometimes see Christoffel symbols described as *being something like a gravitational force*. You can see why from this analysis. The effect that we call gravity comes from the change in the geodesics. In turn, this comes from the extra derivative terms that appear in covariant differentiation and are quantified by the Christoffel symbols. The amount of geodesic deviation (versus, say, a straight line) depends on how fast an object is moving through time and space. This makes sense. If time and space are distorted, then the more time and space an object moves through, the larger the distortion... and the greater the geodesic deviation.

In our crude dung beetle model we ignored any distortion related to moving through space because the dung beetle (definitely) isn't moving at relativistic speeds. Even for a supersonic jet u^x is about 10^{-6}, so $u^x u^x$ is about 10^{-12}. But what about objects moving at relativistic speeds? If the first term in Equation 5.24 matches Newton's gravitational force, then when the second term becomes large enough to be meaningful, there will be a discrepancy between the theories.

Putting it another way, the presence of energy-momentum distorts the geodesics and this manifests in two parts. The first is a *velocity-independent* component that is experienced even if an object is spatially stationary relative to a gravitational field. This is the distortion related to its movement through *time*. It is the gravity we sluggish human's experience. It is Newton's gravity.

In addition, there is a *velocity-dependent* component[4] that is related to the object's movement through *space* towards or away from the gravitational source. It is negligible at low speeds in weak gravitational fields, but for higher speeds in stronger fields, it creates testable differences between the theories.

5.5 Albert Versus Isaac: Seeing the Light

One thing that moves at light speed is, of course, light. What does the geodesic equation tell us? In Chapter 6, we will discuss why light must be affected by gravity (this is a feature of the equivalence principle). However, let me intuitively argue that this must be the case.

Think of the two dung beetles crawling over the sphere in Section 3.4. Their paths were drawn together not because they were walking in curves, but because parallel straight lines on a spherical surface gradually get closer to each other. Any object in motion experiences the same effect. The mass of the object does not enter the calculation. It is irrelevant. It would be weird if this did not also apply to light. In curved space, geodesics are the equivalent of straight paths. The geodesics are the paths taken by undisturbed objects, whether the object is massive or massless.

At first glance, we have a problem using the geodesic equation for light. Notice that Equation 5.24, which evaluates $\frac{dv^x}{d\tau}$, includes u^t and u^x terms, both of which are derivatives with respect to proper time τ. But for light, proper time $d\tau = 0$ as you can see from Equation 5.25 for Minkowski spacetime (shown for an object moving at c in the x direction). Does this mean that there are no geodesics for light? Of course not.

$$c^2 d\tau^2 = c^2 dt^2 - dx^2 = c^2 dt^2 - c^2 dt^2 = 0, \qquad \text{At light speed:} \quad \frac{dx^2}{dt^2} = c^2. \qquad (5.25)$$

With light, we cannot use τ as a coordinate variable because for light $d\tau = 0$, but there is a work-around. We do a simple coordinate change to an arbitrary variable λ, with $\lambda = A\tau + B$, where A and B are non-zero constants. This eliminates the zero-value. As $\frac{d\tau}{d\lambda}$ is a constant, the switch to λ does not change the structure of the geodesic equation. λ is called an *affine parameter* (details on the coordinate switch are in Box 5.3, if you want to see them). This gives Equation 5.26 (using $c = 1$). Using λ as the variable, the geodesic equation also works for light.

$$\frac{d^2 x}{d\tau^2} = -\frac{GM}{x^2}\left(\frac{dt}{d\tau}\frac{dt}{d\tau} - \frac{dx}{d\tau}\frac{dx}{d\tau}\right) \implies \frac{d^2 x}{d\lambda^2} = -\frac{GM}{x^2}\left(\frac{dt}{d\lambda}\frac{dt}{d\lambda} - \frac{dx}{d\lambda}\frac{dx}{d\lambda}\right). \qquad (5.26)$$

If we track the dung beetle in our model, its path through space is negligible compared to through time. On the other hand, for light the paths through space and time are the same. This *doubles* the geodesic deviation as it moves towards or away from a gravitational source. For light, the distortion we calculate from the path's passage through time (the effect of time dilation) and the distortion from its passage through space (the effect of length contraction) must be added together.

This means that light will be deflected in a gravitational field twice as much as you would have expected from Newton's law (applying Newton's law to all energy, not just mass). Einstein realised how important this was and described it eloquently when discussing the deflection of light by the sun. In his words: *according to the theory,*

---

4 This is sometimes referred to as gravitomagnetism because of similarities to magnetism in the electromagnetic force.

half of this deflection is produced by the Newtonian field of attraction of the sun, and the other half by the geometrical modification [curvature] *of space caused by the sun. This result admits of an experimental test... during a total eclipse of the sun.*

A word of warning. Obviously, Einstein did not calculate the deflection of light from a crude 2-D model! The calculation needs to be done in 4-D spacetime with the Schwarzschild metric (this calculation is in Chapter 15). However, the principle is the same. When you deal with fast-moving objects and more intense gravitational fields, Einstein's theory differs from that of Newton.

Box 5.3 Using an affine parameter in the geodesic equation

The essential point is that because $\lambda = A\tau + B$, it means $\frac{d\tau}{d\lambda}$ is a *constant*, so we derive:

$$\frac{d^2x}{d\tau^2} = \frac{d}{d\tau}\left(\frac{dx}{d\lambda}\frac{d\lambda}{d\tau}\right) = \frac{d\lambda}{d\tau}\frac{d}{d\tau}\left(\frac{dx}{d\lambda}\right) = \frac{d\lambda}{d\tau}\frac{d\lambda}{d\tau}\frac{d}{d\lambda}\left(\frac{dx}{d\lambda}\right) = \left(\frac{d\lambda}{d\tau}\right)^2\frac{d^2x}{d\lambda^2}.$$

This gives: $\left(\frac{d\tau}{d\lambda}\right)^2\frac{d^2x}{d\tau^2} = \frac{d^2x}{d\lambda^2}$. We use this result and the geodesic equation from Box 5.2. This switches from variable τ (which is zero for light) to the affine variable λ, without altering the structure of the geodesic equation.

$$\left(\frac{d\tau}{d\lambda}\right)^2\frac{d^2x}{d\tau^2} = -\Gamma^X_{AB}\frac{d\mathbf{A}}{d\tau}\frac{d\mathbf{B}}{d\tau}\left(\frac{d\tau}{d\lambda}\right)^2,$$

$$\frac{d^2x}{d\lambda^2} = -\Gamma^X_{AB}\left(\frac{d\mathbf{A}}{d\tau}\frac{d\tau}{d\lambda}\right)\left(\frac{d\mathbf{B}}{d\tau}\frac{d\tau}{d\lambda}\right) \quad \Longrightarrow \quad \frac{d^2x}{d\lambda^2} = -\Gamma^X_{AB}\frac{d\mathbf{A}}{d\lambda}\frac{d\mathbf{B}}{d\lambda}.$$

5.6 A Victory for Einstein

In May 1919, 4 years after Einstein had published his theory, two British expeditions set out to measure the deviation in the apparent position of stars during a solar eclipse. The project was dubbed the Eddington experiment after its leader, Arthur Eddington. One team headed to Sobral in Brazil and the other to Principe, an island off the coast of West Africa.

In a solar eclipse, the stars in the sky around the sun become visible, allowing their positions to be measured. Any bending of the starlight in the sun's gravitational field changes their apparent position (see Figure 5.2). Einstein predicted an angular deflection of 1.75 arcseconds (an arcsecond is 1/3,600 of a degree). This is twice that expected from Newton's theory.

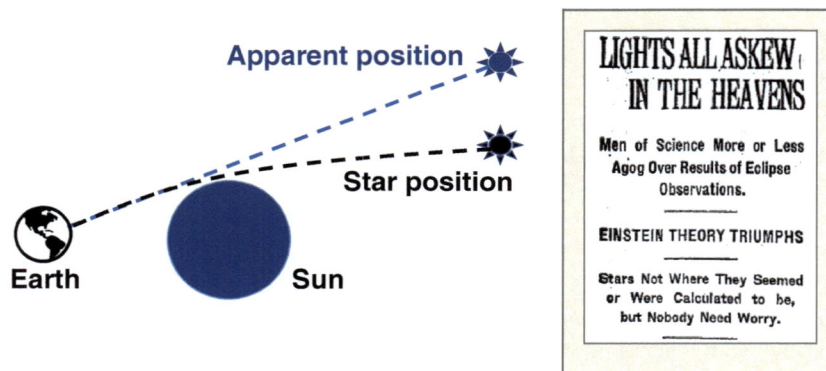

Figure 5.2 The Eddington experiment assessed the curvature of starlight during an eclipse.

The two teams photographed the position of stars during the eclipse and, for comparison, photographed the same stars the following night. They reported findings in line with Einstein's theory. There followed a barrage of publicity, such as the headline of *The New York Times* on 10 November 1919, *Lights All Askew In The Heavens* (see right of Figure 5.2). Einstein became an international celebrity. His reaction at the time was: *the world is a madhouse* (and see Box 5.4).

Some doubts have been raised as to the accuracy of Eddington's original data, but it was accepted as vindication of Einstein's theory and has been confirmed many times since. Three cheers for Einstein... hip hip hooray!

Box 5.4 Einstein's postcard to his mum, 1919

Einstein wrote a postcard to his mother Pauline: *Dear Mother! Happy news today. Lorentz cabled me that the English expeditions really proved the deflection of light in the sun.*

5.7 Time Dilation: Hafele-Keating and GPS

Before accepting Einstein's completely different explanation of gravity, some readers might demand more than a few fuzzy photographs of stars in an eclipse. When Einstein published his theory in 1915, evidence of spacetime curvature and time dilation was hard to come by. The change in the passage of time due to gravity on the surface of the earth is only about one in a billion. That is a change of only 1 second in about 30 years versus the passage of time if there were no gravitational field (see Box 5.5).

However, nowadays clocks are accurate enough to detect the effect. One of the first efforts to directly measure time dilation was the *Hafele-Keating experiment* in 1971 (sadly, 16 years after Einstein had died). The experimental concept and set-up was the simplest imaginable: bung a few super-accurate caesium clocks on a commercial jet aeroplane and fly them around the world! The clock times were then compared with ones that had stayed on the ground. The experiment has been repeated in several forms since, and unambiguously confirms the time dilation of general relativity.

For the clocks in GPS satellites circling some 20,000 kilometres above the earth, compensating for time dilation is essential. Being higher, they experience lower gravity, with the result that their clocks tick faster than those on the earth's surface. The required adjustment is about 40 microseconds per day (a microsecond is 10^{-6} seconds). You probably use a GPS-satnav in your car or on your phone. If there were no adjustment for the time-distortion of general relativity, your maps would drift off target by about 10 kilometres per day.

Box 5.5 Gravitational time dilation at the Earth's surface

We can calculate the gravitational time dilation of a clock at the earth's surface compared with that of a clock outside any gravitational field, using the time dilation in our model metric of Equation 5.3. G is the gravitational constant, M is the mass of the earth, r is the radius of the earth (the distance from the surface to its centre of mass) and c is the speed of light. The time dilation is very small, so we can use the Taylor approximation: $\sqrt{1-x} \approx 1 - \frac{x}{2}$. These approximate numbers show time dilation of $1 : 10^9$, which is about 1 second every 30 years.

$$\frac{\Delta(dt)}{d\tau} \approx \frac{1}{2}\frac{\Delta(dt^2)}{d\tau^2} = \frac{1}{2}\frac{1}{c^2}\frac{2GM}{x} \approx \frac{2\,(7 \times 10^{-11})\,(6 \times 10^{24})}{2\,(9 \times 10^{16})\,(6 \times 10^6)} \approx \frac{10^{13}}{10^{22}} \approx 10^{-9}.$$

(5.27)

$$G = 6.7 \times 10^{-11}\,\mathrm{m^3\,kg^{-1}\,s^{-2}}, \quad M = 6 \times 10^{24}\,\mathrm{kg}, \quad r = 6 \times 10^6\,\mathrm{m}, \quad c = 3 \times 10^8\,\mathrm{m\,s^{-1}}.$$

5.8 Geodesic Summary

My aim with this chapter is to convey the fundamental concept that sits behind general relativity... that the gravity we experience is actually spacetime curvature (the effect is predominantly from time curvature). Rather than gravitational acceleration leading to time dilation (because of increasing speed), it is time dilation of spacetime that leads to gravitational acceleration. To see if this fabulous idea holds up to scrutiny, we used a crude 2-D model. We calculated the time dilation of an object (or clock) accelerating in a gravitational field and built this into the metric.

We determined how the geodesics would be affected. These are the paths followed by unaccelerated objects, the equivalent of straight lines in flat spacetime. With some maths (sorry!), we calculated that the geodesic paths would lead objects to accelerate towards the attracting mass at a rate equivalent to the value of the Christoffel symbol Γ^X_{TT}. In our model, I positioned a dung beetle sitting on the one-dimensional line. What happens to the dung beetle? It accelerates towards the mass at $\frac{GM}{x^2}$. This is exactly the acceleration in Newton's theory. Wow! Einstein's idea really works.

This is a huge step forward. However, just because time dilation *could* be responsible for gravity does not mean that it *is* responsible. Proof of Einstein's theory requires testable differences.

Closer inspection of the geodesic equation revealed that the amount of geodesic deviation depends on an object's passage through time *and* space. The deviation linked to moving through *time* applies even to a spatially stationary object and gives us the gravity that we slow-moving humans are familiar with—Newton's gravity. The deviation linked to moving through *space* is an additional term that does not appear in Newton's theory. We showed that the effect of the time-related and space-related deviations are additive. At higher speeds in strong gravitational fields, this extra additive term creates measurable differences between Einstein and Newton.

As a final flourish, we considered what this means for the geodesic deviation experienced by light. It is affected by its passage through time (Newton's gravity) *and* its passage through space (towards or away from the attracting mass). Travelling at the speed of light, the latter is as large as the former. As a result, in a gravitational field, light deflects twice as much as Newton would have predicted. Eddington showed this effect in 1919. Surely a Nobel Prize for Einstein? Yes and no (see Box 5.6).

When Einstein's theory was published, it was hard to prove the relativistic distortions associated with his ideas. Inevitably there were doubters. Nowadays, with technological improvements, his predictions have been measured directly many times. And if you decide to be a *non-believer* in his theory, you had better adjust the time in your GPS-satnav.

Box 5.6 Einstein's 1921 Nobel Prize

After Eddington's result, there was huge pressure on the Nobel committee to recognise Einstein. He dominated the list of nominees with 8 nominations for 1920, 14 for 1921, and 16 for 1922. The committee resisted. In what appears to be last-ditch resistance in 1921, they declared no candidate suitable and deferred the prize to the following year. Why?

Was it die-hards on the Nobel committee who struggled to accept these new concepts? One member, Allvar Gullstrand, in a report to the committee, concluded that Einstein's theories were *dogma mistaken for facts* and lacked any real *significance for physics*. Another member deemed them *in conflict with the laws of physics*, whatever that means. Or was it growing anti-Jewish sentiment? One prestigious scientist, Philipp Lenard, declared general relativity to be *un-German* and *sickly-Jewish*.

In 1922, the pressure became too much and the committee yielded to the inevitable. Einstein was awarded the 1921 Nobel Prize for physics... 1 year late... and with the compromise that it was specifically *not* for relativity but was awarded for his work on the photoelectric effect completed over 15 years earlier.

On a positive note, some of the prize money went to his ex-wife Mileva as part of a promised divorce settlement. Given Einstein's outrageous letter of demands to her (see Box 4.6), I suspect it was well deserved.

5.9 Tensors: Why...? What...? How...?

Before we move on, I want to tell you a little more about tensors in general. We are going to be using tensors a lot. Equation 5.28 shows you the EFEs written out in full. I'd like to show you that it has three (4 × 4) tensors: the Ricci tensor and the metric tensor (which we have been discussing) on the left and the energy-momentum tensor on the right. Don't worry—you are not meant to understand it yet. But why was Einstein so obsessed with these strange tensor things? What is a tensor? One definition from the Web is: *in mathematics, a tensor is an algebraic object that describes a multilinear relationship between sets of algebraic objects related to a vector space.* But I think you are going to need a bit more than that.

$$R_{\mu\nu} - \frac{1}{2} R g_{\mu\nu} = k T_{\mu\nu},$$

Spacetime curvature $\Longleftrightarrow$ Energy-momentum

$$\begin{bmatrix} R_{TT} & R_{TX} & R_{TY} & R_{TZ} \\ R_{XT} & R_{XX} & R_{XY} & R_{XZ} \\ R_{YT} & R_{YX} & R_{YY} & R_{YZ} \\ R_{ZT} & R_{ZX} & R_{ZY} & R_{ZZ} \end{bmatrix} - \frac{1}{2} R \begin{bmatrix} g_{TT} & g_{TX} & g_{TY} & g_{TZ} \\ g_{XT} & g_{XX} & g_{XY} & g_{XZ} \\ g_{YT} & g_{YX} & g_{YY} & g_{YZ} \\ g_{ZT} & g_{ZX} & g_{ZY} & g_{ZZ} \end{bmatrix} = k \begin{bmatrix} T_{TT} & T_{TX} & T_{TY} & T_{TZ} \\ T_{XT} & T_{XX} & T_{XY} & T_{XZ} \\ T_{YT} & T_{YX} & T_{YY} & T_{YZ} \\ T_{ZT} & T_{ZX} & T_{ZY} & T_{ZZ} \end{bmatrix}. \qquad (5.28)$$

Let me briefly recap the material in Sections 3.1 and 3.2. I explained that, *being tensors, they all transform between coordinate systems in exactly the same way.* This is described as *covariance*. It means that the EFEs work in any coordinate system. Whether we use Cartesian coordinates (t, x, y, z) or spherical coordinates (t, r, θ, ϕ), or any other, they still work. One of the EFEs in Cartesian coordinates is $R_{XY} - \frac{1}{2} R g_{XY} = k T_{XY}$. One in spherical coordinates is $R_{\theta\phi} - \frac{1}{2} R g_{\theta\phi} = k T_{\theta\phi}$. So, one key advantage is that the use of tensors means that the EFEs work in any coordinate system.

Obviously, this simplifies things, but there is more to it than that. Einstein's theory reflects an underlying physical reality (hopefully). If so, surely it should work independent of the coordinate system used. We use coordinates to do the math, but we should be describing something tangible, something real that transcends the coordinate system. To achieve this in curved spacetime you need tensors. Let me try to explain why.

5.10 Where's the Fridge?

Suppose that I am round at your house hoping for a beer, so I want to get to your fridge. You tell me that the fridge is 10 metres away. This is a *scalar* quantity. It has magnitude but no direction.

As I don't have the patience to search everywhere in a 10-metre radius, I question you further. You tell me that I can get to the fridge by going 8 metres east and then 6 metres north. This description is a *vector* quantity. It has magnitude in a particular direction. If I use a coordinate system with the x basis vector pointing east and the y basis vector pointing north, my journey is $8\vec{e}_x + 6\vec{e}_y$. If I wanted to use polar coordinates, then I would need to transform this to approximately $10\vec{e}_r + 0.64\vec{e}_\theta$ (using radians). Whatever coordinates I use, I am describing the same thing. All vectors transform the same way. The vector's components change but what the vector represents is unchanged.

Things get trickier when there is curvature. Suppose that you have a really strange house such that you and your fridge exist on the 2-D surface of a sphere as described in Section 3.5. You have to break the news to me that going $8\vec{e}_x$ followed by $6\vec{e}_y$ is not the same as going $6\vec{e}_y$ followed by $8\vec{e}_x$ (try it on a football if you need to). It is this sort of discrepancy that sits at the heart of how we measure curvature. We will discuss this in depth later. The vector description does not work. Does this mean that the location of the fridge depends on how I get there? Of course not. The fridge has a definite location on the surface. However, to make the trip I must take into account how the basis vectors $\vec{e}_x$ and $\vec{e}_y$ vary across the surface.

In addition, if the coordinates are not orthogonal, I need to know what they mean in terms of each other. For a 2-D surface, I need the value of $\vec{e}_x$ in terms of movement in the x and y directions and the value of $\vec{e}_y$ in the x and y directions at different points on the surface. This is a *tensor*. It is the magnitude of something with direction (in this case a basis vector) that has an effect in more than that one direction. The key point is that there still is an underlying reality and I still need a beer! Changing the coordinates does not change the surface. The tensor represents this reality. In different coordinates, the tensor's components change, but what the tensor represents is unchanged (just as is the case with vectors).

When you extend this from 2-D to spacetime, it means that for the t basis vector you need four entries for what that means in terms of t, x, y and z. Likewise for the x, y and z basis vectors. This gives the (4×4) matrix structure of the spacetime tensors in the EFEs. In the case of the metric tensor $[g_{\mu\nu}]$, its components tell you what each basis vector means in terms of movement (measured in invariant proper time or distance) along each dimension. In the case of the Ricci tensor, it is a measure of the distortion linked to each basis vector moving along each dimension. In the case of the energy-momentum tensor, it is linked with the flow of energy and momentum through each dimension.

That is enough of an introduction to tensors. It is time now to look more closely at the curvature associated with gravity (see Box 5.7 for a lighter view).

Box 5.7 Just for laughs

I really don't like thinking about gravity. *It gets me down.*

Chapter 6

The Equivalence Principle and Ricci Tensor

Let me take a moment to congratulate you. If you have followed the argument so far, you already have grasped the fundamental logic underpinning general relativity. Gravitational acceleration is the result of distortion (curvature) in the spacetime metric. The next question is how the presence of energy-momentum distorts the spacetime metric in the region containing it.

Einstein relied heavily on the *equivalence principle* to guide him. We will use this principle in the hunt for the relationship between energy-momentum, the curvature of spacetime and the distortion of the surrounding geodesic paths. This will lead us to Einstein's conclusion that the presence of energy-momentum reduces the spacetime volume separating geodesics.

I will then introduce the Ricci tensor $[R_{\mu\nu}]$, which you will learn to distinguish from other curvature tensors by its having *two* indices. It measures any change to the spacetime volume between geodesics, therefore its value is related to the presence of energy-momentum. Einstein spotted the central role of the Ricci tensor in general relativity several years before he discovered the correct form of the Einstein Field Equations (EFEs). He got side-tracked for a while, but fortunately figured it out in the end.

The key result is that in the presence of energy-momentum the Ricci tensor $[R_{\mu\nu}]$ is non-zero. And in the absence of energy-momentum (i.e. in a vacuum), the Ricci tensor $[R_{\mu\nu}]$ is zero.[1]

6.1   The Equivalence Principle

Many of you will already have a good idea of what the equivalence principle is, but I should cover the basics anyway. Einstein was working at his desk in the patent office in Berne, which he described as *that worldly cloister where I hatched my most beautiful ideas*. The job was well paid and not overly demanding. He was struck by a revelation giving him his *happiest idea*. In Einstein's words: *If a person falls freely, he will not feel his own weight. I was startled. The simple thought made a deep impression on me. It impelled me toward a theory of gravitation.*

Einstein was pointing out that the weight we feel on the surface of the earth is not the result of us moving in a gravitational field. It is the result of us *not* moving in it. The surface of the earth stops us following the natural path. The electromagnetic repulsion of the ground generates a force that we feel as weight. If there were no force opposing our movement, we would freefall. We would be weightless. We would feel nothing.

As usual, I want to focus on an intuitive example and let that lead us to the mathematical ideal. Consider a group of people in a box (perhaps a lift). The box is released so that it falls in freefall downwards in a gravitational field towards a big rocky planet. There are no windows. What do the people experience? The people and the lift fall together. Both the people and the lift accelerate at exactly the same rate in the gravitational field. As a result, the people are weightless.

Einstein's conclusion was that there is an underlying equivalence between freefall acceleration in a gravitational field and travelling in a straight line with no acceleration (which is an *inertial frame*). The people in the lift do not feel gravity. They have no idea they are accelerating at high speed towards a rocky end.

Let me immediately add that this is a theoretical scenario... the sort of thought-game that Einstein was so good at. Of course, the lift might have a window allowing the occupants to peek out at the approaching lump

---

1 This ignores any vacuum energy (i.e. the cosmological constant), which will be addressed later.

Untangling General Relativity: The Intuitive Self-Study Guide, First Edition. Simon Sherwood.
© 2026 John Wiley & Sons Ltd. Published 2026 by John Wiley & Sons Ltd.

of rock. Or they might do experiments to detect various tidal forces (see Box 6.1).[2] But Einstein's conjecture was that, viewed at the most microscopic *local* level, there is a theoretical equivalence between freefall acceleration in a gravitational field and no acceleration at all. Wow!

Box 6.1 Tidal forces

If you analyse the freefalling lift scenario from a practical perspective, there is a way for the occupants to detect their acceleration. Consider two people slightly apart in the lift. Each will accelerate in a direct line towards the centre of the earth. The distance between them means that these two lines to the centre of the earth will be at slightly different angles. This will lead to a small difference in acceleration.

Differences in position relative to the attracting central mass lead to what are called *tidal forces*. This is illustrated in Figure 6.1a. Two spacewomen are in freefall. Their slightly different positions relative to the centre of the earth lead them to accelerate towards each other, creating an apparent tidal force attracting them together. Using a different example, if an astronaut were to travel through a very strong gravitational field, those parts of his body closer to the mass would be accelerated more than those far away. He would be stretched further and further... now, that would be measurable!

Einstein's genius was the ability to ignore these practical issues and consider two what he called *local* points separated by an infinitesimally small distance.

6.1.1 A Planet with a Hole

I want to be sure you see how weird the equivalence principle is, so I intentionally will make the scenario weirder. We dig a tunnel through the centre of a planet and drop the lift and its occupants in at the north pole. The planet has no atmosphere (no air resistance). Figure 6.1b shows the concept. What happens?

Sitting on the surface, we see the lift fall, accelerating down the hole. Once it is past the centre of the planet, it decelerates because of the gravitational acceleration back towards the centre. The lift reaches the south pole of the planet at which stage all its kinetic energy has been converted back to gravitational potential energy. It then travels back through the planet from the south pole to the north pole... and continues to oscillate from pole to pole.

But, here is the point. The occupants in the lift have no idea that they are oscillating up and down. They are weightless and feel no force (this is the same as the weightlessness of astronauts in orbit, except in order to make the point clearer, my scenario is through the planet rather than around it). From the occupants' perspective, they are doing nothing... no twists, no turns... just following the natural path of the geodesics through spacetime. They oscillate up and down with no clue that they are even moving. Weird, eh?

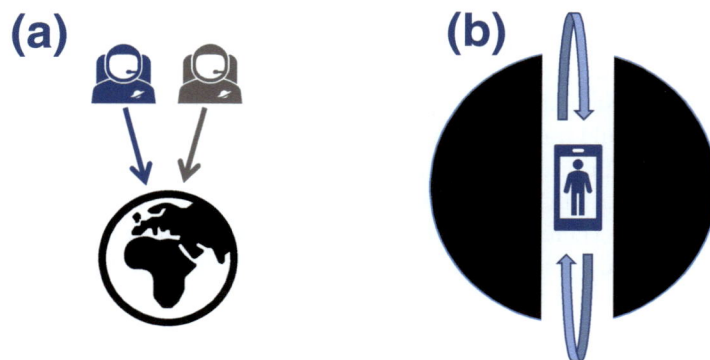

Figure 6.1 Tidal forces and the weirdness of freefall. (a) illustrates tidal forces; (b) is a scenario illustrating the weirdness of freefall.

---

2 Hans Ohanian in his book *Einstein's Mistakes* speculates that Einstein did not understand tidal forces. This strikes me as exceedingly unlikely. Indeed, Einstein's constant insistence that the equivalence principle only works at *local* level is strong evidence that he was fully aware of tidal effects.

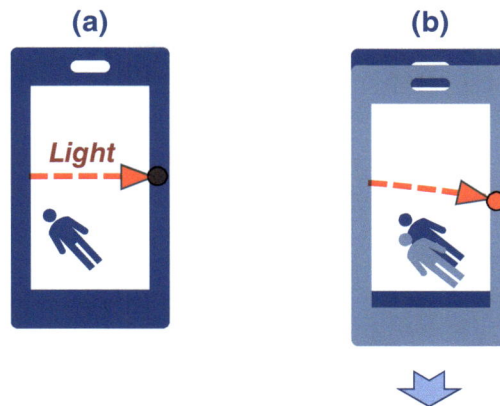

Figure 6.2 Why light must bend in a gravitational field. (a) is the perspective of an observer stationary relative to the falling lift; (b) is the perspective of an observer watching the lift fall.

6.1.2 Light in a Gravitational Field

One of the consequences of the equivalence principle is that light must be affected by a gravitational field. Consider Figure 6.2. Again, the scenario involves occupants in a closed lift. At the moment that the lift is released into freefall, a light is shone from the left side. It travels across and hits the right side of the lift.

Figure 6.2a shows the perspective of, say, a woman in the lift. She is in weightless freefall in a uniform gravitational field. By the equivalence principle, she is in an inertial frame and the laws of physics are the same for all inertial frames. She must see the light go straight across the lift and hit the other side at the same height. This is shown in the figure as a red dotted line from left to right, arriving at a red spot on the right wall of the lift. If there were any deviation in the path, she would know something was different and the equivalence principle would be broken.

Figure 6.2b shows the perspective of an observer outside the lift who is stationary relative to the uniform gravitational field. During the tiny moment of time while the light is crossing, the lift accelerates down (as shown in lighter blue). As a result, this observer must see the light bend slightly downwards (the red dotted line going from left to right) for the equivalence principle to hold. The light must be bent by the gravitational field. For those of you who know something about quantum physics, I should add that the same logic applies to antimatter (see Box 6.2 for CERN's recent work on this).

Box 6.2 Gravity and antimatter

In September 2023, CERN announced results showing that antimatter falls with gravity in the same way that matter does.

So, what is antimatter? Quarks (which make up protons and neutrons) and electrons are the fundamental constituents of atoms. For each type there is an antiparticle, which has the opposite charge. For example, the electron has a negative electric charge while its antiparticle, the positron, is positive. The existence of antiparticles is a requirement of quantum mechanics, predicted by the Dirac equation (see Chapter 18 of my book *Quantum Untangling* for all the details). So, an antiparticle is sort of the opposite of a particle. Why then was it expected to fall in a gravitational field in the same way as a particle? Should it not do the opposite?

Well, if antiparticles fell upwards, it would create a significant difference between falling in a gravitational field and following an unaccelerated path, breaching the equivalence principle. This is why Einstein (and most physicists) expected antimatter to react to gravity in the same way that matter does. And now we know it does. Well done CERN. Thumbs up for your efforts!

One obvious question is what happens if light is moving directly into or out of the gravitational field? The measures of time and space in the metric change as you move through the field. As a result, the frequency and wavelength of the light change. If you shine a light out from a gravitational field, its wavelength will lengthen. It will be redshifted. Its energy will fall. In terms of the conservation of energy, the energy of the light beam falls as its gravitational potential energy grows.

Einstein rationalised this with a neat thought experiment. Imagine we point a light beam away from the earth. We catch the light at a great height and in some way convert it to mass ($E = mc^2$). We drop the mass down. The mass gains kinetic energy falling. For energy to be conserved, the light must lose the same amount of energy in travelling up as its equivalent mass gains in travelling down.

6.2 From Newton's Gravity to Geodesic Separation

By now you should be familiar with the idea that gravity has nothing to do with objects feeling an attractive force and everything to do with the presence of energy-momentum distorting the surrounding spacetime geodesics. Einstein's challenge was to find the nature of the relationship between the presence of the energy-momentum and the geodesic distortion. We can use the equivalence between gravity and freefall as a starting point. Let's see what we can deduce from Newtonian gravity.

Compare what happens to the geodesics in two simple scenarios: when there is energy-momentum present such as in a dust field versus when there is no energy-momentum present anywhere, i.e. in a vacuum (the possibility of vacuum energy is addressed later in the book).

Figure 6.3a illustrates schematically Newtonian gravity in the dust field. It is shown in 2-D, but I want you to think in 3-D. The blue circle is a *sphere* surrounding a volume of dust. The eight black dots are points on the sphere, which you can think of as bits of dust if you wish. Gravity pulls the dust in the cloud closer together over time. The dust within the sphere is evenly spread, so its centre of mass is at the centre of the sphere. The Newtonian view is that the dust at the surface of the sphere (the black dots) feels an attracting force towards the centre (the red arrows). If the black dots are initially stationary, they start to accelerate inwards. After some time, the dust on the surface of the original sphere has moved. It now sits on the surface of a smaller sphere (shown as a grey circle).

Figure 6.3b shows what you might call the Einsteinian view of the same thing (I love to invent words). Again, the blue circle represents an imaginary sphere surrounding part of the dust field. The black dots feel no attracting force. Each piece of dust at the black dots starts stationary. In flat spacetime (a vacuum) the geodesic paths would remain stationary in space while moving through time. We would describe these as parallel spacetime geodesics. However, the presence of the dust inside the sphere distorts the spacetime geodesics. The result is that they don't remain parallel to each other. As time passes, the geodesics draw closer together arriving on the surface of the smaller sphere (the grey circle).

With this simple scenario, we can model how the spacetime geodesics are distorted. In a vacuum, the spacetime geodesics that start at the black dots would remain the same distance apart. The volume of space separating them (the blue sphere) would not change. In the dust field, the spacetime geodesics draw together. The presence of the mass of the dust field gradually reduces the *spatial volume* separating parallel geodesics (from the larger blue sphere down to the smaller grey sphere).

Let's delve a little deeper with a rough back-of-the-envelope calculation of what happens to the geodesic separation based on Newton's law of gravity. I label the volume of the blue sphere V. This is the volume initially separating the spacetime geodesics. The radius of the blue sphere is shown as r. If you remember maths at school, you

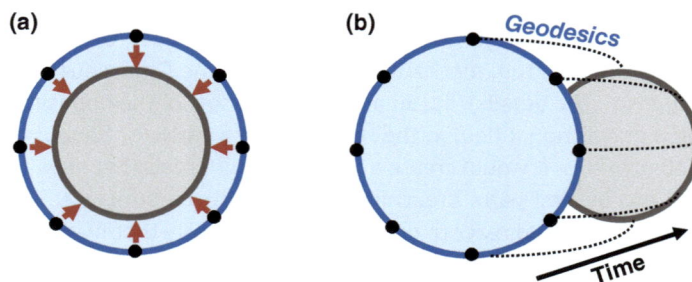

Figure 6.3 Illustrations relating to Ricci curvature. (a) Newtonian; (b) Einstein's general relativity.

will know that the volume of the sphere is $\frac{4}{3}\pi r^3$, so $\frac{\partial V}{\partial r} = 4\pi r^2$. Equation 6.1 develops a relationship between how volume V and radius r change over time.

$$\frac{\partial V}{\partial t} = \frac{\partial V}{\partial r}\frac{\partial r}{\partial t} = 4\pi r^2 \frac{\partial r}{\partial t} = \frac{3V}{r}\frac{\partial r}{\partial t}, \qquad \Longrightarrow \qquad \frac{1}{V}\frac{\partial V}{\partial t} = \frac{3}{r}\frac{\partial r}{\partial t}. \tag{6.1}$$

We take each side from the right of Equation 6.1 and differentiate again (using the slightly laborious process of differentiation by parts) giving the results in Equations 6.2 and 6.3.

$$\frac{\partial}{\partial t}\left(\frac{1}{V}\frac{\partial V}{\partial t}\right) = \frac{\partial}{\partial t}\left(\frac{1}{V}\right)\frac{\partial V}{\partial t} + \frac{1}{V}\frac{\partial^2 V}{\partial t^2},$$
$$= \frac{\partial}{\partial V}\left(\frac{1}{V}\right)\frac{\partial V}{\partial t}\frac{\partial V}{\partial t} + \frac{1}{V}\frac{\partial^2 V}{\partial t^2} \qquad = -\frac{1}{V^2}\left(\frac{\partial V}{\partial t}\right)^2 + \frac{1}{V}\frac{\partial^2 V}{\partial t^2}. \tag{6.2}$$

$$\frac{\partial}{\partial t}\left(\frac{3}{r}\frac{\partial r}{\partial t}\right) = \frac{\partial}{\partial t}\left(\frac{3}{r}\right)\frac{\partial r}{\partial t} + \frac{3}{r}\frac{\partial^2 r}{\partial t^2},$$
$$= \frac{\partial}{\partial r}\left(\frac{3}{r}\right)\frac{\partial r}{\partial t}\frac{\partial r}{\partial t} + \frac{3}{r}\frac{\partial^2 r}{\partial t^2} \qquad = -\frac{3}{r^2}\left(\frac{\partial r}{\partial t}\right)^2 + \frac{3}{r}\frac{\partial^2 r}{\partial t^2}. \tag{6.3}$$

We know from Equation 6.1 that the results of Equations 6.2 and 6.3 are equal as shown below in Equation 6.4.

$$-\frac{1}{V^2}\left(\frac{\partial V}{\partial t}\right)^2 + \frac{1}{V}\frac{\partial^2 V}{\partial t^2} = -\frac{3}{r^2}\left(\frac{\partial r}{\partial t}\right)^2 + \frac{3}{r}\frac{\partial^2 r}{\partial t^2}. \tag{6.4}$$

We want to relate how the spatial volume V separating parallel geodesics changes with mass density. To simplify, let's see what happens if we instantly and magically change the scenario from a vacuum to a dust field of mass density ρ. The dust particles are initially stationary, so $\frac{\partial r}{\partial t}$ and $\frac{\partial V}{\partial t}$ are zero. This gives Equation 6.5, where I have substituted $\frac{\partial^2 r}{\partial t^2}$ with gravitational acceleration (such as Equation 1.6) and mass M with volume and mass density $\frac{4}{3}\pi r^3 \rho$. The result is Equation 6.6, where I have labelled $\frac{\partial^2 V}{\partial t^2}$ as $\ddot{V}$ for clarity.

$$\frac{1}{V}\frac{\partial^2 V}{\partial t^2} = \frac{3}{r}\frac{\partial^2 r}{\partial t^2} = -\frac{3}{r}\frac{GM}{r^2} = -\frac{3G}{r^3}\frac{4}{3}\pi r^3 \rho = -4\pi G\rho, \tag{6.5}$$

$$\Longrightarrow \quad -\frac{\ddot{V}}{V} = 4\pi G\rho. \tag{6.6}$$

What this tells us, is that if we suddenly and magically transform from a vacuum to a dust field, the volume of space separating the spacetime geodesics reduces over time, and that the fractional rate of volume reduction depends on the density of matter present (those of you who have studied Newtonian gravity may notice a similarity with the Poisson equation, which is discussed in detail later in Section 10.4). If the mass density ρ is zero, there is no volume reduction, which is as you would expect.

Note that the fractional rate of change $\frac{\ddot{V}}{V}$ does *not* depend on the volume of the sphere. If we look at the relationship between spacetime geodesics that are closer together, we start with a smaller blue sphere. If we want to consider spacetime geodesics that are infinitesimally close to each other, then the blue sphere is infinitesimally small. None of this changes the fractional rate of change of the volume separating the geodesics. It remains the same.

At this point, I must call on your intuition. Einstein knows that gravitational curvature cannot be quantified using a change of *spatial* volume over time. For one thing, his theory of special relativity shows that space and time are inseparable. For another, his theory of gravity concerns curvature of spacetime, not just of space. I hope you agree that the obvious guess is that there is a relationship between the presence of energy-momentum and a change in the volume of *spacetime* separating the spacetime geodesics.[3] Let's follow this line of thought and see where it leads us.

One thing it highlights is the importance of the *Ricci tensor*. This is the measure of curvature that specifically tracks the change in spacetime volume separating parallel spacetime geodesics.

---

3 Technical point: to be consistent, the definition of the volume element for any spacetime metric is $\sqrt{|g|}\ d^4x$, where $|g|$ is the determinant of the matrix $[g_{\mu\nu}]$. For this book, you need to be aware only that there is a consistent measure.

6.3   The Magnificent Ricci Tensor

The curvature of any surface or spacetime is fully defined by its *Riemann* tensor $[R^{\alpha}{}_{\beta\gamma\delta}]$. If there is any curvature of any nature, the Riemann tensor will pick it up. The *Ricci* tensor $[R_{\mu\nu}]$ is what physicists call a *contraction* of the Riemann tensor, which is a fancy language for adding up some of its components. In curvature, a lot of things are labelled with an *R* so watch out. $[R_{\mu\nu}]$ with two indices is the Ricci tensor. $[R^{\alpha}{}_{\beta\gamma\delta}]$ with four indices is the Riemann tensor. And *R* with no indices is called the Ricci scalar. Aarrgghh! Sorry!

We will address the detailed mathematical structure of these curvature tensors later. They use derivatives of derivatives of the metric, so they are quite complicated. Pretty much everything is complicated in 4-D curved spacetime. First, I want to take you through a more intuitive description of the Ricci tensor $[R_{\mu\nu}]$ and give you a feel for what is happening. I hope this will prepare your brain cells for the ensuing maths.

$$R_{\mu\nu} - \frac{1}{2}Rg_{\mu\nu} = kT_{\mu\nu}, \qquad \textit{Example EFE}: \ R_{TT} - \frac{1}{2}Rg_{TT} = kT_{TT}. \tag{6.7}$$

Spacetime curvature $\iff$ *Energy-momentum*

$$\begin{bmatrix} R_{TT} & R_{TX} & R_{TY} & R_{TZ} \\ R_{XT} & R_{XX} & R_{XY} & R_{XZ} \\ R_{YT} & R_{YX} & R_{YY} & R_{YZ} \\ R_{ZT} & R_{ZX} & R_{ZY} & R_{ZZ} \end{bmatrix} - \frac{1}{2}R \begin{bmatrix} g_{TT} & g_{TX} & g_{TY} & g_{TZ} \\ g_{XT} & g_{XX} & g_{XY} & g_{XZ} \\ g_{YT} & g_{YX} & g_{YY} & g_{YZ} \\ g_{ZT} & g_{ZX} & g_{ZY} & g_{ZZ} \end{bmatrix} = k \begin{bmatrix} T_{TT} & T_{TX} & T_{TY} & T_{TZ} \\ T_{XT} & T_{XX} & T_{XY} & T_{XZ} \\ T_{YT} & T_{YX} & T_{YY} & T_{YZ} \\ T_{ZT} & T_{ZX} & T_{ZY} & T_{ZZ} \end{bmatrix}.$$

Before discussing what the Ricci tensor measures, let's look at the central role it plays in the EFEs as summarised in Equation 6.7. The EFEs show the relationship in spacetime between the presence of energy-momentum (right side) and curvature (left side). $R_{\mu\nu}$ is any component of the Ricci tensor $[R_{\mu\nu}]$. As noted earlier, this is a contraction of the larger Riemann tensor. *R* is a number called the Ricci scalar, which is calculated from components of the Ricci tensor (it is a contraction of the Ricci tensor). $g_{\mu\nu}$ is the corresponding component of the spacetime metric tensor $[g_{\mu\nu}]$, which we discussed in Chapter 3. $T_{\mu\nu}$ is the component of the energy-momentum tensor $[T_{\mu\nu}]$, which we will discuss in Chapter 8. And *k* is just a constant, which we will derive later. It is $\frac{8\pi G}{c^4}$.

There are 16 equations in the EFEs such as, for example: $R_{TT} - \frac{1}{2}Rg_{TT} = kT_{TT}$. At the bottom of Equation 6.7 is the same equation but with the tensors opened to show their components. You can see that the Ricci tensor has (4×4) entries like the other tensors in the equation.

I mentioned back in Chapter 3 that we will be working almost exclusively with diagonal metrics, i.e. where all the off-diagonal entries of $[g_{\mu\nu}]$ are zero (in our studies only rapidly rotating bodies and gravitational waves will involve significant off-diagonal terms). Similarly, in most cases the energy-momentum tensor $[T_{\mu\nu}]$ can be expressed using only diagonal terms. As you can see from the EFEs, if there are only diagonal terms in $[g_{\mu\nu}]$ and $[T_{\mu\nu}]$, the same is true for $[R_{\mu\nu}]$. As a result, we focus on the diagonal terms of the Ricci tensor, such as R_{TT} and R_{XX}.

The structure of the EFEs means that in a vacuum, where there is no energy-momentum (every entry in the $[T_{\mu\nu}]$ matrix is zero), the spacetime is Ricci-0 (every entry in the $[R_{\mu\nu}]$ matrix is zero, and the value of the Ricci scalar is also zero). This is not entirely obvious from the equation, but later when you understand the Ricci scalar *R*, you will see that it is true. The presence of energy-momentum and Ricci curvature come hand in hand. With that background, we can now dig further into why the Ricci tensor is the essential link between energy-momentum and spacetime curvature.

6.4   An Intuitive Explanation of the Ricci Tensor

In Section 6.2, I showed how the equivalence principle allows us to equate Newtonian gravity with a reduction in the volume of *space* separating parallel spacetime geodesics. There then followed an intuitive leap to the relationship for general relativity: that the presence of energy-momentum reduces the volume of *spacetime* separating the geodesics. In turn, this leads to a relationship between the presence of energy-momentum and the relevant curvature measure, the Ricci tensor, which measures any change in the overall spacetime separation between infinitesimally close adjacent parallel spacetime geodesics. That last sentence is far from simple, so let me explain step by

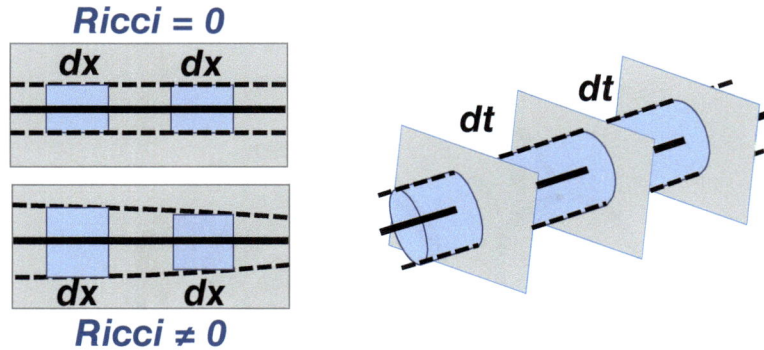

Figure 6.4 Illustrations relating to the Ricci tensor.

step, starting with a 2-D example as shown on the left of Figure 6.4. To avoid a horde of Ricci-loving mathematicians screeching at me, I emphasise this is highly illustrative.

In the example at the top left, the solid black line is a geodesic. The dotted black lines are two nearby geodesics that start parallel with it. As we are working in only two spatial dimensions (x, y), the separation is a tiny distance in the y direction. We can take a small dx length to get some measure of area between geodesics such as that shown in blue. In 2-D, the Ricci tensor measures the *change* in this area as you move along two infinitesimally close parallel geodesics. In this case, as we are moving in the x direction, this is a rough illustration of the component R_{XX}. You can see that the area doesn't change. If the full Ricci tensor $[R_{\mu\nu}]$ is zero, it means the area does not change, whatever direction you track the geodesics ($R_{XX} = 0$, $R_{YY} = 0$, $R_{XY} = 0$ and $R_{YX} = 0$).

Consider now the scenario below it. The geodesics draw together. An example of this sort of curvature is the surface of a sphere (look back at the paths of the two dung beetles in Section 3.4 if needed). The curvature of the surface is such that the area in the blue boxes shrinks as you move along the geodesics. This is a 2-D example of the type of curvature we expect in the presence of energy-momentum. In this case the Ricci tensor is not zero. Note the similarity with the way the geodesics draw together in Figure 6.3. What is more, the Ricci tensor measures the change between *infinitely* close parallel geodesics, so the relationship applies however close they are to each other, which fits with the model in Equation 6.6.

The right side of Figure 6.4 shows a 3-D example (t, x, y), where I have chosen to track the geodesics through time (i.e. in the t direction, so component R_{TT}). The first grey square is an area of space (x, y) at some time. The central geodesic is the solid black running into the page. Around it is a circle of surrounding geodesics (I have shown two as dotted black lines). The second grey square is the same area at a moment of time dt later. The little cylinders illustrate the (t, x, y) volume separating the nearby geodesics. The Ricci tensor measures how this volume changes with progress along the geodesic. In the example, it does not change. If this is true in all directions, then the 3-D spacetime (1 of time, 2 of space) is Ricci-0. This is the sort of pattern we expect for geodesics in a vacuum. Some texts use the label *Ricci-flat* for this sort of spacetime. I find this confusing because it suggests Ricci-0 spacetime is always flat (like Minkowski spacetime). In the case of gravitational curvature in a vacuum, the spacetime is Ricci-0, but definitely not flat Minkowski spacetime.

Let's look more closely at how the Ricci-0 vacuum fits with the equivalence principle, starting with Einstein's own description of the gravitational field in a vacuum: *Now it came to me... in a gravitational field (of small spatial extension) things behave as they do in a space free of gravitation.* Imagine the following theoretical vacuum scenario. There are three small balls stationary relative to each other in a satellite travelling in a gravitational field. This is shown in Figure 6.5. The balls are infinitesimally small, so we can ignore their gravitational effect on each other. They are an infinitesimal distance apart, so we can ignore tidal forces. Providing the tiny volume of space they occupy is a *vacuum*, there is no energy-momentum located between them that would affect each ball differently. Based on Einstein's conjecture, there should be no observation or measurement, actual or theoretical, within the satellite that shows whether or not the balls are accelerating in a gravitational field. The balls should remain stationary relative to each other.

In layman's language, as you peer in closer and closer at parallel geodesics in a vacuum, they should become indistinguishable. Consider again the top left of Figure 6.4. If those little areas are equal, then as you focus in on closer and closer geodesics, the areas will shrink uniformly. In the extreme of an infinitesimally small separation

Figure 6.5 Implications of the equivalence principle for a vacuum.

(which is what the Ricci tensor measures), the geodesics reduce towards the same coordinates. At this extreme, everything follows parallel geodesics as in an inertial frame.

The same applies in 4-D. Ricci-0 spacetime ($[R_{\mu\nu}] = 0$) is exactly what the equivalence principle requires for the vacuum. Infinitesimally close parallel geodesics become indistinguishable. On the other hand, if the blue regions in Figure 6.4 change ($[R_{\mu\nu}] \neq 0$), then however closely you focus in, the adjacent parallel geodesics will never align. This is just what we expect in the presence of energy-momentum based on the dust field model in Section 6.2.

6.5 Vacuum Curvature: An Apparent Paradox

Some readers may be puzzled by what at first appears to be a paradox. If Einstein is right, then the spacetime in a vacuum must be Ricci-0. However, we know there are gravitational forces in a vacuum. For example, the earth orbits the sun. Therefore, there must be spacetime curvature in a vacuum near a mass. How is this consistent with the Ricci tensor being zero in a vacuum? The answer is that for gravity in a vacuum, the *Riemann* tensor, which picks up any and every curvature, must be non-zero (indicating there is spacetime curvature) while the *Ricci* tensor must be zero (to respect the equivalence principle).

I imagine all you dear readers would like me to give you a nice simple 2-D example of how gravity in a vacuum works. Sorry! That is not possible. If there is *any* curvature at all in the manifold (surface), it will show up in its Riemann tensor. In 2-D, if the Riemann tensor shows up any curvature at all, the Ricci tensor cannot be zero. The surface cannot be Ricci-0. This makes sense. How can you draw together, or separate, the geodesics in the flat example (top left) of Figure 6.4 without the size of the blue area gradually changing? You cannot. The same is true in 3-D.

However, it *is* possible in 4-D spacetime to have a curvature that is Ricci-0. This shows up as non-zero components of the Riemann tensor (which detects *all* curvature), even though every component of the Ricci tensor is zero. It turns out that you need at least four dimensions to have Ricci-0 curvature of this sort. What an irony that it is 4-D spacetime (which Einstein himself revealed with special relativity) that allows gravitational curvature to exist in a vacuum.

An example is the Schwarzschild metric, which we will discuss in detail later. It quantifies the spacetime curvature of a vacuum near a spherical mass. If you examine two geodesics some distance apart from each other, you find that they spatially converge or diverge (depending on whether you are measuring towards or away from the mass). They do *not* match up. However, focus in closer and closer and you find that the difference between parallel geodesics decreases until, at an infinitesimal distance apart, they are indistinguishable. This allows gravitational curvature in a vacuum while respecting the equivalence principle.

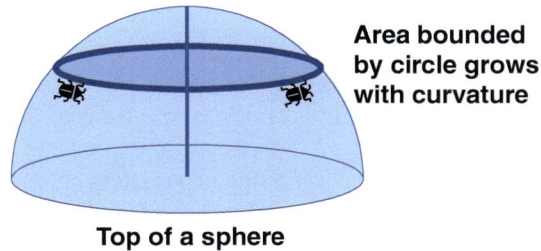

**Area bounded
by circle grows
with curvature**

Top of a sphere

Figure 6.6 The Ricci scalar is a measure of how volumes and areas change with curvature.

6.6 The Ricci Scalar

I have mentioned the Ricci scalar several times, so let me introduce you to it properly. The Ricci scalar is labelled R. Being a scalar, it is a number (no direction) and carries no indices.

The Ricci scalar R can be calculated for a surface or spacetime by adding up components of the Ricci tensor $[R_{\mu\nu}]$. In fact, it is a weighted combination of the principal diagonal terms. In the case of spacetime using orthogonal Cartesian coordinates, this is a weighted combination of R_{TT}, R_{XX}, R_{YY} and R_{ZZ}. If all these components of the Ricci tensor are zero, then the Ricci scalar also is zero.

In the case of spacetime, the Ricci scalar gives a measure of the way that spacetime volumes as a whole are distorted. The Ricci *tensor* components tell you about the distortion as you progress in some particular direction (for example, R_{TT} tells you about change as you move through time). The Ricci *scalar* just adds them all up into a sort of total average. As usual, this is most easily explained with a 2-D example.

We will return yet again to our friendly dung beetles living on the surface of a sphere. They draw a circle as shown in Figure 6.6 (the dark blue line). On a *flat* 2-D surface, the area inside the circle would be the size of the area shaded slightly darker blue. However, as you can see from the figure, in the case of the surface of a sphere, it is the surface area of a dome bounded by the circle. The surface area is *more* than for a flat 2-D surface. The area bounded by the circle increases with positive curvature. The Ricci scalar is a measure of this discrepancy.

For a 2-D surface, the Ricci scalar tells you how the area enclosed by a circle differs from that of a flat surface. In the case of curvature of 3-D space, the Ricci scalar tells you how the volume enclosed in a sphere differs from that of flat 3-D space. With positive spatial curvature, the volume contained inside the sphere will be larger.

But what does this mean for the equivalence principle? If in a vacuum there is no way theoretically to tell the difference locally between freefall in a gravitational field and an inertial frame, then it would be a bit of a stretch (haha) for the spacetime to have a measurable non-zero Ricci scalar. I am not sure if Einstein thought this all through, but if he had, I am sure it could only have increased his confidence that spacetime in a vacuum is Ricci-0.

6.7 Summary

The aim of this chapter has been to introduce you to the Ricci tensor $[R_{\mu\nu}]$ and to give you an intuitive feel of why it plays a crucial role in the link between the presence of energy-momentum and the curvature of spacetime. It is not intended to be mathematically rigorous or a proof. That comes from experiment.

Our starting point was a discussion of the equivalence principle and how central this was to Einstein's thinking with his insatiable appetite for ideas (see Box 6.3). He concluded that, ignoring tidal forces, there is an underlying equivalence between freefall in a gravitational field and being in an inertial frame (i.e. a frame with no acceleration at all). I used the example of passengers in an enclosed lift oscillating up and down in a hole through a planet while having no idea that they are moving. Weird!

Box 6.3 Einstein at 17: ain't got no satisfaction

At 17 years old, Einstein wrote a French essay called *My Future Plans*. In it, he made the following remarkable observation (translating from the French): *A happy man is too satisfied with the present to dwell too much on the future.*

This is a true quote, not folklore, not a rumour. Is it a sign of a restless genius always brain-storming, never satisfied? Or might you and I have written something equally profound... but nobody has bothered to trawl through all our old school papers to find it? I wonder.

I used the simple scenario of a dust field to compare the interpretations of Newton and Einstein. What Newton viewed as an attractive force between masses, Einstein explained as a change to the geodesic pathways. A back-of-the-envelope calculation illustrated that the fractional change in the volume of space separating spacetime geodesics depends on the density of matter present. Importantly, this relationship between geodesic separation and density remains the same, even when the geodesics are only an infinitesimally small distance apart.

With a leap of intuition, I jumped to Einstein's conclusion that the presence of energy-momentum reduces the volume of *spacetime* separating what would have been parallel geodesics in a vacuum. This led to a discussion of the Ricci tensor, which measures any change in the volume of spacetime separating infinitesimally close spacetime geodesics. It therefore will pick up the distortion of geodesics we expect in the dust field. Thus, in the presence of energy-momentum, the Ricci tensor is non-zero ($[R_{\mu\nu}] \neq 0$).

On the other hand, in a vacuum the spacetime is Ricci-0, i.e. every component of the Ricci tensor is zero ($[R_{\mu\nu}] = 0$). This fits with the equivalence principle because as you focus in closer and closer on parallel spacetime geodesics, the spacetime volume separating them will reduce evenly along their paths. At a theoretical infinitesimal level, they tend towards the same coordinates and become indistinguishable.

In a vacuum, the Riemann tensor $[R^{\alpha}_{\beta\gamma\delta}]$ must be non-zero (indicating curvature) while the Ricci tensor $[R_{\mu\nu}]$ must be zero. These two requirements are inconsistent in 2-D or 3-D space, but it turns out this combination is possible in 4-D spacetime.

At the end of the chapter, I introduced you to the Ricci scalar R, which has no indices. This is a number calculated from the Ricci tensor (which is in turn calculated from the Riemann tensor). Curvature in 2-D alters the area contained by a circle. In 3-D it alters the volume contained by a sphere. The Ricci scalar R is a measure of the level of this sort of distortion.

Note that the equivalence principle can equally be described as the inability to tell the difference between being held stationary in a gravitational field and accelerating away from it. Einstein also used this description (see Box 6.4).

Box 6.4 Rindler coordinates: an accelerating reference frame

The equivalence of gravitational freefall with moving in an inertial frame can also be described as the equivalence of being stationary in a gravitational field versus moving in an accelerating frame. Einstein outlined this in a lecture (abridged below) using drugged physicists!

Two physicists, A and B, awake from a drugged sleep and notice they are located in a closed box with opaque walls. They see that bodies they place in the middle of the box and release all fall in the same direction with the same acceleration. Physicist A concludes that the box sits on a planet in its gravitational field. Physicist B takes the view that the box might be maintained in uniform accelerated upward motion. Is there any criterion by which the two physicists can decide who is right? We know of no such criterion.

A uniform accelerated reference frame is often called *Rindler coordinates*. You will find a description of this in textbooks, but personally, I do not find it very helpful in visualising things.

It is now time to dig deeper into the maths of the key measures of curvature: the Riemann tensor $[R^{\alpha}_{\beta\gamma\delta}]$, the Ricci tensor $[R_{\mu\nu}]$ and the Ricci scalar R. I realise all of this is quite hard-going, but hey, in the introduction I promised that general relativity would be intuitive. I never said it would be easy.

Chapter 7

The Maths of Curvature

From this point on, there is no way to discuss curvature without introducing more mathematics, so brew yourself a nice cup of coffee and get ready for a few equations. And if you find the maths befuddling, do not despair... Einstein found it hard, too (see Box 7.1).

Box 7.1 Einstein's mathematical struggle

It took Einstein almost 10 years to figure out the equations of general relativity. So, if you find some of this hard, please take comfort from Einstein's own words: *I have become imbued with a high respect for mathematics, which in its subtler part I had regarded in my simple-mindedness as pure luxury until now. Compared with this problem* [general relativity], *the original* [special relativity] *is child's play.*

7.1   Parallel Transport

A mathematical way to quantify curvature is called *parallel transport*. This involves mathematically manipulating tiny basis vectors around small areas of the surface and seeing whether (and how much) the direction of each basis vector changes en route. Let me demonstrate with the help of our friendly dung beetles. Imagine I give the first beetle an arrow to carry on its back and ask it to walk around a triangle on a flat surface. The left of Figure 7.1 shows this scenario. When it gets to a corner, the beetle puts the arrow down on the ground, turns, then picks up the arrow (without changing the arrow's direction) and walks on in the new direction.

At each corner the beetle turns, but the arrow does not. The arrow after the turn remains parallel to its direction before the turn. As the arrow represents a basis vector, mathematically this is described as *parallel transport of a basis vector.*

Let me use the beetles to show you how and why this measures curvature. The beetle starts at the point labelled *start* (I bet you guessed that). It has the arrow pointing straight forward on its back. At the first corner, it puts down the arrow and turns. It picks up the arrow and walks on. The aim of the beetle is not to change the direction of the arrow while walking. At the next corner, I have relabelled it blue for clarity. It puts down the arrow again, turns, picks it up and walks on. It does the same at the next turn to arrive back where it started (unsurprisingly marked *end*). Lo and behold for the flat surface, if you compare the direction that the arrow is pointing at the start and end, you see that parallel transport doesn't change its direction.

I ask the dung beetle's smelly friend to pace out a similar path, but this time on the curved surface of a sphere, as shown on the right of the figure. It follows the same process but arrives back with its arrow pointing in a different direction. Why? The difference is that it walks up to the north pole and turns a right angle, then down to the equator to turn another right angle, then around the equator and another right angle to head back north. That is three right angles in the triangular path, which is $\frac{3\pi}{2}$ radians or 270°.

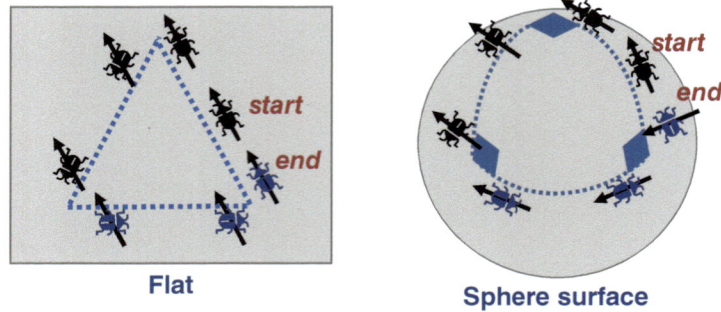

Figure 7.1 Schematic showing the effect of curvature on the parallel transport of a basis vector.

While the internal angles of a triangle on a flat surface add up to π radians or 180°, the internal angles of a triangle on a positively curved surface add up to *more*. The greater the curvature and the larger the triangle, the greater the discrepancy.

With *positive* curvature like the surface of a sphere, the internal angles of a triangle add up to more than on a flat surface. The opposite of positive curvature is *negative* curvature like the surface of a saddle where the internal angles of a triangle add up to less than on a flat surface (see Figure 7.2). If you have any doubts about this, please take a football, measure out a large triangle, dig out that old protractor from school and measure the angles.

The same pattern is true for any quadrilateral (four-sided) path. A quadrilateral can be decomposed into two triangles (connect any two opposing corners with a straight line). On a flat surface, the interior angles of a quadrilateral add up to 2π radians or 360°. On a positively curved surface they add up to more. On a negatively curved surface, they add to less. This brings us to the heart of the maths of curvature and the big beast... the Riemann tensor with its four indices [$R^{\alpha}_{\beta\gamma\delta}$].

7.2 The Riemann Tensor

The Riemann tensor measures the change to the basis vectors in parallel transport. Let me explain with the 2-D example of two beetles on the surface of a sphere as shown in Figure 7.3. The beetles start in the bottom left corner. Each carries an identical arrow (a basis vector that in the figure, you can see is in the y direction so it is $\vec{e}_y$). The black beetle heads north (dy). It then puts down its arrow, turns, picks up the arrow and heads east (dx). The blue beetle first heads east (dx) and then north (dy). They meet and there is a discrepancy between the direction of their arrows because of the curvature of the sphere. The discrepancy in the x direction is shown in red.

This sort of discrepancy is what the Riemann tensor measures. In this example, the part of the discrepancy that is in the x direction is the Riemann tensor component R^{X}_{YXY}. This is the discrepancy in the x direction to the y basis vector when parallel transported dy, then dx, compared to the reverse. Thus, the component R^{A}_{BCD} is the discrepancy in the A direction to the B basis vector when parallel transported, as shown by the last two indices. In this case, they are dD followed by dC (the first step of the two being put last by convention) compared to the reverse.

It is worth pointing out something that allows us quickly to eliminate many Riemann components. Those with the same first two indices such as R^{A}_{ACD} are zero. This is a measure of the directional discrepancy in parallel transporting the A basis vector. Any such discrepancy will not be in the original A direction, but away from it. And

Figure 7.2 Curvature changes the sum of the internal angles of a triangle.

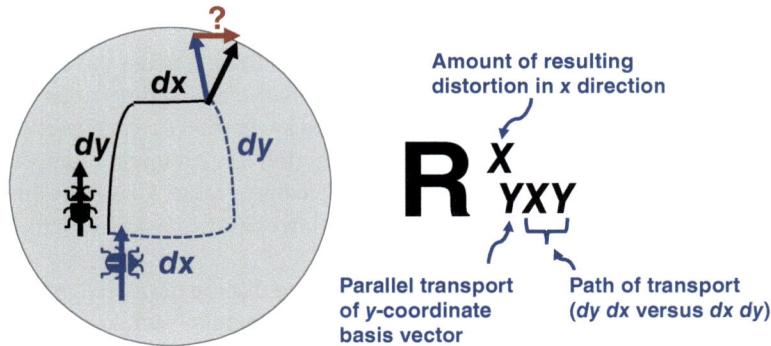

Figure 7.3 Illustration of how the Riemann tensor measures curvature.

any components with the same last two indices such as R^A_{BCC} are zero. For example, think what the component R^X_{YXX} means for the bugs in Figure 7.3. They would carry the basis vector dx followed by dx. The two of them would travel exactly the same route, so there can be no discrepancy at the end. It is also fairly obvious that $R^A_{BCD} = -R^A_{BDC}$. In the case of the bugs, this might be comparing R^X_{YXY} with R^X_{YYX}. The latter just switches the routes of the black and blue bugs so the discrepancy (black minus blue) is the opposite.

The neat thing about the Riemann tensor is that it is not fooled by weird or distorted coordinates. If the Riemann tensor has value, *there is curvature.* Look back at the bug moving along the flat sheet on the left of Figure 7.1. It doesn't matter what coordinates you choose or what crazy route you give the bug, when it arrives back, the direction of the arrow will not have changed. All the components of the Riemann tensor are always zero for a flat surface or flat spacetime.

When working in 4-D spacetime, you will usually see the Riemann tensor written with the conventional Greek symbols, such as $[R^\alpha_{\beta\gamma\delta}]$. I find it harder to follow $(\alpha, \beta, \gamma, \delta, \epsilon, \zeta...)$ than the Latin version $(A, B, C, D, E, F...)$. Therefore, when working with arbitrary components I will stick to regular capital letters such as R^A_{BCD} to keep things as legible as possible.

In 4-D spacetime, every one of the four indices can have four possible values: using Cartesian coordinates, A can be t, x, y or z and the same is true for B, C and D. As examples, the Riemann tensor $[R^\alpha_{\beta\gamma\delta}]$ has components R^T_{XYZ}, R^Z_{YTT}, and R^Y_{YXT}... plus any other combination of the indices. This means in 4-D spacetime that there are 256 possible components $(4 \times 4 \times 4 \times 4)$. To be fair, a lot of these 256 components are zero or repeats but, in total, there are still 20 independent components to the Riemann tensor in 4-D spacetime. Yes, it is that bad. For some historical background on Riemann, check out Box 7.2.

Box 7.2 Bernhard Riemann (1826−1866)

Bernhard Riemann was one of the six children of a poor German Lutheran minister. Riemann was a shy nervous child plagued by ill health. He struggled at school because he was such a perfectionist that he never turned work in on time. However, his teachers spotted that this was a sign of genius not laziness (*note to students: try this line with your professor next time you are late handing in an assignment*).

Riemann worked as a (poorly paid) professor at Göttingen University and could barely afford the paper he needed to write down his calculations. In spite of this, he developed a whole new area of mathematics being one of the first to consider the possibility of additional dimensions and curvature... culminating in the Riemann curvature tensor. In 1866, he had to flee Göttingen when it became a battle zone between the armies of Prussia and Hanover. He made it to Italy, but the trip took its toll and he died of consumption at the age of 39.

In his notes, he wrote about his success in analysing what is now called the Riemann Zeta Function. His conclusion, (the Riemann hypothesis) is of central importance in the analysis of prime numbers. In Riemann's haste to escape Göttingen, he left many notes in his messy office. His housekeeper cleared up and threw the lot out. Mathematicians have tried to construct a proof of the Riemann hypothesis ever since... without success.

7.2.1 Indices of the Riemann Tensor

This is a good moment to return to the topic of upper and lower indices. It is not the most exciting subject, but a basic understanding will prove invaluable later... so top up that coffee cup and try to focus. The Riemann tensor components shown in the last section have one upper and three lower indices, such as $R^T{}_{XYZ}$. Why?

If this is all new to you, the best starting point is to review Section 3.6. An upper index on a vector V such as V^x indicates the component is in what I call vector coordinates. The lower index V_x is in its dual coordinates. We use the upper index coordinates for the vector component and the lower index coordinates for the basis vector so that they combine to an invariant quantity (see also Box 3.6).

Consider the Riemann component $R^T{}_{XYZ}$. It is a measure of the discrepancy when you take the x *basis* vector (lower index) and parallel transport it along the y and then z *basis* vectors (each lower index) versus the reverse. The discrepancy $R^T{}_{XYZ}$ is the *vector component* in the t direction (upper index). Thus, the three lower indices refer to basis vectors (the input basis vector and the basis vectors of parallel transport), and the one upper index refers to the output vector component.

This is not the only way to write components of the Riemann tensor. For example, suppose you are tempted to read a different textbook on general relativity instead of this beautiful oeuvre (heaven forbid), and it gives you R_{TXYZ}. This is exactly the same as $R^T{}_{XYZ}$ *except* the output discrepancy in the t direction is in dual coordinates. If you want to know the output in terms of the vector coordinates, you need to switch coordinates. For the diagonal metrics (no off-diagonal terms) we will work with, you do this by multiplying by g^{TT}, which raises the t index: $g^{TT}R_{TXYZ} = R^T{}_{XYZ}$. Conversely, we lower the index for a diagonal metric with: $g_{TT} R^T{}_{XYZ} = R_{TXYZ}$. Don't forget that, in the case of a diagonal metric, the value of g^{TT} and g_{TT} are related: $g^{TT} = \frac{1}{g_{TT}}$.

If you switch input and output vectors of the Riemann tensor, it reverses the discrepancy measured in dual coordinates: $R_{ABCD} = -R_{BACD}$. From the symmetries discussed earlier, we know that $R_{ABCD} = -R_{ABDC}$, which gives $R_{ABCD} = R_{BADC}$ (all the indices are lower).

$$R_{ABCD} = -R_{BACD} = R_{BADC}, \qquad \textit{another useful Riemann symmetry.} \qquad (7.1)$$

I want to share one final Riemann symmetry that will be important for calculating components of the Ricci tensor. We will use it again and again. The following *applies only to diagonal metrics*. I have explained how to raise and lower indices for *diagonal* metrics such that $g_{AA} R^A{}_{BAB} = R_{ABAB}$ and $g_{BB} R^B{}_{ABA} = R_{BABA}$. The equivalence of the first and third terms in Equation 7.1 gives $R_{ABAB} = R_{BABA}$, leading to Equation 7.2. Multiplying both sides by $g^{AA}g^{BB}$ reveals the very important symmetry for *diagonal* metrics in Equation 7.3.

$$g_{BB} R^B{}_{ABA} = g_{AA} R^A{}_{BAB}, \qquad \textit{note}: \quad g^{AA} = \frac{1}{g_{AA}} \quad g^{BB} = \frac{1}{g_{BB}}, \qquad (7.2)$$

$$g^{AA} R^B{}_{ABA} = g^{BB} R^A{}_{BAB}, \qquad \textit{very important symmetry for diagonal metrics.} \qquad (7.3)$$

7.2.2 Calculating Components of the Riemann Tensor

The process of calculating the Riemann tensor component $R^A{}_{BCD}$ is the following. We take the B dimension basis vector $\vec{e}_B$ and measure how it changes with a step in the D direction, which is its covariant derivative $\nabla_D \vec{e}_B$. We then measure how this changes with the step in the C direction. This is the covariant derivative $\nabla_C (\nabla_D \vec{e}_B)$, which is typically written just as $\nabla_C \nabla_D \vec{e}_B$. In the 2-D example in Figure 7.3, setting $B = y, C = x, D = y$ would give us $\nabla_x \nabla_y \vec{e}_y$, which is the equivalent of the black beetle's route.

To measure the difference, we must subtract the result of the opposite route. In terms of the two beetles in Figure 7.3, this would be $(\nabla_x \nabla_y \vec{e}_y - \nabla_y \nabla_x \vec{e}_y)$. More generally, this gives $(\nabla_C \nabla_D \vec{e}_B - \nabla_D \nabla_C \vec{e}_B)$, which is often written using the commutator symbol as $[\nabla_C, \nabla_D] \vec{e}_B$.

As we are calculating $R^A{}_{BCD}$, we only want the components of this in the $\vec{e}_A$ direction. You can get this by taking the vector dot product with $\vec{e}_A$ or simply picking out the appropriate components.

Deriving the formula for the Riemann tensor is not too tricky, so I show it below in optional Subsection 7.2.3. However, if you are not interested, then you can cheat by using the answer in Box 7.3. You will not need the derivation for what comes next, although you will need the formula in the box now and again.

Box 7.3 Riemann tensor formula   *where* **E** *is a dummy index (shown in bold)*

$$R^A_{\ BCD} = \partial_C \Gamma^A_{BD} - \partial_D \Gamma^A_{BC} + \Gamma^A_{C\mathbf{E}} \Gamma^{\mathbf{E}}_{BD} - \Gamma^A_{D\mathbf{E}} \Gamma^{\mathbf{E}}_{BC}.$$

You must calculate and add together all possible values that the **E** index might be. The symbol ∂C is an abbreviated label for the partial derivative $\frac{\partial}{\partial C}$. Some of the simplest tricks to eliminate calculations are shown below (there are others):

$$R^A_{\ ACD} = R^A_{\ BCC} = 0, \qquad R^A_{\ BCD} = -R^A_{\ BDC}, \qquad R_{ABCD} = -R_{BACD} = R_{BADC}.$$

7.2.3 Deriving the Formula for the Riemann Tensor (Optional)

Let's examine the parallel transport of one basis vector. As usual, for simplicity, I will assume the coordinate axes are orthogonal, so it is a diagonal metric. As described in the previous paragraph, the value of $R^A_{\ BCD}$ is as shown in Equation 7.4. It is the component in the $\vec{e_A}$ direction of the result of parallel transporting the $\vec{e_B}$ basis vector around the C and D directions. Note that we can separate out this component mathematically by taking the vector dot product with $\vec{e_A}$. Don't worry if your knowledge of the dot product is rusty. When we get to this step, you will see that we are just picking out the components in the $\vec{e_A}$ direction.

$$R^A_{\ BCD} = \vec{e_A} \cdot (\nabla_C \nabla_D \vec{e_B} - \nabla_D \nabla_C \vec{e_B}). \tag{7.4}$$

The first step is to express $\nabla_D \vec{e_B}$, as shown in Equation 7.5. As discussed in Section 4.1, in curved spacetime we must account for how the basis vectors change. The derivative of a basis vector is a string of Christoffel symbols, one for each direction (check back to Equation 4.10 if you need to). These can be summarised into one term using the **E** dummy index, which as usual I have put in bold for clarity.

$$\nabla_D \vec{e_B} = \Gamma^A_{BD} \vec{e_A} + \Gamma^B_{BD} \vec{e_B} + \Gamma^C_{BD} \vec{e_C} + \dots = \Gamma^{\mathbf{E}}_{BD} \vec{e_{\mathbf{E}}}, \qquad using\ dummy\ index\ \mathbf{E}. \tag{7.5}$$

The next step is to calculate the $\nabla_C \nabla_D \vec{e_B}$ term that appears in Equation 7.4. This is shown in Equation 7.6 applying the product differentiation rule to the result of Equation 7.5. Don't forget that covariant differentiation includes derivative terms for the basis vectors (check back to Equation 4.6 if needed). This gives us two terms, each of which includes the dummy index.

$$\nabla_C \nabla_D \vec{e_B} = \left(\frac{\partial}{\partial C} \Gamma^{\mathbf{E}}_{BD} \right) \vec{e_{\mathbf{E}}} + \left(\frac{\partial}{\partial C} \vec{e_{\mathbf{E}}} \right) \Gamma^{\mathbf{E}}_{BD}, \tag{7.6}$$

$$= \left(\frac{\partial}{\partial C} \Gamma^A_{BD} \right) \vec{e_A} + \left(\frac{\partial}{\partial C} \Gamma^B_{BD} \right) \vec{e_B} \dots + \left(\Gamma^A_{C\mathbf{E}} \vec{e_A} + \Gamma^B_{C\mathbf{E}} \vec{e_B} \dots \right) \Gamma^{\mathbf{E}}_{BD}, \tag{7.7}$$

$$\vec{e_A} \cdot \left(\nabla_C \nabla_D \vec{e_B} \right) = \frac{\partial}{\partial C} \Gamma^A_{BD} + \Gamma^A_{C\mathbf{E}} \Gamma^{\mathbf{E}}_{BD}. \tag{7.8}$$

These are re-expressed in Equation 7.7. Let's start with the *first* term on the right of Equation 7.6. The dummy index is reopened. This shows its first two components, which are in the $\vec{e_A}$ and $\vec{e_B}$ directions. There are others, but we are calculating $R^A_{\ BCD}$, so we want only the $\vec{e_A}$ component. This becomes the first term in Equation 7.8.

The *last* term of Equation 7.6 takes a little more effort. The derivative of the dummy index basis vector $e_{\mathbf{E}}$ is another string of Christoffel symbols, one for each direction. Two of these terms are shown on the far right in Equation 7.7. Again, there are more, but we are interested only in the $\vec{e_A}$ component. Mathematically, the process is shown in Equation 7.8 as the $\vec{e_A}$ dot product, but all we are doing is picking out the components in the $\vec{e_A}$ direction.

Equation 7.10 pulls it all together. We take the value for $\vec{e_A} \cdot (\nabla_C \nabla_D \vec{e_B})$ from Equation 7.8. We switch the C and D indices to get the answer for $\vec{e_A} \cdot (\nabla_D \nabla_C \vec{e_B})$, and then subtract one from the other. You will often see derivatives written in the form ∂_C instead of $\frac{\partial}{\partial C}$.

$$R^A_{BCD} = \vec{e}_A \cdot (\nabla_C \nabla_D \vec{e}_B - \nabla_D \nabla_C \vec{e}_B), \tag{7.9}$$

$$= \frac{\partial}{\partial C} \Gamma^A_{BD} + \Gamma^A_{CE} \Gamma^E_{BD} - \frac{\partial}{\partial D} \Gamma^A_{BC} - \Gamma^A_{DE} \Gamma^E_{BC},$$

$$R^A_{BCD} = \partial_C \Gamma^A_{BD} - \partial_D \Gamma^A_{BC} + \Gamma^A_{CE} \Gamma^E_{BD} - \Gamma^A_{DE} \Gamma^E_{BC}, \qquad \textit{Riemann tensor formula.} \tag{7.10}$$

To expand this from a single basis vector to a full vector, you simply multiply the result of each basis vector by the relevant vector component, which in the case of Equation 7.9 is V^A. I hope this exercise helps you see what the Riemann tensor is, what it measures and why its formula is this convoluted combination of Cripes-they're-awful symbols.

7.3 Calculating the Ricci Tensor

The Ricci tensor $[R_{\mu\nu}]$ in spacetime is a (4×4) matrix with 16 components as shown on the left of Equation 7.11. In contrast, the Riemann tensor $[R^\alpha{}_{\beta\gamma\delta}]$ in spacetime is a massive $(4 \times 4 \times 4 \times 4)$ monster with 256 entries. The Riemann tensor quantifies any and all the curvature. For the Ricci tensor, you combine some of the components of the Riemann tensor to get a reduced specific measure of the change in spacetime volume between infinitely close parallel geodesics. You may see this described as $R_{AB} = R^C{}_{ACB}$, where the bold C is a dummy variable following the Einstein summation convention. Check back to Box 3.5 if you need a reminder of Einstein summation. Also check out Box 7.4 for a convention warning.

$$[R_{\mu\nu}] = \begin{bmatrix} R_{TT} & R_{TX} & R_{TY} & R_{TZ} \\ R_{XT} & R_{XX} & R_{XY} & R_{XZ} \\ R_{YT} & R_{YX} & R_{YY} & R_{YZ} \\ R_{ZT} & R_{ZX} & R_{ZY} & R_{ZZ} \end{bmatrix}, \qquad R_{TT} = R^T{}_{TTT} + R^X{}_{TXT} + R^Y{}_{TYT} + R^Z{}_{TZT}. \tag{7.11}$$

The right of Equation 7.11 shows the Ricci tensor component R_{TT} in terms of its Riemann tensor components for spacetime using Cartesian coordinates. Note that one of these Riemann tensor components is always zero. In the case shown, it is $R^T{}_{TTT}$ (see symmetries in Box 7.3).

Box 7.4 Ricci tensor convention warning

In this book, I use the convention for the Ricci tensor: $R_{AB} = R^C{}_{ACB}$ where C is the dummy index. Some textbooks use $R_{AB} = R^C{}_{ABC}$. The maths is identical except that $R^C{}_{ACB} = -R^C{}_{ABC}$ so one is the negative of the other. If you ever find a +/− discrepancy in your calculations, then please check the conventions that are being used. Frankly, these different conventions are ruddy irritating... don't blame me... I get as mad as anybody about them.

The best description I can give in words is that R_{TT} is a particular measure of the change in spacetime volume separating geodesics associated with moving the t basis vector through the t dimension (both are basis vectors so lower index). For an intuitive explanation with pretty pictures, refer back to Section 6.4. Similarly, R_{TX} is that from the x basis vector moving through the t dimension. This turns out always to be the same as moving the t basis vector through the x dimension, so $R_{TX} = R_{XT}$.

I should note that there is a general formula for the Ricci tensor. It is shown in Box 7.5 so that you have it if you want it. I *strongly* advise you to calculate the relevant Riemann components individually and add them together because the general Ricci formula has two dummy variables that can lead to mistakes unless you are exceedingly diligent (which I am not). Even more important, is that calculating the individual Riemann components reveals symmetries that reduce the mathematical workload, as I will demonstrate in Subsection 7.5.1.

Box 7.5 Ricci tensor formula   *where* **C** and **D** are both dummy indices

$$R_{AB} = \partial_C \Gamma^C_{AB} - \partial_B \Gamma^C_{AC} + \Gamma^C_{CD} \Gamma^D_{AB} - \Gamma^C_{BD} \Gamma^D_{AC}.$$

You must calculate and add together all possible values that the **C** index and **D** index might be. Personally, I prefer to calculate and add up the individual Riemann components of each Ricci component. Also note that this follows the convention: $R_{AB} = R^C_{ACB}$.

7.4 Calculating the Ricci Scalar

The Ricci scalar is labelled R. You may occasionally see it labelled $R^\mu_{\ \mu}$. It combines components of the Ricci tensor to give what might be described as an average for the Ricci curvature. The Ricci scalar plays an important role in the Einstein Field Equations (EFEs), which we will discuss later. Equation 7.12 shows its formal definition.

$$R = g^{\mu\nu} R_{\mu\nu} = R^\mu_{\ \mu}, \qquad \textit{Ricci Scalar.} \tag{7.12}$$

Equation 7.13 shows an example in two dimensions (x, y). You multiply the components of the Ricci tensor by the appropriate entry in the inverse metric $[g^{\mu\nu}]$ before adding them. This scales them so they match. The telltale sign is that this balances the number of upper and lower indices (one up, one down). Take note that if the Ricci tensor is zero, then so is the Ricci scalar.

$$R = R^X_{\ X} + R^Y_{\ Y} = (g^{XX} R_{XX} + g^{XY} R_{XY}) + (g^{YY} R_{YY} + g^{YX} R_{YX}), \qquad \textit{2-D example.} \tag{7.13}$$

Things are much simpler for the diagonal metrics that we will study. For diagonal metrics, all the off-diagonal components such as g^{XY} are zero, so the formula simplifies as shown in Equation 7.14 using Cartesian coordinates. For diagonal metrics, don't forget that $g^{\mu\nu}$ is simply $\dfrac{1}{g_{\mu\nu}}$ such as, for example, $g^{XX} = \dfrac{1}{g_{XX}}$.

$$R = R^T_{\ T} + R^X_{\ X} + R^Y_{\ Y} + R^Z_{\ Z} = g^{TT} R_{TT} + g^{XX} R_{XX} + g^{YY} R_{YY} + g^{ZZ} R_{ZZ}. \tag{7.14}$$

Before moving on, I want to be sure that you understand what it means when I say that the Ricci scalar is a *scalar*. This simply means that it is a *number*. At any point in spacetime, the Ricci scalar will have some numerical value. It is not a constant, but will be different at different points. Being a scalar, it has a value but no direction, and (importantly) it is Lorentz invariant, meaning that all observers agree on its value (check back to the end of Section 2.3 if you need a refresher). The value of the Ricci scalar tells you about the overall distortion of the spacetime volume at that point (see Section 6.6 for an intuitive description).

As an analogy, consider making some measurements at an airport to check it is safe for planes to land. First you measure temperature. For each point, you note the temperature down as a number. Temperature is a *scalar* quantity. Contrast this with the wind speeds at different points in the airport, which involve a number (speed) and a direction. Wind velocity is a *vector* quantity.

7.5 Example Calculations: Aarrgghh!

Before unleashing the horror of a proper calculation on you, I am going to start with an easy example: Minkowski spacetime. The metric using (ct, x, y, z) coordinate units (which match time with space) is shown again as Equation 7.15 (this is Equation 3.4 from earlier).

$$\textit{Minkowski metric } (ct, x, y, z) : [g_{\mu\nu}] = \begin{bmatrix} 1 & 0 & 0 & 0 \\ 0 & -1 & 0 & 0 \\ 0 & 0 & -1 & 0 \\ 0 & 0 & 0 & -1 \end{bmatrix}. \tag{7.15}$$

As there are no variables at all in the metric (it is just the constants 1 and −1), the value of every Christoffel symbol is zero. If you have any doubts, check back to the cheat sheet in Box 4.4. Every Christoffel value comes from derivatives of components of the metric tensor. As the components of the metric tensor are all constants, the derivatives are all zero and the value of every Christoffel symbol is zero. The Riemann tensor is calculated using Christoffel symbols (see the formula in Box 7.3). Therefore, every component of the Riemann tensor is zero, which confirms that Minkowski spacetime is flat. Both the Ricci tensor and scalar are combinations of components of the Riemann tensor so they are, of course, also zero. Wow, that was easy. Unfortunately, calculating these things is rarely such a breeze. Even Einstein found it hard (see Box 7.6).

Box 7.6 Help!

Again and again Einstein had difficulties with the maths of curvature. In 1912, some 3 years before arriving at his final theory, he beseeched his classmate and friend, the Swiss mathematician Marcel Grossman:
> *You must help me, or else I'll go crazy.*

7.5.1   Symmetry Shortcut for Diagonal Metrics

Let me show you an important shortcut for calculating the Ricci tensor and scalar. This *works only for diagonal metrics*. The definition of the Ricci scalar for a diagonal metric is shown as Equation 7.16 (as from Equation 7.14, but in 2-D).

$$R = g^{XX}R_{XX} + g^{YY}R_{YY}, \qquad \text{\textit{Ricci scalar (2-D), diagonal metric.}} \tag{7.16}$$

Equation 7.17 expands each term to show its underlying Riemann components using the definition of the Ricci tensor for a diagonal metric, on the right of Equation 7.11. From the first Riemann symmetries at the bottom of Box 7.3 we know that R^X_{XXX} and R^Y_{YYY} are zero. From the symmetry of Equation 7.3 for diagonal metrics, we know that $g^{XX}R^Y_{XYX} = g^{YY}R^X_{YXY}$, so I have labelled both of them a, and the value of the Ricci scalar is $2a$ (from Equation 7.16).

$$g^{XX}R_{XX} = g^{XX}R^X_{XXX} + g^{XX}R^Y_{XYX} = 0 + a, \qquad \text{\textit{Ricci Scalar (2-D) = 2a,}}$$
$$g^{YY}R_{YY} = g^{YY}R^X_{YXY} + g^{YY}R^Y_{YYY} = a + 0, \qquad \text{\textit{Diagonal metrics only.}} \tag{7.17}$$

Whenever you calculate the Ricci tensor or scalar of a diagonal metric, I urge you to lay out the underlying Riemann tensor components in this way. Not only does it ease the maths, it also reveals the underlying symmetries of the Riemann curvature hidden within the Ricci tensor. This will prove crucial when we discuss more complicated 4-D spacetime metrics such as the Schwarzschild and FRW metrics. And if you need the actual Ricci tensor values, these are easily extracted by multiplying with the relevant component from $[g_{\mu\nu}]$. For example: $g_{YY}a = g_{YY}g^{YY}R_{YY} = R_{YY}$.

As for the Ricci components with mixed indices, both are zero for any 2-D diagonal metric. We know from symmetry that $R_{XY} = R_{YX}$ (Section 7.3). Expanding to $R_{XY} = R^X_{XXY} + R^Y_{XYY}$, both terms are zero based on the symmetries at the bottom of Box 7.3.

7.5.2   Flat Space with Polar Coordinates

My first serious example is 2-D flat space but using polar coordinates. Can we fool the curvature tensors or will they see through the confusing coordinates and spot that the underlying space is flat? Ooh! How exciting! The answer is of course that the curvature tensors will do their job. Here goes. The 2-D spatial metric is shown in Equation 7.18 (as from Equation 3.11).

$$[g_{ij}] = \begin{bmatrix} 1 & 0 \\ 0 & r^2 \end{bmatrix}, \qquad \text{2-D, flat, polar coordinates } (r,\theta): \quad g_{RR} = 1 \quad g_{\theta\theta} = r^2. \tag{7.18}$$

The Christoffel symbols of the metric are shown as Equation 7.19. I showed how to calculate four of these Christoffel symbols (look back to Equation 4.14). Use the cheat sheet formulas in Box 4.4 to check the others if you wish (the metric is diagonal).

$$
\begin{aligned}
\Gamma^R_{RR} &= 0, & \Gamma^\theta_{RR} &= 0, \\
\Gamma^R_{\theta\theta} &= -r, & \Gamma^\theta_{\theta\theta} &= 0, \\
\Gamma^R_{R\theta} &= \Gamma^R_{\theta R} = 0, & \Gamma^\theta_{R\theta} &= \Gamma^\theta_{\theta R} = \frac{1}{r}.
\end{aligned} \tag{7.19}
$$

The 2-D metric is diagonal, so we lay out the underlying Riemann tensor components as in Equation 7.17. This is shown with the coordinates we are using as Equation 7.20. I have again used the label a to show the symmetry. The value of a will of course vary depending on the metric.

$$
\begin{aligned}
g^{RR}R_{RR} &= g^{RR}R^R_{RRR} + g^{RR}R^\theta_{R\theta R} = 0 + a, & \textit{Ricci Scalar (2-D)} &= 2a, \\
g^{\theta\theta}R_{\theta\theta} &= g^{\theta\theta}R^R_{\theta R\theta} + g^{\theta\theta}R^\theta_{\theta\theta\theta} = a + 0, & \textit{Polar coordinates.}
\end{aligned} \tag{7.20}
$$

To calculate R_{RR} and $R_{\theta\theta}$ we will use the Riemann tensor component $R^\theta_{R\theta R}$ based on the formula from Box 7.3 shown again as Equation 7.21. I will go through each step slowly. You rarely will need this sort of calculation, but best you know how. In Equation 7.22, I have written the formula substituting the variables we want. In Equation 7.23, I drop the first term because it is zero. It is a derivative of a Christoffel symbol with respect to θ and none of the Christoffel symbols contain that variable as you can see from the list in Equation 7.19. At the same time, I have expanded out the dummy variable **E** to cover both r and θ. In Equation 7.24 I have deleted all terms with zero-value. I have then substituted in the Christoffel values from 7.19, which leaves an easy calculation in 7.25. This shows that the value of $R^\theta_{R\theta R}$ and therefore also of R_{RR} is zero as expected.

$$R^A_{BCD} = \partial_C \Gamma^A_{BD} - \partial_D \Gamma^A_{BC} + \Gamma^A_{CE}\Gamma^E_{BD} - \Gamma^A_{DE}\Gamma^E_{BC}. \tag{7.21}$$

$$R^\theta_{R\theta R} = \partial_\theta \Gamma^\theta_{RR} - \partial_R \Gamma^\theta_{R\theta} + \Gamma^\theta_{\theta E}\Gamma^E_{RR} - \Gamma^\theta_{RE}\Gamma^E_{R\theta}, \tag{7.22}$$

$$= -\partial_R \Gamma^\theta_{R\theta} + \Gamma^\theta_{\theta R}\Gamma^R_{RR} + \Gamma^\theta_{\theta\theta}\Gamma^\theta_{RR} - \Gamma^\theta_{RR}\Gamma^R_{R\theta} - \Gamma^\theta_{R\theta}\Gamma^\theta_{R\theta}, \tag{7.23}$$

$$= -\partial_R \Gamma^\theta_{R\theta} - \Gamma^\theta_{R\theta}\Gamma^\theta_{R\theta}, \tag{7.24}$$

$$= -\partial_R \left(\frac{1}{r}\right) - \frac{1}{r}\frac{1}{r} \tag{7.25}$$

$$= \frac{1}{r^2} - \frac{1}{r^2} = 0, \qquad \Rightarrow R_{RR} = 0.$$

We need to do no more calculations. The symmetries give us the rest. We know that $R^\theta_{R\theta R} = 0$, which means $a = 0$ in Equation 7.20 and therefore $R_{\theta\theta}$ and the Ricci scalar are both zero. As you can see, the Riemann and Ricci tensors are not fooled by the polar coordinates. Whatever the coordinates, flat space will always give this zero result.

If you are a bit of a weirdo like me and enjoy a mathematical challenge, you can calculate the Ricci tensor for the curved 2-D surface of a unit sphere. I have outlined this challenge in Box 7.7. Let me warn you that this is *not* easy. If you get it right first time, you should be pleased with yourself. Don't forget to lay out the underlying Riemann tensor components in the format of Equation 7.17. The full correct calculation is shown at the end of this chapter in Box 7.9.

Again, let me repeat that you will rarely need to do this sort of laborious calculation. So, be not afraid. Most of our interest in the Ricci tensor will focus on those occasions when $[R_{\mu\nu}] = 0$.

Box 7.7 Calculating the Ricci tensor for the surface of a unit sphere   Optional!

For the brave readers who want to try this challenge, the aim is to calculate the Ricci curvature for a bug living on the surface of a unit sphere (radius = 1). The scenario is shown earlier in Figure 3.3. The metric is shown as Equation 7.26 (the same as Equation 3.17).

$$ds^2 = d\theta^2 + \sin^2\theta \, d\phi^2,$$

$$\text{Surface of a sphere, } (r = 1), \ (\theta, \phi) : [g_{ij}] = \begin{bmatrix} 1 & 0 \\ 0 & \sin^2\theta \end{bmatrix}. \tag{7.26}$$

To help you along, the answers are below so that you can check your work. The challenge is to produce the calculation that shows this:

$$R_{\theta\theta} = 1, \qquad R_{\phi\phi} = \sin^2\theta, \qquad R_{\theta\phi} = R_{\phi\theta} = 0.$$

Hint: calculate $R^{\phi}_{\ \theta\phi\theta}$ and use symmetries. These trigonometric identities may help you:

$$\frac{\partial \sin^2\theta}{\partial\theta} = 2\sin\theta\cos\theta, \qquad \frac{\partial}{\partial\theta}\left(\frac{\cos\theta}{\sin\theta}\right) = -\frac{1}{\sin^2\theta}.$$

The detailed calculation is shown at the end of this chapter in Box 7.9.

7.6 Hunting for Vacuum Solutions

That is definitely enough detailed maths for this chapter. It is time to consider what this tells us about gravity in a vacuum (check out Box 7.8 for some vacuous humour). Einstein's conclusion was that the curvature of spacetime in a vacuum must be Ricci-0, so every component of the Ricci tensor must be zero ($[R_{\mu\nu}] = 0$). On the other hand, gravity comes from curvature so, if there is a gravitational field, the Riemann tensor of the metric cannot be zero ($[R^{\alpha}_{\ \beta\gamma\delta}] \neq 0$). After all, if the Riemann tensor is zero, then the metric is perfectly flat and there is no curvature at all. In short, to account for the gravitational effects that we experience in a vacuum, the Ricci tensor of the metric must be zero, but the Riemann tensor of the metric must have non-zero components.

Box 7.8 Just for laughs

I was going to tell you some vacuum jokes... but they all suck.

In Section 6.5, I told you that this can only happen in four dimensions or higher. Based on what you have learned in this chapter, I can show you why it is mathematically impossible for a 2-D surface to have a non-zero Riemann tensor alongside a zero-value Ricci tensor.

Let me use x and y as the 2-D coordinates (which could be anything). Riemann symmetries tell us that any Riemann component with the same two indices at start or end is zero (bottom of Box 7.3). In 2-D, this leaves only four components of the Riemann tensor as potentially non-zero value: $R^X_{\ YXY}$, $R^X_{\ YYX}$, $R^Y_{\ XYX}$ and $R^Y_{\ XXY}$. If one of these is non-zero, it indicates curvature such that parallel transport distorts a vector. If there is distortion in the upper-index vector component, there must be distortion also in the lower-index version, i.e. if $R^X_{\ YXY}$ is non-zero, then so is R_{XYXY}.

We know from the symmetries on the right at the bottom of Box 7.3 that: $R_{XYXY} = -R_{XYYX} = R_{YXYX} = -R_{YXXY}$, which means that if one of the potentially non-zero 2-D Riemann components is non-zero, then all four must be non-zero.

If you check back to Equation 7.17, you will see that if $R^Y_{\ XYX}$ is non-zero, then R_{XX} is non-zero, and if $R^X_{\ YXY}$ is non-zero then R_{YY} is non-zero. The point is that if any of the potential non-zero components of the 2-D Riemann tensor is non-zero, then the Ricci tensor must also be non-zero. Ricci-0 curvature in 2-D is impossible.

You can repeat this exercise in three dimensions. It is a bit more complicated, but the result is the same. If there is *any* Riemann curvature in 2-D or 3-D, the Ricci tensor cannot be zero. Vacuum gravity as described in Einstein's theory of general relativity cannot exist in only two or three dimensions.

Thankfully for Einstein, Ricci-0 curvature is possible in four dimensions. Phew! His idea works in our 4-D spacetime. However, the Riemann symmetries put huge constraints on possible solutions for the metric. Indeed, in Chapter 12 we will use these symmetries to guide us to the most famous vacuum solution: the Schwarzschild metric.

Box 7.9 Calculating the Ricci tensor for the surface of a unit sphere   **Optional!**

2-D Ricci component template for a diagonal metric

$$g^{\theta\theta}R_{\theta\theta} = g^{\theta\theta}R^{\theta}{}_{\theta\theta\theta} + g^{\theta\theta}R^{\phi}{}_{\theta\phi\theta} \qquad = 0 + a, \qquad Ricci\ Scalar\ (2\text{-}D) = 2a,$$

$$g^{\phi\phi}R_{\phi\phi} = g^{\phi\phi}R^{\theta}{}_{\phi\theta\phi} + g^{\phi\phi}R^{\phi}{}_{\phi\phi\phi} \qquad = a + 0.$$

Christoffel symbols you need (formulas in Box 4.4)

$$\Gamma^{\theta}_{\theta\theta} = \Gamma^{\theta}_{\phi\theta} = \Gamma^{\theta}_{\theta\phi} = \Gamma^{\phi}_{\theta\theta} = \Gamma^{\phi}_{\phi\phi} = 0,$$

$$\Gamma^{\phi}_{\theta\phi} = \Gamma^{\phi}_{\phi\theta} = \frac{1}{2}g^{\phi\phi}\frac{\partial g_{\phi\phi}}{\partial\theta} = \frac{1}{2\sin^2\theta}\frac{\partial\sin^2\theta}{\partial\theta} = \frac{2\sin\theta\cos\theta}{2\sin^2\theta} = \frac{\cos\theta}{\sin\theta}.$$

Calculation of $R^{\phi}{}_{\theta\phi\theta}$ (formula in Box 7.3)

$$R^{\phi}{}_{\theta\phi\theta} = \frac{\partial}{\partial\phi}\Gamma^{\phi}_{\theta\theta} - \frac{\partial}{\partial\theta}\Gamma^{\phi}_{\theta\phi} + \Gamma^{\phi}_{E\phi}\Gamma^{E}_{\theta\theta} - \Gamma^{\phi}_{E\theta}\Gamma^{E}_{\theta\phi},$$

$$= -\frac{\partial}{\partial\theta}\Gamma^{\phi}_{\theta\phi} - \Gamma^{\phi}_{\theta\theta}\Gamma^{\phi}_{\theta\phi} = -\frac{\partial}{\partial\theta}\left(\frac{\cos\theta}{\sin\theta}\right) - \left(\frac{\cos\theta}{\sin\theta}\right)^2,$$

$$= \frac{1}{\sin^2\theta} - \frac{\cos^2\theta}{\sin^2\theta} = \frac{1-\cos^2\theta}{\sin^2\theta} = 1, \qquad \Longrightarrow \qquad a = g^{\theta\theta}(1) = \frac{1}{g_{\theta\theta}}(1) = 1.$$

Ricci tensor

$$R_{\theta\theta} = 1, \qquad R_{\phi\phi} = g_{\phi\phi}\,a = \sin^2\theta, \qquad R = 2a = 2,$$

$$R_{\theta\phi} = R_{\phi\theta} = 0, \qquad \text{(see last paragraph of Section 7.5.1).}$$

More generally, if you do the calculation for the surface of a sphere of radius *r* (rather than the unit sphere in this calculation), the Ricci scalar is $\frac{2}{r^2}$. This makes sense because as the radius *r* of the sphere increases, the surface flattens. At the limit of $r = \infty$, the Ricci scalar *R* tends to zero, as the surface becomes progressively flatter.

7.7 Summary

Many of you will be pleased to hear that we have covered all the specialised curvature maths we will need. You can breathe a sigh of relief. You now have the full toolbox for your study of general relativity: the geodesic equation, Cripes-they're-awful symbols, the Riemann tensor $[R^{\alpha}{}_{\beta\gamma\delta}]$, the Ricci tensor $[R_{\mu\nu}]$ and the Ricci scalar *R*. Hearty congratulations! Give yourself a pat on the back.

Let me briefly recap the contents of this chapter. The first step was to look at how the angles inside a figure like a triangle are changed by curvature. If the surface has positive curvature, such as is the case for the surface of a sphere, then the interior angles of the triangle add up to more than π (180°). With negative curvature, such as a saddle shape, they add up to less. This means a vector that is parallel transported around the triangle will arrive back pointing in a different direction from its starting point.

If the Riemann tensor $[R^{\alpha}_{\beta\gamma\delta}]$ is zero, the surface is flat. If the Riemann tensor has any non-zero components, there is curvature. The value of each Riemann component is calculated by parallel transporting a basis vector around two axes and seeing if there is a discrepancy that depends on the order of the axes. Mathematically, this involves taking covariant derivatives and results in the formula for the Riemann tensor shown in Box 7.3.

We can use the Riemann tensor $[R^{\alpha}_{\beta\gamma\delta}]$ to calculate the Ricci tensor $[R_{\mu\nu}]$. I showed you how this is most easily done using the symmetries inherent in the Riemann components that make up the Ricci tensor. As a reminder, the Ricci tensor is important because it measures the change in the volume of spacetime separating infinitely close parallel geodesics.

The next module of this book is all about Ricci-0 curvature. This means the Riemann tensor is non-zero while the Ricci tensor is zero ($[R^{\alpha}_{\beta\gamma\delta}] \neq 0$, $[R_{\mu\nu}] = 0$). This is impossible in 2-D or 3-D, but can and does occur in 4-D spacetime. What metrics are possible? The most famous is the Schwarzschild metric, which you will learn is the *only* Ricci-0 solution with spherical symmetry. It will tell us all sorts of weird stuff about intense gravitational fields and black holes.

But first, we will turn our attention to the other side of the EFEs. What happens when energy-momentum is present? In this case, the spacetime is *not* Ricci-0 ($[R_{\mu\nu}] \neq 0$). How does the presence of energy-momentum affect the spacetime metric? What is its relationship with the value of the Ricci tensor? It is time to discuss the *energy-momentum tensor*.

Chapter 8

The Energy-Momentum Tensor

Einstein's theory is based on linking energy-momentum with spacetime curvature. In Chapter 5, I showed you how gravitational acceleration can be explained as the result of curvature in the spacetime metric. Using a simple 2-D model, I demonstrated how time dilation alters the geodesics and so changes an object's natural path of motion (in the case of the model, the poor old dung beetle ended up hurtling towards the attracting mass).

In Chapter 6, we dug further into this distortion. I used Einstein's equivalence principle to link the presence of energy-momentum with a reduction in the spacetime volume separating what would normally be parallel infinitesimally close geodesics. This led in turn to a link between the presence of energy-momentum and the value of the spacetime's Ricci tensor $[R_{\mu\nu}]$.

Up to now, we have focussed on the left-hand side of the Einstein Field Equations (EFEs), shown (yet) again as Equation 8.1. In order to accurately link the presence of energy-momentum with spacetime curvature, we must quantify it. It is time to discuss the right-hand side: the energy-momentum tensor $[T_{\mu\nu}]$.

$$R_{\mu\nu} - \frac{1}{2} R\, g_{\mu\nu} = k\, T_{\mu\nu},$$

Spacetime curvature $\Longleftrightarrow$ *Energy-momentum,*

$$
\begin{bmatrix}
R_{TT} & R_{TX} & R_{TY} & R_{TZ} \\
R_{XT} & R_{XX} & R_{XY} & R_{XZ} \\
R_{YT} & R_{YX} & R_{YY} & R_{YZ} \\
R_{ZT} & R_{ZX} & R_{ZY} & R_{ZZ}
\end{bmatrix}
- \frac{1}{2}R
\begin{bmatrix}
g_{TT} & g_{TX} & g_{TY} & g_{TZ} \\
g_{XT} & g_{XX} & g_{XY} & g_{XZ} \\
g_{YT} & g_{YX} & g_{YY} & g_{YZ} \\
g_{ZT} & g_{ZX} & g_{ZY} & g_{ZZ}
\end{bmatrix}
= k
\begin{bmatrix}
T_{TT} & T_{TX} & T_{TY} & T_{TZ} \\
T_{XT} & T_{XX} & T_{XY} & T_{XZ} \\
T_{YT} & T_{YX} & T_{YY} & T_{YZ} \\
T_{ZT} & T_{ZX} & T_{ZY} & T_{ZZ}
\end{bmatrix}.
\tag{8.1}
$$

8.1 Tensor Indices

Before we address the energy-momentum tensor, I need to give more background on the indices that tensors carry. All the ones shown in Equation 8.1 are lower indices, but there can, of course, also be upper indices (as we discussed earlier for the larger Riemann tensor). The upper/lower distinction is exactly the same as for vectors. An upper index indicates that the coordinates used are vector coordinates. A lower index indicates that it is in dual coordinates. The combination of one upper index vector coordinate with one lower one, such as $V^x V_x$, gives a measure in proportion with the invariant interval. If you need to, please check back to Section 3.6 before continuing.

Each component of a *vector* is associated with *one* dimension. I am using the word *tensor* in this book for those with components associated with *two*.[1] If the tensor component has lower indices such as g_{TT} or R_{TT}, then it's in dual coordinates. Throughout this book, I will show the EFEs with lower indices because it puts the metric tensor component $g_{\mu\nu}$ in the familiar form, matching the invariant interval equation (see Section 3.1). It also appeals to

---

1 Technical point: Tensors can be associated with any number of dimensions. A vector sometimes is described as a one-tensor, something like the Ricci tensor as a two-tensor, the Riemann tensor as a four-tensor and so forth.

Untangling General Relativity: The Intuitive Self-Study Guide, First Edition. Simon Sherwood.
© 2026 John Wiley & Sons Ltd. Published 2026 by John Wiley & Sons Ltd.

me as the most natural form for the Ricci tensor. For example, R_{TT} gives a measure of the distortion linked with moving the t basis vector along the t dimension... again a basis vector, so both lower index.

For the diagonal metrics we will study, you raise an index by multiplying with the appropriate component of the metric tensor (Section 3.6). To convert a tensor from say R_{TT} to R^{TT}, both indices must be raised, so the process must be repeated twice, giving: $R^{TT} = g^{TT}g^{TT}R_{TT}$ for a diagonal metric. This is true for all the tensors, so the EFEs can be written with either upper or lower indices. An example of how this works is shown as Equation 8.2. I have simplified the calculation by assuming all the tensors are diagonal, but the result applies generally. The EFEs work with both upper and lower indices.

$$R_{YY} - \frac{1}{2} R g_{YY} = k T_{YY}, \qquad \text{\textit{Lower index EFE,}}$$
$$g^{YY}g^{YY} \left(R_{YY} - \frac{1}{2} R g_{YY} \right) = g^{YY}g^{YY} k T_{YY},$$
$$g^{YY}g^{YY} R_{YY} - \frac{1}{2} R g^{YY}g^{YY} g_{YY} = k g^{YY}g^{YY} T_{YY}, \qquad (8.2)$$
$$\implies \quad R^{YY} - \frac{1}{2} R g^{YY} = k T^{YY}, \qquad \text{\textit{Upper index EFE.}}$$

In a moment we will be working on the energy-momentum tensor. It is the measure of the flow of energy-momentum through spacetime. In my explanation, I derive $[T^{\mu\nu}]$ (upper index) rather than $[T_{\mu\nu}]$ (lower index) because I find the derivation easier to follow. It involves velocity vectors, so working with vector coordinates will appear more familiar to you than using dual coordinates. Don't forget that $[T^{\mu\nu}]$ and $[T_{\mu\nu}]$ are the *same* tensor, just expressed with different units (check back to Section 5.10 if you need to). All the symmetries and features of the energy-momentum tensor belong to the underlying tensor regardless of whether it is expressed as $[T^{\mu\nu}]$ or $[T_{\mu\nu}]$.

8.2 Introduction to the Energy-Momentum Tensor

The energy-momentum tensor $[T^{\mu\nu}]$ is also known as the *stress-energy* tensor. This is the (4×4) matrix tensor on the right of the EFEs, which describes the flow of energy-momentum through spacetime. Why are there 16 entries in the matrix? You might think you could quantify energy-momentum with only four numbers, describing how much of it is in a lump of spacetime: one for energy E and three more for momentum in the three dimensions p_x, p_y and p_z. However, when you are dealing with spacetime, there is a twist. You cannot look at how much energy-momentum is *in it*... rather, you have to look at how energy-momentum *flows through it*.

Why? If you are dealing with a space (not spacetime), you can talk about something being *in it*. Let me assume you are sitting in a room reading this fine book. Providing nobody moves the book, you can say that it is *in* a defined volume of space (the room). Spacetime is very different. It is defined not only by a small chunk of space, but also by a small chunk of time. Even if nobody moves the book spatially, it must flow out of any given chunk of spacetime into an adjacent chunk of spacetime... which is at a later time. Things cannot stay *in* spacetime. They *flow through* it.

Consider the energy, perhaps in the form of particles, in spacetime. The spacetime has four dimensions: Δt, Δx, Δy and Δz. Any particles that remain in the *spatial* volume ($\Delta x \Delta y \Delta z$) over the time period, we describe as flowing via the time dimension t. These particles don't move in terms of space. They flow through time.

Particles (and their energy) may also flow through the x, y or z directions. For example, a particle might move in the x direction. Therefore, in order to fully account for the flow of energy through the spacetime volume, we have to consider the energy flow through the t, x, y and z dimensions. That is four entries just to describe the flow of energy. You need another four entries to describe the flow of momentum p_x through spacetime, four more for p_y and four more for p_z. In total, that is 16 entries in the form of a (4×4) matrix to fully quantify the flow of energy-momentum. It turns out that six of the 16 entries are the same. For example, the flow of p_x in the time direction is the same as the flow of energy in the x dimension. I will explain why in a moment.

The flow of energy and momentum through spacetime is hard to visualise. Any readers who find this chapter challenging may be relieved to learn that there is a short *need-to-know* summary of the essentials at the end of the chapter in Box 8.4.

8.3 Mass Density Flow of Dust

I think the easiest way to explain the energy-momentum tensor is with what physicists call a *dust* model. Imagine we are out in space in the middle of a huge dust field. The density of the dust may vary at different places in the field, but it is made up of non-interacting identical small particles that are static relative to each other. We need to assess how the energy-momentum of the dust flows through space and time. In terms of the invariant interval, time Δt^2 has the opposite sign from the equivalent in space Δx^2. This makes it harder to visualise a volume of spacetime as some sort of simple hypercube (simple?) with sides $\Delta t, \Delta x, \Delta y, \Delta z$. Fortunately, we can easily visualise a volume of space, so let's start with that.

The scenario is illustrated in Figure 8.1a. To the bottom left is part of the dust field with the dust particles shown as black specks. A small chunk of space is shown in slightly darker blue as $\Delta x, \Delta y, \Delta z$. We want to know how the dust flows during a short period of time. The dust that stays in the space volume $\Delta x, \Delta y, \Delta z$ flows through the *time dimension* to the same space at a later time. This flow is shown as a blue arrow. Alternatively, dust can flow during the time period into an adjacent volume of space. This flow is shown as the green arrow. There are actually three separate green flows because the dust can flow through the *x, y* or *z* dimensions into adjacent spatial volumes. The energy-momentum tensor is the measure at *a point in spacetime* of the net flows through each of the four dimensions *t, x, y* and *z*.

To keep things comparable, we set the sides of the spatial cube and the time period to be the same, measured in terms of the invariant interval τ. This gives 1:1 comparable measurements when comparing the flow through space and time. In terms of the metres and seconds you use in normal life, if we were to consider a spatial volume of one cubic metre ($1 \times 1 \times 1$), then we would assess the flow over a time period of only $1/(3 \times 10^8)$ seconds of proper time τ. To make things even simpler, I am going to work with natural units where the speed of light $c = 1$. This gets rid of all the confusing c terms in the equations. It means that any velocity is measured as a fraction of the speed of light. It will allow you better to see the symmetries that sit within the energy-momentum tensor. We will add the c factors back only when we need them later.

What does an observer who is stationary relative to the dust field see? The particles are stationary, so all the dust flows through time (blue arrow) but not through space (green arrow). Let's call the rest mass density of the field at a particular location ρ_m (the value of ρ_m may vary with location in the dust field). Let me remind you that in this book (and any decent physics book) any reference to mass is *rest mass*.

For a stationary observer, we say that the flow of mass *density* through the time dimension (blue arrow) is ρ_m and the flow through each of the three spatial dimensions (represented by one green arrow) is zero. It may help you to think of a full $1 \times 1 \times 1$ cube of dust flowing in time period $\Delta \tau$ through the time dimension.

Now let's consider the perspective of an observer who is moving relative to the dust field (or equivalently, the dust field is moving relative to the observer). For this observer, dust is going to move into adjacent spatial volumes (green arrow) during the time interval. We can imagine building a barrier at a certain value of *x*. How much of the $1 \times 1 \times 1$ cube of dust will cross the barrier in the time interval? If we were to include the c factors, the cube moves in the *x* direction at velocity $\frac{dx}{d\tau}$ for $\frac{1}{c}$ seconds of proper time, which is $\frac{1}{c}\frac{dx}{d\tau}$. In natural units ($c = 1$) this is

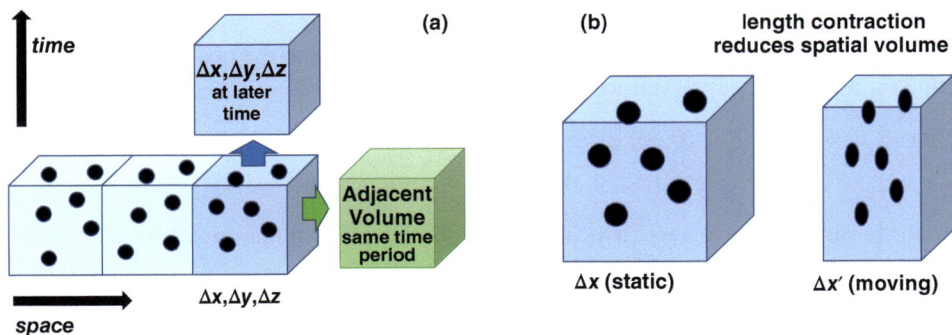

Figure 8.1 Energy-momentum flows through a spacetime volume (a) illustrates the flow of energy-momentum through spacetime; (b) illustrates the effect of length contraction.

simply $\frac{dx}{d\tau}$. Therefore, the volume crossing the barrier is $\frac{dx}{d\tau} \times 1 \times 1$. The flow of mass density in the x direction is $\rho_m \frac{dx}{d\tau}$. Velocity $\frac{dx}{d\tau}$ is typically labelled u^x. This means that the flow of mass density for the dust field is $\rho_m u^x$ through the x dimension. Similarly, it is $\rho_m u^y$ and $\rho_m u^z$ through the y and z dimensions.

Box 8.1 Einstein and Charlie Chaplin

Einstein attended a movie premiere with Charlie Chaplin in Los Angeles. The crowd went wild when they appeared together. The following exchange is reported to have occurred.

Einstein: *What I most admire about your art, is your universality. You don't say a word, yet the world understands you!*

Chaplin: *True. But your glory is even greater! The whole world admires you, even though they don't understand a word of what you say.*

I suspect some readers are thinking: *this is obvious and boring* (check out Box 8.1 for an entertaining interlude). But, we have not yet addressed an important question. For the moving observer, what is the mass density flow through the time dimension? Your instinctive answer may be that it is the same ρ_m as for the stationary observer, because any dust exiting in the x direction (to the right in Figure 8.1a) will be replaced by dust coming in from the left, so the flow through time will be unchanged. Wrong!

The correct answer is that the flow of dust across the time t dimension depends on how fast the dust is moving through time t, in exactly the same way as the flow across the x dimension depends on its speed through x. The mass density flow through time for the moving observer is $\rho_m \frac{dt}{d\tau} = \rho_m u^t$. The intuitive explanation is that the moving observer's measurement of time u^t is different from the u^t measured by the stationary observer. The quantity u^t is $\frac{dt}{d\tau}$, which is γ: time dilation (see Section 2.4). For a stationary observer, the value is one, but for a moving observer, it is more than one.

At this moment, I can imagine a lot of smart readers pounding the table. For a moving observer, u^t is more than one, which means the mass density flow $\rho_m u^t$ is greater than ρ_m. But if the moving observer sees a higher flow of mass density during one time interval, the same will be true of the next moment in time... and the next. The moving observer will measure the dust field to have higher mass density than the stationary observer, which seems crazy. We seem to have conjured up extra dust particles from nowhere!

The answer to this apparent paradox is shown in Figure 8.1b. On the left is the viewpoint of the stationary observer. To the right is that of the moving observer. For the moving observer, there is length contraction in the direction of motion by a factor of γ. This decreases the spatial volume by a factor of γ and thus increases the density by a factor of γ (which is u^t). The higher density is not due to extra particles or mass. It is because, for the moving observer, the same mass is in a smaller volume of space, which means higher density. We discussed in Section 2.5 why length contraction is essential for special relativity to be mathematically consistent. The flow of mass density in a dust field gives you another example of how length contraction goes hand in hand with time dilation.

This brings us to the end of the first step in our analysis: the flow of mass density in a dust field through space and time. The result is summarised in Equation 8.3 for any observer (including a moving one), which becomes Equation 8.4 for a stationary observer, i.e. the dust is stationary relative to the observer. Don't forget that there is no time dilation for a stationary observer, so $u^t = 1$.

$$\text{Mass density flow} \, (t, x, y, z) = (\rho_m u^t, \, \rho_m u^x, \, \rho_m u^y, \, \rho_m u^z), \qquad generally, \qquad (8.3)$$

$$\implies \; (\rho_m, \, 0, \, 0, \, 0), \qquad stationary \; dust/observer. \qquad (8.4)$$

To summarise the story so far, we need to measure the *flow* of energy-momentum through spacetime. Our starting point has been to analyse the flow of mass density in a very simple dust field. For a stationary observer, the mass density ρ_m all travels through time as shown in Equation 8.4. For a moving observer, there is flow through the spatial dimensions and the flow through the time dimension actually *increases* because of the time dilation factor u^t as shown in Equation 8.3. This apparent paradox is resolved when you take into account length contraction. For a moving observer, the volume of space is reduced and therefore the mass density of the dust field is higher than that measured by a stationary observer.

8.4 Energy-Momentum Tensor of Dust

We have done the heavy groundwork and can now analyse what this means for the flow of energy and momentum through spacetime. Personally, I find this next step of analysis rather fun. It has a pretty symmetry. I hope you like it as much as I do.

We need to evaluate what the flow of mass density means in terms of energy and momentum. Note that if we were to include the c factor, we would need to balance the units of energy with momentum as E with $p_x c$, $p_y c$ and $p_z c$, i.e. increase all the momentum terms by a factor of c (see Box 8.2 if you are unfamiliar with the energy-momentum four-vector). Working with natural units ($c = 1$) gives the simpler combination E, p_x, p_y and p_z.

Box 8.2 Energy-momentum four-vector

The four-vector of a particle is defined such that its length is the same for all observers. In spacetime, we achieve this by balancing the units of time and space such as using (ct, x, y, z). The equivalent for energy-momentum is $(E, p_x c, p_y c, p_z c)$ and is often labelled as $\vec{p}$. As shown in Equation 8.5, this matches the units of energy and momentum so that the length of the four-vector $\vec{p} \cdot \vec{p}$ is invariant for all observers (check back to Equation 2.14 if you need to).

$$\vec{p} \cdot \vec{p} = E^2 - p_x^2 c^2 - p_y^2 c^2 - p_z^2 c^2 = E^2 - p_{total}^2 c^2 = m^2 c^4. \tag{8.5}$$

I need you to cast your minds all the way back to Section 2.9 where we showed what relativistic energy and momentum are in terms of movement through time and space. The results from that section are shown again in Equation 8.6 (based on Equations 2.16 and 2.17). Equation 8.7 shows them in the natural units we will use.

$$E = mc^2 \frac{dt}{d\tau}, \qquad p_x = m \frac{dx}{d\tau}, \qquad (t, x, y, z), \tag{8.6}$$

$$E = m \frac{dt}{d\tau} = mu^t, \qquad p_x = m \frac{dx}{d\tau} = mu^x, \qquad \text{natural units } (c = 1). \tag{8.7}$$

Let's look first at the flow of the dust's *energy* through spacetime. Energy is mu^t so the energy-density is $\rho_m u^t$. We combine this definition with the result from Equation 8.3 to give Equation 8.8, which shows the energy flow for the dust field across the time t dimension, the x dimension, the y dimension and the z dimension.

$$\text{Mass density flow } (t, x, y, z) = (\rho_m u^t, \rho_m u^x, \rho_m u^y, \rho_m u^z).$$

$$\text{Energy } E \text{ density flow } (t, x, y, z) = (\rho_m u^t u^t, \rho_m u^t u^x, \rho_m u^t u^y, \rho_m u^t u^z), \tag{8.8}$$

$$= \rho_m (u^t u^t, u^t u^x, u^t u^y, u^t u^z).$$

We can perform the same calculation for *momentum*. How much x-direction momentum p_x does the flowing mass carry through spacetime? We know that $p_x = mu^x$ so x-momentum density is $\rho_m u^x$. This gives the flow of p_x momentum density shown in Equation 8.9, which also has the result for p_y and p_z momentum flow.

$$\text{Mass density flow } (t, x, y, z) = (\rho_m u^t, \rho_m u^x, \rho_m u^y, \rho_m u^z).$$

$$\text{Momentum } p_x \text{ density flow } (t, x, y, z) = (\rho_m u^x u^t, \rho_m u^x u^x, \rho_m u^x u^y, \rho_m u^x u^z),$$

$$= \rho_m (u^x u^t, u^x u^x, u^x u^y, u^x u^z). \tag{8.9}$$

$$\text{Momentum } p_y \text{ density flow } (t, x, y, z) = \rho_m (u^y u^t, u^y u^x, u^y u^y, u^y u^z).$$

$$\text{Momentum } p_z \text{ density flow } (t, x, y, z) = \rho_m (u^z u^t, u^z u^x, u^z u^y, u^z u^z).$$

We now have a full description of the energy-momentum flow through spacetime for the dust field. This result is expressed as a (4×4) matrix as shown in Equation 8.10 and is called the energy-momentum or stress-energy tensor $[T^{\mu\nu}]$. The top line is the energy flow E through each dimension of spacetime. The following lines are the flow of momentum through the spacetime: p_x, p_y and p_z. Equation 8.11 shows $[T^{\mu\nu}]$ in the stationary frame. In this case $u^x = u^y = u^z = 0$ and $u^t = 1$ because there is no time dilation, in which case there is no momentum and all the energy flow is through the time dimension and appears as the entry T^{TT}. I have relabelled this ρ_E to show it is the energy-density, albeit in natural units ($c = 1$) it is the same as mass-density.

Energy-momentum tensor of a dust field (natural units, $c = 1$)

$$[T^{\mu\nu}] = \begin{bmatrix} E\ flow\ (t,x,y,z) \\ p_x\ flow\ (t,x,y,z) \\ p_y\ flow\ (t,x,y,z) \\ p_z\ flow\ (t,x,y,z) \end{bmatrix} = \rho_m \begin{bmatrix} u^t u^t & u^t u^x & u^t u^y & u^t u^z \\ u^x u^t & u^x u^x & u^x u^y & u^x u^z \\ u^y u^t & u^y u^x & u^y u^y & u^y u^z \\ u^z u^t & u^z u^x & u^z u^y & u^z u^z \end{bmatrix} = \rho_m\, u^\mu u^\nu, \tag{8.10}$$

$$\Longrightarrow stationary\,[T^{\mu\nu}] = stationary\,[T_{\mu\nu}] = \begin{bmatrix} \rho_E & 0 & 0 & 0 \\ 0 & 0 & 0 & 0 \\ 0 & 0 & 0 & 0 \\ 0 & 0 & 0 & 0 \end{bmatrix}. \tag{8.11}$$

The only non-zero component of $[T^{\mu\nu}]$ in Equation 8.11 is T^{TT}. This scenario is a good approximation for what you might call day-to-day gravity. For non-relativistic objects, *rest mass energy* is so dominant that we treat it as the only significant component, i.e. if only $T^{TT} \neq 0$ you have an approximate *rest-mass-energy-dominated* model.

As shown in Equation 8.11, ρ_E also is the value of T_{TT}. The components of $[T_{\mu\nu}]$ are the same as $[T^{\mu\nu}]$ in Equation 8.10 except using dual coordinates instead of the vector coordinates i.e. $[T_{\mu\nu}] = \rho u_\mu u_\nu$. In the case of the stationary dust field, this gives: $T_{TT} = g_{TT}g_{TT}T^{TT} = T^{TT}$, because we are using natural units ($c = 1$) and working with the Minkowski spacetime metric (an accurate approximation for anything except extreme gravitational fields).

8.5 Symmetry of the Energy-Momentum Tensor

The first thing to notice is that the tensor components are symmetrical about the diagonal: $T^{\mu\nu} = T^{\nu\mu}$. This is true for all energy-momentum tensors, not just for the dust model. Why? Let me try to explain using the example $T^{TX} = T^{XT}$.

T^{TX} is the flow of energy through the x dimension. Energy is the velocity of mass through the time dimension $\rho_m u^t$. Therefore, T^{TX} is the velocity of $\rho_m u^t$ through the x dimension, which is $\rho_m u^t u^x$. It is a measure of the velocity through space of the velocity of mass through time.

T^{XT} is the flow of x momentum through the t dimension. What is x momentum? It is the velocity of mass through the x dimension $\rho_m u^x$. Therefore, T^{XT} is the velocity of $\rho_m u^x$ through the t dimension, which is $\rho_m u^x u^t$. It is a measure of the velocity through time of the velocity of mass through space.

Hopefully you can see intuitively that T^{TX} and T^{XT} amount to the same thing and must be equal. The same applies for any other combination of dimensions. For example, T^{XY} is the flow through the y dimension of p_x, which is the flow of mass through the x dimension. Therefore, T^{XY} is the flow through the y dimension of the flow of mass through the x dimension. T^{XY} is the flow through the x dimension of the flow of mass through the y dimension. This is the same quantity. For dust, it is $\rho_m u^x u^y = \rho_m u^y u^x$, so the energy-momentum tensor is symmetrical about its diagonal.

8.6 Covariant Derivative of the Energy-Momentum Tensor

The covariant derivative of the energy-momentum tensor is zero. This is a consequence of the local conservation of energy and momentum. You will learn in Chapter 9 that this was an important factor that led Einstein to his EFEs. It is typically expressed as $\nabla_\nu T^{\mu\nu} = 0$ or $\nabla_\mu T^{\mu\nu} = 0$. They are the same thing because $T^{\mu\nu} = T^{\nu\mu}$.

Don't forget that using Einstein summation notation, $\nabla_\nu\, T^{\mu\nu} = 0$ represents four distinct equations. The single μ term means it can take any value, while we sum all possible values of the repeated ν term (check back to Box 3.5 if you need a refresher on Einstein summation). For example, in Equation 8.12 μ takes the value t and we sum over all possible values of ν. In Equation 8.13, we repeat the process setting μ as x, and so forth. As a reminder, ∇_x means the covariant derivative with respect to x.

$$\nabla_\nu\, T^{\mu\nu} = 0, \qquad\qquad Summary\ form,$$

$$\nabla_t T^{TT} + \nabla_x T^{TX} + \nabla_y T^{TY} + \nabla_z T^{TZ} = 0, \qquad Conservation\ of\ energy\ E, \qquad (8.12)$$

$$\nabla_t T^{XT} + \nabla_x T^{XX} + \nabla_y T^{XY} + \nabla_z T^{XZ} = 0, \qquad Conservation\ of\ momentum\ p_x, \qquad (8.13)$$

$$\nabla_t T^{YT} + \nabla_x T^{YX} + \nabla_y T^{YY} + \nabla_z T^{YZ} = 0, \qquad Conservation\ of\ momentum\ p_y,$$

$$\nabla_t T^{ZT} + \nabla_x T^{ZX} + \nabla_y T^{ZY} + \nabla_z T^{ZZ} = 0, \qquad Conservation\ of\ momentum\ p_z.$$

Let me explain. Consider the conservation of energy E in Equation 8.12. Imagine a small volume of spacetime. As described earlier, nothing can *be in* it. Energy flows *through* it; if not spatially, then through the time dimension. Assuming energy is locally conserved, the amount of energy flowing into it must match the amount flowing out.

Any net inflow or outflow through the x sides depends on the flow rate T^{TX} being different on the two x sides of the spacetime volume, which I will label x and $x + dx$. This is the change in flow T^{TX} with respect to x, which is the covariant derivative $\nabla_x T^{TX}$. Similarly, net inflow or outflow through the y and z or t sides is $\nabla_y T^{TY}$ and $\nabla_z T^{TZ}$ and $\nabla_t T^{TT}$. As energy is conserved, the net inflow or outflow across the four sides must total zero.

To repeat this important point, if $\nabla_t T^{TT} + \nabla_x T^{TX} + \nabla_y T^{TY} + \nabla_z T^{TZ} \neq 0$, it would mean that some energy magically appeared or disappeared at that point in spacetime. Therefore, the conservation of energy gives Equation 8.12.

The same logic using conservation of momentum leads to Equation 8.13 for p_x and the equations for p_y and p_z. Taken together, these four results based on conservation of energy and momentum can be summarised with Einstein notation as: $\nabla_\nu\, T^{\mu\nu} = 0$.

8.7 Energy-Momentum Tensor of a Perfect Fluid

What is a perfect fluid? For me it is a delicious glass of wine at dinner. For some it is a malt whisky or a pint of beer and for others it is a relaxing mug of tea. Unfortunately, general relativity textbooks make no mention of any of these. Instead, you will learn that for physicists a perfect fluid is a rather drab model of interacting particles. How sad is that?

In the dust model the particles don't interact with each other. Obviously, this is not a good model for something like the compressed hydrogen gas that makes up a star. The perfect fluid model caters for this, but only in the simplest of ways. It models into the energy-momentum tensor the pressure that is exerted outwards by each small volume of gas or liquid (yes—a perfect fluid can be a gas). Let's take a brief look at how this is done.

Imagine the particles inside the cube volume in Figure 8.1a are bouncing madly to and fro in the x direction. You can think of them smashing into particles in adjacent volumes and bouncing back. This changes their momentum from mv^x to $-mv^x$, creating a change in momentum at the boundary. This creates an outward force. The force F at the surface is the amount of momentum delivered per second. Many of you will know the classical formula $F = \dfrac{dp}{dt}$.

A real fluid can have viscosity. This is the stickiness that drags against any flow and causes shearing forces if the fluid is moving. It also can conduct heat. The perfect fluid of physics has none of this. The model only allows for pressure at the x boundary to be in the same x direction.

In the energy-momentum tensor, T^{XX} is the density of flow of p_x momentum across the x dimension. It is the amount of momentum transferred per unit of time. This is $\dfrac{\Delta p_x}{\Delta t}$ across the x boundary, which is therefore the force F on it. This force is spread over the surface, which has an area $\Delta y \Delta z$. This measure $\dfrac{F}{\Delta y \Delta z}$ is force/area, which is *pressure*. This means that T^{XX} is the pressure in the x direction.

Similarly, T^{YY} and T^{ZZ} are pressure in the y and z direction. In the perfect fluid model, the pressure in all three dimensions is the same and is labelled P, so $T^{XX} = T^{YY} = T^{ZZ} = P$ (be careful not to confuse this capital P with the labels for density or momentum). This gives the energy-momentum tensor for a stationary perfect fluid shown as

Equation 8.14. In summary, T^{TT} is the energy-density, which in this case is the same as the mass density because ($c = 1$). And the other diagonal entries T^{XX}, T^{YY} and T^{ZZ} are all the same pressure P.

This is based on the Minkowski metric (all tensors transform in the same way, so a change to the coordinates of the spacetime metric identically changes the energy-momentum tensor). We will need the lower index version $[T_{\mu\nu}]$ of this later when we discuss cosmology. In the case of energy-momentum tensors based on the Minkowski metric ($c = 1$), $[T_{\mu\nu}] = [T^{\mu\nu}]$, as can be easily shown for this diagonal metric because: $T_{TT} = g_{TT}g_{TT}T^{TT} = (1)(1)T^{TT}$ and $T_{XX} = g_{XX}g_{XX}T^{XX} = (-1)(-1)T^{XX}$.

Energy-momentum tensor of a stationary perfect fluid (Minkowski metric, $c = 1$)

$$[T^{\mu\nu}] = [T_{\mu\nu}] = \begin{bmatrix} \rho_E & 0 & 0 & 0 \\ 0 & P & 0 & 0 \\ 0 & 0 & P & 0 \\ 0 & 0 & 0 & P \end{bmatrix} , \quad \rho_E \text{ is energy-density, } P \text{ is pressure.} \tag{8.14}$$

While the perfect fluid is not moving in the wider sense, we know that the pressure P from the frantic motion of the particles within it is significant. If the constraints on the cube volume were removed, the particles would fly out. For example, if the gravitational field of the sun were suddenly to disappear, the pressure in the hydrogen would hurtle it outwards at close to light speed and we would all be burnt to cinders some 10 minutes later. Ouch!

Equation 8.14 can be generalised for any observer in Minkowski spacetime as shown in Equation 8.15. You can derive $[T^{\mu\nu}]$ as in Equation 8.14 by setting $u^t = 1$, $u^x = u^y = u^z = 0$ and $g^{\mu\nu}$ to the components of the matrix $[g^{\mu\nu}]$ in Equation 3.22 (setting $c = 1$).

$$T^{\mu\nu} = (\rho_E + P)u^\mu u^\nu - Pg^{\mu\nu}, \quad \text{Signature}: \quad <+--->\quad (c = 1). \tag{8.15}$$

Some ambitious readers may wonder why physicists do not use a more complicated model. The perfect fluid seems a bit simplistic. The truth is that the EFEs are so hard to work with that you have to keep things simple. More complicated energy-momentum tensors are generally impossible to handle without the help of computers. We are like the drunk man who has lost his keys at night. He looks under the street light because that is the only place worth exploring. In the same way, the study of general relativity is limited to those scenarios we can realistically assess. That is enough about perfect fluids for now... except perhaps for one last glass of wine.

8.8 Summary

I started this chapter with a brief discussion on upper indices such as T^{TT} and lower indices such as T_{TT}. The former indicates that the tensor component is in vector coordinates, the latter that it is in dual coordinates. There are two indices because each tensor component is associated with two dimensions. We can toggle between the two using the metric tensor, once for each index. A simple example for our diagonal metric is: $T_{TT} = g_{TT}g_{TT}T^{TT}$. The underlying tensor remains the same. All the key features of the tensor are unaffected.

I explained the structure of the energy-momentum tensor using the model of a dust field. If the dust is stationary, we describe it as flowing through the time dimension. It can also flow through each spatial dimension. The next step was to quantify this flow of mass in terms of the flow of energy and momentum. This gives us the 16 components of the energy-momentum tensor and is summarised on the left of Figure 8.2. In textbooks, the labelling is often 0 for time and 1–3 for the spatial dimensions, in which case for example T^{TY} is T^{02}.

The top line (darkest blue) shows the energy flow across each dimension. You could describe T^{TT} as the flow of mass/energy-density through time (which is energy E), through the time dimension i.e. time...time. Similarly, you can describe T^{TX} as the flow of mass density through time (which is energy), through the x dimension. On the next line down T^{XT} is the flow of mass density through x (which is p_x), through the time dimension. This means that T^{TX} is the flow of mass through time, through space... while T^{XT} is the flow of mass through space, through time. The result is that the energy-momentum tensor is symmetrical: $T^{\mu\nu} = T^{\nu\mu}$.

Another important feature of the energy-momentum tensor is that its covariant derivative is zero due to local conservation laws. The conservation of energy means that the following covariant derivatives of the E-flow line in Figure 8.2 sum to zero: $\nabla_t T^{TT}$, $\nabla_x T^{TX}$, $\nabla_y T^{TY}$ and $\nabla_z T^{TZ}$. These measure the net inflow or outflow (across each dimension) of energy into a spacetime volume. Any energy flowing into the spacetime volume also flows out, so

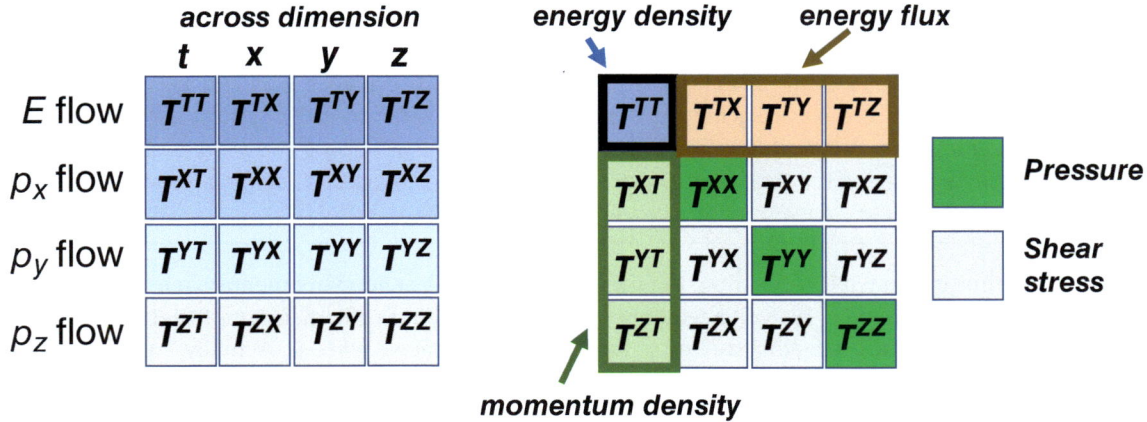

Figure 8.2 Components of energy-momentum tensor $[T^{\mu\nu}]$.

they must total to zero. Taken together with the conservation of momentum, this leads to the relationships shown in Equations 8.12 and 8.13, which can be expressed in summary as: $\nabla_\nu T^{\mu\nu} = 0$.

We derived the energy-momentum tensors for two simple models: a dust field and a perfect fluid. Both are shown in Equation 8.16, measured in the stationary reference frame. In the case of the dust field, the dust is stationary and non-interacting, so there is no flow of energy or momentum through the spatial dimensions. Indeed, there is no momentum at all. The only non-zero component of the energy-momentum tensor is T^{TT}, which is the energy-density ρ_E of the dust's rest mass energy flowing through time while the dust remains spatially static. We will use this later as a general model for non-relativistic *rest-mass-energy-dominated* scenarios.

$$
\begin{array}{cc}
\textit{Dust field} & \textit{Perfect fluid} \\[4pt]
\begin{bmatrix} \rho_E & 0 & 0 & 0 \\ 0 & 0 & 0 & 0 \\ 0 & 0 & 0 & 0 \\ 0 & 0 & 0 & 0 \end{bmatrix}, &
\begin{bmatrix} \rho_E & 0 & 0 & 0 \\ 0 & P & 0 & 0 \\ 0 & 0 & P & 0 \\ 0 & 0 & 0 & P \end{bmatrix},
\end{array}
\qquad
\begin{array}{l}
\textit{(stationary reference frame)} \\[12pt]
\textit{Minkowski metric } (ct, x, y, z).
\end{array}
\tag{8.16}
$$

The perfect fluid allows also for internal pressure P. This is from the momentum of the particles within the fluid. Although the fluid is stationary in the broader sense, these internal interactions exert an outward pressure, which is shown as the T^{XX}, T^{YY} and T^{ZZ} components. This will become relevant later in the book when we will model the entire universe as a perfect fluid. This model is particularly relevant in the early stages when the universe was an interacting plasma of radiation and particles. Until we get to the module on cosmology, don't worry about it.

An overall summary of the components of the energy-momentum tensor is shown on the right of Figure 8.2. The top left component T^{TT} is the *energy-density*, which as noted a moment ago includes the rest mass energy. The three dark green boxes on the diagonal T^{XX}, T^{YY} and T^{ZZ} are the *pressure* terms discussed in the perfect fluid model. Virtually all the energy-momentum tensors we work with feature only these four diagonal components (i.e. all off-diagonal components are zero).

To the right of the T^{TT} component is the energy flow through the spatial dimensions, which is often called *energy flux*. The boxes vertically below T^{TT} are the *momentum density*: this is the momentum passing through time but remaining in the same spatial volume. The symmetry of the tensor means that the energy flux and momentum density are equal (you may see them labelled the other way round). The light grey squares show *shear stresses*. These are momentum/pressure terms that are not directly outwards from the volume. For example, the force on the x side of the volume may have a component in the y or z directions. I tell you this only for interest. Shear forces will not be relevant in this book.

So, what then is really important? If we assess the energy-momentum of any massive object, it is always easiest to work in the stationary reference frame. The T^{TT} component is likely to be the overwhelmingly dominant component of the tensor for a massive body because its rest mass energy is so large relative to the other components (which, for example, might include pressure and rotational energy).

Box 8.3 Professor Eddington's third person

Eddington returned in 1919 from his eclipse trip (Section 5.6) with evidence vindicating Einstein's theory. He presented his data to the Royal Society. At the end of the presentation, the physicist Ludwik Silberstein commented to Eddington: *You must be one of the three persons in the world who understands general relativity.* When Eddington hesitated at the thought, Silberstein added: *Don't be modest, Eddington.* To which Eddington replied: *On the contrary, I am trying to think who the third person is!*

And now it is time for a terrible confession. Detailed knowledge of the energy-momentum tensor is *not* required to follow the rest of this module. We will need only the T^{TT} component in Chapters 9 and 10. I can imagine you gnashing your teeth with fury at having studied so hard for so little that we will use. Actually, I am exaggerating a bit. We will also use the fact that the covariant derivative of the energy-momentum tensor is zero. Box 8.4 highlights the need-to-know essentials for what is to come. In any case, think how much you have learned that so few people know (see Box 8.3). And think how much fun this chapter has been.

Box 8.4 Energy-momentum tensor need-to-know essentials

- The energy-momentum tensor is a 4×4 matrix quantifying the flow of energy and momentum through spacetime.
- The components are symmetrical about the diagonal, e.g. $T^{TX} = T^{XT}$.
- The top four components of the matrix (such as $T^{TT}, T^{TX}, T^{TY}, T^{TZ}$) are the flow of energy across time and each of the three spatial dimensions.
- For a non-relativistic object the dominant component is T^{TT}. In weak gravitational fields, $T^{TT} \approx T_{TT}$ ($c = 1$). This is the rest mass energy-density ρ_E of the object.
- The covariant derivative of the energy-momentum tensor is zero: $\nabla_\nu T^{\mu\nu} = 0$.

Chapter 9

Deriving the Einstein Field Equations

In this chapter, we will derive the structure of the Einstein Field Equations (EFEs) shown in Equation 9.1 (I will continue to ignore the cosmological constant throughout this chapter). Before embarking on this mission, I will try to give you an intuitive fuzzy feel why it is not perhaps so weird that there is a link between the presence of energy-momentum and spacetime curvature.

$$R_{\mu\nu} - \frac{1}{2}g_{\mu\nu}R = \frac{8\pi G}{c^4}T_{\mu\nu},$$

Spacetime curvature $\Longleftrightarrow$ *Energy-momentum.*

(9.1)

For Einstein, it proved a major struggle to deduce the somewhat complicated-looking form of the equation on the left side of the EFEs. I will show you why the EFEs cannot have a simpler relationship. Then I will use some of what we learned in Chapter 8 about the energy-momentum tensor to derive this particular mix of the Ricci tensor component $R_{\mu\nu}$ and the Ricci scalar R. I will end by comparing it all to a hippo, so you have much to look forward to.

9.1 Why Does Energy-Momentum Curve Spacetime?

Why? Most wise professors will say this is just the way it is... and stop asking stupid questions. While I cannot answer the question, let me offer a few reminders of how spacetime and energy-momentum are linked.

Energy-momentum and the spacetime metric are not independent entities. In Section 2.8 I showed you that $[g_{\mu\nu}]$ underpins the very definition of energy and momentum. Indeed, we derived the relationship between mass, energy and momentum ($m^2c^4 = E^2 - p^2c^2$) directly from the invariant interval of Minkowski spacetime. And later in Chapter 14, I will give an example of how these conserved quantities, which we call energy and momentum, change when the metric is not Minkowski spacetime.

Let's dig a bit deeper. Think back to the analysis in the last chapter of the structure of the energy-momentum tensor $[T^{\mu\nu}]$. I showed you that energy-density T^{TT} is the rate of flow of mass through time, through time, which can be expressed as the mass density multiplied by $u^t u^t$ (its speed through time squared). Similarly, the momentum density T^{TX} is the rate of flow of mass through space, through time $u^x u^t$, and so forth for the other components of $[T^{\mu\nu}]$ (see Section 8.4 and Equation 8.10). The way that the spacetime dimensions combine... t, x, y, z... affects these velocities and thus affects the flow of energy-momentum through the spacetime volume.

My final point on this topic is for readers familiar with the maths of quantum mechanics. The quantum mechanical wave function can be fully defined in terms of spacetime or in terms of energy-momentum. It turns out that the two are different ways of representing the same thing (you Fourier transform between them). You could say that they are two sides of the same coin.

Let me emphasise that nothing here explains why energy-momentum curves spacetime. I simply want to remind you how tightly they are intertwined in the hope that this makes you feel a bit more comfortable with Einstein's theory. Anyway, that is enough fluff for now. It is time to get back on more solid ground.

Untangling General Relativity: The Intuitive Self-Study Guide, First Edition. Simon Sherwood.
© 2026 John Wiley & Sons Ltd. Published 2026 by John Wiley & Sons Ltd.

9.2 Generalising Coordinates

I have mentioned several times that the EFEs are covariant, which means they work with any coordinates. For our analysis, the main two spacetime coordinate systems are Cartesian coordinates (ct, x, y, z) and spherical coordinates (ct, r, θ, ϕ). To emphasise the covariant nature of the EFEs, I will start to use the more general coordinate labels $(0, 1, 2, 3)$, where 0 is the time dimension and 1, 2 and 3 are the three spatial dimensions. For example, the metric tensor component g_{12} refers to g_{XY} if we are working in Cartesian coordinates or $g_{R\theta}$ if we are using spherical coordinates. Let me provide a refresher on spherical coordinates for those of you who may be less familiar with them.

The left of Figure 9.1 shows how spherical coordinates work. Rather than defining a point in terms of its distance (x, y, z) from the origin, we define it by its distance radially from the origin (r), the angle down from the vertical pole (θ) and the angle around (ϕ). Spherical coordinates are very useful when handling scenarios with spherical symmetry, such as the gravitational effect around a static star, which varies only with r (the distance from the star) and not with θ or ϕ. The box in the left of Figure 9.1 shows the way to switch between coordinates (x, y, z) and (r, θ, ϕ). Looks hard to translate across, eh? The good news is that you will not need to. The covariance of the EFEs means that they work equivalently in either coordinates.

The right side of Figure 9.1 shows what this means in terms of the invariant interval. Clearly, a miniscule change dr in the r coordinate is a physical displacement of dr. A miniscule change $d\theta$ in the θ coordinate is a physical displacement of $r\,d\theta$ because it is a small move around the circumference of a circle of radius r (note that we are working in radians as described earlier in Box 3.3). A miniscule change $d\phi$ in the ϕ coordinate is a physical displacement of $r\sin\theta\,d\phi$ because it is a small move around the circumference of a circle of radius $r\sin\theta$. For Minkowski spacetime, this gives the metric shown in Equation 9.2. The time component is positive and the spatial component is negative in line with the $< + - - - >$ signature used throughout this book.

$$\text{Minkowski metric } (ct, r, \theta, \phi): \quad [g_{\mu\nu}] = \begin{bmatrix} 1 & 0 & 0 & 0 \\ 0 & -1 & 0 & 0 \\ 0 & 0 & -r^2 & 0 \\ 0 & 0 & 0 & -r^2\sin^2\theta \end{bmatrix}. \tag{9.2}$$

For most of this chapter I will be using coordinate labels $(0, 1, 2, 3)$. It may be easier for you to think in terms of (ct, x, y, z) or you may prefer to think (ct, r, θ, ϕ). The choice is yours! That is the beauty of the covariant EFEs.

That is enough background on spherical coordinates. It is time to take a closer look at the symmetries lurking within the Ricci tensor.

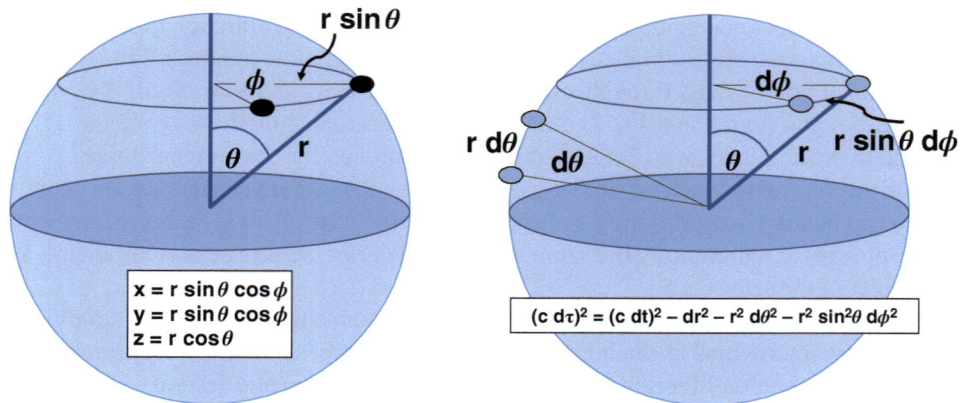

Figure 9.1 Spherical coordinates.

9.3   The Ricci Tensor: Why So Complicated?

The left side of Equation 9.1 has the component of the Ricci tensor $[R_{\mu\nu}]$ combined with half the Ricci scalar R multiplied by the same component of the metric tensor $[g_{\mu\nu}]$. I mean... what? Why not a simple relationship such as $R_{\mu\nu} = kT_{\mu\nu}$? Actually, Einstein (working with the mathematician Marcel Grossman) tried to use this simple formula as a basis for his theory, but he had to abandon it. The structure of the Ricci tensor and its symmetries mean that something this simple cannot possibly work. Let me explain why.

9.3.1   Riemann Symmetries Within the Ricci Tensor

To understand the problem, you need to appreciate the symmetries of the underlying Riemann tensor components within the Ricci tensor. Here is how you do that. The key components of the Ricci tensor for our purposes are R_{TT} plus the three spatial equivalents such as R_{XX}, R_{YY}, R_{ZZ} or, for example, the spherical coordinates R_{RR}, $R_{\theta\theta}$, $R_{\phi\phi}$ (see Box 9.1 for why mixed-index components such as R_{TX} are not significant in most cases).

Whatever the coordinates, if you want to calculate and/or understand the Ricci tensor for any 4-D diagonal metric, I recommend the layout discussed in Subsection 7.5.1 for 2-D metrics. It works for any diagonal metric (i.e. no off-diagonal components). It shows the underlying Riemann symmetries within the Ricci tensor and eases the maths.

Just as we did for the 2-D example in Subsection 7.5.1, we start from the definition of the Ricci scalar, shown in Equation 9.3 for any 4-D diagonal metric using coordinate numbers $(0, 1, 2, 3)$, which could be any orthogonal coordinates.

$$R = g^{00}R_{00} + g^{11}R_{11} + g^{22}R_{22} + g^{33}R_{33}, \qquad \textit{Ricci scalar for any 4-D diagonal metric.} \qquad (9.3)$$

Equation 9.4 expands each term from Equation 9.3 to show the Riemann tensor components. This is a similar exercise to Equation 7.17 for 2-D diagonal metrics, so check back to that if you need to. As in 2-D earlier, I have applied the Riemann symmetry $g^{AA} R^{B}{}_{ABA} = g^{BB} R^{A}{}_{BAB}$ (see Equation 7.3) and assigned labels a, b, c, d, e and f to matching values. Each component has a twin. For example, $g^{00}R^{1}{}_{010} = g^{11}R^{0}{}_{101}$, so I have labelled both a.

Underlying symmetries for any 4-D diagonal metric $\hspace{4cm}$ (9.4)

$$
\begin{aligned}
g^{00}R_{00} &= g^{00}R^{0}{}_{000} + g^{00}R^{1}{}_{010} + g^{00}R^{2}{}_{020} + g^{00}R^{3}{}_{030} &= \quad 0 \quad +a \quad +b \quad +c, \\
g^{11}R_{11} &= g^{11}R^{0}{}_{101} + g^{11}R^{1}{}_{111} + g^{11}R^{2}{}_{121} + g^{11}R^{3}{}_{131} &= \quad a \quad +0 \quad +d \quad +e, \\
g^{22}R_{22} &= g^{22}R^{0}{}_{202} + g^{22}R^{1}{}_{212} + g^{22}R^{2}{}_{222} + g^{22}R^{3}{}_{232} &= \quad b \quad +d \quad +0 \quad +f, \\
g^{33}R_{33} &= g^{33}R^{0}{}_{303} + g^{33}R^{1}{}_{313} + g^{33}R^{2}{}_{323} + g^{33}R^{3}{}_{333} &= \quad c \quad +e \quad +f \quad +0.
\end{aligned}
$$

The symmetries between underlying Riemann components in the Ricci tensor will be useful later when we work with more complicated diagonal spacetime metrics such as the Schwarzschild metric.

Box 9.1 Mixed-index components of the Ricci tensor

You don't need to worry about mixed-index Ricci components such as R_{TX} until we discuss non-diagonal metrics such as the Kerr metric later (much later).

Usually all the mixed-index Ricci components of a diagonal metric are *zero*. Indeed, there can only be non-zero mixed-index components if the metric depends on *both* the variables, e.g. for there to be a non-zero R_{TX} value of a diagonal metric, the metric must vary with both x and t. This will become important later. If you are the sort of person who wants a proof of this, see Box 9.3.

9.3.2 The Symmetries Complicate Things

Let's return to our original topic. Einstein was searching for a simple relationship between the energy-momentum tensor $[T_{\mu\nu}]$ and the Ricci tensor $[R_{\mu\nu}]$. He would have been particularly interested in the relationship between T_{00} and R_{00}, because T_{00} contains rest mass energy (typically the dominant term), and R_{00} is the Ricci component linked with distortions in time (strongly implicated in gravitational effects).

Dear old Albert struggled with this for years. What makes things so challenging? Take a long look at the symmetries shown on the right of Equation 9.4. Do you see the problem? Before reading on, pause for a while and think about it.

Now, here is the punchline. A change to T_{00} (such as an increase in rest mass energy-density) *cannot affect only* R_{00}, because it would involve changing the value of a, b or c (or all three) in Equation 9.4. This would have a knock-on effect because each underlying Riemann tensor component has a twin. For example, any change to a also affects R_{11}. If T_{00} affects R_{00}, it must affect at least one of the spatial Ricci components R_{11}, R_{22} and R_{33}. Einstein's dream solution $R_{\mu\nu} = kT_{\mu\nu}$ cannot possibly work. The presence of rest mass energy T_{00} cannot affect *only* R_{00}.

The story is similar if something alters the value of R_{11}. This entails changing a, d or e in Equation 9.4, which in turn affects the values of R_{00}, R_{22} or R_{33}. The Riemann tensor symmetries that lurk within the Ricci tensor mean that the Ricci components are not independent. Change one and it affects others.

Box 9.2 Einstein's loss of confidence

Even Einstein had doubts that he could find the right EFE formula. In 1914, exasperated and frustrated, he commented after a dinner with colleagues: *In Berlin they speculate on me like a prize hen but I don't know whether I can still lay eggs.*

What a mess. Aarrgghh! You can see why Einstein's hair was so wild! We want to find a relationship between some function of a component $R_{\mu\nu}$ of the Ricci tensor and the corresponding component $T_{\mu\nu}$ of the energy-momentum tensor. How on earth can we figure out what is happening? Fortunately, Einstein (with help from Grossman and others) found a way forward.

9.4 Deriving the Ricci Relationship

For some of you, this section will prove challenging. Even Einstein found it hard (see Box 9.2). Einstein's silver bullet was to use $\nabla^{\mu}T_{\mu\nu} = 0$. In other words, the covariant derivative of the energy-momentum tensor is zero. We showed this in Section 8.6. It is required for local conservation of energy and momentum. Einstein was happy to assume that this basic conservation law must also be true for curved spacetime.

The argument is simple. Whatever the correct formula is involving $R_{\mu\nu}$ that matches up with $T_{\mu\nu}$, the covariant derivative of that formula also must be zero. The calculation of the covariant derivative of the Ricci tensor is reasonably complicated as you will see in a moment, but it turns out there is one combination of components of the Ricci tensor $R_{\mu\nu}$ combined with the Ricci scalar R for which this is true. This allowed Einstein, after years of searching, to derive the structure of the EFEs in Equation 9.5.

$$\nabla^{\mu}\left(R_{\mu\nu} - \frac{1}{2}g_{\mu\nu}R\right) = 0, \quad \Longrightarrow \quad R_{\mu\nu} - \frac{1}{2}g_{\mu\nu}R = kT_{\mu\nu}. \tag{9.5}$$

9.4.1 Finding the Right Ricci Function: Tensor Contraction (Optional)

This is one of the few times we will manipulate tensors. I will step through it slowly because some of you may need to use tensor calculus in future work. Feel free to skip this calculation if it is too much. The starting point is a relationship called the *Second Bianchi Identity for the Riemann Tensor*. It is a well-known rule based on the symmetries of the Riemann tensor. I am not going to prove the rule. That is a step too far. Anyway, Einstein used this shortcut, so I am following in some famous footsteps.

The identity is shown as Equation 9.6 using Einstein notation, and then below, where I have switched to capital letters for the indices so that the maths is easier to follow. The relationship works for any indices. They do not have to be different. Indeed, as there are only four spacetime dimensions but five indices in the equation, we know at least two of the indices must be the same. You will notice that I have used F and G indices for α and β. This is because I will be changing them to A and B in a moment. As shown in Equation 9.7, the first step is to shift around the indices using the Riemann symmetries (see Box 7.3).

$$0 = \nabla_\sigma R_{\alpha\beta\mu\nu} + \nabla_\nu R_{\alpha\beta\sigma\mu} + \nabla_\mu R_{\alpha\beta\nu\sigma}, \qquad \text{\textit{Second Bianchi identity,}} \qquad (9.6)$$

$$0 = \nabla_E R_{FGCD} + \nabla_D R_{FGEC} + \nabla_C R_{FGDE}, \qquad \text{\textit{with clearer indices,}}$$

$$= \nabla_E R_{FGCD} - \nabla_D R_{FGCE} - \nabla_C R_{GFDE}, \qquad \text{\textit{switch using symmetries.}} \qquad (9.7)$$

The second step is to raise the F and G indices of all the terms. In doing so, I switch the index from F to A and from G to B using the rule $V^\nu = g^{\mu\nu} V_\mu$ from Equation 3.29. In the case of the F index, this is $V^A = g^{FA} V_F$, summing over all possible values of F (the Bianchi relationship holds for all coordinates, so it also works for the sum). We can move g^{FA} *inside* the covariant derivative because any covariant derivative of the metric tensor is zero i.e. $\nabla_\alpha g^{\mu\nu} = 0$, where α, μ and ν can be any coordinates (see Box 4.7 if needed). This means that $\nabla(gR) = R\nabla g + g\nabla R = 0 + g\nabla R$ and therefore $g\nabla R = \nabla(gR)$.

At the same time, we can follow exactly the same process using g^{GB}, creating an upper B index for every term, giving Equation 9.8. Intuitively, you can think of this as just raising the first two indices on all the terms. I have changed the index (F to A and G to B) to be a stickler for the correct rules of tensor calculus. After all, we don't want some professor reading this and seething with rage.

$$0 = \nabla_E \left(g^{FA} g^{GB} R_{\mathbf{F}GCD} \right) - \nabla_D \left(g^{FA} g^{GB} R_{\mathbf{F}GCE} \right) - \nabla_C \left(g^{FA} g^{GB} R_{\mathbf{G}FDE} \right),$$
$$= \nabla_E R^{AB}{}_{CD} - \nabla_D R^{AB}{}_{CE} - \nabla_C R^{BA}{}_{DE}. \qquad (9.8)$$

Now comes the clever bit, which is called *tensor contraction*. The expression in Equation 9.8 is completely general. It works for any set of indices, so we can choose $A = C$. At the same time, we make A into a dummy index, which I will label in bold as $\mathbf{A}$ giving Equation 9.9. Instead of the 3 terms in Equation 9.8, there are 12 terms in Equation 9.9 because we add the version of Equation 9.8 where $A = t$ to those where $A = x$, $A = y$ and $A = z$ (or whatever coordinates). Each one adds up to zero, so the total in Equation 9.9 is still zero.

$$0 = \nabla_E R^{\mathbf{A}B}{}_{\mathbf{A}D} - \nabla_D R^{\mathbf{A}B}{}_{\mathbf{A}E} - \nabla_{\mathbf{A}} R^{B\mathbf{A}}{}_{DE}, \qquad (9.9)$$

$$= \nabla_E R^{B}{}_{D} - \nabla_D R^{B}{}_{E} - \nabla_{\mathbf{A}} R^{B\mathbf{A}}{}_{DE}. \qquad (9.10)$$

In Equation 9.10, the first and second terms $R^{B}{}_{D}$ and $R^{B}{}_{E}$ are no longer Riemann tensor components. They are Ricci tensor components with one upper and one lower index. As an example, Equation 9.11 shows how this works using Cartesian coordinates. You can compare back with Equation 7.11. This process is called tensor contraction because it reduces the number of indices on the tensor. We have contracted four Riemann tensor components into one Ricci tensor component. In a moment we will contract again by creating a dummy index on the Ricci tensor, which then results in the Ricci scalar (as in Section 7.4). This is shown in Equation 9.12.

$$R^{\mathbf{A}B}{}_{\mathbf{A}D} = R^{TB}{}_{TD} + R^{XB}{}_{XD} + R^{YB}{}_{YD} + R^{ZB}{}_{ZD} = R^{B}{}_{D}, \qquad \text{\textit{Contraction to Ricci tensor,}} \qquad (9.11)$$

$$R^{\mathbf{B}}{}_{\mathbf{B}} = R^{T}{}_{T} + R^{X}{}_{X} + R^{Y}{}_{Y} + R^{Z}{}_{Z} = R, \qquad \text{\textit{Contraction to Ricci scalar.}} \qquad (9.12)$$

Starting from Equation 9.10, we now elect that the indices B and D are the same, so $B = D$. We are free to do this because the Bianchi identity works for any combination of indices. At the same time, we make B into a dummy index $\mathbf{B}$. Again, this means we are adding up the four versions of Equation 9.10 where B takes on each possible index. Every one of the four versions adds to zero, so the total also adds to zero. This is Equation 9.13, which results in Equation 9.14. The first term has now contracted to the Ricci scalar, which has no index. The last term also contracts to a Ricci tensor component.

$$0 = \nabla_E R^{\mathbf{B}}{}_{\mathbf{B}} - \nabla_{\mathbf{B}} R^{\mathbf{B}}{}_{E} - \nabla_{\mathbf{A}} R^{\mathbf{B}\mathbf{A}}{}_{\mathbf{B}E}, \qquad (9.13)$$

$$= \nabla_E R - \nabla_{\mathbf{B}} R^{\mathbf{B}}{}_{E} - \nabla_{\mathbf{A}} R^{\mathbf{A}}{}_{E}, \qquad \text{\textit{R is the Ricci scalar.}} \qquad (9.14)$$

The last step (phew!) is to raise the E index of the Ricci tensors so they have two upper indices. This is done by multiplying through with the metric tensor g^{EH} as shown in Equation 9.15 (I am using H because we used F and G earlier). Don't forget that the covariant derivative of the metric tensor is zero, so this term can move inside the ∇ signs. We add together all index possibilities for E, making it a dummy variable (this actually happens automatically in Einstein notation, but I show the dummy variables in bold anyway for clarity). The result is Equation 9.16.

Look carefully at Equation 9.16. We now have three dummy variables **A**, **B** and **E**, each acting on a separate term and doing exactly the same thing (summing over all possible values of the index). Therefore, we can use one symbol for all three—I have chosen **E**. This gives Equation 9.17. After that we factorise, divide both sides by -2 and switch back to the usual spacetime Greek indices (μ for the dummy variable **E** and ν for H) to give the final answer in Equation 9.18. As shown, the upper and lower indices are interchangeable.

$$0 = \nabla_{\mathbf{E}}\, g^{EH} R - \nabla_{\mathbf{B}}\, g^{EH} R^{\mathbf{B}}_{\,\mathbf{E}} - \nabla_{\mathbf{A}}\, g^{EH} R^{\mathbf{A}}_{\,\mathbf{E}}, \tag{9.15}$$

$$= \nabla_{\mathbf{E}}\, g^{EH} R - \nabla_{\mathbf{B}}\, R^{\mathbf{B}H} - \nabla_{\mathbf{A}}\, R^{\mathbf{A}H}, \tag{9.16}$$

$$= \nabla_{\mathbf{E}}\, g^{EH} R - \nabla_{\mathbf{E}}\, R^{EH} - \nabla_{\mathbf{E}}\, R^{EH}, \tag{9.17}$$

$$= \nabla_{\mathbf{E}} \left(g^{EH} R - 2R^{EH} \right),$$

$$= \nabla_{\mathbf{E}} \left(R^{EH} - \frac{1}{2} g^{EH} R \right),$$

$$= \nabla_{\mu} \left(R^{\mu\nu} - \frac{1}{2} g^{\mu\nu} R \right) = \nabla^{\mu} \left(R_{\mu\nu} - \frac{1}{2} g_{\mu\nu} R \right). \tag{9.18}$$

This is exactly the relationship that we were looking for (see the left side of Equation 9.5). Well done. You have followed in Einstein's footsteps. And don't forget this took him several years to figure out! For many of you, this whole calculation is overkill. However, those of you who study general relativity in depth will appreciate this introduction to tensor contraction, which can be very confusing if you don't understand exactly what is going on.

9.5   What Does This Tell Us About Spacetime Curvature?

Welcome back to those who skipped the last section. I hope you enjoyed a nice cup of tea while I explained some heavy maths. Good decision! Let me repeat in words what we just did. Einstein was looking for some function including a component of the Ricci tensor to match up with the corresponding component of the energy-momentum tensor. He was confident that the covariant derivative of the energy-momentum tensor is zero. Therefore, he reasoned, the covariant derivative of the corresponding Ricci function also must be zero. This led him to the following structure of the EFEs.

$$R_{\mu\nu} - \frac{1}{2} g_{\mu\nu} R = k\, T_{\mu\nu}. \tag{9.19}$$

We combine the Ricci tensor, Ricci scalar and metric tensor to make a function with zero covariant derivative. We now can use this to investigate how the presence of energy-momentum affects the curvature of spacetime.

9.5.1   Vacuum

Scenario: $[T_{\mu\nu}] = 0$. You will soon see that Equation 9.19 is a sneakily complicated relationship that reflects how the components of the Ricci tensor entwine in a merry dance. However, first we need a reality check. Einstein surmised that spacetime curvature in a vacuum is Ricci-0 (Section 6.4), which means $[R_{\mu\nu}] = 0$ if $[T_{\mu\nu}] = 0$. Let's check if this works. Our first scenario is to set component $T_{\mu\nu} = 0$ and multiply both sides by $g_{\mu\nu}$, which means because of the multiple indices of μ and ν that we are summing over all possible values of μ and ν (Einstein summation). A little mathematical manipulation (remembering the sum of all possible values for $g^{\mu\nu} g_{\mu\nu} = 4$ in 4-D spacetime—see paragraph above Equation 3.22) gives Equation 9.21.

$$R_{\mu\nu} - \frac{1}{2} g_{\mu\nu} R = 0, \quad \implies \quad g^{\mu\nu} R_{\mu\nu} - \frac{1}{2} g^{\mu\nu} g_{\mu\nu} R = 0, \qquad \text{if } T_{\mu\nu} = 0, \tag{9.20}$$

$$\implies \quad g^{\mu\nu} R_{\mu\nu} = 2R, \qquad \text{if } T_{\mu\nu} = 0. \tag{9.21}$$

Equation 9.21 is in disagreement with the definition of the Ricci scalar, $R = g^{\mu\nu}R_{\mu\nu}$, unless both $R = 0$ and $[R_{\mu\nu}] = 0$. Therefore, we have shown that in a vacuum, $[T_{\mu\nu}] = 0$, $[R_{\mu\nu}] = 0$ and $R = 0$ (I continue to ignore the cosmological constant for now).

9.5.2 Rest Mass Energy

Scenario: $T_{00} \neq 0$ but all the other components of $[T_{\mu\nu}]$ equal zero. This is a good approximation for any non-relativistic massive object in the stationary reference frame, because rest mass energy overwhelmingly dominates its energy-momentum tensor. Our actual scenario is the profile of the energy-momentum tensor inside a dust cloud, observed stationary relative to the particles. The only energy-momentum is the rest mass energy of the dust. We will continue to simplify by assuming the metric is diagonal. As only T_{00} is non-zero, we know $T_{11} = T_{22} = T_{33} = 0$. Starting from the EFE for T_{11}, we get Equation 9.22 for the first spatial dimension (index 1). Don't forget that being an individual component, $g^{11}g_{11} = 1$.

$$T_{11} = R_{11} - \frac{1}{2}g_{11}R, \quad \implies \quad g^{11}T_{11} = g^{11}R_{11} - \frac{1}{2}R = 0. \tag{9.22}$$

If we multiply by -2 to get rid of the awkward half factor and substitute R with the components for a diagonal metric (see Equation 7.14), we arrive at Equation 9.23, which also must be zero.

$$-2g^{11}T_{11} = -2g^{11}R_{11} + (g^{00}R_{00} + g^{11}R_{11} + g^{22}R_{22} + g^{33}R_{33}),$$
$$= g^{00}R_{00} - g^{11}R_{11} + g^{22}R_{22} + g^{33}R_{33} = 0. \tag{9.23}$$

Below, shown as Equation 9.24, I have repeated the identical calculation for the other two spatial indices. All three equations must be true if only T_{00} is non-zero.

$$g^{00}R_{00} - g^{11}R_{11} + g^{22}R_{22} + g^{33}R_{33} = 0,$$
$$g^{00}R_{00} + g^{11}R_{11} - g^{22}R_{22} + g^{33}R_{33} = 0, \tag{9.24}$$
$$g^{00}R_{00} + g^{11}R_{11} + g^{22}R_{22} - g^{33}R_{33} = 0.$$

If you subtract the first line of Equation 9.24 from the second, you get $g^{11}R_{11} = g^{22}R_{22}$. Subtracting the third from the second gives $g^{33}R_{33} = g^{22}R_{22}$. Overall $g^{11}R_{11} = g^{22}R_{22} = g^{33}R_{33}$. Substituting this result back into Equation 9.24 gives Equation 9.25, and the result is $g^{11}R_{11} = -g^{00}R_{00}$. Note the minus sign.

$$g^{00}R_{00} - g^{11}R_{11} + g^{11}R_{11} + g^{11}R_{11} = 0,$$
$$\implies \quad g^{11}R_{11} = -g^{00}R_{00}, \tag{9.25}$$
$$\implies \quad g^{11}R_{11} = g^{22}R_{22} = g^{33}R_{33} = -g^{00}R_{00}, \quad \text{when only } T_{00} \neq 0.$$

Equation 9.25 clearly shows that the presence of rest mass energy affects not only R_{00}. It spreads through all the diagonal components of the Ricci tensor. In fact, it creates as much distortion in each of the spatial Ricci components as it does in the time component (bear in mind the $< + - - - >$ signature of Minkowski spacetime). Note carefully that Equation 9.25 does *not* mean $R_{11} = R_{22} = R_{33}$ because the values of g^{11}, g^{22} and g^{33} may well differ. For example, using spherical coordinates in Minkowski spacetime $g^{11} = -1$, $g^{22} = -\frac{1}{r^2}$ and $g^{33} = -\frac{1}{r^2\sin^2\theta}$.

9.5.3 Relating the Ricci Tensor to Energy-Density

T_{00} typically is the dominant component of the energy-momentum tensor because it contains the rest mass energy-density in the spacetime. Therefore, we model what you might describe as the day-to-day gravity we experience by setting $T_{00} = \rho_E$ where ρ_E is energy-density (check back to Equation 8.11) and assuming that all the other components of the $[T_{\mu\nu}]$ are negligible.

For day-to-day gravity, this means we can approximate that $T_{\mu\nu} = 0$ *except* for component $T_{00} = \rho_E$. Starting with the EFE formula for T_{00}, we substitute for R based on Equation 9.25 and then relate the Ricci component R_{00} with the level of energy-density ρ_E present. The result, shown in Equation 9.26, will prove useful later when we quantify the value of the Einstein constant k.

$$k\rho_E = R_{00} - \frac{1}{2} g_{00} R, \qquad (kT_{00} = k\rho_E) \qquad\qquad \text{EFE from Equation 9.19,}$$

$$= R_{00} - \frac{1}{2} g_{00} (g^{00} R_{00} + g^{11} R_{11} + g^{22} R_{22} + g^{33} R_{33}), \qquad \text{substitute Equation 9.25,} \qquad (9.26)$$

$$= R_{00} - \frac{1}{2} (-2R_{00}) = 2R_{00}, \quad \Longrightarrow \quad R_{00} = \frac{k\rho_E}{2}, \qquad \text{when only } T_{00} \neq 0.$$

9.6   Curvature Footprints

I want to look more closely at the pattern of Riemann curvature in the case of what I am calling day-to-day gravity (i.e. when the only significant component of the energy-momentum tensor is T_{00}) and also in the case of a vacuum ($[T_{\mu\nu} = 0]$).

The Riemann tensor $[R^{\alpha}{}_{\beta\gamma\delta}]$ is a mammoth ($4 \times 4 \times 4 \times 4$) beast, so we will continue to focus on the components of the Riemann tensor that contribute to the diagonal components of the Ricci tensor $[R_{\mu\nu}]$. The starting point is Equation 9.27, which shows the symmetries that exist between these Riemann components for *all* 4-D diagonal metrics (check back to Equation 9.4).

Underlying symmetries of every 4-D diagonal metric $\hspace{4cm}$ (9.27)

$$
\begin{aligned}
g^{00} R_{00} &= g^{00} R^0{}_{000} + g^{00} R^1{}_{010} + g^{00} R^2{}_{020} + g^{00} R^3{}_{030} = 0 \ \ +a \ \ +b \ \ +c, \\
g^{11} R_{11} &= g^{11} R^0{}_{101} + g^{11} R^1{}_{111} + g^{11} R^2{}_{121} + g^{11} R^3{}_{131} = a \ \ +0 \ \ +d \ \ +e, \\
g^{22} R_{22} &= g^{22} R^0{}_{202} + g^{22} R^1{}_{212} + g^{22} R^2{}_{222} + g^{22} R^3{}_{232} = b \ \ +d \ \ +0 \ \ +f, \\
g^{33} R_{33} &= g^{33} R^0{}_{303} + g^{33} R^1{}_{313} + g^{33} R^2{}_{323} + g^{33} R^3{}_{333} = c \ \ +e \ \ +f \ \ +0.
\end{aligned}
$$

From Equation 9.25, we know that if the only non-zero component of $[T_{\mu\nu}]$ is $T_{00} \neq 0$, then it follows: $-g^{00} R_{00} = g^{11} R_{11} = g^{22} R_{22} = g^{33} R_{33}$. Combining this with the values from the right of Equation 9.27 gives the three equalities in Equation 9.28. A bit of maths gives d, e and f in terms of a, b and c as shown in Equation 9.29.

$$(1) -a-b-c = a+d+e, \quad (2) -a-b-c = b+d+f, \quad (3) -a-b-c = c+e+f,$$

$$(1) \ \ d+e = -2a-b-c, \quad (2) \ \ d+f = -a-2b-c, \quad (3) \ \ e+f = -a-b-2c, \qquad (9.28)$$

from which: $\ (1) + (2) - (3) \implies 2d = -2a - 2b,$

$$\implies \qquad d = -a-b, \qquad e = -a-c, \qquad f = -b-c. \qquad\qquad (9.29)$$

Substituting this into Equation 9.27 gives the pattern of Riemann curvature in Equation 9.30. You can think of this pattern as being an approximate footprint of the curvature created when rest mass is the dominant term in the energy-momentum tensor.

Rest-mass-energy-dominated model: only T_{00} is non-zero (diagonal metric) $\hspace{2cm}$ (9.30)

$$
\begin{aligned}
g^{00} R_{00} &= g^{00} R^0{}_{000} + g^{00} R^1{}_{010} + g^{00} R^2{}_{020} + g^{00} R^3{}_{030} = 0 \ \ \ +a \ \ \ +b \ \ \ +c, \\
g^{11} R_{11} &= g^{11} R^0{}_{101} + g^{11} R^1{}_{111} + g^{11} R^2{}_{121} + g^{11} R^3{}_{131} = a \ \ \ +0 \ \ \ -a-b \ \ \ -a-c, \\
g^{22} R_{22} &= g^{22} R^0{}_{202} + g^{22} R^1{}_{212} + g^{22} R^2{}_{222} + g^{22} R^3{}_{232} = b \ \ \ -a-b \ \ \ +0 \ \ \ -b-c, \\
g^{33} R_{33} &= g^{33} R^0{}_{303} + g^{33} R^1{}_{313} + g^{33} R^2{}_{323} + g^{33} R^3{}_{333} = c \ \ \ -a-c \ \ \ -b-c \ \ \ +0.
\end{aligned}
$$

Take a moment to appreciate how the components of the Ricci tensor link together. Changes to a, b or c affect many Ricci tensor components. This curvature footprint will prove useful in Chapter 10 when we discuss a gravitational model of the spacetime metric $[g_{\mu\nu}]$.

We can perform exactly the same exercise for gravitational curvature in a *vacuum*. In this case, every component of the Ricci tensor $[R_{\mu\nu}]$ is zero, so $g^{00} R_{00} = g^{11} R_{11} = g^{22} R_{22} = g^{33} R_{33} = 0$. Referring back to the general symmetries for all diagonal metrics in Equation 9.27, we know immediately for a vacuum that $(a + b + c = 0)$ and therefore that: $a = -(b + c)$.

The other three lines in Equation 9.27 must also be zero, which gives us the three equalities below. Combining them as shown gives $d = c$. Feeding this back into equality (1) gives $e = b$, and simple maths shows $f = -b - c$.

$$(1) -b -c + d + e = 0, \quad (2)\ b + d + f = 0, \quad (3)\ c + e + f = 0,$$

$$\textit{from which:} \quad (1) + (2) - (3) \implies -2c + 2d = 0 \implies d = c, \tag{9.31}$$

$$\implies \qquad d = c, \qquad e = b, \qquad f = -b - c.$$

This gives the curvature footprint shown as Equation 9.32. The only stipulation is that it is for a diagonal metric, which means the coordinate axes must be orthogonal. The rest comes from the symmetries of the Riemann tensor and the requirement that the Ricci tensor for a vacuum is zero: $[R_{\mu\nu}] = 0$.

Underlying symmetries of vacuum curvature (diagonal metric) (9.32)

$$
\begin{aligned}
g^{00}R^0_{\ 000} + g^{00}R^1_{\ 010} + g^{00}R^2_{\ 020} + g^{00}R^3_{\ 030} &= & 0 \quad & -b-c \quad & +b \quad & +c \quad &= 0,\\
g^{11}R^0_{\ 101} + g^{11}R^1_{\ 111} + g^{11}R^2_{\ 121} + g^{11}R^3_{\ 131} &= & -b-c \quad & +0 \quad & +c \quad & +b \quad &= 0,\\
g^{22}R^0_{\ 202} + g^{22}R^1_{\ 212} + g^{22}R^2_{\ 222} + g^{22}R^3_{\ 232} &= & b \quad & +c \quad & +0 \quad & -b-c \quad &= 0,\\
g^{33}R^0_{\ 303} + g^{33}R^1_{\ 313} + g^{33}R^2_{\ 323} + g^{33}R^3_{\ 333} &= & c \quad & +b \quad & -b-c \quad & +0 \quad &= 0.
\end{aligned}
$$

This footprint of vacuum curvature will prove helpful when we discuss the Schwarzschild metric in Chapter 12.

9.7 Summary

Einstein worked for almost 10 years on the theory of general relativity and I suspect it is the contents of this chapter that troubled him the most. In Chapter 10, we still have some work to do to figure out all the details of the EFEs (specifically, calculating the value of the constant k in Equation 9.33 and the possible inclusion of a cosmological constant), but we have broken the back of the problem.

$$R_{\mu\nu} - \frac{1}{2}g_{\mu\nu}R = kT_{\mu\nu}, \tag{9.33}$$

Spacetime curvature $\Longleftrightarrow$ Energy-momentum.

At the start of the chapter, I tried to give you an intuitive feel (*not* a proof) for why there is a link between the presence of energy-momentum and the curvature of spacetime. The spacetime metric $[g_{\mu\nu}]$ quantifies the relationship between mass, energy and momentum. Therefore, it is not surprising that $[g_{\mu\nu}]$ and the presence of energy-momentum are intertwined.

We discussed the symmetries of the Ricci tensor. I showed you that it is impossible to have a simple relationship between T_{00} and R_{00} because any change to R_{00} must affect one or more of the spatial components of the Ricci tensor, such as R_{11}.

This created a mathematical headache for Einstein, which he solved using local conservation of energy and momentum. The covariant derivative $\nabla^\mu T_{\mu\nu}$ is zero. Einstein reasoned the same must be true for the correct Ricci function on the left side of his EFEs. Otherwise, the two sides of the EFEs cannot be equal. I walked you through the mathematics of this (at least those of you who did not skip out of it). The result is the combination of components shown on the left of Equation 9.33, called the Einstein tensor, labelled $[G_{\mu\nu}]$. It includes an additional term based on the Ricci scalar R.

This relationship does not change the vacuum solution. It is consistent with Einstein's supposition that in a vacuum ($[T_{\mu\nu}] = 0$), spacetime is Ricci-0 ($[R_{\mu\nu}] = 0$). However, it signals that the presence of energy-momentum affects the spatial components of the Ricci tensor, R_{11}, R_{22} and R_{33}, as well as R_{00}.

I amuse myself by thinking of the Einstein tensor $[G_{\mu\nu}]$ as being a bit like a hippo. The relationship $G_{\mu\nu} = kT_{\mu\nu}$ looks simple and innocent like a hippo floating in a river. But when you delve more deeply into $[G_{\mu\nu}]$, the symmetries of the Ricci tensor and the addition of the Ricci scalar term spread around the distortion, like the hippo's legs beneath the water's surface thrashing around, stirring things up (Figure 9.2).

Figure 9.2 The Einstein tensor—seemingly calm but stirring things up beneath. *Source:* Bernard Dupont / Wikimedia Commons / CC BY-SA 2.0.

Don't forget that this complicated relationship involving the Ricci tensor and Ricci scalar is not some strange feature of gravitational curvature nor is it a mathematical coincidence. It is driven by the symmetries inherent in the Riemann tensor and by local conservation of energy and momentum.

Based on the structure of the EFEs in Equation 9.33, I showed you that when $[T_{\mu\nu}] = 0$ (a vacuum), the Ricci tensor is zero: $[R_{\mu\nu}] = 0$. This is exactly as Einstein expected (ignoring the cosmological constant).

I also showed you some implications of the EFEs in terms of the pattern of Riemann curvature (the curvature footprint) of spacetime, both when energy-momentum is present (the rest-mass-energy-dominated scenario) and in the case of a vacuum.

In Chapter 10, we will use the structure of the EFEs to build an approximate model of what the spacetime metric looks like in a gravitational field.

Box 9.3 Mixed-index components of the Ricci tensor: proof

All the mixed-index Ricci components *must* be zero for any diagonal metric $[g_{\mu\nu}]$ unless the metric depends on *both* the variables, e.g. for non-zero R_{TX}, the diagonal metric must vary both with t and x. Note that the following calculation does *not* use Einstein summation for repeated indices, but shows dummy indices in bold.

Labelling the four different spacetime indices as A, B, C and D, we evaluate R_{AB}. $R^A{}_{AAB}$ and $R^B{}_{ABB}$ are zero because the first or last two indices are the same. Equation 9.34 shows $R^C{}_{ACB}$. As it is a diagonal metric, any Christoffel symbols with three different indices such as Γ^A_{BC} are zero resulting in Equation 9.35. Every term involves derivatives with respect to *both A* and *B*. By symmetry, the same is true for $R^D{}_{ADB}$. It is therefore true of R_{AB}. The indices A and B are arbitrary, so the rule applies generally for any mixed-index component of $[R_{\mu\nu}]$.

$$R_{AB} = R^A{}_{AAB} + R^B{}_{ABB} + R^C{}_{ACB} + R^D{}_{ADB} = R^C{}_{ACB} + R^D{}_{ADB}, \quad (\textit{index symmetry} - \textit{see Box 7.3}),$$

$$R^C{}_{ACB} = \frac{\partial}{\partial C}\Gamma^C_{AB} - \frac{\partial}{\partial B}\Gamma^C_{AC} + \Gamma^C_{\mathbf{E}C}\Gamma^{\mathbf{E}}_{AB} - \Gamma^C_{\mathbf{E}B}\Gamma^{\mathbf{E}}_{AC}, \qquad (\textbf{E } \textit{is a dummy index}), \qquad (9.34)$$

$$= -\frac{\partial}{\partial B}\Gamma^C_{AC} + \Gamma^C_{AC}\Gamma^A_{AB} + \Gamma^C_{BC}\Gamma^B_{AB} - \Gamma^C_{CB}\Gamma^C_{AC}, \qquad (9.35)$$

$$= -\frac{\partial}{\partial B}\left(\frac{1}{2}g^{CC}\frac{\partial g_{CC}}{\partial A}\right) + \frac{1}{2}g^{CC}\frac{\partial g_{CC}}{\partial A}\frac{1}{2}g^{AA}\frac{\partial g_{AA}}{\partial B} + \frac{1}{2}g^{CC}\frac{\partial g_{CC}}{\partial B}\frac{1}{2}g^{BB}\frac{\partial g_{BB}}{\partial A}\cdots$$

$$\cdots - \frac{1}{2}g^{CC}\frac{\partial g_{CC}}{\partial B}\frac{1}{2}g^{CC}\frac{\partial g_{CC}}{\partial A}.$$

$$R^C{}_{ACB} \text{ and } R^D{}_{ADB} = 0 \implies R_{AB} = 0 \qquad \textit{unless } \frac{\partial g_{\mu\nu}}{\partial A} \textit{ and } \frac{\partial g_{\mu\nu}}{\partial B} \textit{ are BOTH non-zero.}$$

Chapter 10

Einstein Field Equations: The Full Story

In Chapter 9, we derived the detailed structure of the Einstein Field Equations (EFEs) and discussed the possible curvature of spacetime under two scenarios: when there is no energy-momentum present (a vacuum, $[T_{\mu\nu}] = 0$), and when there is energy-momentum present that is rest-mass-energy-dominated (T_{00} is the only significant component of $[T_{\mu\nu}]$). We calculated for each scenario, the components of its Ricci tensor, revealing for each a pattern of curvature, which I call its curvature footprint.

This is all useful stuff, but we still need to link this curvature to the spacetime metric $[g_{\mu\nu}]$. You might think that given $[R_{\mu\nu}]$ and the curvature footprint, it would be easy to calculate $[g_{\mu\nu}]$. Oh, how wrong you would be! The process of calculating from $[g_{\mu\nu}]$ to $[R_{\mu\nu}]$ is fairly straightforwards albeit at times mathematically cumbersome. The reverse is not.

Einstein worked out his theory of general relativity *without* having any gravitational spacetime metric $[g_{\mu\nu}]$ that exactly matches any of our EFE scenarios. It was only after he published his theory that the first exact solution (the Schwarzschild vacuum metric) was found.

What Einstein did have was a model spacetime metric $[g_{\mu\nu}]$ called the *Weak Field* metric that *approximately* matches the vacuum and rest-mass-energy-dominated scenarios. In this chapter, I will show you this metric and demonstrate that it does indeed approximately match the curvature footprints discussed in Chapter 9. We will then use the Weak Field metric to establish the value of the Einstein constant k in the EFEs.

10.1 Einstein's Weak Field Metric

The invariant interval of the Weak Field metric is shown as Equation 10.1. The symbol Φ is the gravitational potential as described in Newton's theory. This is a *negative* quantity, so g_{TT} is less than one (as you would expect from time dilation). One great thing about the Weak Field metric is that it uses Cartesian coordinates, so a step in any spatial direction is the same in terms of the invariant interval. This makes mathematical computations less taxing. Three cheers for Einstein!

$$c^2\,d\tau^2 = \left(1 + \frac{2\Phi}{c^2}\right)c^2\,dt^2 - \left(1 - \frac{2\Phi}{c^2}\right)(dx^2 + dy^2 + dz^2), \qquad \textit{Weak Field metric.} \tag{10.1}$$

Equation 10.2 shows the metric in matrix form. Please note that I have used the usual (ct, x, y, z) coordinate units, which match time to space. As we will be comparing the Weak Field metric with Newtonian gravity, I will occasionally switch to units (t, x, y, z) in order to fit with the metres and seconds you are used to day to day. I will give good warning when I do.

$$\textit{Weak Field } (ct,x,y,z) \ : \ [g_{\mu\nu}] = \begin{bmatrix} 1 + \dfrac{2\Phi}{c^2} & 0 & 0 & 0 \\ 0 & -\left(1 - \dfrac{2\Phi}{c^2}\right) & 0 & 0 \\ 0 & 0 & -\left(1 - \dfrac{2\Phi}{c^2}\right) & 0 \\ 0 & 0 & 0 & -\left(1 - \dfrac{2\Phi}{c^2}\right) \end{bmatrix}. \tag{10.2}$$

Untangling General Relativity: The Intuitive Self-Study Guide, First Edition. Simon Sherwood.
© 2026 John Wiley & Sons Ltd. Published 2026 by John Wiley & Sons Ltd.

10.1.1 Refresher: Gravitational Potential

To calculate the value of the Einstein constant k, we will match the predictions of general relativity with the classical Newtonian description of gravity. For this, you need to understand the Newtonian classical concept of *gravitational potential* Φ. Let me explain what it is.

At any given position in space, Φ is the amount of gravitational potential energy a test particle of unit mass ($m = 1$) would have due to the gravitational fields around it. Let's start with the simplest example: when there is only one attracting mass M. We set the value of Φ as zero for a test particle at an infinite distance from the mass. What happens if we allow the test particle to move to a point that is closer to mass M? The test particle loses gravitational potential energy and, in this scenario, gains kinetic energy (total energy is conserved). The gravitational potential energy is Φ (it would be $m\Phi$ but we have set the mass of the test particle to $m = 1$). Φ is a *negative* quantity because the potential energy is always lower than at an infinite distance away (which we set as zero).

We aren't limited to one attracting mass M. The scenario can cater for any number of attracting masses. The logic remains the same. Moving a test particle (of unit mass) from an infinite distance away to a point (x, y, z) leads to a drop in gravitational potential energy. That negative change in energy is $\Phi(x, y, z)$.

I also need to introduce you to (or remind you of) the relationship between gravitational potential Φ and classic Newtonian gravitational force and acceleration, which I will label F and a. Energy is a measure of force multiplied by distance. If you move a metre against an opposing force of one newton, that requires one joule of energy (joules are measured in newton-metres). Of course, the force may change as you move, so to calculate the overall energy change, you need to integrate the opposing force $-F$ over the distance moved. In the same way, to calculate gravitational potential energy $m\Phi$ of an object of mass m distance r from attracting mass M, we *integrate* the gravitational force $-F$ towards the attracting mass starting infinitely far from the mass and ending at the point we want: $m\Phi = -\int_\infty^r F dr$.

Conversely, if we know $m\Phi$ and want to know F, we must *differentiate* $-m\Phi$ with respect to distance r from the attracting mass M. This is shown in Equation 10.3 using only one dimension (r). As expected, you can see that gravitational acceleration a does not depend on the mass m of the test object. It is the same for all objects, whatever their mass.

$$m\Phi = -\frac{GMm}{r}, \qquad F = ma = -\frac{GMm}{r^2} \qquad a = -\frac{d\Phi}{dr}, \qquad (10.3)$$

$$\vec{a} = -\left(\frac{\partial\Phi}{\partial x}\vec{e_x} + \frac{\partial\Phi}{\partial y}\vec{e_y} + \frac{\partial\Phi}{\partial z}\vec{e_z}\right), \qquad \textit{Cartesian coordinates.} \qquad (10.4)$$

Equation 10.4 shows the relationship between a and Φ in three spatial dimensions using the Cartesian coordinates of the Weak Field metric. In words, the acceleration is $-\frac{\partial\Phi}{\partial x}$ in the x direction (hence the basis vector $\vec{e_x}$) plus the equivalent in the y and z directions. We say the net gravitational acceleration at any point is the negative of what is called the *gradient* of Φ at that point.

10.1.2 Weak Field Geodesic Equation

For a single attracting mass M at distance r, the value of Φ is $-\frac{GM}{r}$ (see Equation 10.3). Therefore in the presence of a single mass, the Weak Field invariant interval in Equation 10.1 can be expressed as shown in Equation 10.5.

$$c^2 d\tau^2 = \left(1 - \frac{2GM}{rc^2}\right)c^2 dt^2 - \left(1 + \frac{2GM}{rc^2}\right)(dx^2 + dy^2 + dz^2), \quad \textit{mass M at distance r.} \qquad (10.5)$$

Compare this with the model metric of Chapter 5 that incorporated the time dilation of gravitational acceleration into the metric (Section 5.1). For a single attracting mass M, the time dilation factor g_{TT} in the Weak Field metric (attracting mass distance r away) is identical to g_{TT} in the 2-D model metric (attracting mass distance x away). In Section 5.3, I used the geodesic equation to show you this distortion creates the equivalent of Newton's gravitational acceleration (see Equation 5.19). Therefore, the distorting effect of the Weak Field metric matches the type of day-to-day gravitational acceleration experienced by non-relativistic objects in a weak gravitational field. By weak, I mean $\frac{2\Phi}{c^2} \ll 1$. On the surface of the earth $\frac{2\Phi}{c^2} \approx 10^{-9}$ (see Box 5.5).

I think it is worth quickly stepping through this same geodesic calculation using the Weak Field metric. Starting with the geodesic Equation 10.6 (see Equation 5.13 earlier), we can discount all except the $v^t v^t$ term because, for day-to-day objects, the values of v^x, v^y and v^z are negligible. For non-relativistic objects $g_{XX} \approx -1$ and also $t \approx \tau$, so $v^t = \dfrac{dt}{d\tau} \approx 1$ (check back to Equation 5.18 if that helps). To keep things simple, the calculation in Equation 10.7 uses the value of g_{TT} in units of (t, x) rather than (ct, x), so that it matches the metres and seconds we use day to day.

$$\frac{dv^x}{d\tau} = -\Gamma^X_{AB} v^A v^B, \qquad\qquad\qquad \textit{Geodesic equation,} \qquad\qquad (10.6)$$

$$\approx -\Gamma^X_{TT} v^t v^t \approx -\Gamma^X_{TT}, \qquad\qquad v^x \ll v^t, \quad v^y \ll v^t, \quad v^z \ll v^t,$$

$$= \frac{1}{2g_{XX}}\left(\frac{\partial g_{TT}}{\partial x}\right), \qquad\qquad t \approx \tau, \qquad g_{XX} \approx -1,$$

$$\frac{dv^x}{dt} \approx -\frac{1}{2}\frac{\partial}{\partial x}\left(c^2 + 2\Phi\right) = -\frac{\partial \Phi}{\partial x}, \qquad g_{TT} = c^2 + 2\Phi, \quad (t, x, y, z). \qquad (10.7)$$

The total gravitational acceleration at non-relativistic speeds is the sum of the change in velocity along all three axes as shown in Equation 10.8. The approximation is accurate for non-relativistic objects in weak gravitational fields. As you can see, the gravitational acceleration of the Weak Field metric is a good match to that of classical Newtonian gravity in Equation 10.4.

$$\vec{a} = \frac{dv^x}{dt}\vec{e}_x + \frac{dv^y}{dt}\vec{e}_y + \frac{dv^z}{dt}\vec{e}_z \approx -\left(\frac{\partial \Phi}{\partial x}\vec{e}_x + \frac{\partial \Phi}{\partial y}\vec{e}_y + \frac{\partial \Phi}{\partial z}\vec{e}_z\right). \qquad (10.8)$$

10.1.3 Weak Field Ricci Tensor

The next step is to calculate the Ricci tensor for the Weak Field metric. This leads to the value of the Einstein constant k. It is also useful to check how the curvature compares with the curvature footprints we studied in Section 9.6. The metric is diagonal, so we can use the symmetries from Equation 9.27, shown again as Equation 10.9. In the case of the Weak Field metric, the coordinates $(0, 1, 2, 3)$ are (ct, x, y, z) and the symmetry between spatial coordinates eases the mathematics.

$$\textit{Underlying Symmetries for any 4-D Diagonal Metric} \qquad\qquad (10.9)$$

$$g^{00}R_{00} = g^{00}R^0_{000} + g^{00}R^1_{010} + g^{00}R^2_{020} + g^{00}R^3_{030} = 0 \quad +a \quad +b \quad +c,$$

$$g^{11}R_{11} = g^{11}R^0_{101} + g^{11}R^1_{111} + g^{11}R^2_{121} + g^{11}R^3_{131} = a \quad +0 \quad +d \quad +e,$$

$$g^{22}R_{22} = g^{22}R^0_{202} + g^{22}R^1_{212} + g^{22}R^2_{222} + g^{22}R^3_{232} = b \quad +d \quad +0 \quad +f,$$

$$g^{33}R_{33} = g^{33}R^0_{303} + g^{33}R^1_{313} + g^{33}R^2_{323} + g^{33}R^3_{333} = c \quad +e \quad +f \quad +0.$$

There are two simplifying assumptions that help further. The first assumption is that Φ is very small so, except when differentiating, we approximate with $g_{TT} \approx 1$ and $g_{XX} \approx -1$, which also gives $g^{TT} \approx 1$ and $g^{XX} \approx -1$. The second assumption is that changes in the gravitational potential of a weak gravitational field are smooth, which means that derivatives of Φ are small, and we can ignore multiple or squared terms such as $\left(\dfrac{\partial \Phi}{\partial x}\dfrac{\partial \Phi}{\partial y}\right)$ or $\left(\dfrac{\partial \Phi}{\partial t}\right)^2$.

This second assumption is particularly important. It reduces the formula for components of the Riemann tensor in Equation 10.10 to that in Equation 10.11 (the original formula is from Box 7.3 earlier). Each Christoffel symbol involves a derivative of Φ, so we ignore the $\Gamma\,\Gamma$ terms.

$$R^A_{BCD} = \partial_C \Gamma^A_{BD} - \partial_D \Gamma^A_{BC} + \Gamma^A_{CE}\Gamma^E_{BD} - \Gamma^A_{DE}\Gamma^E_{BC}, \qquad \textbf{E} \textit{ is a dummy index,} \qquad (10.10)$$

$$\approx \partial_C \Gamma^A_{BD} - \partial_D \Gamma^A_{BC}, \qquad\qquad\qquad \textit{Weak Field assumption.} \qquad (10.11)$$

The calculation of R^X_{TXT} is shown in Equation 10.12. This is R^1_{010} in the curvature footprint of Equation 10.9. The metric is not time t dependent, so only the first term in Equation 10.12 has value. As $g_{TT} \approx 1$, the result is the value of symbol a in Equation 10.9. By symmetry, to get the values of b and c, we substitute for x with y and z respectively.

$$R^X_{TXT} \approx \partial_x \Gamma^X_{TT} - \partial_t \Gamma^X_{TX} = \partial_x \Gamma^X_{TT} = \partial_x\left(-\frac{1}{2}g^{XX}\frac{\partial g_{TT}}{\partial x}\right),$$

$$\qquad\qquad\qquad\qquad (10.12)$$

$$= \partial_x\left(-\frac{1}{2}(-1)\frac{2}{c^2}\frac{\partial \Phi}{\partial x}\right) = \frac{1}{c^2}\frac{\partial^2 \Phi}{\partial x^2}.$$

The calculation of R^Y_{XYX} is shown as Equation 10.13. This is R^2_{121} in the curvature footprint of Equation 10.9. As $g_{XX} \approx -1$, the negative of this result is d. To get R^Y_{ZYZ} we simply substitute z for x, which gives the value of f. And to get R^Z_{XZX} we substitute z for y giving the value of e.

$$R^Y_{XYX} \approx \partial_y \Gamma^Y_{XX} - \partial_x \Gamma^Y_{XY},$$

$$= \partial_y \left(-\frac{1}{2} g^{YY} \frac{\partial g_{XX}}{\partial y} \right) - \partial_x \left(\frac{1}{2} g^{XX} \frac{\partial g_{YY}}{\partial x} \right),$$

$$= \partial_y \left(-\frac{1}{2}(-1)\frac{2}{c^2}\frac{\partial \Phi}{\partial y} \right) - \partial_x \left(\frac{1}{2}(-1)\frac{2}{c^2}\frac{\partial \Phi}{\partial x} \right),$$

$$= +\frac{1}{c^2}\frac{\partial^2 \Phi}{\partial x^2} + \frac{1}{c^2}\frac{\partial^2 \Phi}{\partial y^2}.$$

(10.13)

We can substitute these values into the general footprint of Equation 10.9 to give the approximate curvature footprint of the Weak Field metric shown as Equation 10.14. Don't forget the simplifying assumptions we made. It is *approximate* because it applies only to non-relativistic objects in weak, smoothly changing gravitational fields.

Approximate Curvature Footprint of Weak Field Metric (10.14)

$$g^{00}R_{00} = g^{TT}R_{TT} = \frac{1}{c^2} \left(\quad 0 \quad +\frac{\partial^2\Phi}{\partial x^2} \quad +\frac{\partial^2\Phi}{\partial y^2} \quad +\frac{\partial^2\Phi}{\partial z^2} \quad \right),$$

$$g^{11}R_{11} = g^{XX}R_{XX} = \frac{1}{c^2} \left(\frac{\partial^2\Phi}{\partial x^2} \quad +0 \quad -\frac{\partial^2\Phi}{\partial x^2}-\frac{\partial^2\Phi}{\partial y^2} \quad -\frac{\partial^2\Phi}{\partial x^2}-\frac{\partial^2\Phi}{\partial z^2} \right),$$

$$g^{22}R_{22} = g^{YY}R_{YY} = \frac{1}{c^2} \left(\frac{\partial^2\Phi}{\partial y^2} \quad -\frac{\partial^2\Phi}{\partial x^2}-\frac{\partial^2\Phi}{\partial y^2} \quad +0 \quad -\frac{\partial^2\Phi}{\partial y^2}-\frac{\partial^2\Phi}{\partial z^2} \right),$$

$$g^{33}R_{33} = g^{ZZ}R_{ZZ} = \frac{1}{c^2} \left(\frac{\partial^2\Phi}{\partial x^2} \quad -\frac{\partial^2\Phi}{\partial x^2}-\frac{\partial^2\Phi}{\partial z^2} \quad -\frac{\partial^2\Phi}{\partial y^2}-\frac{\partial^2\Phi}{\partial z^2} \quad +0 \right).$$

Do take a moment to compare this with the curvature footprint of rest-mass-energy-dominated spacetime shown as Equation 9.30 back in Section 9.6, that is the curvature dictated by the EFEs when the only non-zero component of the energy-momentum tensor is T_{00}. You will see that the approximate curvature footprint of the Weak Field metric is identical. To compare substitute a, b and c variables in Equation 9.30 with $\frac{\partial^2\Phi}{\partial x^2}$, $\frac{\partial^2\Phi}{\partial y^2}$ and $\frac{\partial^2\Phi}{\partial z^2}$ respectively.

Compare also with the curvature footprint of the vacuum shown as Equation 9.32 (also in Section 9.6). You will see it is identical when the total in Equation 10.15 is zero. As shown, this total typically is labelled $\nabla^2\Phi$ (be careful not to confuse this with the covariant derivative that happens to use a similar symbol). I will explain this in more detail in a moment, and will show you that in Newtonian gravity, it *is zero for a vacuum and non-zero when mass is present*.

$$\nabla^2\Phi = \frac{\partial^2\Phi}{\partial x^2} + \frac{\partial^2\Phi}{\partial y^2} + \frac{\partial^2\Phi}{\partial z^2} = 0, \qquad \textit{for a vacuum.} \tag{10.15}$$

To summarise, we have made a couple of important simplifying assumptions: that $\frac{2\Phi}{c^2}$ is very small and that changes in Φ are smooth enough to ignore multiples of derivative terms, such as $\frac{\partial\Phi}{\partial x}\frac{\partial\Phi}{\partial x}$ and $\frac{\partial\Phi}{\partial x}\frac{\partial\Phi}{\partial y}$. Given these assumptions, the Ricci tensor and associated curvature of the Weak Field metric match what the EFEs tell us about both rest-matter-energy-dominated and vacuum curvature in general relativity:

- In Newton's theory (as I will show you later) $\nabla^2\Phi \neq 0$ in the presence of mass. In this case, the *approximate* curvature of the Weak Field metric matches the rest-mass-energy-dominated curvature required by the EFEs (Equation 9.30).
- In Newton's theory $\nabla^2\Phi = 0$ in a vacuum. In this case the *approximate* curvature of the Weak Field metric matches the vacuum curvature required by the EFEs (Equation 9.32).

Please note my continuing emphasis of *approximate*. The Weak Field metric is not an exact solution to the EFEs for either scenario. However, this all builds confidence that it is an excellent non-relativistic approximation of the spacetime metric $[g_{\mu\nu}]$ when $\frac{2\Phi}{c^2} \ll 1$.

10.2   Energy-Momentum, Curvature and the Vacuum

Let me take a moment to drop the maths and address an obvious question. In the rest-mass-energy-dominated scenario, the presence of energy-momentum requires Ricci curvature, and Ricci curvature requires the presence of energy-momentum. To use an English phrase, they are two sides of the same coin. The two are inseparable.

In the case of a vacuum, there is no energy-momentum and the Ricci tensor is zero. It is Ricci-0. However, there still can be Riemann curvature. Ricci-0 spacetime can vary from Minkowski spacetime (which is flat) to the curvature of spacetime around the earth, which leads satellites to orbit, to the intense distortion of spacetime around black holes, which we will discuss in the next module. In all cases, there is no energy-momentum present in the spacetime itself (albeit there may be some nearby), so what is it that governs the pattern?

The answer is that the pattern of curvature of the vacuum metric is governed by the presence of any energy-momentum in the spacetime surrounding it. This should not surprise you. If the presence of energy-momentum distorts the spacetime containing it, then you would not expect the level of distortion to suddenly disappear at the border. If it did, you would have a sharp transition.

To steal a phrase from my other book, *Quantum Untangling*, nature shouldn't have sharp edges. Any discontinuity in the curvature of spacetime is difficult to fathom. An example that we will discuss is the singularity inside a black hole. The mathematician Roger Penrose considered the physical consequences to be so distasteful that there must be a law of physics shielding the rest of the universe from any singularity. In Stephen Hawking's words: *God abhors a naked singularity.*

The presence of energy-momentum distorts the spacetime containing it *and* the surrounding spacetime. In the case of a vacuum, the distortion is Ricci-0 curvature that smoothly fades away with distance from the source of energy-momentum.

10.3   Calculating the Value of Einstein's Gravitational Constant

We still need to derive the value of Einstein's gravitational constant k in the EFEs. The answer is shown on the right of Equation 10.16.

$$R_{\mu\nu} - \frac{1}{2}g_{\mu\nu}R = k\,T_{\mu\nu}, \qquad k = \frac{8\pi G}{c^4},$$

Spacetime curvature $\Longleftrightarrow$ Energy-momentum.

$$(10.16)$$

The way we approach this is disarmingly simple. In Chapter 9 (see Equation 9.26), we showed that if matter dominates, then T_{00} is the only significant component in $[T_{\mu\nu}]$ and $R_{TT} = \frac{k\rho_E}{2}$. In order to compare with classical Newtonian gravity, we restate this energy-density in terms of matter density ρ_m using $E = mc^2$:

$$\textit{EFE curvature if only } T_{00} \neq 0 \qquad R_{TT} = \frac{k\rho_E}{2} = \frac{k\rho_m c^2}{2}. \qquad (10.17)$$

The Weak Field metric, which we have shown is a good (weak field) approximation of the distortion created by matter, also gives us a value of R_{TT}. This is shown in Equation 10.18 (see the top line of the curvature footprint shown as Equation 10.14 using $g^{TT} \approx 1$).

$$\textit{Weak Field}: \quad R_{TT} \approx \frac{1}{c^2}\frac{\partial^2\Phi}{\partial x^2} + \frac{1}{c^2}\frac{\partial^2\Phi}{\partial y^2} + \frac{1}{c^2}\frac{\partial^2\Phi}{\partial z^2} = \frac{1}{c^2}\nabla^2\Phi. \qquad (10.18)$$

There is a well-established relationship in Newtonian gravity between the gravitational potential and the presence of matter. This is called the *Poisson equation*: $\nabla^2\Phi = 4\pi G\rho_m$. Equating Equation 10.17 to Equation 10.18 and substituting it with the Poisson formula gives the value of constant k as shown in Equation 10.19. Please note that I define $T_{00} = \rho_E$ (energy-density). Very rarely, texts use $T_{00} = \rho_m$ (mass density), which is smaller by a factor of c^2, in which case $k = \frac{8\pi G}{c^2}$.

$$\frac{k\rho_m c^2}{2} \approx \frac{1}{c^2}\nabla^2\Phi = \frac{4\pi G\rho_m}{c^2}, \qquad \Longrightarrow \qquad k = \frac{8\pi G}{c^4}. \qquad (10.19)$$

Note also that the Poisson equation $\nabla^2\Phi = 4\pi G\rho_m$ means that $\nabla^2\Phi$ is zero for a vacuum and non-zero when mass is present. This is exactly what we required earlier (Equation 10.15) to match the curvature footprint of the Weak Field metric with what the EFEs tell us about rest-matter-energy-dominated and vacuum curvature.

So that is that. Everybody happy? I am guessing not. For those of you less familiar with the works of Baron Simeon Poisson (who came up with the Poisson equation, see Box 10.1), I offer an optional explanation/refresher in Section 10.4. For experienced physicists who are deeply familiar with Newton's theory, this will be a piece of cake. However, mere mortals might want to munch down a piece of cake before embarking on it.

Box 10.1 Baron Simeon Denis Poisson (1781–1840)

Poisson was smart. Really smart. While at university, he wrote such an impressive paper that the university published it and graduated Poisson without even bothering to have him attend the final exam. I know a few students who would like that!

He grew up during the French Revolution, but rather sensibly eschewed politics to focus on his maths. It was definitely a time to remain calm and, as they say, keep your head while others lose theirs... which several famous scientists tragically did.

In addition to discovering the Poisson equation, he worked on electromagnetism and statistics. You may have heard of the Poisson distribution, which is also named after him. A few years before his death, he was made an honorary Baron for his contributions to science. He is reputed to have said: *Life is good for only two things: doing mathematics and teaching it.*

10.4 The Poisson Equation (Optional Refresher)

The maths isn't that complicated, but I will be working with gradient and divergence, so it may be a bit much for some of you. I have tried to add sufficient explanation alongside the maths, so that everybody can follow the logical argument.

The Poisson equation comes from the classical theory of gravity. This is based on Newton's inverse-square law, so we are stepping back in time and putting aside for a moment the concept of curved spacetime.

Classically, gravity is treated as a field that is mathematically very similar to an electric field. We visualise these fields by imagining *field lines* that, in the case of the electric field, run into an electric charge and, in the case of gravity, run into a mass. You can think of each gravitational field line as being one unit of gravitational acceleration on a test object. The density of field lines tells you the overall acceleration of the test object. The higher the density of field lines, the higher the gravitational acceleration.

Figure 10.1a shows the field lines ending on an attracting mass M. The figure is only in 2-D but imagine surrounding the mass with a sphere, illustrated by the orange circle in the figure. Increasing the radius of this surrounding sphere does not change the number of field lines running into it but does change its surface area, which is $4\pi r^2$. Therefore, as the radius r of the sphere increases, the density of the field lines at the surface decreases

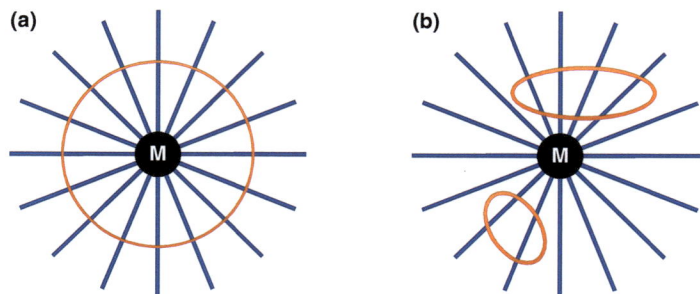

(a)   **(b)**

Figure 10.1 Classical representation of gravity as a field; *M* is the attracting mass. (a) the orange line represents a sphere containing the mass M, (b) the orange lines represent volumes containing no mass.

with r^2. As each field line is a unit of acceleration, this means that the total acceleration a (along with the force) follows an inverse-square law, which is what we want.

Figure 10.1b illustrates an important point that I hope is fairly obvious. There are two areas shown surrounded by orange lines. I want you to think in 3-D of these as volumes. If there is no mass in the volume, then the number of field lines going into the volume must be the same as the number of field lines coming out. This is because the field lines only ever end when they reach a mass. Therefore, the number of field lines ending inside a volume depends on the amount of mass in that volume. This is the basis of Poisson's equation.

Poisson imagined a sphere centred around the centre-of-mass of a bunch of matter. We need to calculate the number of field lines ending in the sphere. We do this by treating the gravitational acceleration a as a vector field. At any point, the net gravitational acceleration a is the sum of three components: $\vec{a} = a^x \vec{e}_x + a^y \vec{e}_y + a^z \vec{e}_z$.

The key measure is the *divergence* of $\vec{a}$, which is a *number* at each point. It has no direction. The symbol for the divergence of $\vec{a}$ is $\nabla \cdot a$ (yet another use of the ∇ symbol). The formula is shown as Equation 10.20. It adds the change in a^x as you move in the x direction, plus the change in a^y moving in the y direction, plus the change in a^z moving in the z direction. As a is an attractive acceleration, any increase in gravitational effect makes it more negative, i.e. it *decreases*.

Imagine that we analyse a tiny region. Suppose a^x decreases as you move in the x direction. This would indicate that you are encountering more field lines. Therefore, more field lines are entering the region from the x direction in which you are moving, than are exiting it from the other side. Either these field lines stop in the region, or veer off to create a net surplus in the y or z directions.

An overall decrease in all directions (negative divergence) means field lines are ending in the region. With gravity, if there is no mass in a region, then the same number of field lines enter and exit and the divergence is zero. We want to measure the mass present, so we use the total net number of field lines ending in the sphere, which is the negative of the integral of divergence over the volume of the sphere as shown in Equation 10.21.

$$\text{Divergence}: \quad \nabla \cdot a = \frac{\partial a^x}{\partial x} + \frac{\partial a^y}{\partial y} + \frac{\partial a^z}{\partial z}, \qquad \textit{Net change in field lines,} \tag{10.20}$$

$$\text{Total sphere}: \quad -\oint_{Vol} \nabla \cdot a, \qquad \textit{Field lines ending inside the sphere.} \tag{10.21}$$

Poisson now turned his attention to the surface of the sphere. We have centred it on the centre-of-mass so the attractive acceleration is the same at every point and is by Newton's law $a = -\frac{GM}{r^2}$ where r is the radius of the sphere, which is of course the distance of the surrounding surface from the centre-of-mass. This means the density of field lines entering the surface of the sphere and ending in it must match this acceleration. The net number of field lines ending in the sphere must be the inflow density $\frac{GM}{r^2}$ multiplied by the total surface area of the sphere $4\pi r^2$, which gives $4\pi GM$.

Now the clever twist. The overall mass M of the sphere can be expressed as the mass density ρ_m at each point integrated over the volume of the sphere, which gives Equation 10.22 ($4\pi G$ is a constant so can move inside the integral).

$$\text{Total sphere}: \quad 4\pi G \oint_{Vol} \rho_m = \oint_{Vol} 4\pi G \rho_m, \qquad \textit{Mass inside the sphere.} \tag{10.22}$$

If all field lines end on mass, then the overall mass must also be a measure of the total number of field lines ending in the sphere, so we can equate the answer in Equation 10.22 with the one we derived in Equation 10.21. The contents of the integral must be the same, so: $4\pi G \rho_m = -\nabla \cdot a$. The last step (thank heavens) is to substitute using the relationship that gravitational acceleration a is the negative of the gradient of the gravitational potential Φ, as shown earlier in Equation 10.4. You can see, for example, that the x component of a is $a^x = -\frac{\partial \Phi}{\partial x}$.

The result in Equation 10.23 is the Poisson equation that we used back in Section 10.3 to calculate the value of Einstein's gravitational constant $k = \frac{8\pi G}{c^4}$.

$$4\pi G \rho_m = -\nabla \cdot a = -\frac{\partial}{\partial x}\left(-\frac{\partial \Phi}{\partial x}\right) - \frac{\partial}{\partial y}\left(-\frac{\partial \Phi}{\partial y}\right) - \frac{\partial}{\partial z}\left(-\frac{\partial \Phi}{\partial z}\right),$$
$$= \frac{\partial^2 \Phi}{\partial x^2} + \frac{\partial^2 \Phi}{\partial y^2} + \frac{\partial^2 \Phi}{\partial z^2} = \nabla^2 \Phi, \qquad \textit{The Poisson equation.} \tag{10.23}$$

I expect a round of applause after all that! To summarise briefly in words, our dear friend Baron Poisson considered a spherical volume containing mass. He worked out in two ways the number of field lines ending in it. The first was to treat the gravitational acceleration as a vector field and calculate its divergence. The second was to consider the gravitational acceleration at the surface of the sphere, which he knew from Newton's laws. Equating the two gives Poisson's famous equation.

10.5 The EFEs in Full Form (Almost!)

Establishing the value of k gives Equation 10.24, which is the EFEs *almost* in full form. I say almost because I have not yet considered the cosmological constant. Before I address that complication, let's quickly review where we stand.

$$\textit{General form}: \quad R_{\mu\nu} - \frac{1}{2}R g_{\mu\nu} = \frac{8\pi G}{c^4} T_{\mu\nu}.$$

$$\textit{Vacuum form}: \quad R_{\mu\nu} = 0, \qquad because \quad T_{\mu\nu} = 0.$$

(10.24)

The EFEs are *covariant*. This means that they apply for any coordinate system. The only glitch is the value of k. As mentioned earlier, I define $[T_{\mu\nu}]$ in terms of *energy-density*. Very rarely, it is defined as *mass density*, which is smaller by a factor of c^2 in which case $k = \frac{8\pi G}{c^2}$.

The top *General form* shows the EFEs in the presence of energy-momentum. We modelled the effect of spacetime containing matter. Both the time and space components of the spacetime metric are distorted by Ricci curvature, which is a change in the spacetime volume separating infinitely close parallel geodesics.

The *Vacuum form* has no Ricci curvature. The spacetime is Ricci-0. However, it can have Riemann curvature because the vacuum metric can be distorted by the presence of energy-momentum in surrounding spacetime.

10.6 The Cosmological Constant

So far, I have ignored Einstein's final tweak to his field equations: the *cosmological constant*, generally labelled Λ. In terms of what you might call *local* gravitational effects (the sun, earth, moon and especially black holes) it is so small that, even if it exists, it is negligible. However, it has a major effect on what the EFEs say about the development of the universe as a whole. The Λ term may be tiny at the local level, but its combined effect is significant over billions of years of time and billions of light years of space. As shown in Equation 10.25 (which you may want to compare with Equation 10.24), it appears in the field equations as $\Lambda g_{\mu\nu}$ where Λ is the cosmological constant. Note that the Λ term appears positive in some texts. This depends on the conventions used, including the metric signature, which is $< + - - - >$ in this book.

$$R_{\mu\nu} - \frac{1}{2}R g_{\mu\nu} - \Lambda g_{\mu\nu} = \frac{8\pi G}{c^4} T_{\mu\nu}.$$

(10.25)

The cosmological constant has a rather ignoble origin. Einstein calculated from his field equations that the universe must either be expanding or contracting. At that time, the whole scientific community believed the universe was static, so dear old Albert came up with a fudge factor. He introduced a term containing the constant Λ, which could be set (he thought) to balance out the net expansion or contraction of the universe... leaving it static. In his words, he introduced it *only for the purpose of making possible a quasi-static distribution of matter*. There were two problems with this. First, in 1929, the astronomer Edwin Hubble showed the universe is not static. It is expanding. Second, the cosmological constant does not actually lead to a static universe because Einstein's proposed equilibrium is unstable. Einstein abandoned the cosmological constant, reportedly describing it as *my biggest blunder*.

There is another twist in this tale. Einstein's Λ term is consistent mathematically with his field equations so it *can* exist. And there is now some strong evidence from cosmological observations that it *does* exist.

Let me explain why Λ is consistent with the EFEs. In Section 9.4, we derived the left side of the EFEs by requiring that its covariant derivative be zero (as is the case for $[T_{\mu\nu}]$). I told you that there is only one mix of $[R_{\mu\nu}]$ and R that works. However, any covariant derivative of the spacetime metric $[g_{\mu\nu}]$ also is zero (see Box 4.7), so we can add this term (multiplied by any constant) to the equation, and the left side of the EFEs will still have zero covariant derivative.

What about the equivalence principle, which we discussed in Section 1.3? Measuring on a local basis, you cannot tell in a vacuum if you are being accelerated by a gravitational field or just trundling along unaffected by gravity (our scenario was freefall in a box with the windows closed). Therefore, to avoid any measurable change in the trajectory of nearby particles, we concluded that the vacuum *with* a gravitational field must be Ricci-0 because the vacuum *without* any gravitational field is Ricci-0.

Can you spot the assumption? As gravity is curvature, we have assumed that a spacetime vacuum in the absence of any gravitational field has no curvature of any kind, that it is bereft of energy and therefore utterly *flat*. That makes sense based on Newton's laws of motion, our day-to-day experience and good old common sense. But what if it is wrong? Could all spacetime have what we might describe as a slight amount of background Ricci curvature? This possibility is what the Λ term in the EFEs represents.

And what of the equivalence principle? Well it still works, providing the same level of background energy/curvature applies to *all* spacetime (including, for example, gravitational curvature in a vacuum). There would still be underlying local equivalence between freefall acceleration in a gravitational field and no acceleration at all.

If you follow the science press, you will have read that this is a big deal. This is because of the relationship between energy and Ricci curvature. I encourage you not to think of this as $(A \rightarrow B)$ with energy (A) causing Ricci curvature (B), but more as the two being equivalent $(A = B)$. The presence of energy in a volume of spacetime indicates that it has Ricci curvature. And, conversely, any Ricci curvature of spacetime indicates the presence of energy. The two are inextricably linked.

If the cosmological constant has a value as we currently believe, then vacuum spacetime contains energy. Physicists call this *dark energy* or *vacuum energy*. The universe is very very large and there is lots of vacuum relative to the presence of galaxies and stars so, even though negligible except on cosmological scales, vacuum energy represents well over half of the total energy in the universe based on current estimates (in fact, about 70% of it). Wow! We will discuss this in detail later.

10.7 Summary

I will keep this short because Chapter 11 has a full summary of the whole module. We used Einstein's Weak Field metric as a model of the spacetime metric $[g_{\mu\nu}]$ for a weak gravitational field, i.e. when: $\frac{2\Phi}{c^2} \ll 1$.

The Weak Field metric has similarities with the simple 2-D vacuum model of gravitational curvature that we used earlier in the book. We calculated the geodesic distortion of the Weak Field metric showing that the accelerating effect fits well with that of Newtonian gravity. We calculated (approximately) its Ricci tensor and curvature footprint, which match the curvature for rest-mass-energy-dominated spacetime and vacuum spacetime determined from the EFEs.

The next step was to use the Ricci curvature of the Weak Field metric to calculate Einstein's gravitational constant k using the Poisson equation. My derivation of the Poisson equation may have tested some readers' tolerance to maths!

I ended the chapter by discussing Einstein's cosmological constant Λ. It can legitimately be added to the EFEs in the term $-\Lambda g_{\mu\nu}$ because any covariant derivative of the metric is zero. Although Einstein subsequently dismissed it as his biggest blunder, there is cosmological evidence that it exists. The term becomes important only when considering the effects of gravity at a cosmological scale.

Chapter 11

Module Summary and Conventions

11.1 Module Summary

It is time to trawl back through the story so far and review what we have learned. In Chapter 2 on *Special Relativity*, we discussed the impact on physics of Einstein's proposal that the speed of light is constant for all observers. For this to be mathematically consistent, there is time dilation and length contraction for objects that are observed in motion, plus a change in synchronisation of clocks. It leads us to a linkage between the dimensions of time and space: Minkowski spacetime. We derived Equation 11.1 (the invariant interval), and Equation 11.2 (the relativistic relationship between energy and momentum), which in turn gives Einstein's famous result for a stationary object: $E = mc^2$.

$$c^2\, d\tau^2 = c^2\, dt^2 - dx^2 - dy^2 - dz^2, \qquad \textit{Invariant interval of Minkowski spacetime,} \qquad (11.1)$$

$$m^2 c^4 = E^2 - p_x^2 c^2 - p_y^2 c^2 - p_z^2 c^2, \qquad \textit{Energy-momentum relationship.} \qquad (11.2)$$

In Chapter 3, I showed how the invariant interval can be expressed in the form of a (4×4) matrix called the metric $[g_{\mu\nu}]$. Equation 11.3 shows this for Minkowski spacetime. The metric is the fundamental measure of distances on curved surfaces. As an example, we examined the metric of the surface of a sphere and how two little dung beetles living on the surface would be drawn towards each other while walking on parallel paths. I fantasised that, if unaware they were living on a curved surface, they might assume there was some force drawing them together, creating an identical acceleration for all objects... a bit like gravity.

$$\textit{Minkowski metric }(ct, x, y, z): \quad [g_{\mu\nu}] = \begin{bmatrix} 1 & 0 & 0 & 0 \\ 0 & -1 & 0 & 0 \\ 0 & 0 & -1 & 0 \\ 0 & 0 & 0 & -1 \end{bmatrix}, \quad \textit{often labelled } [\eta_{\mu\nu}]. \qquad (11.3)$$

At the end of the chapter, I pointed out that this perceived attractive force between the dung beetles disappears if they stop moving. This sounds different from gravity, which is ever-present. However, this simply is because nothing can ever stop moving through *spacetime*... still as a stone in terms of space, everything still travels on through time.

In Chapter 4, I introduced some of the mathematics required to analyse how an undisturbed object (no external force on it) behaves in the presence of curvature. With curvature, we must calculate covariant derivatives that take into account the change in the basis vectors on the surface. We use Christoffel symbols to quantify this. If you want a reminder, the formula for Christoffel symbols is shown later in this chapter in Box 11.5.

In Chapter 5, we used the mathematics of curvature to have some fun with one of our friendly dung beetles. We know how time dilates for an object travelling in a gravitational field towards a mass: a certain profile of acceleration leads to a certain profile of time dilation. We built this time dilation into the metric for a stationary clock in order to examine Einstein's crazy idea: if a profile of acceleration can lead to time dilation, might a profile of background time dilation lead to acceleration? We used Equation 11.4 for geodesics to calculate what would

Untangling General Relativity: The Intuitive Self-Study Guide, First Edition. Simon Sherwood.
© 2026 John Wiley & Sons Ltd. Published 2026 by John Wiley & Sons Ltd.

happen to the dung beetle following (from its perspective) an unaccelerated path. The result was that our stinky friend would accelerate towards the attracting mass at a rate matching that of gravity. Einstein's idea works!

$$\frac{d^2x}{d\tau^2} = -\Gamma^X_{AB} \, v^A \, v^B, \qquad \textit{Geodesic equation.} \qquad (11.4)$$

However, I imagined a smart student asking Einstein: *Listen Albert, this curvature effect does look the same as the gravitational force. Your idea is really cute, but my old buddy Isaac had a cute idea too. Something about an apple falling on his head. How can I know which of you two is right?*

It turns out that, in addition to the gravitational distortion to an object's movement through time (time dilation), there is gravitational distortion to its movement through space, towards or away from the gravitational source (length contraction). These combine additively in altering an object's natural path of travel, albeit the spatial component is significant only at relativistic speeds.

As a result, there are measurable differences between the theories of Einstein and Newton, differences that were observed in the deflection of starlight during an eclipse and hailed as a vindication of Einstein's theory. Anyway, nowadays we have more direct evidence because clocks are accurate enough to detect gravitational time dilation. For fun, I have interspersed in this summary some reactions to Einstein's success (Boxes 11.1, 11.2 and 11.4).

Box 11.1 Reactions to Einstein's theory of general relativity (1)

The theory of relativity of Einstein, quite apart from its validity, cannot but be regarded as a magnificent work of art. Ernest Rutherford (Nobel Prize, Chemistry).

By his theory of relativity, Albert Einstein has provoked a revolution of thought in physical science. Einstein stands above his contemporaries even as Newton did. Arthur Eddington (astronomer who made eclipse observations, see Section 5.6).

In Chapter 6, we turned to another of Einstein's brainwaves, the equivalence principle, and what it tells us about the curvature of spacetime in a vacuum. Extending from the observation that we feel no acceleration in gravitational freefall, he concluded: *now it came to me... in a gravitational field (of small spatial extension), things behave as they do in a space free of gravitation.*

I illustrated with the scenario of a dust field how the presence of energy-momentum draws together spacetime geodesics. Einstein concluded that the true relationship is between the presence of energy-momentum in a region and a reduction in the *spacetime* volume separating what would be infinitesimally close parallel spacetime geodesics in flat spacetime.

The Ricci tensor is important because it measures any change in the spacetime volume separating infinitesimally close parallel geodesics. In the presence of energy-momentum, the Ricci tensor is non-zero ($[R_{\mu\nu}] \neq 0$), while in a vacuum (ignoring the possible existence of a cosmological constant) the Ricci tensor is zero ($[R_{\mu\nu}] = 0$).

Chapter 7 examined how we measure curvature using parallel transport. We compare the change to a basis vector transported around two dimensions (say, t followed by x), versus the opposite (x followed by t). Any difference indicates curvature and is quantified in the Riemann tensor. Combining several of its components gives the Ricci tensor (the formulas are in Box 11.5). In 2-D and 3-D, it turns out that if the Ricci tensor is zero, then the Riemann tensor is zero. Ricci-0 metrics with gravitational curvature cannot exist in 2-D or 3-D. However, it *is* possible in 4-D spacetime, which means there can be gravitational curvature while respecting the equivalence principle. In the next module, we will discuss this in more detail.

In Chapter 8, we turned our attention from the left side of the Einstein Field Equations (EFEs) (curvature) to the right side (energy-momentum). I showed why and how the presence of energy-momentum is quantified as *flowing* through the four dimensions of spacetime using the (4×4) energy-momentum tensor $[T_{\mu\nu}]$. I explained why $[T_{\mu\nu}]$ is a symmetric tensor: for example, the flow of energy through the x dimension (flow of mass through t, through x) is the same as the flow of x-momentum through time (flow of mass through x, through t). I showed you that its covariant derivative $\nabla^\nu T_{\mu\nu}$ must be zero to respect local conservation of energy and momentum, an important result in deriving the EFEs.

Box 11.2 Reactions to Einstein's theory of general relativity (2)

Probably the greatest scientific discovery ever made. Paul Dirac (a pioneer of quantum theory).

One of the greatest examples of the power of speculative thought. Hermann Weyl (mathematician who helped solve Schrödinger's equation for hydrogen).

Probably the most beautiful of all existing physical theories. Lev Landau (Nobel Prize winner for work on superfluids).

We began to pull together the structure of the EFEs in Chapter 9. I highlighted the symmetries of the Ricci tensor. It is not possible to have a simple relationship such as T_{TT} affecting just R_{TT}. Altering any of the Riemann components that make up the Ricci tensor affects others. The relationship is complicated. This gave Einstein a major headache and it took him several years to find the solution.

The answer came from considering that the covariant derivative of the energy-momentum tensor $\nabla^\nu T_{\mu\nu}$ is zero. Therefore, the covariant derivative of the curvature expression on the left side of the EFEs also must be zero. To achieve this, the Ricci tensor and scalar must be combined in a specific way: $R_{\mu\nu} - \frac{1}{2}g_{\mu\nu}R$. As a result, even the simplest scenario of a stationary dust field (only $T_{TT} \neq 0$), affects all the components of the Ricci tensor. There is distortion across all the dimensions.

In Chapter 10, which I jokingly titled *The Full Story*, we built a more complete picture of how the presence of energy-momentum distorts the spacetime metric. We saw how Einstein's Weak Field metric creates geodesic distortion in line with the gravitational acceleration in Newton's theory, and how its curvature footprint approximately matches the requirements of the EFEs for both rest-mass-energy-dominated and vacuum spacetime.

We derived the value of the constant k using the Poisson equation, which relates mass density with second derivatives of the gravitational potential Φ.

At the end of the chapter, we discussed the cosmological constant Λ, which can appear in the EFEs as $-\Lambda g_{\mu\nu}$. It can be included because any covariant derivative of the metric tensor $\nabla^\alpha g_{\mu\nu}$ is zero (see Box 4.7). Einstein added it because he thought (incorrectly) that it would give a static model of the universe. He abandoned it when Hubble showed the universe is expanding. However, there is astronomical evidence that it really exists. If so, it means that there is some underlying curvature in the vacuum. In turn, that indicates some underlying source of energy that physicists call dark energy or vacuum energy. This will become relevant only when we study the development of the universe as a whole in the module on cosmology.

11.2 The Field Equations in Full Form (Finally!)

Box 11.3 Einstein Field Equations (EFEs)

$$\text{General form}: \qquad R_{\mu\nu} - \frac{1}{2}R g_{\mu\nu} = \frac{8\pi G}{c^4}T_{\mu\nu}, \qquad\qquad (11.5)$$

$$\text{Vacuum form}: \qquad R_{\mu\nu} = 0, \qquad \text{because } T_{\mu\nu} = 0, \qquad (11.6)$$

$$\text{Full form with } \Lambda: \qquad R_{\mu\nu} - \frac{1}{2}R g_{\mu\nu} - \Lambda g_{\mu\nu} = \frac{8\pi G}{c^4}T_{\mu\nu}. \qquad (11.7)$$

$R_{\mu\nu}$ is any component of the (4 × 4) Ricci tensor $[R_{\mu\nu}]$. Similarly, $T_{\mu\nu}$ and $g_{\mu\nu}$ are the corresponding components of the energy-momentum tensor $[T_{\mu\nu}]$, and the spacetime metric tensor $[g_{\mu\nu}]$. R is the Ricci scalar (a number).

Λ is the cosmological constant. G is Newton's gravitational constant. c is the speed of light.

Conventions: Metric: $< + - - - >$ Ricci tensor: $R_{AB} = R^C{}_{ACB}$

Box 11.3 summarises the EFEs. The three equations shown are the same EFEs but with slightly different emphasis. The first, Equation 11.5, has zero cosmological constant ($\Lambda = 0$). It is the one we use for modelling the curvature of regions that contain energy-momentum, such as the dust field model, because the cosmological constant is negligible except on vast scales. The second, Equation 11.6, is the EFEs for a vacuum ($[T_{\mu\nu}] = 0$) and will be our reference point when we discuss *Vacuum Curvature* in the next module. The third, Equation 11.7, recognises a possible cosmological constant, which will be relevant when we discuss the development of the universe as a whole in the module on *Cosmology*.

All three versions reflect the relationship between spacetime curvature (on the left) and the presence of energy-momentum (on the right). Don't forget that the EFEs are actually 16 different equations, 10 of which are independent. This being said, the ones most familiar to students will be those containing the diagonal terms such as T_{TT} and T_{XX}. The former is particularly important because, typically, the energy-momentum tensor of a region containing non-relativistic matter will be dominated by the T_{TT} term, which contains rest mass energy.

The EFE's are *covariant*, which means that they work in any coordinates. Whether you work with the Cartesian coordinates I have tended to use (ct, x, y, z), the spherical coordinates (ct, r, θ, ϕ), which we will be using in the next module, or any other weird coordinate system, the EFEs always apply. This is because all tensors transform in the same way. Switching coordinates transforms the tensors on one side of the EFEs, but at the same time it transforms those on the other side in exactly the same way.

11.3   Why Does Energy-Momentum Distort Spacetime?

Why does energy-momentum distort the spacetime metric $[g_{\mu\nu}]$? I again admit that I cannot provide a philosophical justification. Without doubt, my best answer is that this is just the way it is. However, in Section 9.1, I explained that energy-momentum and spacetime are not independent quantities.

In Section 2.8, we derived the relationship between mass, energy and momentum directly from the Minkowski metric. Furthermore, the energy-momentum tensor, which is a key variable in the EFEs, measures the flow of energy-momentum through spacetime. Clearly, it is affected by the spacetime metric. Indeed later (in Chapter 14), I will show you that the definitions of energy and momentum as conserved quantities entirely depend on the structure of $[g_{\mu\nu}]$.

While this doesn't answer the fundamental question, it points to tight links between energy-momentum and the structure of spacetime. This is the best I can do. If you remain unconvinced, remember that the laws of physics don't have to explain themselves!

11.4   Conventions (Optional)

You need to read the following sections on *Conventions* only if you will be comparing the contents of this book with other textbooks or lecture notes. In any comparison, there lurks a problem in spite of the covariant flexibility of the EFEs. I am sorry to say that physicists have done all they can to mess things up by being unable to agree on basic conventions! This will lead to much gnashing of teeth and tearing of hair if you are not careful. You are going to find time and time again that there is a $+/-$ discrepancy between texts or the odd factor of c missing. Aarrgghh!

My solution is two-fold. I try my utmost to highlight the conventions that I use (bottom of Box 11.3). I also offer you the following description of the convention problem in the hope that it will save you some frustration.

Box 11.4 Reactions to Einstein's theory of general relativity (3)

The foundation of general relativity appeared to me then, and still does as the greatest feat of human thinking about Nature, the most amazing combination of philosophical penetration, physical intuition and mathematical skill. Max Born (Nobel Prize for work on quantum theory).

One of the highest achievements of human thought, although I have to confess that no one has yet succeeded in stating in clear language what the theory of Einstein really is. J J Thomson (Nobel Prize for discovering the electron).

11.4.1 Conventions (1): Metric Signature

The choice of metric signature is the biggest convention wobbler. The square of time dt^2 and space dx^2 offset each other in the invariant interval equation (Section 3.1), so we need to treat one as positive and one as negative. Broadly speaking, I sense a favouritism for $< + - - - >$ (time positive) among quantum/particle physicists and a bias towards $< - + + + >$ (space positive) for those who focus on general relativity, but you will find a healthy mixture of opinion everywhere, so watch out. I have chosen to use $< + - - - >$ (time positive, space negative) because this matches my first book, *Quantum Untangling*. It is often called the West-coast convention, versus the $< - + + + >$ East-coast option. I guess I am a laid back West-coast kind of guy.

Let's start with good news. Changing the metric signature changes the sign of $[g_{\mu\nu}]$ and $[g^{\mu\nu}]$ (it reverses the sign of all the components). However, this does *not* change the Christoffel symbols (the change in sign cancels out in the calculation). Therefore, it does *not* affect the value of the components of the Riemann tensor or the sign of the Ricci tensor $[R_{\mu\nu}]$ in the EFEs.

Also $[T_{\mu\nu}]$ is typically unaffected because, by convention (another one), the T_{TT} component is almost always shown positive, such as $T_{TT} = \rho_E$ (so that rest mass energy is positive). On rare occasions when the metric signature is $< - + + + >$, you may come across $T_{TT} = -\rho_E$ to match (energy negative). In this case, the right side of the EFEs becomes $-kT_{\mu\nu}$. However, let me stress that this is rare.

So what does a change in metric signature affect? The total term $-\frac{1}{2}Rg_{\mu\nu}$ does not change. However, the sign of the Ricci scalar R changes. It is calculated as $g^{\mu\nu}R_{\mu\nu}$. As noted earlier, the Ricci tensor components such as R_{TT} do not change, but the inverse metric tensor terms such as g^{TT} do change, so the sign of R must change.

Switching the metric signature changes the minus sign in front of the cosmological constant, from $-\Lambda g_{\mu\nu}$ for $< + - - - >$ to $+\Lambda g_{\mu\nu}$ for $< - + + + >$. This is because it is convention to associate a *positive* value of Λ with an *accelerating* effect on the expansion of the universe. You might feel there is a slight inconsistency here. Both the R and Λ terms are number multiples of $g_{\mu\nu}$ in the EFEs. When you switch metric signature, the actual value of R changes sign, but the value of Λ does not. You just switch from the $-$ to $+$ sign in front of Λ. Umm... err....

Let's summarise this SNAFU. If you switch metric signature from $< + - - - >$ to $< - + + + >$, $[R_{\mu\nu}]$ does not change and nor do $[T_{\mu\nu}]$ or k unless you are a weirdo. The overall sign in front of the term $-\frac{1}{2}Rg_{\mu\nu}$ stays the same because the sign of R changes and is offset by the change in sign of $g_{\mu\nu}$. The overall sign in front of the cosmological constant term changes from $-\Lambda g_{\mu\nu}$ to $+\Lambda g_{\mu\nu}$ because we refuse to change the actual value of Λ and must offset the sign change in $g_{\mu\nu}$. All clear? Stop crying and don't shout. This is convention. Don't blame me.

11.4.2 Conventions (2): Definition of Curvature Tensors

We are not done with this mess yet. There can be different conventions for the measures of curvature. Box 11.5 summarises the definitions of curvature in this book. Pretty much everybody seems to agree about Equations 11.8 and 11.9 as the definition of the Christoffel symbol and Riemann tensor, but if you find a $+/-$ discrepancy in another book, please check these definitions. There *is* some variation in the definition of the Ricci tensor. The version in Equation 11.10 has become prevalent, but in a few texts the definition is $R_{AB} = R^C_{ABC}$, which *changes the sign* of the Ricci tensor and Ricci scalar with knock-on effects on other signs in the EFEs (I am told this is the case in Rindler's well-known textbook).

Box 11.5 Curvature definitions in this book

$$\Gamma^A_{BC} = \frac{1}{2}g^{AD}\left(\frac{\partial g_{DC}}{\partial B} + \frac{\partial g_{DB}}{\partial C} - \frac{\partial g_{BC}}{\partial D}\right), \qquad \text{Christoffel symbol.} \qquad (11.8)$$

$$R^A_{BCD} = \partial_C\Gamma^A_{BD} - \partial_D\Gamma^A_{BC} + \Gamma^A_{CE}\Gamma^E_{BD} - \Gamma^A_{DE}\Gamma^E_{BC}, \qquad \text{Riemann tensor.} \qquad (11.9)$$

$$R_{AB} = R^C_{ACB}, \qquad \text{Ricci tensor.} \qquad (11.10)$$

11.4.3 Conventions (3): Definition of Energy-Momentum Tensor

The other thing to watch out for is the choice of units for $[T_{\mu\nu}]$. I use energy-density. However, in some textbooks you may see components of $[T_{\mu\nu}]$ written in terms of mass density, such as $T_{00} = \rho_m$. Mass density is smaller by a factor of c^2. As a result the value of k is larger by a factor of c^2 to compensate. If $[T_{\mu\nu}]$ is measured in terms of mass density, then $k = \frac{8\pi G}{c^2}$ instead of $k = \frac{8\pi G}{c^4}$.

11.5 Stationary Action Derivation of the EFEs

Leaving the horror of conventions behind us, I want to make you aware that there is an alternative approach to deriving the EFEs that uses the principle of *stationary action* or *least action*. Let me warn you that it requires a decent background knowledge of physics. It is not intuitive for a beginner and involves some tough advanced calculations. I am not a fan myself, but I know some who swear by it (yes, Gareth Williams, I am referring to you).

The theory of least or stationary action is broadly applied in classical physics. Let me introduce it using the flight of a cannonball through the air (this actually was one of the original applications). The concept is that a stable repeatable pattern indicates some measure along the path is being maximised or minimised (in this case minimised). Why? Because something is holding the cannonball to that pathway. Any variation away from it must be detrimental on some measure, or the pattern would not be stable. The value at each point on the path is called the *Lagrangian*, which in the case of the cannonball is the difference between the ball's kinetic and potential energy (*k.e. − p.e.*). You integrate the Lagrangian over the pathway to get the total *action*, which is at a minimum for the stable pathway. From the action, you can use what is called the *Euler–Lagrange equation* to derive the cannonball's equations of motion.

The cunning trick is to use this concept to derive the EFEs. I will not go through the maths, but I will explain the basics. Let's start with a region of vacuum spacetime. What is the metric $[g_{\mu\nu}]$ of this vacuum region? Assuming it is a stable solution, we infer that it is at a minimum or maximum for some measure (the action). If this were not the case, the solution for $[g_{\mu\nu}]$ would not be stable.

But what is this measure of action that gives a stable solution from which $[g_{\mu\nu}]$ will not vary? The conventional approach is to assume it is the *Einstein–Hilbert action* shown as Equation 11.11 and then to prove this leads to the EFEs, which is not an entirely satisfactory argument. However, there is some intuitive logic behind Equation 11.11. The action must be an integral over d^4x (this symbol means the integration is over all four dimensions with time and space treated equally), and $\sqrt{|g|}$ is needed as a normalisation factor ($|g|$ is the determinant of the matrix $[g_{\mu\nu}]$) to match up the size of the spacetime volume element in curved coordinate systems (trust me on this). These elements are a must.

$$\int R\sqrt{|g|}\, d^4x, \qquad \textit{Hilbert–Einstein action.} \tag{11.11}$$

Therefore, the only query relates to R, the Ricci scalar. Why R? Well, that term should be a scalar so that we are integrating a *number* over each point in spacetime. If it were not a number, it would be coordinate-dependent and the physics would be affected by coordinate choice, which is unattractive (we want a covariant solution). Also, the scalar must depend only on $[g_{\mu\nu}]$ because the spacetime metric fully defines the spacetime. The simplest scalar that depends solely on the metric is the Ricci scalar R, which is a mix of Riemann tensor components that are, in turn, a complicated mix of second derivatives of $[g_{\mu\nu}]$ (see Chapter 7).

Hey presto, you have the Einstein–Hilbert action! You take this action, apply Euler–Lagrange, go through some tough calculations (well, I think they are very tough) and arrive at the EFEs for a vacuum. You can incorporate additional terms into the Einstein–Hilbert action for the presence of energy-momentum and even the cosmological constant. So there you are. You can get the rest on the Web. Umm... err... as I said, probably not for beginners (but you may be interested to check out Box 11.6 for Hilbert's role in developing the EFEs).

While I am not a big fan of using the Einstein–Hilbert action to derive the EFEs, it is worth taking a moment to think about its simplicity. The Einstein–Hilbert action is much simpler than the somewhat complicated structure of the EFEs. This simplicity adds further beauty to Einstein's extraordinary theory.

Let me remind you again that neither the EFEs nor the Einstein–Hilbert action mean there is a unique solution for the metric of a region with a certain density of energy-momentum in it. Not at all! Boundary conditions are important (i.e. the metric of spacetime surrounding the region). An obvious example is the metric of spacetime in a vacuum ($[R_{\mu\nu}] = 0$), which can vary from flat Minkowski spacetime to the strong gravitational curvature of the vacuum near a black hole.

Box 11.6 Should they be HFEs not EFEs?

Should we be discussing the Hilbert Field Equations? It is not disputed that David Hilbert (1862–1943) developed the Einstein–Hilbert action and published it along with his work on the EFEs five days before Einstein made his official announcement. Consider the following sequence of events in November 1915:

- *November 11th*: Einstein submits an incorrect version of EFEs to the Prussian Academy.
- *November 18th*: Einstein has an advance copy of Hilbert's paper but claims he had already discovered the correct EFEs (having submitted incorrect ones only a week earlier).
- *November 20th*: Hilbert presents his equations at Göttingen Academy.
- *November 25th*: Einstein finally submits a correct version of the EFEs.

Things get weirder. Although Einstein on November 18th appears to acknowledge that Hilbert's paper contained the EFEs, we find that Hilbert's original paper has been mutilated by someone cutting off the key section that would show the EFEs. Very suspicious! In spite of this, Hilbert never contested Einstein's priority and, in fairness, the whole idea of general relativity was the output of Einstein's brain and his 10 years of concentrated effort. So, whether or not Hilbert was the first to derive the detailed EFEs, I feel Einstein fully deserves the credit he is given.

11.6 Final Thoughts on This Module

You now have a full description of Einstein's theory. I hope you feel you have a good grasp of it, and that I have delivered on the promise of the book's title (if you do, please leave a review for others). For those readers who want more, I offer the next two modules, which probe further into the subject than you might expect in this sort of book.

The modules cover *Vacuum Curvature* (with a focus on black holes) and *Cosmology*. Some of the topics covered are very advanced. To keep the material accessible, I offer intuitive overviews alongside *optional* sections containing mathematical detail for those who can cope with it. Even so, there may be pieces that you find challenging. If and when this happens, please bear in mind that many of these topics would only appear in a postgraduate physics syllabus.

If you decide to read on, I must address a teeny-weeny problem with the EFEs. Indeed they are beautiful, but for the most part they are insoluble! The field equations are horribly tricky because they are non-linear. Einstein restricted himself to working with approximations. He was most surprised when, a month after publishing his theory, a front-line soldier, Karl Schwarzschild, presented him with an exact solution, albeit for the simplest case where $[T_{\mu\nu}] = 0$ and $\Lambda = 0$ (gravitational curvature in a vacuum with no cosmological constant).

Imagine that you are lucky enough to know the value of every one of the components in the energy-momentum tensor $[T_{\mu\nu}]$. This tells you all you need to know about the right side of the field equations. You want to know what this means in terms of the structure of spacetime, so you need to calculate the value of the components in the spacetime metric tensor $[g_{\mu\nu}]$. Easy? No! There is no easy (or even difficult) route to calculating $[g_{\mu\nu}]$.

Take a look at the terms that make up the left side of the field equations (shown in Box 11.3). As you have seen, the value of $[R_{\mu\nu}]$ is based on second derivatives of a bunch of components in the $[g_{\mu\nu}]$ tensor, the value of R is based on a different bunch of its second derivatives and, on top of this, the weighting of the component $R_{\mu\nu}$ versus R depends on components of $[g_{\mu\nu}]$, which is what we are trying to calculate. It is a spaghetti-mess that makes it virtually impossible to get an exact picture of what is happening.

How is it then that Einstein's theory can teach us so much? The answer is that we can use approximations and focus on very simple scenarios. For example, Schwarzschild's solution used the simplest scenario ($[T_{\mu\nu}] = 0$ so $[R_{\mu\nu}] = 0$) *and* perfect spherical symmetry (point-like stationary uncharged mass). Although idealised, it tells us a huge amount about how gravity and spacetime curvature works. This and other vacuum solutions will be our focus in the next module.

I end this chapter with Box 11.7. Not everybody was a fan of Einstein. In Germany in the 1930s, there were, in some quarters, strong feelings against him and his theories... Nazi sentiment against the theories of the most famous scientist of the era... who was a Jew... go figure.

Box 11.7 Not yet hanged

The Nazi movement described relativity as *Jewish physics* and organised conferences and events to counter Einstein's theory. After all, there is no more effective way of making a scientific argument than holding a good old book burning. In 1931, they published *One Hundred Authors Against Einstein*. Reputedly, Einstein quipped: *If I were wrong, it would take only one.*

In the following years, Einstein's bank accounts were seized, and a Nazi magazine printed an article with a photo of him under a title *Not Yet Hanged*! Then, in August 1933, assassins shot one of his colleagues, Theodor Lessing, who had also appeared in that article.

By this time, Einstein had fled from Germany. With a price on his head, he was forced to hide in a holiday hut near Cromer on the east coast of England. There he stayed, guarded by armed locals, until papers came through allowing him to travel further away from the Nazi menace. Before leaving for the United States, he told a reporter:

I could not believe that it was possible that such spontaneous affection could be extended to one who is a wanderer on the face of the earth. The kindness of your people has touched my heart so deeply that I cannot find words to express in English what I feel. I shall leave England for America at the end of the week, but no matter how long I live, I shall never forget the kindness which I have received from the people of England.

11.7   Module Memory Jogger

Below is a list of some of the key topics we have covered so far. Hopefully, it can act as a memory-jogger if you look back at this book some day in the dim and distant future. And Box 11.8 provides details of some further resources that you might find helpful.

- *A moving object's time changes (ageing more slowly).*
- *A moving object's length contracts in the direction of motion.*
- *A moving object's clocks fall out of sync (leading clocks lag).*
- *Invariant interval of Minkowski spacetime:* $c^2 \, d\tau^2 = c^2 \, dt^2 - dx^2 - dy^2 - dz^2.$
- *Energy-momentum relationship:* $m^2 c^4 = E^2 - p_x^2 c^2 - p_y^2 c^2 - p_z^2 c^2.$
- *Minkowski metric* (ct, x, y, z): $[g_{\mu\nu}] = \begin{bmatrix} 1 & 0 & 0 & 0 \\ 0 & -1 & 0 & 0 \\ 0 & 0 & -1 & 0 \\ 0 & 0 & 0 & -1 \end{bmatrix}$, *often labelled* $[\eta_{\mu\nu}]$.
- *Index lowering:* $g_{\mu\nu} V^\mu = V_\nu$, *Index raising:* $g^{\mu\nu} V_\mu = V^\nu.$
- *Covariant derivative of* $\vec{V}^x$ *with respect to x:* $\nabla_x \vec{V}^x = \frac{\partial V^x}{\partial x} \vec{e_x} + V^x \frac{\partial \vec{e_x}}{\partial x}.$
- *Covariant derivative of metric tensor* $\nabla^\alpha g_{\mu\nu}$ *is zero.*
- *Christoffel symbol:* $\Gamma^A_{BC} = \frac{1}{2} g^{AD} \left(\frac{\partial g_{DC}}{\partial B} + \frac{\partial g_{DB}}{\partial C} - \frac{\partial g_{BC}}{\partial D} \right).$

- *Christoffel cheat sheet for diagonal metrics:* $A \neq B \neq C$,

$$\Gamma^A_{AA} = \frac{1}{2g_{AA}}\left(\frac{\partial g_{AA}}{\partial A}\right), \qquad \Gamma^A_{BB} = -\frac{1}{2g_{AA}}\left(\frac{\partial g_{BB}}{\partial A}\right),$$

$$\Gamma^A_{AB} = \Gamma^A_{BA} = \frac{1}{2g_{AA}}\left(\frac{\partial g_{AA}}{\partial B}\right), \qquad \Gamma^A_{BC} = 0.$$

- *Geodesic equation:* $\frac{d^2x}{d\tau^2} = \frac{dv^x}{d\tau} = -\Gamma^X_{\mathbf{AB}}\frac{d\mathbf{A}}{d\tau}\frac{d\mathbf{B}}{d\tau} = -\Gamma^X_{\mathbf{AB}} v^{\mathbf{A}} v^{\mathbf{B}}.$
- *Riemann tensor:* $R^A_{\ BCD} = \partial_C \Gamma^A_{BD} - \partial_D \Gamma^A_{BC} + \Gamma^A_{\mathbf{CE}}\Gamma^{\mathbf{E}}_{BD} - \Gamma^A_{\mathbf{DE}}\Gamma^{\mathbf{E}}_{BC}.$
- *Ricci tensor (4-D), example:* $R_{TT} = R^T_{\ TTT} + R^X_{\ TXT} + R^Y_{\ TYT} + R^Z_{\ TZT}.$
- *Ricci scalar (4-D), diagonal metric* $R = g^{TT}R_{TT} + g^{XX}R_{XX} + g^{YY}R_{YY} + g^{ZZ}R_{ZZ}.$
- *Spacetime curvature in a vacuum is Ricci-0 (ignoring a possible cosmological constant).*
- *Energy-momentum tensor of a dust field (natural units, $c = 1$).*

$$[T^{\mu\nu}] = \begin{bmatrix} E\ flow\ (t,x,y,z) \\ p_x\ flow\ (t,x,y,z) \\ p_y\ flow\ (t,x,y,z) \\ p_z\ flow\ (t,x,y,z) \end{bmatrix} = \rho_m \begin{bmatrix} u^t u^t & u^t u^x & u^t u^y & u^t u^z \\ u^x u^t & u^x u^x & u^x u^y & u^x u^z \\ u^y u^t & u^y u^x & u^y u^y & u^y u^z \\ u^z u^t & u^z u^x & u^z u^y & u^z u^z \end{bmatrix} = \rho_m\, u^{\mu}u^{\nu},$$

$$\Rightarrow stationary\ [T^{\mu\nu}] = stationary\ [T_{\mu\nu}] = \begin{bmatrix} \rho_E & 0 & 0 & 0 \\ 0 & 0 & 0 & 0 \\ 0 & 0 & 0 & 0 \\ 0 & 0 & 0 & 0 \end{bmatrix}.$$

- *Covariant derivative of energy-momentum tensor* $\nabla^{\nu} T_{\mu\nu}$ *is zero.*
- *Einstein Weak Field metric:*

$$[g_{\mu\nu}] = \begin{bmatrix} 1 + \frac{2\Phi}{c^2} & 0 & 0 & 0 \\ 0 & -\left(1 - \frac{2\Phi}{c^2}\right) & 0 & 0 \\ 0 & 0 & -\left(1 - \frac{2\Phi}{c^2}\right) & 0 \\ 0 & 0 & 0 & -\left(1 - \frac{2\Phi}{c^2}\right) \end{bmatrix}.$$

- *Poisson equation:* $\nabla^2\Phi = 4\pi G\rho_m$.
- *Einstein Field Equations (full form):* $R_{\mu\nu} - \frac{1}{2}R g_{\mu\nu} - \Lambda g_{\mu\nu} = \frac{8\pi G}{c^4}T_{\mu\nu}.$
- *Einstein–Hilbert action:* $\int R\sqrt{|g|}\,d^4x.$
- *Module Conventions in this book:*

Metric: $< + - - - >$ \qquad Ricci tensor: $R_{AB} = R^{\mathbf{C}}_{\ ACB}$

Box 11.8 Module 1: further resources (web links as at February 2025)

Check out Leonard Susskind's online course on general relativity. It has less mathematical detail than this book, but moves at a leisurely pace and gives a helpful overview on many topics:
https://theoreticalminimum.com/courses/general-relativity/2012/fall

For more mathematical rigour and detail, there are Sean Carroll's lecture notes. In my opinion, they require a particularly strong background in mathematics:
https://preposterousuniverse.com/wp-content/uploads/grnotes-one.pdf

If you want more detail on how the mathematics of curvature works (the Riemann tensor, Ricci tensor, Ricci scalar), I recommend Eigenchris's videos:
https://www.youtube.com/playlist?list=PLJHszsWbB6hqlw73QjgZcFh4DrkQLSCQa

Module II

Vacuum Curvature

Chapter 12

The Schwarzschild Metric: Derivation

In this module, we will be looking at the curvature of spacetime in a vacuum. In order to have gravitational effects, there must be spacetime curvature, because that is what causes the gravitational effect. This means that the Riemann tensor cannot be zero, because it picks up all and any spacetime curvature (Section 7.2). However, if one ignores the tiny effect from a possible cosmological constant, then in a vacuum the Ricci tensor must be zero (see Section 6.4 if this is not clear). Therefore, for gravitational effects in a vacuum around a mass, such as an orbiting satellite, the spacetime of the vacuum must have *non-zero* Riemann tensor but *zero* Ricci tensor (which I label Ricci-0). This constrains the possible structure of the spacetime metric $[g_{\mu\nu}]$.

In this chapter, I will show that if we add in the requirements that the metric is time-independent and has spherical symmetry, there turns out to be only one possible solution, the *Schwarzschild metric* (see Box 12.1 for some history).

The Schwarzschild derivation that I have developed for this chapter is different from that used in other texts. The initial focus is on symmetries. This approach has two benefits. First, and most importantly, it will help you see intuitively the huge constraints that the symmetries put on possible solutions (indeed, there is only one possible solution). Second, the maths is simpler, which appeals to a simple chap like me.

Let's start with the basics. Schwarzschild looked for a solution to the EFEs that had the following characteristics:

1. Vacuum solution and therefore Ricci-0: $[R_{\mu\nu}] = 0$ (cosmological constant $\Lambda = 0$).
2. Gravitational effect present, so the Riemann tensor is non-zero.
3. Static and stationary solution (not spinning): metric is not affected by changes in time t coordinate or by time reversal (i.e. no change if dt becomes $-dt$).
4. Spherically symmetrical: in spherical coordinates $[g_{\mu\nu}]$ depends only on the value of r, and is not affected by changes to the angular coordinates θ or ϕ.

You may be wondering what sort of scenario is described by these constraints. The first three of the constraints apply to the gravitational field surrounding any stationary massive object (note that the cosmological constant Λ is negligible except at cosmological scales). The fourth requires that the object is spherically symmetrical and not spinning (I will address spinning objects in Chapter 17). Therefore, the Schwarzschild metric describes conditions surrounding planets, stars and even black holes. In reality, everything in astrophysics is spinning, but the Schwarzschild metric provides a good, albeit not perfect, description for most scenarios (although later we will need to use the Kerr metric to account for the rapid spin of black holes).

Untangling General Relativity: The Intuitive Self-Study Guide, First Edition. Simon Sherwood.
© 2026 John Wiley & Sons Ltd. Published 2026 by John Wiley & Sons Ltd.

Box 12.1   Black shield

Karl Schwarzschild (1873–1916) was a German astronomer. He worked with Hilbert and Minkowski before signing up at the start of the First World War and heading to the Eastern Front as an artillery officer. In spite of freezing conditions, he stayed in touch with scientific developments and received a copy of Einstein's paper on the Einstein Field Equations (EFEs) in December 1915, a month after its publication.

Within a few weeks, he had found the exact spherically symmetrical solution that bears his name. He sent it to Einstein in late December 1915, commenting: *The war treated me kindly enough, in spite of the heavy gunfire, to allow me to get away from it all and take this walk in the land of your ideas.*

Einstein was surprised by the simplicity of the solution and responded: *I had not expected that the exact solution to the problem could be formulated so simply. Your analytic treatment of the problem appears to me splendid.*

Schwarzschild was already ill with pemphigus, an incurable skin disease. He was invalided back from the front in March 1916 and died 2 months later, never knowing the importance of his discovery. One exciting implication of his metric is the existence of black holes. As fate would have it, the name Schwarzschild translates into English as: *black shield.*

12.1   Metric Symmetries: A Diagonal Metric

As we are investigating a spherically symmetric solution to the EFEs, it will not surprise you that we will be using spherical coordinates (ct, r, θ, ϕ). For a reminder of how spherical coordinates work, check back to Section 9.2, which contains Figure 9.1. The first important step in our analysis is to establish that Schwarzschild's symmetry requirements give a diagonal metric in spherical coordinates (no off-diagonal terms). This is important because the maths of metrics with off-diagonal terms is much more complicated (see, for example, the calculation of Christoffel symbols in Subsection 4.2.2).

Let's consider what factors we can vary in the Minkowski spacetime metric while respecting the symmetries we desire. The invariant interval of the Minkowski metric is shown in spherical coordinates as Equation 12.1 (the same as on the right of Figure 9.1 earlier).

$$(c\,d\tau)^2 = (c\,dt)^2 - dr^2 - r^2\,d\theta^2 - r^2\sin^2\theta\,d\phi^2, \qquad Minkowski\ metric. \qquad (12.1)$$

Suppose we include an off-diagonal component in the metric involving $d\theta$. This would result in an additional term in the invariant interval: $dt\,d\theta$, $dr\,d\theta$ or $d\theta\,d\phi$. This would not respect spherical symmetry because, under spherical symmetry, it makes no difference to the metric if we consider a movement in the $+\theta$ or $-\theta$ direction. They are equivalent. Thus, we must be able to change from $+d\theta$ to $-d\theta$ without changing the metric. This change doesn't affect the diagonal components: dt^2, dr^2, $d\theta^2$ and $d\phi^2$ (the $d\theta^2$ term is squared and so is unchanged). However if there were, for example, a $dr\,d\theta$ term, the sign would be changed, and that would change the metric. An identical argument (changing $+d\phi$ to $-d\phi$) disallows any off-diagonal component that includes $d\phi$.

Thus, spherical symmetry eliminates the possibility of any off-diagonal component that includes $d\theta$ or $d\phi$. That leaves only the possible $dt\,dr$ off-diagonal element. However, we can eliminate this because the metric is static. As there is nothing spinning or moving that would be affected by the direction of the flow of time, the invariant interval should not change if we switch dt to $-dt$. This would switch any $dt\,dr$ element in the invariant interval to $-dt\,dr$, which would change the metric. That would not match a static solution.

There is an additional intuitive step to simplify our search. Consider the spacetime interval between two points at constant t and r. They have the same value of t and r, and sit on a sphere radius r away from the origin. As there is spherical symmetry, we can always choose angular coordinates such that the spacetime interval between them is still $(-r^2\,d\theta^2 - r^2\sin^2\theta\,d\phi^2)$. The presence of a distant mass may affect dr, but it cannot change this relationship without destroying the spherical symmetry.[1]

---

1 You can find rigorous proofs that this is the general form of such a metric in textbooks such as *A First Course in General Relativity*, Bernard Schutz pages 256–8.

To summarise, the spherical symmetry and static nature of the metric allow us to eliminate all possible off-diagonal elements ($dt\,dr$, $dt\,d\theta$, $dt\,d\phi$, $dr\,d\theta$, $dr\,d\phi$, $d\theta\,d\phi$). The spherical symmetry also dictates the form of the $g_{\theta\theta}$ and $g_{\phi\phi}$ elements. The result is shown in Equation 12.2. In addition, we know that both g_{TT} and g_{RR} are only r dependent. They cannot depend on the values of θ or ϕ because of spherical symmetry, and they cannot depend on time t because the metric is stationary and static. The next step in our challenge is to figure out the values of g_{TT} and g_{RR}... but we are not through with symmetries yet.

$$\text{Static with spherical symmetry}: [g_{\mu\nu}] = \begin{bmatrix} g_{TT} & 0 & 0 & 0 \\ 0 & g_{RR} & 0 & 0 \\ 0 & 0 & -r^2 & 0 \\ 0 & 0 & 0 & -r^2\sin^2\theta \end{bmatrix}. \tag{12.2}$$

12.2 Ricci and Riemann Symmetries

The Ricci and Riemann symmetries are going to be a great help. Let me remind you of the underlying symmetries, which I describe as curvature footprints, that we discussed in Chapter 9. Equation 12.3 lays out the curvature symmetries common to all 4-D diagonal metrics. This is based on the Riemann symmetry $g^{AA}R^{B}_{\ ABA} = g^{BB}R^{A}_{\ BAB}$ (see Equation 7.3).

$$\textit{Underlying Symmetries for any 4-D Diagonal Metric} \tag{12.3}$$

$$
\begin{aligned}
g^{00}R_{00} &= g^{00}R^{0}_{\ 000} + g^{00}R^{1}_{\ 010} + g^{00}R^{2}_{\ 020} + g^{00}R^{3}_{\ 030} = & 0 & +a & +b & +c, \\
g^{11}R_{11} &= g^{11}R^{0}_{\ 101} + g^{11}R^{1}_{\ 111} + g^{11}R^{2}_{\ 121} + g^{11}R^{3}_{\ 131} = & a & +0 & +d & +e, \\
g^{22}R_{22} &= g^{22}R^{0}_{\ 202} + g^{22}R^{1}_{\ 212} + g^{22}R^{2}_{\ 222} + g^{22}R^{3}_{\ 232} = & b & +d & +0 & +f, \\
g^{33}R_{33} &= g^{33}R^{0}_{\ 303} + g^{33}R^{1}_{\ 313} + g^{33}R^{2}_{\ 323} + g^{33}R^{3}_{\ 333} = & c & +e & +f & +0.
\end{aligned}
$$

In Section 9.6, I showed you that for vacuum curvature we know $[R_{\mu\nu}] = 0$, which means for our diagonal metric: $R_{00} = R_{11} = R_{22} = R_{33} = 0$. This leads to the curvature footprint shown again as Equation 12.4. This was shown earlier as Equation 9.32 where you can find the supporting maths if you need a reminder.

$$\textit{Underlying Symmetries of Vacuum Curvature (Diagonal metric)} \tag{12.4}$$

$$
\begin{aligned}
g^{00}R^{0}_{\ 000} + g^{00}R^{1}_{\ 010} + g^{00}R^{2}_{\ 020} + g^{00}R^{3}_{\ 030} = & \quad 0 & -b-c & +b & +c & = 0, \\
g^{11}R^{0}_{\ 101} + g^{11}R^{1}_{\ 111} + g^{11}R^{2}_{\ 121} + g^{11}R^{3}_{\ 131} = & -b-c & +0 & +c & +b & = 0, \\
g^{22}R^{0}_{\ 202} + g^{22}R^{1}_{\ 212} + g^{22}R^{2}_{\ 222} + g^{22}R^{3}_{\ 232} = & \quad b & +c & +0 & -b-c & = 0, \\
g^{33}R^{0}_{\ 303} + g^{33}R^{1}_{\ 313} + g^{33}R^{2}_{\ 323} + g^{33}R^{3}_{\ 333} = & \quad c & +b & -b-c & +0 & = 0.
\end{aligned}
$$

With Schwarzschild, we work in spherical coordinates so $(0, 1, 2, 3)$ is (ct, r, θ, ϕ). Let me remind you again that I use capital letters for the tensor coordinates (such as T for t) because they are easier to read.

Spherical symmetry tells us that (measured on the same scale), curvature over the θ and ϕ directions must be equivalent. Bear in mind that we could switch the θ or ϕ directions by rotating the sphere. This means $R^{\theta}_{\ T\theta T} = R^{\phi}_{\ T\phi T}$. The first has one upper and one lower θ index, and the second has one upper and one lower ϕ index, which means they are the same scale in units of invariant interval (see Box 12.2).

Box 12.2   Coordinate symmetry and components of the Riemann tensor

We can use coordinate symmetries to simplify the calculation of the Riemann components that make up the Ricci tensor. For example, in the case of the Schwarzschild metric, there is spherical symmetry between the θ and ϕ coordinates, so $R^{\theta}{}_{T\theta T} = R^{\phi}{}_{T\phi T}$ and $R^{\theta}{}_{R\theta R} = R^{\phi}{}_{R\phi R}$.

Each Riemann tensor component quantifies the distortion from parallel transport of a basis vector (Section 7.2). The useful thing about the Riemann tensor is that it measures real underlying curvature and is not fooled by weird or wonky coordinates. Therefore, given the spherical symmetry of the Schwarzschild metric, the distortion must be the same whether assessed in the θ or ϕ coordinate directions, *providing* the results are scaled to be equivalent (measured in units of the invariant interval).

$R^{\theta}{}_{T\theta T}$ measures the distortion of the t basis vector in the θ direction from parallel transport around the t and θ axes (see Figure 7.3). Let's focus on the measurements relating to the θ coordinate. Parallel transport along the θ axis is scaled with dual coordinates (lower index), and then the resulting distortion in the θ direction is scaled with vector coordinates (upper index). This combination of one lower and one upper index means the magnitude matches the invariant interval (see Section 3.6).

The same is true for the ϕ coordinate in $R^{\phi}{}_{T\phi T}$. As this measurement matches that of $R^{\theta}{}_{T\theta T}$ in terms of the invariant interval and there is spherical symmetry between θ and ϕ, we know the distortion must be the same, so $R^{\theta}{}_{T\theta T} = R^{\phi}{}_{T\phi T}$ and similarly $R^{\theta}{}_{R\theta R} = R^{\phi}{}_{R\phi R}$.

If you struggle with this intuitively, do not worry, I will prove it later. In Equation 12.4, $R^{\theta}{}_{T\theta T} = R^{\phi}{}_{T\phi T}$ is $R^{2}{}_{020} = R^{3}{}_{030}$, which means $b = c$. Substituting this into Equation 12.4 gives the curvature footprint for a spherically symmetrical diagonal metric shown as Equation 12.5. I have labelled $b = c$ as S (I have chosen S for spherical).

Spherically Symmetric Vacuum Solution (Diagonal metric) $\hspace{2cm}$ (12.5)

$$g^{TT}R_{TT} = \quad g^{TT}R^{T}{}_{TTT} + g^{TT}R^{R}{}_{TRT} + g^{TT}R^{\theta}{}_{T\theta T} + g^{TT}R^{\phi}{}_{T\phi T} = \quad 0 \quad -2S \quad +S \quad +S \quad = 0,$$

$$g^{RR}R_{RR} = \quad g^{RR}R^{T}{}_{RTR} + g^{RR}R^{R}{}_{RRR} + g^{RR}R^{\theta}{}_{R\theta R} + g^{RR}R^{\phi}{}_{R\phi R} = \quad -2S \quad +0 \quad +S \quad +S \quad = 0,$$

$$g^{\theta\theta}R_{\theta\theta} = \quad g^{\theta\theta}R^{T}{}_{\theta T\theta} + g^{\theta\theta}R^{R}{}_{\theta R\theta} + g^{\theta\theta}R^{\theta}{}_{\theta\theta\theta} + g^{\theta\theta}R^{\phi}{}_{\theta\phi\theta} = \quad S \quad +S \quad +0 \quad -2S \quad = 0,$$

$$g^{\phi\phi}R_{\phi\phi} = \quad g^{\phi\phi}R^{T}{}_{\phi T\phi} + g^{\phi\phi}R^{R}{}_{\phi R\phi} + g^{\phi\phi}R^{\theta}{}_{\phi\theta\phi} + g^{\phi\phi}R^{\phi}{}_{\phi\phi\phi} = \quad S \quad +S \quad -2S \quad +0 \quad = 0.$$

Let me briefly remind you of the symmetries that this reflects. It is the pattern of curvature for a diagonal spherically symmetrical metric that is a vacuum ($[R_{\mu\nu}] = 0$). Do you see how much the symmetries constrain the pattern? I was puzzled when I first learned that Schwarzschild's is the unique solution. However, when you consider the effect that all these symmetries have on curvature, you can understand why. Only one pattern of curvature works.

It is time to move on and derive the Schwarzschild metric. The symmetries give us a shortcut to the answer, although it still involves a little mathematical legwork. In the next few pages, we will be using the two relationships on the right of Equations 12.6 and 12.7. These come directly from the symmetries in Equation 12.5.

$$g^{TT}R^{\theta}{}_{T\theta T} = g^{RR}R^{\theta}{}_{R\theta R} = S, \hspace{2cm} \implies \hspace{0.5cm} g^{TT}R^{\theta}{}_{T\theta T} = g^{RR}R^{\theta}{}_{R\theta R}. \hspace{1cm} (12.6)$$

$$g^{TT}R^{\theta}{}_{T\theta T} = S, \quad g^{\theta\theta}R^{\phi}{}_{\theta\phi\theta} = -2S, \hspace{1cm} \implies \hspace{0.5cm} g^{TT}R^{\theta}{}_{T\theta T} = -\frac{1}{2}g^{\theta\theta}R^{\phi}{}_{\theta\phi\theta}. \hspace{1cm} (12.7)$$

12.3   Simon's Ricci Cheat Sheet

We are going to need to calculate some of the Riemann components that add together for the Ricci tensor. It will always be for diagonal metrics (no off-diagonal terms) and of the form $R^{A}{}_{BAB}$. To make life easy, I am going to give you another shortcut. I know... I spoil you.

Let me label whatever the coordinates are as (A, B, C, D). Based on this I have used the formula for the Riemann tensor (Box 7.3) to give Equation 12.8 for the Riemann component R^A_{BAB}. The bold index $\mathbf{E}$ is a dummy index that can be any coordinate. For any dummy index (shown in bold in this book), we must sum over all possible coordinates.

I then expand this noting that $\mathbf{E}$ can be any one of the four coordinates (A, B, C, D) in the third term, but can be only A or B in the fourth term because the value of any Christoffel symbol with three different indices is zero for diagonal metrics (see Box 4.5). This leaves eight terms. I have shifted the two C index and D index terms to the end.

$$
\begin{aligned}
R^A_{BAB} &= \partial_A \Gamma^A_{BB} - \partial_B \Gamma^A_{BA} + \Gamma^A_{AE} \Gamma^E_{BB} - \Gamma^A_{BE} \Gamma^E_{BA}, \\
&= \partial_A \Gamma^A_{BB} - \partial_B \Gamma^A_{BA} + \Gamma^A_{AA} \Gamma^A_{BB} + \Gamma^A_{AB} \Gamma^B_{BB} - \Gamma^A_{BA} \Gamma^A_{BA} - \Gamma^A_{BB} \Gamma^B_{BA} + \Gamma^A_{AC} \Gamma^C_{BB} + \Gamma^A_{AD} \Gamma^D_{BB}.
\end{aligned}
\tag{12.8}
$$

The next step is to insert the formula of each Christoffel symbol (again see Box 4.5). As an example, the formula for the first term in Equation 12.8 is shown as Equation 12.9.

$$
\partial_A \Gamma^A_{BB} = -\frac{\partial}{\partial A} \left(\frac{g^{AA}}{2} \frac{\partial g_{BB}}{\partial A} \right).
\tag{12.9}
$$

All that is left to do is to tidy up the formulas to give the eight terms in *Simon's Ricci cheat sheet* in Box 12.3, which works for any diagonal spacetime metric. I know this looks cumbersome, but you will sing my praises when you use it.

Box 12.3   Simon's Ricci cheat sheet for 4-D diagonal metrics

The eight-term formula below helps in calculating the Ricci tensor of any diagonal spacetime metric (no off-diagonal components). The dimensions are labelled arbitrarily (A, B, C, D). You set A and B to match the coordinates of R^A_{BAB}, then set C and D to the remaining two coordinates.

$$
R^A_{BAB} = -\frac{\partial}{\partial A} \left(\frac{g^{AA}}{2} \frac{\partial g_{BB}}{\partial A} \right) - \frac{\partial}{\partial B} \left(\frac{g^{AA}}{2} \frac{\partial g_{AA}}{\partial B} \right) - \frac{g^{AA} g^{AA}}{4} \frac{\partial g_{AA}}{\partial A} \frac{\partial g_{BB}}{\partial A} + \frac{g^{AA} g^{BB}}{4} \frac{\partial g_{AA}}{\partial B} \frac{\partial g_{BB}}{\partial B}
$$

$$
\dots - \frac{g^{AA} g^{AA}}{4} \left(\frac{\partial g_{AA}}{\partial B} \right)^2 + \frac{g^{AA} g^{BB}}{4} \left(\frac{\partial g_{BB}}{\partial A} \right)^2 - \frac{g^{AA} g^{CC}}{4} \frac{\partial g_{AA}}{\partial C} \frac{\partial g_{BB}}{\partial C} - \frac{g^{AA} g^{DD}}{4} \frac{\partial g_{AA}}{\partial D} \frac{\partial g_{BB}}{\partial D}.
$$

$$
R^A_{BAB} = [1] + [2] + [3] + [4] + [5] + [6] + [7] + [8], \qquad \text{terms numbered for easy reference.}
$$

Note that if $A = B$, you don't need the formula because any Riemann component with its first or last two components the same is zero, e.g. $R^A_{AAA} = 0$ (see Section 7.3 for Riemann symmetries).

The best way to explain is with a couple of examples. I promised earlier to prove to you that, based on the spherical symmetry we are studying, quantities such as $R^\theta_{T\theta T}$ and $R^\phi_{T\phi T}$ are equivalent. Let's start by calculating the value of $R^\theta_{T\theta T}$, which means in the cheat sheet that $A = \theta$, $B = t$ and we can set $C = r$ and $D = \phi$ (C and D could also be vice versa). The metric is shown again below as Equation 12.10. To keep things as clear as possible, I have labelled g_{TT} as bold $\mathbf{T}$ and g_{RR} as bold $\mathbf{R}$. Both are only r dependent. They depend only on the radial distance away from, for example, the mass such as a star. I will label $\frac{\partial g_{TT}}{\partial r}$ as $\mathbf{T}'$ and $\frac{\partial g_{RR}}{\partial r}$ as $\mathbf{R}'$.

$$
\text{Time-independent, spherical symmetry} : [g_{\mu\nu}] = \begin{bmatrix} \mathbf{T} & 0 & 0 & 0 \\ 0 & \mathbf{R} & 0 & 0 \\ 0 & 0 & -r^2 & 0 \\ 0 & 0 & 0 & -r^2 \sin^2\theta \end{bmatrix}.
\tag{12.10}
$$

As this is the first time we are using the cheat sheet, I have written the formula out in full from Box 12.3, substituting in the coordinates $A = \theta$, $B = t$, $C = r$ and $D = \phi$.

$$R^{\theta}{}_{T\theta T} = -\frac{\partial}{\partial \theta}\left(\frac{g^{\theta\theta}}{2}\frac{\partial g_{TT}}{\partial \theta}\right) - \frac{\partial}{\partial t}\left(\frac{g^{\theta\theta}}{2}\frac{\partial g_{\theta\theta}}{\partial t}\right) - \frac{g^{\theta\theta}g^{\theta\theta}}{4}\frac{\partial g_{\theta\theta}}{\partial \theta}\frac{\partial g_{TT}}{\partial \theta} + \frac{g^{\theta\theta}g^{TT}}{4}\frac{\partial g_{\theta\theta}}{\partial t}\frac{\partial g_{TT}}{\partial t}$$

$$\ldots - \frac{g^{\theta\theta}g^{\theta\theta}}{4}\left(\frac{\partial g_{\theta\theta}}{\partial t}\right)^2 + \frac{g^{\theta\theta}g^{TT}}{4}\left(\frac{\partial g_{TT}}{\partial \theta}\right)^2 - \frac{g^{\theta\theta}g^{RR}}{4}\frac{\partial g_{\theta\theta}}{\partial r}\frac{\partial g_{TT}}{\partial r} - \frac{g^{\theta\theta}g^{\phi\phi}}{4}\frac{\partial g_{\theta\theta}}{\partial \phi}\frac{\partial g_{TT}}{\partial \phi}.$$

To calculate $R^{\theta}{}_{T\theta T}$, note that neither g_{TT} or $g_{\theta\theta}$ depend on t or θ, so the derivatives in the first six terms ([1],[2],[3],[4],[5],[6]) are all zero. g_{TT} and $g_{\theta\theta}$ also do not depend on ϕ so term [8] is zero. Only term [7] is non-zero. We can immediately write down and solve giving the result in Equation 12.11. Note that $\frac{\partial g_{\theta\theta}}{\partial r} = -2r$. Also $g^{\theta\theta} = \frac{1}{g_{\theta\theta}}$ and $g^{RR} = \frac{1}{g_{RR}}$ for any diagonal metric.

$$R^{\theta}{}_{T\theta T} = -\frac{g^{\theta\theta}g^{RR}}{4}\frac{\partial g_{\theta\theta}}{\partial r}\frac{\partial g_{TT}}{\partial r} = \frac{1}{4r^2}\frac{1}{\mathbf{R}}(-2r)\,\mathbf{T}' = -\frac{1}{2r}\frac{\mathbf{T}'}{\mathbf{R}}. \tag{12.11}$$

To calculate $R^{\phi}{}_{T\phi T}$, we repeat the process with $A = \phi$, $B = t$, $C = r$ and $D = \theta$. I recommend you write out the eight terms in the cheat sheet formula, substituting in these coordinates. Again, the first six terms are zero because neither g_{TT} nor $g_{\phi\phi}$ are t or ϕ dependent. Also, term [8] is zero because g_{TT} is not θ dependent. Only term [7] is non-zero. You can see from the result in Equation 12.12 that the values of $R^{\theta}{}_{T\theta T}$ and $R^{\phi}{}_{T\phi T}$ are equivalent, whatever the values of $\mathbf{T}$ and $\mathbf{R}$.

$$R^{\phi}{}_{T\phi T} = -\frac{g^{\phi\phi}g^{RR}}{4}\frac{\partial g_{\phi\phi}}{\partial r}\frac{\partial g_{TT}}{\partial r} = \frac{1}{4r^2\sin^2\theta}\frac{1}{\mathbf{R}}(-2r\sin^2\theta)\,\mathbf{T}' = -\frac{1}{2r}\frac{\mathbf{T}'}{\mathbf{R}}. \tag{12.12}$$

We will use the cheat sheet for further calculations in this chapter and in the *Cosmology* module. If you think it is tricky, then be my guest and try the calculations above without it!

12.4 Deriving the Schwarzschild Metric: Relating Time and Space

Our goal is to derive the relationship between $\mathbf{T}$ and $\mathbf{R}$, which are the labels I am using for g_{TT} and g_{RR}. I will spoil things by telling you in advance that $\mathbf{R} = -\frac{1}{\mathbf{T}}$. We will use the relationship $g^{TT}R^{\theta}{}_{T\theta T} = g^{RR}R^{\theta}{}_{R\theta R}$ from Equation 12.6 to show this. We use the cheat sheet for $R^{\theta}{}_{R\theta R}$ setting $A = \theta$, $B = r$, $C = t$ and $D = \phi$. If you run through the derivatives, you will see that only terms [2], [4] and [5] are non-zero. These are calculated to give the result in Equation 12.13.

$$R^{\theta}{}_{R\theta R} = -\frac{\partial}{\partial r}\left(\frac{g^{\theta\theta}}{2}\frac{\partial g_{\theta\theta}}{\partial r}\right) + \frac{g^{\theta\theta}g^{RR}}{4}\frac{\partial g_{\theta\theta}}{\partial r}\frac{\partial g_{RR}}{\partial r} - \frac{g^{\theta\theta}g^{\theta\theta}}{4}\left(\frac{\partial g_{\theta\theta}}{\partial r}\right)^2,$$

$$= -\frac{\partial}{\partial r}\left(\frac{1}{r}\right) + \frac{1}{2r}\frac{\mathbf{R}'}{\mathbf{R}} - \frac{1}{4r^4}(-2r)^2 = -\frac{1}{r^2} + \frac{1}{2r}\frac{\mathbf{R}'}{\mathbf{R}} + \frac{1}{r^2}, \tag{12.13}$$

$$= \frac{1}{2r}\frac{\mathbf{R}'}{\mathbf{R}}.$$

We already calculated the value of $g^{TT}R^{\theta}{}_{T\theta T}$ (Equation 12.11), so, we can now express in terms of the metric, the relationship $g^{TT}R^{\theta}{}_{T\theta T} = g^{RR}R^{\theta}{}_{R\theta R}$ (Equation 12.6), which we established through symmetries in the curvature footprint.

$$g^{TT}R^{\theta}{}_{T\theta T} = g^{RR}R^{\theta}{}_{R\theta R}, \qquad \text{\textit{using Equations 12.11 and 12.13,}}$$

$$\frac{1}{\mathbf{T}}\left(-\frac{1}{2r}\frac{\mathbf{T}'}{\mathbf{R}}\right) = \frac{1}{\mathbf{R}}\left(\frac{1}{2r}\frac{\mathbf{R}'}{\mathbf{R}}\right), \qquad \text{\textit{multiply both sides by} : } 2r\mathbf{T}\mathbf{R}^2,$$

$$-\mathbf{R}\mathbf{T}' = \mathbf{T}\mathbf{R}', \qquad \implies \qquad \mathbf{T}\mathbf{R}' + \mathbf{R}\mathbf{T}' = 0. \tag{12.14}$$

The result on the right of Equation 12.14 means that **TR** is a constant (positive or negative). We know this because using differentiation by parts $\mathbf{TR}' + \mathbf{RT}' = \frac{\partial}{\partial r}(\mathbf{TR}) = 0$. As **TR** is a constant, it must have the same value in all scenarios. At an infinite distance from the mass ($r \rightarrow \infty$), the metric must tend towards flat spacetime, which means $\mathbf{T} = 1$ and $\mathbf{R} = -1$. Therefore, it must always be the case that $\mathbf{TR} = -1$, which shows that $\mathbf{R} = -\frac{1}{\mathbf{T}}$, or using the usual labels $g_{RR} = -\frac{1}{g_{TT}}$.

12.5 Birkhoff's Theorem

The next step is to determine the structure of g_{TT} and g_{RR}. For this, we need the value of $R^\phi{}_{\theta\phi\theta}$. Using the cheat sheet in Box 12.3, we set $A = \phi$, $B = \theta$, $C = t$ and $D = r$. Looking through the derivatives, only terms [2], [5] and [8] are non-zero, leading to Equation 12.15. To help less experienced readers who may, like me, need to look up these trigonometric horrors, I offer the following tips: $\frac{\partial g_{\phi\phi}}{\partial \theta} = -2r^2 \sin\theta \cos\theta$ and $\frac{\partial}{\partial\theta}\left(\frac{\cos\theta}{\sin\theta}\right) = -\frac{1}{\sin^2\theta}$.

$$R^\phi{}_{\theta\phi\theta} = -\frac{\partial}{\partial\theta}\left(\frac{g^{\phi\phi}}{2}\frac{\partial g_{\phi\phi}}{\partial\theta}\right) - \frac{g^{\phi\phi}g^{\phi\phi}}{4}\left(\frac{\partial g_{\phi\phi}}{\partial\theta}\right)^2 - \frac{g^{\phi\phi}g^{RR}}{4}\frac{\partial g_{\phi\phi}}{\partial r}\frac{\partial g_{\theta\theta}}{\partial r}, \tag{12.15}$$

$$= -\frac{\partial}{\partial\theta}\left(\frac{2r^2 \sin\theta\cos\theta}{2r^2\sin^2\theta}\right) - \frac{4r^4\sin^2\theta\cos^2\theta}{4r^4\sin^4\theta} - \frac{1}{\mathbf{R}}\frac{4r^2\sin^2\theta}{-4r^2\sin^2\theta},$$

$$= -\frac{\partial}{\partial\theta}\left(\frac{\cos\theta}{\sin\theta}\right) - \frac{\cos^2\theta}{\sin^2\theta} + \frac{1}{\mathbf{R}} = \frac{1}{\sin^2\theta} - \frac{\cos^2\theta}{\sin^2\theta} + \frac{1}{\mathbf{R}},$$

$$= 1 + \frac{1}{\mathbf{R}}, \qquad\qquad \textit{note that}: 1 - \cos^2\theta = \sin^2\theta. \tag{12.16}$$

The result of this grisly calculation is the simple expression $1 + \frac{1}{\mathbf{R}}$ in Equation 12.16. From the symmetry in Equation 12.7, we know $g^{TT}R^\theta{}_{T\theta T} = -\frac{1}{2}g^{\theta\theta}R^\phi{}_{\theta\phi\theta}$. We now substitute with the values from Equations 12.11 and 12.16. For the symmetries to hold, this gives the result in Equation 12.17. From this we can deduce that $\mathbf{T} = 1 - \frac{C}{r}$, where C is a constant. We can prove this by differentiating this expression to give $\mathbf{T}' = \frac{C}{r^2}$, and then substituting in the value $C = (1 - \mathbf{T})r$ from the original expression for **T**. This is shown as Equation 12.18.

$$g^{TT}R^\theta{}_{T\theta T} = -\frac{1}{2}g^{\theta\theta}R^\phi{}_{\theta\phi\theta},$$

$$\frac{1}{\mathbf{T}}\left(-\frac{1}{2r}\frac{\mathbf{T}'}{\mathbf{R}}\right) = \frac{1}{2r^2}\left(1 + \frac{1}{\mathbf{R}}\right), \qquad \textit{note that}: \frac{1}{\mathbf{TR}} = -1 \textit{ and } \frac{1}{\mathbf{R}} = -\mathbf{T},$$

$$\mathbf{T}' = \frac{1 - \mathbf{T}}{r}, \tag{12.17}$$

$$\implies \quad \mathbf{T} = 1 - \frac{C}{r}, \quad \textit{because} \quad \mathbf{T}' = \frac{C}{r^2} = \frac{1 - \mathbf{T}}{r}, \quad (C \textit{ is a constant}). \tag{12.18}$$

The outcome of all our effort is that we have shown that the Schwarzschild metric must have this $g_{TT} = 1 - \frac{C}{r}$ structure and that it is the only Ricci-0 solution with spherical symmetry. This is known as *Birkhoff's theorem*. To be fair, Birkhoff's theorem shows there is only one Ricci-0 spherically symmetrical solution, *and* it shows that it must be time-independent... so he covered a little more ground than I have here.

12.6 The Schwarzschild Metric

Many of you may be fed up with all this maths and want the answer. Fortunately, we have all the information we need to pull together the solution. We know what level of time dilation leads to the gravitational acceleration we experience. It is $\mathbf{T} = 1 - \frac{2GM}{c^2 r}$ where r is the distance from the central mass M, G is the gravitational constant and c is the speed of light. In Section 5.1, we used the geodesic equation to prove this. You can also see it in Equation 10.5, where we discussed the Weak Field metric.

Now that we know $\mathbf{R} = -\frac{1}{\mathbf{T}}$ and have shown that $\mathbf{T} = 1 - \frac{C}{r}$, we can complete the Schwarzschild solution. The metric is Equation 12.19, and its equivalent in terms of the invariant interval is Equation 12.20. You will often see the abbreviation $R_S = \frac{2GM}{c^2}$ where R_S is called the *Schwarzschild radius*.

Schwarzschild metric and invariant interval

$$[g_{\mu\nu}] = \begin{bmatrix} 1 - \dfrac{2GM}{c^2 r} & 0 & 0 & 0 \\ 0 & -(1 - \dfrac{2GM}{c^2 r})^{-1} & 0 & 0 \\ 0 & 0 & -r^2 & 0 \\ 0 & 0 & 0 & -r^2\sin^2\theta \end{bmatrix}, \quad (ct, r, \theta, \phi), \tag{12.19}$$

$$(cd\tau)^2 = \left(1 - \frac{2GM}{c^2 r}\right)(c\,dt)^2 - \left(1 - \frac{2GM}{c^2 r}\right)^{-1} dr^2 - r^2 d\theta^2 - r^2\sin^2\theta d\phi^2. \tag{12.20}$$

We can calculate the value of S in the curvature footprint of Equation 12.5. This is shown in Equation 12.21. Using the relationship $g^{TT}R^\theta{}_{T\theta T} = S$ and substituting with the result in Equation 12.11 gives $S = \frac{GM}{c^2 r^3}$.

$$S = g^{TT}R^\theta{}_{T\theta T} = -\frac{1}{2r}\frac{\mathbf{T}'}{\mathbf{TR}} = \frac{1}{2r}\mathbf{T}' = \frac{1}{2r}\frac{\partial}{\partial r}\left(1 - \frac{2GM}{c^2 r}\right) = \frac{GM}{c^2 r^3}. \tag{12.21}$$

The total of each component of the Ricci tensor must be zero, but knowing the value of its Riemann components will prove useful later.

12.7 Summary

Some of you will have found these mathematical steps cumbersome, so let me review what we have covered. In Section 12.1, I showed you that the characteristics of the Schwarzschild solution (static, stationary and spherically symmetric) lead to a diagonal metric in spherical coordinates (no off-diagonal terms).

In Section 12.2, we applied two simple constraints from the Schwarzschild metric: the Ricci tensor $[R_{\mu\nu}] = 0$ because it is a vacuum solution with no cosmological constant and spherical symmetry. These constraints allow only one pattern of curvature footprint.

In this curvature footprint, $g^{TT}R^\theta{}_{T\theta T} = g^{RR}R^\theta{}_{R\theta R}$. Calculating the value of these Riemann components reveals that for the Schwarzschild metric: $g_{RR} = -\dfrac{1}{g_{TT}}$.

Then I used the symmetry $g^{TT}R^\theta{}_{T\theta T} = -\frac{1}{2}g^{\theta\theta}R^\phi{}_{\theta\phi\theta}$ to show that the *unique* solution is of the form $g_{TT} = 1 - \frac{C}{r}$ where C is a constant.

Comparing with the weak field gravity that we know, leads to the Schwarzschild metric and invariant interval shown as Equations 12.19 and 12.20.

The calculation in Equation 12.21 of the value of S in the underlying Riemann components gives the curvature footprint of the Schwarzschild metric shown below.

Curvature Footprint of Schwarzschild Metric $\qquad S = \dfrac{GM}{c^2 r^3}$ $\qquad\qquad$ (12.22)

$$g^{TT}R_{TT} = \quad g^{TT}R^T{}_{TTT} + g^{TT}R^R{}_{TRT} + g^{TT}R^\theta{}_{T\theta T} + g^{TT}R^\phi{}_{T\phi T} = \quad 0 \quad -2S \quad +S \quad +S \quad = 0,$$

$$g^{RR}R_{RR} = \quad g^{RR}R^T{}_{RTR} + g^{RR}R^R{}_{RRR} + g^{RR}R^\theta{}_{R\theta R} + g^{RR}R^\phi{}_{R\phi R} = \quad -2S \quad +0 \quad +S \quad +S \quad = 0,$$

$$g^{\theta\theta}R_{\theta\theta} = \quad g^{\theta\theta}R^T{}_{\theta T\theta} + g^{\theta\theta}R^R{}_{\theta R\theta} + g^{\theta\theta}R^\theta{}_{\theta\theta\theta} + g^{\theta\theta}R^\phi{}_{\theta\phi\theta} = \quad S \quad +S \quad +0 \quad -2S \quad = 0,$$

$$g^{\phi\phi}R_{\phi\phi} = \quad g^{\phi\phi}R^T{}_{\phi T\phi} + g^{\phi\phi}R^R{}_{\phi R\phi} + g^{\phi\phi}R^\theta{}_{\phi\theta\phi} + g^{\phi\phi}R^\phi{}_{\phi\phi\phi} = \quad S \quad +S \quad -2S \quad +0 \quad = 0.$$

The Schwarzschild metric is the only Ricci-0 solution for spacetime with spherical symmetry. I should note that there are other Ricci-0 solutions if the metric is *not* spherically symmetrical, such as the Kerr solution (of which more later). However, let me repeat for emphasis what we have shown in this chapter: if you combine the limitations imposed from being Ricci-0 with those from spherical symmetry, the Schwarzschild metric is the

Figure 12.1 Our pet cat Roxy.

only possible stationary (non-rotating) solution. Now that you have seen the constraints imposed by the various symmetries, I hope this does not surprise you.

12.8 Why Do We Care?

The unique nature of the Schwarzschild metric makes it a powerful tool. Take a look at the metric as displayed in Equation 12.19. If there is no mass present, then the $\left(\frac{2GM}{c^2 r}\right)$ terms disappear and it becomes flat Minkowski spacetime. You can compare with Equation 12.1 if needed. And the Schwarzschild solution tends to Minkowski spacetime as you get further away from the mass because as r tends to infinity, the $\left(\frac{2GM}{c^2 r}\right)$ terms tend to zero.

Perhaps more interesting is what happens in extremely strong gravitational fields. Consider the metric if $r = \frac{2GM}{c^2}$. In terms of invariant proper time, the $(c\,dt)^2$ term becomes zero and the dr^2 term becomes infinite. This is the Schwarzschild radius and is the *event horizon* of a black hole. We will be discussing the implications in detail in Chapter 13.

The important point is the following. We *cannot ignore* what the Schwarzschild metric says about extreme gravitational conditions. We cannot hypothesise that there is an alternative solution. There is not. If we accept that spacetime in a vacuum is Ricci-0, then the Schwarzschild metric *must* be a decent model of intense gravity. This is why we care. And it brings us to Chapter 13, where we will take a closer look at the implications. Get your little brain cells ready to learn about *black holes*.

As there has been little by way of light relief in this chapter, I offer Figure 12.1, which is a photo of our pet cat Roxy as the Easter bunny. I have absolutely no justification for this except that I hear books containing cat pictures sell rather well... so, tell your friends all about it.

12.9 Schwarzschild with Cosmological Constant (Optional)

The cosmological constant is too small to have a noticeable effect when considering stars or black holes in our universe. However, I am going to show how it fits into the Schwarzschild metric, just in case any readers need it. The good news is that you don't have to read this section unless you have some special interest...

The cosmological constant has an equivalent effect everywhere, so it doesn't change the spherical symmetry of the Schwarzschild solution. I will show you that the components labelled $(+S)$ in the curvature footprint shown earlier as Equation 12.5 become $(S - \frac{\Lambda}{2})$ as shown in Equation 12.23. Let me explain how this matches up with the EFEs including Λ, by assuming this answer and demonstrating that it produces the correct result.

Spherically Symmetric Vacuum Solution With Cosmological Constant (Diagonal metric) (12.23)

$$g^{TT}R_{TT} = \quad 0 \quad\quad -2S \quad +(S - \tfrac{\Lambda}{2}) \quad +(S - \tfrac{\Lambda}{2}) \quad = -\Lambda,$$

$$g^{RR}R_{RR} = \quad -2S \quad\quad +0 \quad +(S - \tfrac{\Lambda}{2}) \quad +(S - \tfrac{\Lambda}{2}) \quad = -\Lambda,$$

$$g^{\theta\theta}R_{\theta\theta} = \quad (S - \tfrac{\Lambda}{2}) \quad +(S - \tfrac{\Lambda}{2}) \quad +0 \quad -2S \quad = -\Lambda,$$

$$g^{\phi\phi}R_{\phi\phi} = \quad (S - \tfrac{\Lambda}{2}) \quad +(S - \tfrac{\Lambda}{2}) \quad -2S \quad +0 \quad = -\Lambda,$$

$$R = g^{TT}R_{TT} + g^{RR}R_{RR} + g^{\theta\theta}R_{\theta\theta} + g^{\phi\phi}R_{\phi\phi} = -4\Lambda.$$

Starting from the general form of the EFEs on the left of Equation 12.24 for g_{TT} (you can use any component), I multiply through by g^{TT} to give the answer on the right. The value of $g^{TT}g_{TT}$ is 1. In Equation 12.25, I substitute in values using the assumed solution of the curvature footprint in Equation 12.23. As you can see, the result is correct. The assumed solution is consistent with the vacuum EFEs including Λ, while respecting all the required Schwarzschild symmetries.

$$R_{TT} - \frac{1}{2}R g_{TT} - \Lambda g_{TT} = 0, \quad\Longrightarrow\quad g^{TT}R_{TT} - \frac{1}{2}R - \Lambda = 0, \tag{12.24}$$

$$\Longrightarrow \quad -\Lambda + 2\Lambda - \Lambda = 0, \text{ Q.E.D.} \tag{12.25}$$

Turning to the relationships in Equations 12.6 and 12.7, the first is unchanged as shown on the right of Equation 12.26, while the second is changed by $\frac{\Lambda}{2}$ as shown on the right of Equation 12.27. We can use these to derive the effect of a non-zero cosmological constant on the Schwarzschild metric.

$$g^{TT}R^{\theta}_{T\theta T} = g^{RR}R^{\theta}_{R\theta R} = S - \frac{\Lambda}{2}, \quad\quad \Longrightarrow g^{TT}R^{\theta}_{T\theta T} = g^{RR}R^{\theta}_{R\theta R}. \tag{12.26}$$

$$g^{TT}R^{\theta}_{T\theta T} = S - \frac{\Lambda}{2}, \quad g^{\theta\theta}R^{\phi}_{\theta\phi\theta} = -2S, \quad \Longrightarrow g^{TT}R^{\theta}_{T\theta T} = -\frac{1}{2}g^{\theta\theta}R^{\phi}_{\theta\phi\theta} - \frac{\Lambda}{2}. \tag{12.27}$$

As Equation 12.26 is the same as Equation 12.6, the calculation in Section 12.4 is unchanged, giving the relationship: $g_{RR} = -\frac{1}{g_{TT}}$, just as is the case for the Schwarzschild metric without Λ.

The change in value on the right of Equation 12.27 means that we must adjust the relationship in Equation 12.17 by $\frac{\Lambda}{2}$ as shown in Equation 12.28. This gives Equation 12.29.

$$g^{TT}R^{\theta}_{T\theta T} = -\frac{1}{2}g^{\theta\theta}R^{\phi}_{\theta\phi\theta} - \frac{\Lambda}{2}, \tag{12.28}$$

$$\frac{1}{\mathbf{T}}\left(-\frac{1}{2r}\frac{\mathbf{T}'}{\mathbf{R}}\right) = \frac{1}{2r^2}\left(1 + \frac{1}{\mathbf{R}}\right) - \frac{\Lambda}{2}, \quad\quad note\ that: \frac{1}{\mathbf{TR}} = -1 \ and \ \frac{1}{\mathbf{R}} = -\mathbf{T},$$

$$\frac{\mathbf{T}'}{2r} = \frac{1}{2r^2}(1 - \mathbf{T}) - \frac{\Lambda}{2},$$

$$\mathbf{T}' = \frac{1 - \mathbf{T}}{r} - \Lambda r. \tag{12.29}$$

As $\mathbf{T}'$ is the derivative, we can deduce that the presence of Λ adds an r^2 term to $\mathbf{T}$. This changes the expression we calculated for $\mathbf{T}$ at the end of Equation 12.17 to that on the left of Equation 12.30, where C and D are constants. Differentiating this gives the expression for $\mathbf{T}'$ on the right.

$$\mathbf{T} = 1 - \frac{C}{r} + Dr^2, \quad\quad \Longrightarrow \quad\quad \mathbf{T}' = \frac{C}{r^2} + 2Dr. \tag{12.30}$$

We now substitute these values for $\mathbf{T}$ and $\mathbf{T}'$ into Equation 12.29 to calculate the value of the new Dr^2 term as shown in Equation 12.31 to give: $D = -\frac{\Lambda}{3}$.

$$\frac{C}{r^2} + 2Dr = \frac{1}{r}\left(\frac{C}{r} - Dr^2\right) - \Lambda r, \quad \Longrightarrow \quad 3Dr = -\Lambda r, \quad \Longrightarrow \quad D = -\frac{\Lambda}{3}. \tag{12.31}$$

These results show that a non-zero Λ doesn't change the relationship $g_{RR} = -\dfrac{1}{g_{TT}}$, but adds a $-\dfrac{\Lambda}{3}r^2$ term as shown in Equation 12.32. For simplicity, I use natural units (speed of light $c = 1$) in this equation. The result is the Schwarzschild metric including Λ. If you set $\Lambda = 0$, you get (as you must) the original Schwarzschild solution.

Schwarzschild metric with non $-$ zero Λ $\qquad (c = 1)$

$$d\tau^2 = \left(1 - \frac{2GM}{r} - \frac{\Lambda}{3}r^2\right) dt^2 - \frac{1}{\left(1 - \frac{2GM}{r} - \frac{\Lambda}{3}r^2\right)}dr^2 - r^2 d\theta^2 - r^2\sin^2\theta \, d\phi^2. \qquad (12.32)$$

Note that at an infinite distance from the mass ($r \to \infty$), the metric does *not* tend towards the metric of flat spacetime. This is because the presence of a non-zero cosmological constant means that there is Ricci-curvature even in a vacuum ($[R_{\mu\nu}] \neq 0$). The non-zero cosmological constant also affects the position of the event horizon of black holes. However, in our universe this effect is negligible because the value of the cosmological constant is very small.

Chapter 13

Schwarzschild and Black Holes

The main topic of this chapter is what the Schwarzschild metric tells us about black holes. I suspect (and hope) that you are familiar with the concept. Their gravitational pull is so strong that even light cannot escape their clutch. Black holes are weird and befuddling (see Box 13.1), so I begin the chapter with a precis of what is to come. This is to prepare you, rather like those announcements at the start of films that warn you they contain explicit or graphic content.

Once the gravitational pull on an object becomes strong enough, nothing can stop it collapsing in on itself. Physicists believe this happened early in the development of the universe as vast amounts of gas coalesced. There also are *stellar-mass black holes* formed in the aftermath of supernovae. Stars are fuelled by nuclear fusion in which hydrogen fuses to form helium, which fuses in turn to form heavier elements all the way up to iron. Once this process is complete, some stars explode in a supernova. If the remaining mass of the star after the explosion is over about three solar masses, the result is a black hole.

Box 13.1 Just for laughs: black holes

How did the black hole stay healthy? *It ate light.*

You think black holes aren't important? *You don't understand the gravity of the situation.*

We will not spend a lot of time discussing what happens at the very centre of black holes for the simple reason that we do not know much. The Schwarzschild metric becomes a mess at the centre because $r = 0$ and spacetime curvature becomes infinite. General relativity appears to predict that the black hole mass crunches down on itself without limit to a point-like singularity. Quantum mechanics says this is not possible because it would run counter to Heisenberg's uncertainty principle, which is a key tenet of quantum theory. What is more, the centre of a black hole is hidden from our view. Ho-hum.

Instead, our focus will be on the theoretical limit of what is observable: the *event horizon*, which is also called the *Schwarzschild radius* of the black hole. At the event horizon, it would take an infinite amount of acceleration to resist falling into the black hole (we will prove this). Once inside the event horizon, nothing, not even light, can escape plunging to the centre.

From the perspective of an observer outside the event horizon, strange things happen. Our analysis using the Schwarzschild metric will show that an approaching clock appears (to the observer) to tick exponentially slower and slower as it nears the event horizon. As a result, they will never see the clock reach it. From their perspective, nothing ever reaches the singularity at the centre. All the material falling into the black hole is observed to remain in the vicinity of, but always outside of, the event horizon, forming what is called the *accretion disk*.

The perspective is very different for a black hole explorer who freefalls in, crossing the event horizon. From his or her perspective, nothing strange happens at the event horizon if the black hole is big enough. Everything is hunky dory until the explorer is what physicists describe as *spaghettified*: torn apart by tidal forces.

Educators should note that I intentionally delay introducing alternative coordinate systems until as late as possible. Although they are a useful way to peer inside the event horizon, they are neither simple nor intuitive. So, I beg patience from any reader desperate to jump into Eddington–Filkenstein coordinates and such-like.

Untangling General Relativity: The Intuitive Self-Study Guide, First Edition. Simon Sherwood.
© 2026 John Wiley & Sons Ltd. Published 2026 by John Wiley & Sons Ltd.

13.1 Schwarzschild Revisited

Box 13.2 The Schwarzschild metric

$$[g_{\mu\nu}] = \begin{bmatrix} 1 - \dfrac{2GM}{c^2 r} & 0 & 0 & 0 \\ 0 & -(1 - \dfrac{2GM}{c^2 r})^{-1} & 0 & 0 \\ 0 & 0 & -r^2 & 0 \\ 0 & 0 & 0 & -r^2 \sin^2\theta \end{bmatrix}, \quad (ct, r, \theta, \phi),$$

$$(cd\tau)^2 = \left(1 - \frac{R_S}{r}\right)(c\,dt)^2 - \frac{1}{\left(1 - \frac{R_S}{r}\right)}dr^2 - r^2 d\theta^2 - r^2 \sin^2\theta\, d\phi^2. \tag{13.1}$$

$R_S = \dfrac{2GM}{c^2}$,   *G*: Newton's gravitational constant   *M*: central mass   *c*: speed of light

---

I would love to assume that all my readers have picked through every detail of the Schwarzschild derivation in the last chapter... but I know some of you will have dozed off. Therefore, I will start from scratch to give the lazier contingent a chance to catch up.

The Schwarzschild metric is the *unique* spherically symmetric solution of the EFEs in a vacuum ($[R_{\mu\nu}] = 0$). It describes the vacuum spacetime distortion around a spherically symmetric mass. While it does not account for rotation, it is a decent model for anything that is not spinning at relativistic speeds, so it describes much of the gravity we know, such as our solar system (the sun rotates around its axis, but it takes about 30 days).

The invariant interval equation of the Schwarzschild metric is shown as Equation 13.1 in Box 13.2. We simplify by labelling $R_S = \frac{2GM}{c^2}$. R_S is called the *Schwarzschild radius*. On the left side of the invariant interval Equation 13.1, $d\tau$ is proper time. On the right side, dt, dr, $d\theta$ and $d\phi$ are the coordinates as measured by an observer in flat Minkowski spacetime. It may help you to think of the observer being an infinite distance from the mass. When $r = \infty$, the $\left(\frac{R_S}{r}\right)$ terms tend to zero and the Schwarzschild invariant interval equation becomes the same as that of Minkowski spacetime.

As the Schwarzschild metric is the unique solution of its sort for the EFEs, we need to take its implications seriously for both weak and strong gravitational fields. One obvious question is what happens when $r = R_S$. You can see that the factor alongside the $(cdt)^2$ term becomes zero. Oops!

13.2 Black Holes: Overview

When $r = R_S$, you are at the *event horizon* of a *black hole*. As a clock approaches the event horizon, an outside observer will see the clock progressively slow down. Before digging into further detail, I want to get things in perspective. What does $r = R_S$ mean quantitatively? After all, we don't regularly stumble across areas where we see clocks slow dramatically.

As an example, consider the Schwarzschild radius of a mass such as the earth. The calculation is shown in Equation 13.2. The answer is that R_S for the mass of the earth is slightly less than one centimetre. This means that for the earth to be a black hole, its entire mass would need to be compressed into less than the volume of a sphere of radius one centimetre. For the earth's mass to fit into this small volume, the density would need to be about 10^{30} kg m^{-3}. Compare this with the density of lead at 10^4 kg m^{-3}. Clearly, you don't have to worry that a black hole will suddenly appear in your backyard.

$$R_S = \frac{2GM}{c^2} \approx \frac{2\,(7 \times 10^{-11})\,(6 \times 10^{24})}{9 \times 10^{16}} \approx 10^{-2} \text{ metres,}$$
$$G \approx 6.7 \times 10^{-11} \text{m}^3\,\text{kg}^{-2}\,\text{s}^{-1}, \qquad \textit{Earth mass} \approx 6 \times 10^{24} \text{ kg}, \qquad c \approx 3 \times 10^8 \text{ m s}^{-1}. \tag{13.2}$$

Way back in 1783, the remarkable mathematician Rector John Michell postulated that black holes might exist. He imagined *dark stars* with too strong a gravitational field for light to escape, calculating this to be true for any star with the same average density as the sun but with a radius 500 times bigger (a bizarrely accurate result as discussed in Box 13.3).

On the other hand, Einstein doubted that black holes exist. In a paper he published in 1939, he set out to prove that they cannot. But dear old Albert was wrong! We now know that super-massive black holes, millions or even billions times more massive than our sun, sit at the centre of most galaxies. Of course, *seeing is believing,* so I offer Figure 13.1, which shows images of two black holes, both recorded by the *Event Horizon Telescope* detecting in the microwave range.

On the left is M87, which is about 55 million light years away in the constellation of Virgo and is estimated to weigh in at over five billion solar masses... yes, that is a mass equivalent to five *billion* suns! The event horizon of such a super-massive black hole extends some 15 billion kilometres from its centre. To put this in perspective, it is about 100 times the distance from the earth to the sun. On the right of the figure is Sagittarius A*, which sits around 25,000 light years away at the heart of our Milky Way galaxy with about four million solar masses. The central dark disc in each image is the shadow of the black hole. The surrounding glowing radiation is emitted by gas and dust orbiting the black holes.

Box 13.3   Rector John Michell (1724–1793): father of the black hole

Rector John Michell theorised that light is a stream of particles subject to the effect of gravity. Even in the 1700s, there were fairly decent estimates of the speed of light (calculated by studying small changes in the timing of such things as the eclipses of Jupiter at different distances from the earth). He equated the speed of light with the required escape velocity of a gravitational field, concluding that light would not be able to escape from the surface of a star with the density of the sun, but with a radius 500 times greater.

Let's use modern methods to undertake a similar calculation. The mass of the sun is about 2×10^{30} kg, so the mass of Michell's dark star is $(500)^3$ greater at about 2.5×10^{38} kg. For this mass, we calculate the Schwarzschild radius as shown in Equation 13.3 to be about 4×10^8 km.

The radius of the sun is 7×10^5 km. Multiplying this by 500× gives about 4×10^8 km. Thus, the surface of Mitchell's dark star would indeed be a black hole event horizon. I should note that any star the size of Michell's dark star would collapse under its gravitational force. However, it is still impressive that Michell's 1783 result was right, at least in theory. Wow!

$$R_S = \frac{2GM}{c^2} \approx \frac{2\,(7 \times 10^{-11})\,(2.5 \times 10^{38})}{9 \times 10^{16}} \approx 4 \times 10^{11} \text{ metres} = 4 \times 10^8 \text{ km.} \tag{13.3}$$

Michell presented his findings to the Royal Society, but nobody seems to have taken much notice. What was he like? With no portrait, we have only the following description: *a little short man, of a black complexion...* (i.e. sad)... *and fat*. Not much recognition for a genius who not only predicted the existence of black holes, but even proposed how they might be spotted:

If there really should exist in nature any bodies... [dark stars]... we could have no information from light; yet if any other luminous bodies should happen to revolve about them, we might still perhaps from the motions of these revolving bodies infer the existence of the central ones.

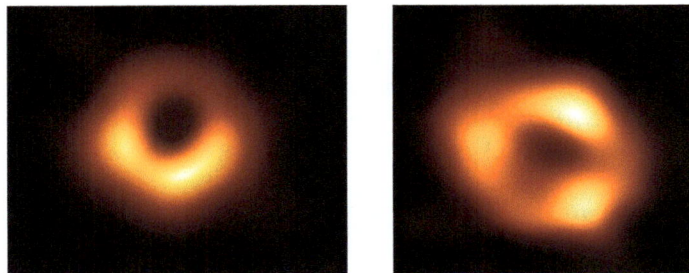

Figure 13.1   Black holes. On the left: M87. *Source*: Event Horizon Telescope / Wikimedia Commons / CC BY 4.0. On the right: Sagittarius A* in our Milky Way galaxy. The black hole is the dark centre. The glow is from material in the surrounding accretion disk. *Source*: EHT Collaboration / Wikimedia Commons / CC BY 4.0.

13.3 Minky and Schwart

To help our discussion, I want to introduce you to two imaginary adventurous black hole explorers: Minky and her boyfriend Schwart. Minky stays well away from the black hole. For her, the value of $\frac{R_S}{r}$ is approximately zero. If you look at the Schwarzschild invariant interval shown again as Equation 13.4, you will see that this means Minky's spacetime metric is very similar to flat Minkowski spacetime (*Minky* in *Mink*owski spacetime - geddit?) while her boyfriend Schwart will head into black holes. He will experience the Schwarzschild metric (*Schwart* in *Schwar*zschild spacetime). Given what is going to happen to poor old Schwart in the next few pages, I have some doubts about the level of Minky's affection for him. How unpredictable love can be!

Before we get into the detailed analysis, I want to highlight the importance of determining who is the observer in any scenario. Time is not absolute. If we say there is time dilation it must be *relative* to some observer. In most of the following scenarios, we will have Minky sitting stationary at a great distance observing Schwart near to the black hole (or other gravitational source). Let me show you in general terms what the Schwarzschild metric tells us.

We assume Minky is far enough away that $\frac{R_S}{r}$ is negligible and for the Schwarzschild invariant interval, shown again as Equation 13.4, we can use $\frac{R_S}{r} = 0$ for her clock. Being stationary relative to the black hole $dr = d\theta = d\phi = 0$. This leads to $d\tau^2 = dt^2$. The length of a tick dt on Minky's clock matches a tick of invariant time $d\tau$.

$$(cd\tau)^2 = \left(1 - \frac{R_S}{r}\right)(c\,dt)^2 - \frac{1}{\left(1 - \frac{R_S}{r}\right)}dr^2 - r^2 d\theta^2 - r^2\sin^2\theta\,d\phi^2. \tag{13.4}$$

Let's analyse Minky's observation of Schwart's clock. How fast does Minky observe his clock to tick? It is easier to use numbers, so imagine that Schwart is hovering stationary at a distance of 50× the Schwarzschild radius from the centre of the black hole i.e. $1 - \frac{R_S}{r} = 1 - 0.02$.

As Schwart is stationary relative to the black hole $dr = d\theta = d\phi = 0$, we can calculate the disparity between Minky's clock tick (matching $d\tau$) and Schwart's clock tick, which is shown as dt in Equation 13.5. The calculation divides both sides by c and uses the Taylor approximation $\sqrt{1-x} \approx 1 - \frac{x}{2}$ for small x.

$$d\tau^2 = (1 - 0.02)\,dt^2, \implies d\tau \approx (1 - 0.01)\,dt = 0.99dt\,, \quad \textit{Stationary.} \tag{13.5}$$

The result is that $d\tau \approx 0.99\,dt$. We know that $d\tau$ matches the tick on Minky's own clock. For the right side of the equation to match the left, Minky must observe the tick length of Schwart's clock (dt) to be slightly longer than hers ($d\tau$). Minky observes Schwart's clock to tick more slowly than that of hers. This is gravitational time dilation.

Now let's think what happens if Schwart is moving relative to the black hole. Movement only in the θ direction gives the invariant interval in Equation 13.6. The movement further reduces the right side of the equation. For the two sides to match, Minky must observe Schwart's clock's tick dt to be slightly longer than it was when Schwart was stationary. This is time dilation of a moving clock, as discussed in Section 2.4 on special relativity. The attracting mass M affects the value of R_S, but this term doesn't appear in $g_{\theta\theta}$ or $g_{\phi\phi}$. The gravitational field doesn't change the time dilation relating to movement in the θ or ϕ directions.

$$(cd\tau)^2 = 0.98(c\,dt)^2 - r^2 d\theta^2, \quad \textit{Movement in } \theta \textit{ direction.} \tag{13.6}$$

Finally, Equation 13.7 shows the invariant interval if Schwart moves in the radial r direction. As before, this reduces the right side of the equation. For the sides to match, Minky must observe Schwart's clock tick dt to be slightly longer than it was when Schwart was stationary. This *is affected* by the gravitational field. It increases the overall value of the dr^2 term. The stronger the field, the bigger the reduction in value on the right-hand side and therefore the greater the time dilation Minky observes in the ticks of Schwart's clock.

$$(cd\tau)^2 = 0.98(c\,dt)^2 - \frac{1}{0.98}dr^2, \quad \textit{Movement in } r \textit{ radial direction.} \tag{13.7}$$

In summary, you might describe there being two sources of the gravitational time dilation Minky observes on Schwart's clock. The first is due to the distortion in g_{TT}. The second is due to the distortion of the spatial g_{RR} component of the metric, which increases the time dilation linked to radial movement. The presence of the c terms in Equation 13.7 means that the radial effect is much the smaller unless an object is moving at a substantial proportion of light speed.

I will show you in optional Chapter 15 how this radial distortion approximately doubles the deflection of light in a weak gravitational field, which Einstein described with a quote that I showed earlier (Section 5.5): *according to the theory, half of this deflection is produced by the Newtonian field of attraction of the sun, and the other half by the geometrical modification* [curvature] *of space caused by the sun. This result admits of an experimental test... during a total eclipse of the sun.* The geometrical modification of space that Einstein refers to is the change to g_{RR}.

13.4 Proper Acceleration

After that introduction, let's get started on the detailed analysis by calculating the accelerating effect of the Schwarzschild metric on any object in its vicinity. We will work with *proper acceleration* (which I will label α). It is an invariant quantity that all observers agree upon. This is the acceleration measured locally on an object that is stationary relative to the gravitational field. You can think of it as the acceleration at the very moment an object is released into freefall or, perhaps more intuitively, as the resisting acceleration needed to hold an object stationary (i.e. hovering) at some point in the gravitational field. I expect you are familiar with the proper acceleration on the surface of the earth, which is $9.8 \, \mathrm{ms}^{-2}$.

When gravity is weak, an approximate calculation for the Schwarzschild metric is simple because we have already done it. It turns out to be the very same calculation we used when we were abusing dung beetles back in Chapter 5. However, we must step through it carefully to prepare for the more extreme scenario of black holes.

Our storyline is to send the intrepid explorer Schwart in his spaceship and have him measure the acceleration he needs to hover stationary relative to the black hole. To assess his proper acceleration, we must measure the distortion to the geodesics. We use the geodesic equation shown again in Box 13.4 (if you need a reminder, check back to Box 5.2). Using (ct, r, θ, ϕ) coordinates, I label the rate of change over proper time τ as velocity u (the same label used when we were discussing the energy-momentum tensor). Thus u^t is $\frac{dt}{d\tau}$ and u^r is $\frac{dr}{d\tau}$ etc.

Box 13.4 Geodesic equation where A and B are dummy indices

$$\frac{du^x}{d\tau} = -\Gamma^X_{AB} \frac{dA}{d\tau} \frac{dB}{d\tau} = -\Gamma^X_{AB} u^A u^B.$$

The first step is to calculate the radial acceleration $\alpha^r = \frac{du^r}{d\tau}$. As Schwart is hovering stationary, we know that $u^r = u^\theta = u^\phi = 0$. However, we must remember that Schwart is moving through time so $u^t \neq 0$. As a result, the geodesic equation in Box 13.4 is only non-zero if $\mathbf{A} = \mathbf{B} = t$ as shown in Equation 13.8 (you can check the Christoffel symbol formula using Box 4.4). As Schwart is hovering stationary, we know $dr = d\theta = d\phi = 0$ in Equation 13.1, giving $u^t = \frac{dt}{d\tau} = \left(1 - \frac{R_S}{r}\right)^{-\frac{1}{2}}$. The Schwarzschild metric is diagonal so the value of g^{RR} is $\frac{1}{g_{RR}} = -\left(1 - \frac{R_S}{r}\right)$.

$$\alpha^r = \frac{du^r}{d\tau} = -\Gamma^R_{TT} u^t u^t = \frac{1}{2} g^{RR} \frac{\partial g_{TT}}{\partial r} u^t u^t = -\frac{1}{2}\left(1 - \frac{R_S}{r}\right)\left(\frac{R_S}{r^2}\right) u^t u^t, \tag{13.8}$$

$$= -\frac{R_S}{2r^2}\left(1 - \frac{R_S}{r}\right)\left(1 - \frac{R_S}{r}\right)^{-\frac{1}{2}}\left(1 - \frac{R_S}{r}\right)^{-\frac{1}{2}},$$

$$= -\frac{R_S}{2r^2}. \tag{13.9}$$

Although it seems intuitively obvious that Schwart must accelerate directly away from the black hole (i.e. radially), it is worth confirming that he does not require any other components of acceleration to hover (i.e. in directions other than radial). You can see from Equation 13.8 that for any coordinate μ, α^μ is zero if $\frac{\partial g_{TT}}{\partial \mu}$ is zero. In the Schwarzschild metric, g_{TT} does not depend on t, θ or ϕ. Therefore, we know $\alpha^t = \alpha^\theta = \alpha^\phi = 0$ and Schwart can hover by accelerating only in the radial direction exactly as common sense would suggest.

We can summarise how Schwart hovers using four-vectors (see Box 8.2 for a refresher on how these work). Equation 13.10 shows Schwart's velocity four-vector, moving through time but stationary in space. Equation 13.11 shows Schwart's acceleration four-vector, accelerating only in the radial direction. Obviously, Schwart's spaceship acceleration would need to be equal and opposite to this in order to hover. Note that both these four-vectors are shown in vector coordinates (upper index).

$$u^\mu = \left(\left(1 - \frac{R_S}{r} \right)^{-\frac{1}{2}}, \ 0, \ 0, \ 0 \right), \qquad (ct, r, \theta, \phi), \qquad (13.10)$$

$$\alpha^\mu = \left(0, -\frac{R_S}{2r^2}, \ 0, \ 0 \right), \qquad (ct, r, \theta, \phi). \qquad (13.11)$$

To calculate the magnitude of Schwart's proper acceleration, we must be careful to scale it to the invariant interval by combining the upper and lower index (check back to Section 3.6 if you need to): $|\alpha| = |\sqrt{\alpha^r \alpha_r}|$ (the symbol $|\alpha|$ means the absolute value of α whether positive or negative). The Schwarzschild metric is diagonal so $\alpha_r = g_{RR} \alpha^r$. This leads to the final result in Equation 13.12.

$$|\alpha| = |\sqrt{\alpha^r \alpha_r}| = |\sqrt{g_{RR} (\alpha^r)^2}| = |\alpha^r \sqrt{g_{RR}}|,$$
$$= \frac{R_S}{2r^2} \sqrt{\frac{1}{1 - \frac{R_S}{r}}}, \qquad \textit{Proper acceleration} \qquad (ct, r, \theta, \phi). \qquad (13.12)$$

This looks more complicated than the gravity we know. Let's start with a reality check in weak gravitational fields. At the earth's surface, $\frac{R_S}{r}$ is only about 10^{-9} so the second term in Equation 13.12 is almost exactly one giving the formula in Equation 13.13 (accurate to about one in a billion).

Equation 13.13 is the result using coordinates (ct, r, θ, ϕ). The advantage of these coordinates is that they match up the units of time and space, removing the requirement to have a c^2 factor in g_{TT} (as explained in Section 3.1). In Equation 13.13 I have shown the result using the more familiar coordinates (t, r, θ, ϕ). When you do the calculation using these coordinates, the result is greater by a factor of c^2 because the value of $\frac{\partial g_{TT}}{\partial r}$ in Equation 13.8 is greater by a factor of c^2. You can see that the result in Equation 13.14 matches Newton's formula for gravitational acceleration (check back to Equation 1.6 if needed). This is a comforting check, but not surprising seeing as we used the weak field comparison to define the value of R_S in the Schwarzschild metric.

$$|\alpha| \approx \frac{R_S}{2r^2} = \frac{GM}{c^2 r^2} \quad (ct, r, \theta, \phi), \qquad \textit{Weak Field,} \qquad (13.13)$$
$$= \frac{GM}{r^2} \quad (t, r, \theta, \phi). \qquad (13.14)$$

I imagine readers yowling at so much work for so little reward. I beg patience of you yapping learners. The calculation gets to the heart of what a black hole is... well, not actually the heart, which is a singularity... but the event horizon where $R_S = r$. As Schwart approaches the event horizon, the proper acceleration he needs to hover (Equation 13.12) increases until, at the event horizon itself, $\frac{R_S}{r} = 1$ and the second term becomes $\frac{1}{0} = \infty$. At this point, he requires an *infinite* amount of acceleration to resist the pull. Nothing, not even light, can escape. This is the event horizon of a black hole.

13.5 White Dwarfs and Neutron Stars

The maths of general relativity and the geodesic equation is clear. In the scenario where the surface of a star sits within its Schwarzschild radius, then that surface would need an infinite acceleration outwards to resist moving towards the centre. Nothing whatsoever can produce an infinite acceleration. The star must and will collapse towards its centre, forming a black hole. Clearly, black holes can exist in theory. The next question is how and under what conditions might a black hole form?

The calculation similar to Rector Michell's shown in Box 13.3 indicates that a star the same density as the sun would have to be about 500^3 (125 million) solar masses to form a black hole. Current estimates are that the largest sun-like stars in the universe are only about 150 solar masses.[1] However, having a higher density might allow less massive stars to become black holes. How dense can a star become?

Our sun is powered by nuclear fusion, so quantum mechanics enters into our story. For those with a copy of my book *Quantum Untangling*, there are cross-references in the footnotes. Put in the simplest terms, the intense gravitational pressure in the sun leads nuclei of hydrogen (one proton) to fuse into helium (two protons, two neutrons). For larger stars, this fusion process can continue up to the production of iron (26 protons, 30 neutrons). Nuclear fusion produces energy, creating an outward pressure in the star that counteracts the inward gravitational pressure.

Once the fusion process nears conclusion, this outward pressure reduces and the star collapses inwards. If the star is large enough, it explodes in a supernova. Fusion to elements heavier than iron *requires* energy.[2] They are forged in these stellar explosions, so if you own any gold or silver jewellery, you can be confident that it all started with a bang. The end of life for most stars is to become what are called *white dwarfs*. They emit progressively less radiation and finally fade into obscurity as *black dwarfs*.

The gravitational pressure in white and black dwarfs is offset by *electron degeneracy pressure*. If you crush together the atoms in the core of a star, there is resistance. The electrons in the atoms are *fermions*, which cannot be in the same state, because of the *Pauli exclusion principle*.[3] In layman's language, each electron (actually, pair of electrons) sort of needs its own space. The tighter you constrain an electron, the shorter its wavelength and the higher its kinetic energy.[4] The energy required to constrain electrons creates what we call electron degeneracy pressure. Is it possible for the gravitational pressure of a massive star to crush the resistance of these defiant electrons?

The first person to address this question was Subrahmanyan Chandrasekhar, an Indian astronomer. He calculated that the maximum mass of a white dwarf is only 1.4 solar masses (yes, that is 1.4, not 14), now known as the *Chandrasekhar limit*. Beyond this mass, gravitational pressure wins out. The implication in terms of black holes was clear. He presented his result to the Royal Society in 1935. It was not well-received by some of his peers (see Box 13.5), but today it is still accepted as an accurate result.

If the residual mass after a supernova explosion is greater than the Chandrasekhar limit, it will collapse to a *neutron star*. The basic process is that, given the intense gravitational pressure in a star of over 1.4 solar masses that has burnt through its fusion fuels, the negatively charged electrons combine with the positively charged atomic protons to form electrically neutral neutrons.[5] In layman's language, the star crunches from a tight cluster of atoms down to a much tighter cluster of atomic nuclei (but all neutrons). The neutrons are also fermions and have their own degeneracy pressure. However, most of you will know that the nucleus is only a tiny component of an atom. In fact, the nucleus represents about 10^{-15} of the total atomic volume. Therefore, the difference in density between a white dwarf and a neutron star is enormous.

I will now show you why Chandrasekhar's conclusion so alarmed those physicists who did not accept the existence of black holes. Let's do a rough calculation to illustrate what size of neutron star creates a black hole. The mass of a star of radius r is its average density ρ multiplied by its volume, which is $\rho \frac{4}{3}\pi r^3$. If its physical radius is exactly the same as its Schwarzschild radius R_S (so it is just massive enough to become a black hole), then its mass is: $M = \rho \frac{4}{3}\pi R_S^3$. From this and the definition of the Schwarzschild radius, we can calculate its mass in kilograms versus its density ρ as shown in Equation 13.15.

$$M = \rho \frac{4\pi}{3} R_S^3 = \rho \frac{4\pi}{3} \left(\frac{2GM}{c^2}\right)^3 \approx \rho \frac{10\pi G^3}{c^6} M^3,$$

(13.15)

$$\implies M^2 = \left(\frac{1}{\rho}\right) \frac{c^6}{10\pi G^3} \approx \left(\frac{1}{\rho}\right) \frac{(3 \times 10^8)^6}{10\pi (7 \times 10^{-11})^3} \approx \left(\frac{1}{\rho}\right) 10^{80}.$$

---

1 See for example D. Figer, *An upper limit for the masses of stars, Nature*, 8 March 2005.
2 *Quantum Untangling*, Chapter 16, Section 16.7.
3 *Quantum Untangling*, Chapter 17, Section 17.2.
4 *Quantum Untangling*, Chapter 7, Section 7.4.
5 Technical note: neutrinos are emitted in this process.

We can approximate the density of a neutron star using the mass of a neutron m_n (1.6×10^{-27} kg) and its radius r_n (10^{-15} m), giving density $\rho \approx 10^{18}$ kg m^{-3} as in Equation 13.16. Substituting this into Equation 13.15 gives $M^2 \approx 10^{62}$, which means $M \approx 10^{31}$ kg.

$$\rho \approx m_n \frac{3}{4\pi r_n^3} \approx 10^{18}\,\text{kg}\,\text{m}^{-3} \qquad \Longrightarrow \qquad M \approx 10^{31}\,\text{kg}. \tag{13.16}$$

Based on our rough calculation, any neutron star with mass above 10^{31} kg will become a black hole. The mass of our sun is about 2×10^{30} kg, so this result is only a few solar masses... and our sun is by no means abnormally large for a star. In summary, once a star is all out of fuel for fusion it collapses. If the result is an object of more than 1.4 solar masses, gravity will overcome the electron degeneracy pressure (as Chandrasekhar showed) and it will crunch down to a neutron core. If that core is more than a few solar masses, it will form a black hole. A more accurate calculation than our rough estimate arrives at a mass limit of between 2 and 3 solar masses for neutron stars (called the *Tolman–Oppenheimer–Volkoff limit*). There are many many billions of stars in the universe much larger than the sun. Once Chandrasekhar had shown how easy (well, relatively easy) it is for gravity to overcome electron degeneracy pressure, there was no realistic obstacle to the formation of black holes.

Box 13.5 Chandrasekhar and Eddington: simple heresy

The 24-year-old Subrahmanyan Chandrasekhar (1910–1995) presented his work on the collapse of white dwarf stars to the Royal Society in 1935. Also present was his 52-year-old friend and mentor Arthur Eddington (1882–1944), renowned for the eclipse observations that underpinned Einstein's theory (see Section 5.6). To Chandrasekhar's horror, Eddington stood up immediately after his presentation and brutally attacked and rebutted his work. Eddington was unable to accept the idea that stars might collapse into black holes; that in his words:

The star has to go on radiating and radiating and contracting and contracting until, I suppose, it gets to a few kilometres' radius, when gravity becomes strong enough to hold in the radiation, and the star can at last find peace...
I think there should be a law of Nature to prevent a star from behaving in this absurd way!

At another meeting, Eddington described it as *simple heresy*. He was highly respected, and his opposition proved a huge drag to the acceptance of the theory. Finally, almost 50 years after his presentation and 40 years after Eddington's death, Chandrasekhar won the 1983 Nobel Prize. In his prize acceptance, he quoted the following poem:

Where the mind is without fear and the head is held high;
Where knowledge is free;
Where words come out from the depth of truth;
Where tireless striving stretches its arms towards perfection;
Where the clear stream of reason has not lost its way into the dreary desert sand of dead habit;
Into that haven of freedom, let me awake. R. Tagore

Now you know how black holes can form, let's return to our analysis of the Schwarzschild metric and the story of our intrepid astronaut friend, Schwart. You may remember that we left him hovering near a black hole.

13.6 Falling into a Black Hole

My readers, being super-intelligent, may have figured out that hovering near the event horizon of a black hole is unlikely to be good for Schwart's health. As he approaches it, he needs to apply more and more acceleration. Before he reaches the event horizon, the acceleration required to resist the gravitational pull will squash him to a pancake.

But what if Minky, in a romantic effort to save her boyfriend from being pancaked, tells Schwart to forget the experiment and turn off his spaceship's engines? Rather than being pancaked, Schwart will no longer feel any accelerating force. He will be weightless as he freefalls into the black hole.

If this confuses you, think of the equivalence principle (Chapter 6). Gravitational acceleration is not the result of an external accelerating force. It is distortion of the geodesics. Each particle in Schwart's body will freefall, floating along its geodesic towards the centre of the black hole. At the event horizon where it would take an infinite amount of acceleration to hover, Schwart will feel no acceleration at all measured locally (this caveat is important, as I will clarify in a moment). He will float across the event horizon happily un-pancaked, heading towards the centre of the black hole, blissfully unaware of any acceleration... except for tidal forces.

If Schwart's body occupied a single infinitely small point, our story would be complete. However, this is not the case. Every particle in Schwart's body follows its own geodesic. The separation between the particles means these geodesics differ. Assuming Schwart chooses to float in feet first, his feet will be closer to the gravitational source than his head, creating a relative acceleration between the two that will progressively stretch his body and tear it apart (spaghettification). The only comfort I can offer is that, for a large enough black hole, this will happen inside the event horizon, so his girlfriend Minky will never see it.

Let's have a closer look at what happens to the metric near the event horizon where $r = R_S$. If we consider an object falling radially towards the black hole, then $d\theta = d\phi = 0$. The Schwarzschild metric becomes that shown on the left of Equation 13.17 and tends towards the value on the right as the object closes in on the event horizon. What the heck kind of a metric is that?

$$(c d\tau)^2 = \left(1 - \frac{R_S}{r}\right)(c\,dt)^2 - \frac{1}{\left(1 - \frac{R_S}{r}\right)}dr^2, \implies (c d\tau)^2 = (0)\,(c\,dt)^2 - (\infty)\,dr^2. \tag{13.17}$$

At the event horizon, the time coordinate tends to zero in terms of the invariant interval, and the space coordinate tends to infinity. That sounds like a singularity of some sort. How can we reconcile this with Schwart's smooth freefall, which is required by the equivalence principle (ignoring those pesky spaghettifying tidal forces)?

The answer is that this is a *coordinate singularity*, not a fundamental curvature singularity. As an example of a coordinate singularity, consider the metric for the surface of a unit sphere in Chapter 3. You may want to refer back to Figure 3.3 and the metric in Equation 3.17. If we consider the surface at $\theta = 0$ (the north pole), then $\sin^2\theta = 0$ and the $d\phi$ term disappears in the invariant interval. Does this mean that the north pole is a special point on the surface; some sort of singularity? No! Of course not. The north pole has the same curvature as any other point on the sphere's surface. The north pole is a coordinate singularity, not a curvature singularity.

We can check for curvature singularities in the Schwarzschild metric by looking at its underlying Riemann curvature. The Schwarzschild curvature footprint is shown again as Equation 13.18 (this is Equation 12.22 from Chapter 12). The Riemann curvature causing the gravitational effect is always a multiple of $\frac{GM}{c^2 r^3}$ (as calculated earlier in Equation 12.21). First let's evaluate the point $r = 0$, which is the centre of the black hole. As r tends to zero, the curvature tends to infinity. Clearly, the centre of a black hole is a *curvature singularity*.

$$\text{Curvature Footprint of Schwarzschild Metric} \qquad S = \frac{GM}{c^2 r^3} \tag{13.18}$$

$$
\begin{array}{llllllll}
g^{TT}R_{TT} = & g^{TT}R^T_{TTT} + g^{TT}R^R_{TRT} + g^{TT}R^\theta_{T\theta T} + g^{TT}R^\phi_{T\phi T} = & 0 & -2S & +S & +S & = 0, \\
g^{RR}R_{RR} = & g^{RR}R^T_{RTR} + g^{RR}R^R_{RRR} + g^{RR}R^\theta_{R\theta R} + g^{RR}R^\phi_{R\phi R} = & -2S & +0 & +S & +S & = 0, \\
g^{\theta\theta}R_{\theta\theta} = & g^{\theta\theta}R^T_{\theta T\theta} + g^{\theta\theta}R^R_{\theta R\theta} + g^{\theta\theta}R^\theta_{\theta\theta\theta} + g^{\theta\theta}R^\phi_{\theta\phi\theta} = & S & +S & +0 & -2S & = 0, \\
g^{\phi\phi}R_{\phi\phi} = & g^{\phi\phi}R^T_{\phi T\phi} + g^{\phi\phi}R^R_{\phi R\phi} + g^{\phi\phi}R^\theta_{\phi\theta\phi} + g^{\phi\phi}R^\phi_{\phi\phi\phi} = & S & +S & -2S & +0 & = 0.
\end{array}
$$

On the other hand, at the event horizon where $r = R_S$, we can substitute in the definition of the Schwarzschild radius $R_S = \frac{2GM}{c^2}$. As shown in Equation 13.19, there is no weird singularity. The Riemann curvature remains a well-behaved measurable quantity. Its value does not veer off to infinity. This means that the event horizon is a *coordinate singularity*.

$$r = \frac{2GM}{c^2}, \quad \implies \quad \frac{GM}{c^2 r^3} = \frac{c^4}{8\,G^2 M^2}. \tag{13.19}$$

In summary, the underlying Riemann curvature varies smoothly at the event horizon. There is no spike in the distortion of the geodesics, so Schwart will freefall across, unaware of any sudden local change. The weird metric at the event horizon shown on the right of Equation 13.17 is the result of the coordinates we are using. Later, I will introduce some alternative coordinate systems that help display what happens at the event horizon and beyond. However, these different coordinate systems are complicated and far from intuitive. So, for the time being, we will stick with coordinates that are based on how things appear to a distant observer. Let's examine what Minky observes.

13.7   Time: For Minky the Clock Stops

The simplest scenario is to compare the time of Schwart near the event horizon of a black hole with the time of Minky observing from afar. We will assume both observers are stationary relative to the black hole ($dr = d\theta = d\phi = 0$). The Schwarzschild metric simplifies to that shown on the left of Equation 13.20. In this scenario, Minky's clock tick is $d\tau$ because $\frac{R_S}{r}$ is very small, as explained in Section 13.3.

$$(cd\tau)^2 = \left(1 - \frac{R_S}{r}\right)(c\,dt)^2, \quad \implies \quad (cd\tau)^2 = (0)\,(c\,dt)^2, \quad \textit{at the event horizon.} \tag{13.20}$$

Minky observes Schwart's clock to tick more slowly than hers. This should not surprise you. Schwart is in a gravitational field, and so his clock is affected by gravitational time dilation. I showed you in Section 13.3 that if Schwart hovers at 50× the Schwarzschild radius, his clock would run about 1% slower than Minky's. If he stayed there for 20 years and then headed back to Minky to renew their relationship, he would find she had aged about 2 months more than he had.[6] In Section 5.7, I described how GPS satellite times must be adjusted to take account of this effect.

What happens if Schwart hovers closer and closer to the black hole? As Schwart hovers ever closer to the event horizon, the invariant interval equation tends to the equation shown to the right of Equation 13.20. The metric tells us that Minky, a distant observer in flat spacetime, will see Schwart's clock progressively slow down (in theory it would stop ticking altogether at the event horizon). As a result, Minky will never see Schwart reach the event horizon. And what if Minky moves closer to the black hole so she too is in a gravitational field? If she positioned herself stationary at $\frac{R_S}{r} = 0.5$ (twice the Schwarzschild radius), she would see Schwart's clock tick faster than before, but Schwart's clock would still tick ever slower for her as he approaches the event horizon. Minky's move does not change the location of the event horizon, nor the fact that she will never see him reach it.

A clever student might point out that the left of Equation 13.20 is just $g_{TT} = \left(1 - \frac{2GM}{c^2}\right)$, which is nothing new. It has come up again and again in this book. We had it from time dilation back in Section 5.1 and again in the Weak Field metric of Section 10.1. Why then the surprise? The difference is that those metrics were created by matching up with the conditions we experience in a *weak* gravitational field. There was no assumption that this would tell the whole story in extreme conditions. As the Schwarzschild metric is the *only* spherically symmetrical vacuum solution to the EFEs, it indicates that this extreme time dilation really does occur near the event horizon.

Let me summarise with a more complete scenario. Minky accelerates her spaceship to hold steady in the black hole's gravitational field. She kicks Schwart out of the airlock (they must have argued) and watches him head in towards the event horizon. Initially, she will see him accelerate. However, as he approaches the event horizon, she will see him go slower and slower because of the time dilation. She will never see him reach the event horizon. As his time dilates, the light from him will redshift and become fainter and fainter. The redshift is because high frequency light based on his clock becomes progressively lower frequency for her (see Box 13.6). The light becomes fainter because, for example, one photon per second emission from him will become one photon per week, then one photon per year, then one photon per million years for her. From her perspective, he will never cross the event

---

6 Schwart's journeys to and from Minky also would affect relative ageing: see the twin paradox, Section 2.7.

horizon even if she watches for eternity. He will remain there, in what is called the *accretion disk* of gas around the event horizon, and fade from view, a somewhat sad end to their romantic involvement.

Tying back to the black hole images in Figure 13.1, we distant observers see a weak redshifted version of the high-intensity radiation from gas and dust particles crashing together and swirling down to the accretion disk at the event horizon... which does not sound much fun for Schwart.

Box 13.6 Gravitational redshift

Light experiences a redshift when leaving a gravitational field. This is because clocks within the gravitational field run more slowly than clocks outside it. Consider the time between wave crests of a light beam emitted by Schwart close to the black hole. This time is longer for Minky (because her clock runs faster). The speed of light is constant, so the wavelength λ of the emitted light is longer when it reaches Minky. The frequency of the light is inversely proportional to its wavelength $\left(f \propto \frac{1}{\lambda}\right)$, so the frequency of the light is lower when it reaches Minky.

Based on the relationship in Equation 13.20, we can compare the frequency of light emitted at r within the gravitational field (labelled f_r) with the frequency of light arriving at an observer who is an infinite distance from the gravitational source (labelled f_∞). The redshift, typically labelled z, is the proportional change in frequency. The right of Equation 13.21 shows how to calculate the resulting redshift of the light.

$$\frac{dt}{d\tau} = \left(1 - \frac{R_S}{r}\right)^{-\frac{1}{2}}, \qquad \Longrightarrow \qquad (1 + z) = \frac{f_r}{f_\infty} = \left(1 - \frac{R_S}{r}\right)^{-\frac{1}{2}}. \tag{13.21}$$

13.8 A Simple Illustrative Model

The fact that Minky never sees Schwart cross the event horizon befuddled me the first time I heard of it. It raises all sorts of questions. If Minky sees Schwart's clock slow towards a stop, does that mean Schwart, looking back, would see the whole future of the universe flash in a moment? The answer to that is *no*, as I will explain. To properly understand how objects and light behave near a black hole, we later will use alternative coordinate systems. However, these different coordinates can be confusing, so let me give you an intuitive feel of what is going on. I must warn serious physicists that I am about to break some rules... but I think the emergent illustration is worth it. We will be using *spacetime diagrams*, so Box 13.7 offers a brief refresher if you need it.

Box 13.7 Spacetime diagrams

The left of Figure 13.2 shows an example. Time is shown passing vertically, and space is shown horizontally. Typically, only one dimension of space is shown. Time and space are scaled to be equivalent, just as in this book with the units (ct, x, y, z). If, for example, we measure time in seconds, then space is measured in units of 3×10^8 metres (light seconds of distance). When scaled this way, the path of light is always a 45° line across the diagram (shown as the speed of light c in the figure).

Referring again to the left of Figure 13.2, the path of an object is shown in red. We know at once that it is a massive object because it is not moving at 45° so it is at sub-light speed. From any point (*a*) on its path, nothing can reach outside the subtended light paths. This is because nothing moves faster than the speed of light. The accessible region is called the *future light cone*. The name comes from spacetime diagrams that use two spatial dimensions where the speed of light paths forms a cone from point (*a*). Just as there is a future light cone, there is a *past light cone* that the object must have come from.

Let's return to our distant hovering observer Minky watching Schwart head towards the event horizon of a black hole. One is tempted to create a spacetime diagram along the lines of that shown on the right of Figure 13.2. This is completely *wrong*, but let me explain it. Minky (*M*) and Schwart (*S*) start the same distance away from the black

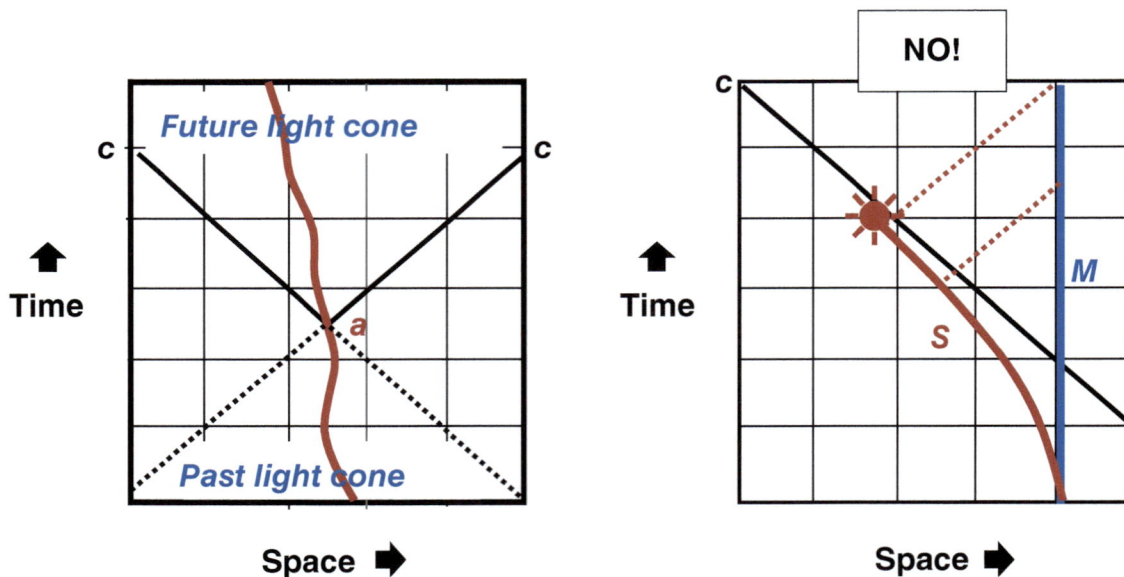

Figure 13.2 An example spacetime diagram is shown on the left. The right shows an incorrect diagram of the journey into a black hole.

hole. Minky (*M*) hovers, remaining at that distance (illustrated by a perpendicular line travelling through time but not space). Schwart (*S*) is kicked out of the airlock and heads towards the black hole, which is to the left of Minky in the diagram. Schwart accelerates up to (but never quite reaches) light speed. He crashes into the black hole singularity (shown by a red explosion mark). The dotted red lines show the light from Schwart reaching Minky (diagonal as at light speed). Based on this *wrong* spacetime diagram, it appears that Minky will be able to watch Schwart's journey all the way into the singularity, albeit that it will take a little time for the light to reach her.

Of course, the mistake is that we must use the curved Schwarzschild metric. Figure 13.2 uses the flat Minkowski metric. Oops! However, we can take advantage of the equivalence principle to build an illustrative model in Minkowski spacetime that I hope will help you understand what is going on. The equivalence principle tells us that, from Schwart's perspective in freefall along a geodesic, spacetime appears locally flat. In this frame of reference, Schwart is stationary (he has no engine and feels no acceleration). Minky is accelerating away from him (her engines must be on to hover). As for the black hole, Schwart sees it and its event horizon accelerating towards him. This is shown in Figure 13.3. On the left, you can see that Schwart (*S*) is stationary and so moves straight up the spacetime diagram. The equivalence principle says that Minky's position, hovering stationary in a gravitational field, is equivalent to accelerating away from the attracting mass. Therefore, Minky (*M*) is shown accelerating away from Schwart. Minky will travel closer and closer to light speed but can never actually reach it. The path in the figure tracks her up to 0.85*c*, but she continues to accelerate beyond that. This path is a *hyperbola*. The black diagonal in the figure is a speed-of-light path, which is the *asymptote* of Minky's hyperbola.[7] Her path gets ever closer to this asymptote without ever crossing it. While Minky never crosses this diagonal, Schwart does and this is the moment he crosses the event horizon. Stating this more correctly in our reference frame, it is the moment that the event horizon, coming in from the left, passes Schwart, whom we view as stationary. Behind the event horizon is the black hole singularity, which hits Schwart a short time later.

After the event horizon has passed Schwart (i.e. he is inside it), the left of the figure shows a dotted red line representing the path of light from Schwart travelling in Minky's direction. You can see that this light can never reach her. As proper acceleration is infinite at the event horizon (see Section 13.4), Schwart sees the event horizon pass him at light speed. No light that he produces can ever escape beyond it. This, after all, is what the event horizon is.

---

7 To calculate the distance *D* travelled by time *t* at acceleration *a*, the formula is: $D = \frac{c^2}{a}\left(\sqrt{1 + \frac{a^2 t^2}{c^2}} - 1\right)$.

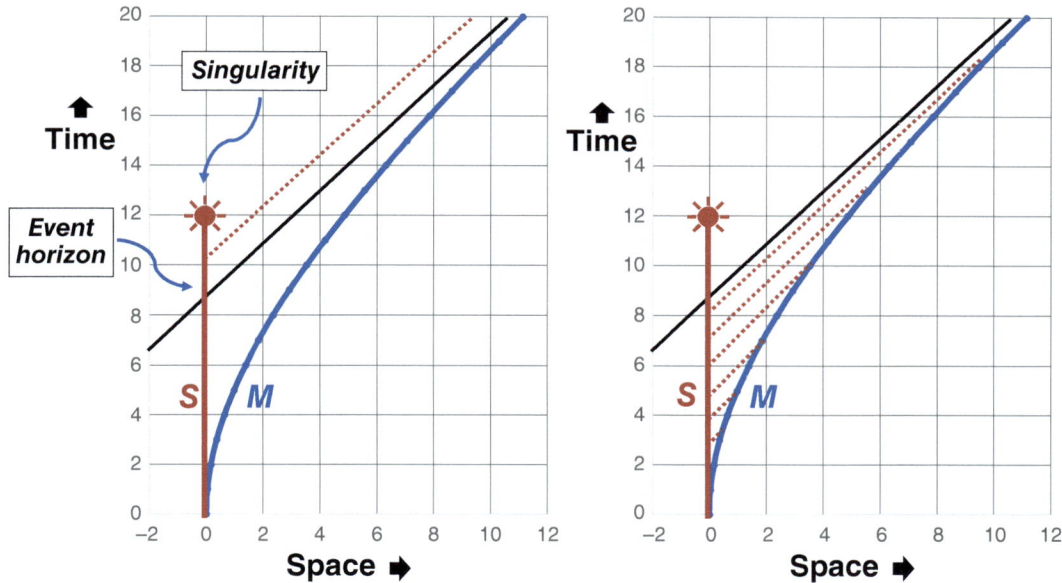

Figure 13.3 Changing perspective: Schwart follows a geodesic, Minky is accelerating.

Now turn to the right of Figure 13.3, which shows as dotted red lines the paths of light from Schwart to Minky before he crosses the event horizon. For illustration, let's suppose that the unit of time is measured in days (it could be anything just as long as the unit of space matches). What does Minky see as Schwart approaches the event horizon? Roughly reading off the figure, on Day 5, she sees Schwart's Day 4. On Day 13, she sees Schwart's Day 7. On Day 18, she sees Schwart's Day 8... and so on. She can wait forever, but will never see him cross the event horizon because that is the asymptote to her hyperbola path. She will see Schwart slow down and finally stop. His image will be redshifted by the gravitational field from visible light to infrared to microwave to radio wave... and will fade away from view as fewer and fewer photons per day reach her. However in theory, given some miraculous ideal viewing device, Minky would be able to detect Schwart at the event horizon for eternity.

The last step in this analysis is to review what Schwart sees as he reaches the event horizon. If Minky sees Schwart's clock slow for eternity towards a stop, then does Schwart see Minky's clock speed up such that he sees the whole future of the universe flash by in an instant? For Minky's light to reach Schwart, it must be in his *past light cone* (see Box 13.7). This means that light from Minky would travel up diagonally left in Figure 13.3 to reach Schwart. I have not shown this explicitly in the figure, but if you draw light lines up diagonally to the left from Minky to Schwart, you can see that nothing changes at the event horizon. Schwart continues to see her until he hits the singularity. However, he is *not* seeing her future. For example, if you draw a dotted light line from Minky on Day 12, you can see it will not reach Schwart, who has been consumed in the singularity.

Let me end this section with a warning. I have spent several pages discussing this model because I find the equivalence principle gives a useful shortcut to a helpful picture. However, it is at best a rough illustration. The correct analysis must be done using the Schwarzschild metric in alternate coordinates. We will cover this in Chapter 16.

13.9 Space: Schwarzschild Radial Coordinate

It is time to turn our attention to the radial coordinate r. To save you referring back, Equation 13.22 shows the Schwarzschild metric, and Equation 13.23 gives the metric for a radial path at the event horizon (as earlier in Equation 13.17). As you can see, the radial measure dr^2 also becomes weirdly distorted at the event horizon of a black hole.

$$(cd\tau)^2 = \left(1 - \frac{R_S}{r}\right)(c\,dt)^2 - \frac{1}{\left(1 - \frac{R_S}{r}\right)}dr^2 - r^2 d\theta^2 - r^2\sin^2\theta\,d\phi^2, \tag{13.22}$$

$$\implies (cd\tau)^2 = (0)\,(c\,dt)^2 - (\infty)\,dr^2, \qquad \textit{radial path at event horizon.} \tag{13.23}$$

Some of you will be surprised to learn that the radial coordinate *r* is *not* the radial distance out from the centre of the black hole. What then is it? It is the radial distance out from the centre *if and only if* the spacetime is flat. The presence of a black hole, or indeed to a lesser extent any massive body, will distort spacetime. The radial distance from the centre will no longer be *r*.

What then is *r*? At a spacetime point with a coordinate value *r*, a circle around the origin in the θ direction has circumference $2\pi r$. For any given value of *r*, the value (in terms of the invariant interval) of a move in the $d\theta$ or $d\phi$ directions are identical in the Schwarzschild metric and the Minkowski metric (compare the last two terms of Equations 13.22 and 12.1). Thus, a rocket travelling near the black hole could in theory calculate the value of coordinate *r* by circling around the black hole and dividing the length of the circumference by 2π. For obvious reasons, this coordinate *r* is called the *reduced circumference*.

It took a while for my brain to get a fix on this, so let me explain with the following scenario. Minky from afar observes two brothers, Schwart-1 and Schwart-2 on a direct line between her and a central point of origin. They are respectively at radial coordinates r_1 and r_2, as shown in Figure 13.4. In the absence of any gravitational field (the left of the figure), I hope it is obvious that Minky sees the clocks of Schwart-1 and Schwart-2 tick at the same speed, and she measures the distance *D* between them as $D = r_2 - r_1$.

Imagine now that a central mass or black hole appears out of nowhere at the origin. We allow a moment for the spacetime curvature to settle (gravitational curvature must obey the light speed limit). As shown on the right of the figure, Minky will measure no change in the circumference of the circular paths of Schwart-1 and Schwart-2 around the origin. You already know that Minky will measure the clock of Schwart-1 to tick more slowly than that of Schwart-2 (both will tick more slowly than hers). In addition as shown on the right of the figure, Minky will measure the distance between Schwart-1 and Schwart-2 to be *greater* than the distance between them that she would measure in flat space (gravitational curvature means that $D > r_2 - r_1$).

The curvature created by the mass or black hole does not change the circumference of the rockets' circular paths, but it does change Minky's measurement of radial distances from the central mass. The value of the Schwarzschild *dr* in terms of the invariant interval grows versus that in flat Minkowski spacetime. The disparity depends on the value of coordinate *r* and becomes ever larger as you approach the event horizon. Figure 13.5 is called the *Flamm paraboloid*. It shows the relationship at an instant in time (i.e. $dt = 0$). At the event horizon, the schematic becomes vertical, and however great the distance of travel measured by Minky, she observes no radial progress towards the central mass (i.e. no reduction in coordinate *r*).

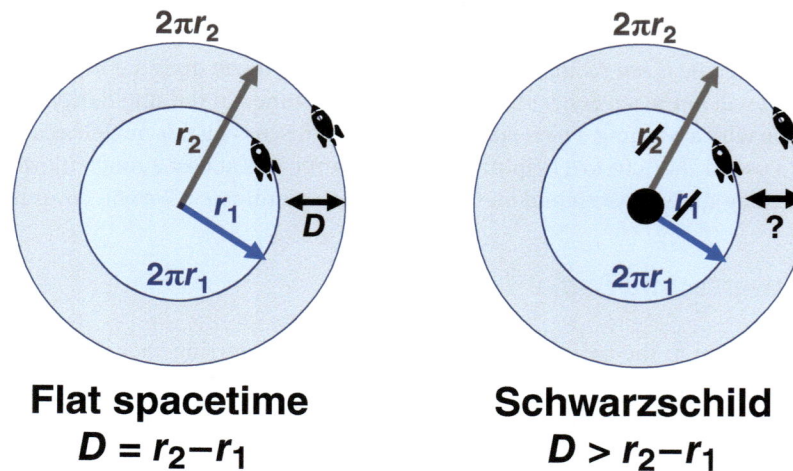

Flat spacetime
$D = r_2{-}r_1$

Schwarzschild
$D > r_2{-}r_1$

Figure 13.4 The presence of a central mass changes a distant observer's measure of the radial distance between two objects.

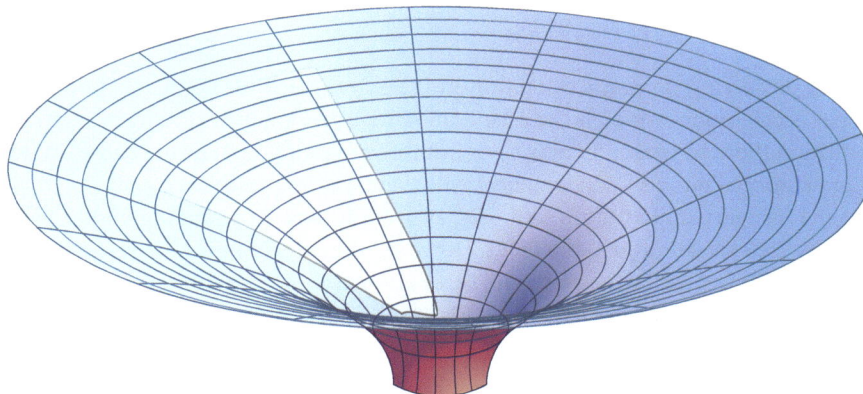

Figure 13.5 The Flamm paraboloid.

The trend to the (0) $(c\,dt)^2$ term in Equation 13.23 means that Minky measures each step forwards in time on Schwart's clock (each tick) as an ever-growing passage of time on her clock. Conversely, the (∞) dr^2 term means that Minky measures Schwart's radial progress through space (in terms of the value of radial coordinate r) as being ever less significant... until from her perspective, he is making no radial progress at all in spite of the fact that, from his perspective, he is moving at high speed towards the black hole singularity (or if you prefer, the black hole is moving at high speed towards him). She gets what you might loosely describe as a foreshortened view of his radial progress. In Minky's reference frame, the time and spatial distortions combine additively to slow Schwart's progress towards the event horizon.

This should sound vaguely familiar. You might want to check back to Section 5.4, leading to the metric in Equation 5.21. You will see that it is a similar scenario to the Schwarzschild metric, albeit for a weak field, not a black hole. In that section, I went on to derive the geodesic deviation and show the time and spatial curvature combine (Equation 5.23) such that a fast-moving object, such as light, experiences the combined deviation of its passage through space and time. There is a detailed discussion of the deflection of light in optional Chapter 15.

13.10 Inside the Event Horizon

Any inquisitive reader is going to wonder what is happening to the Schwarzschild metric inside the event horizon. When $r < R_S$, then both the dt^2 and dr^2 terms in the invariant interval switch sign (see Equation 13.22). It is as if time and space switch roles. Let me give you a tentative hand-wavy sort of introduction to this subject, which will become an issue for further discussion in Chapter 16.

Let's compare the properties of time and space. Outside the event horizon, Schwart can (at least in theory) start his engines and blast in a spatial direction away from the black hole. The progression in value of r can increase or decrease. It can become anything in theory. In contrast, his progress through time is one-way only.

Inside the event horizon things are different. Schwart in freefall follows a geodesic. In theory, he can start his engines and blast in any direction on to a different geodesic, but every path and every direction leads to the same singularity. In the reference frame of the model where Schwart is stationary, the singularity is approaching him from *every direction*. His radial progress through space now is one way only... like the passage of time outside the event horizon.

What can we say about where the singularity is? Measured in terms of space, the coordinate point $r = 0$ is in every direction. Radial progress dr no longer aligns with a certain spatial direction. It aligns with progress through time dt. The roles have reversed.

And what happens at the actual singularity $r = 0$? We know that any singularity conflicts with the laws of quantum mechanics, but taking the Schwarzschild metric at face value, the forward progression of time dt, measured using the invariant interval of proper time, becomes infinitely negative. Time as we know it ceases to exist.

13.11 Summary

I started this chapter discussing proper acceleration in the Schwarzschild metric. Using the geodesic equation, I showed that proper acceleration, which is the acceleration required to hover stationary relative to a gravitational field, becomes infinite at the Schwarzschild radius. As nothing can produce an infinite acceleration, the conclusion must be that the gravitational pull on such an object will overcome any resistance.

We have to take the Schwarzschild metric seriously, even for intense gravitational fields, because it is the only spherically symmetric vacuum solution to the EFEs. However, the Schwarzschild radius is small for objects in our solar system: about one centimetre for the earth and three kilometres for the sun, so you don't have to worry about being sucked into a nearby black hole (see cartoon in Figure 13.6). Initially physicists could argue that, while black holes might exist in theory, there is no physical way to form such a dense body.

Along came Chandrasekhar to calculate what happens when a star runs out of fusion fuel and explodes in a supernova. He demonstrated that the electron degeneracy pressure that underpins the resistance of atoms to collapse will be overcome by the gravitational pressure if it forms a white dwarf of more than 1.4 solar masses. In this case, the atomic electrons and protons combine into neutrons. The star collapses into a neutron star, and its density increases to a level similar to that of the nucleus of an atom. We used a rough calculation to demonstrate that, given its high density, a neutron star of more than a few solar masses will have a radius less than its Schwarzschild radius. It will collapse and form a black hole. It is now known that there are black holes at the centre of most galaxies, including one of about four million solar masses at the centre of our Milky Way. I even provided a couple of images (Figure 13.1).

The focus of the rest of this chapter has been what the Schwarzschild metric tells us about the curvature of spacetime near a black hole. At first glance, there seems to be a curvature singularity at the Schwarzschild metric ($r = R_S$). However, examining the underlying Riemann curvature (the curvature footprint in Equation 13.18) reveals that this is a singularity from the perspective of a distant observer. It is a coordinate singularity, not a curvature singularity.

We compared the viewpoints of two astronauts, Schwart and Minky. Schwart travels across the event horizon while Minky observes from afar. Schwart freefalls across the event horizon unaware of any sudden change.

Figure 13.6 Beware of the black hole. *Source*: BAUSCHRON / Wikimedia commons / CC BY 4.0.

However, Minky sees Schwart's clock tick ever slower as he approaches the event horizon. His image redshifts and fades away. She never sees him cross it.

I tried to illustrate with a spacetime diagram that treats Schwart as stationary (freefall following a geodesic) and Minky as accelerating away from him (rocket engine on to counter the gravitational field). The event horizon is the asymptote to Minky's accelerating hyperbola path (Figure 13.3). Once Schwart has crossed it, his light can never reach her. She will only ever see light from him before he reached the event horizon. And although Minky sees Schwart's clock tick slower and slower for eternity, Schwart does not see the eternity of Minky's future. See Box 13.8 for a discussion of whether all this might best be described as illusion or reality.

There is also distortion to the radial component of the metric dr. The radial coordinate r at any point is the reduced circumference of a circular path around the centre. The curvature of the Schwarzschild metric does not affect this measure, but it does increase the radial separation between objects as measured by a distant observer (Figures 13.4 and 13.5). The effect on the geodesics of this spatial distortion adds to that of the time dilation in the sense that Minky, from her perspective, observes both Schwart's clock to slow and his radial progress to decrease... the combined effect slowing his inward fall so she never sees him reach the event horizon.

We briefly discussed the strange change to the Schwarzschild metric inside the event horizon. All geodesics lead to the singularity. Whatever direction Schwart might decide to move leads to the singularity. Let me repeat that we cannot say much about conditions at the singularity. That requires a theory of quantum gravity, which we do not yet have.

Box 13.8 Reality or illusion?

Some people ponder whether time dilation and space distortion are best described as *real* or just as *illusions* due to perspective. You can ask the same question of special relativity. For example, Observer A sees the clock of a passing Observer B tick more slowly (time dilation). And yet Observer B, considering herself stationary, similarly sees the clock of Observer A tick more slowly. That seems like an *illusion* due to perspective. However, if one of these observers accelerates off and then returns, less time will have passed on her clock than on the other's (the famous twin paradox). Now, that seems *real*.

At the black hole horizon, Minky sees Schwart's clock slow towards a stop. However, Schwart does not see the eternity of Minky's future flash by. That seems like an *illusion* of perspective. However, if Schwart leaves Minky, travels towards the black hole, and then returns, less time will have passed on his clock than will have on hers. That seems *real*. Personally, I find it easier to think of these things as *real*, but it is a question of semantics. To quote the amazing theoretical physicist Kip Thorne in his book *Black Holes and Time Warps*:

Since the two viewpoints agree on the results of all experiments, they are physically equivalent. Which viewpoint tells the "real truth" is irrelevant for experiments; it is a matter for philosophers to debate, not physicists.

Chapter 14

Orbits and Conserved Quantities

Conserved quantities in the Schwarzschild metric, equivalent to energy and angular momentum, reveal differences between general relativity and Newton's gravitational theory. Newtonian physics allows for bound orbits at any distance from the gravitational source, which would allow stuff to swirl around in orbit close to the black hole singularity. The Schwarzschild metric doesn't allow this. Once an object has crossed the event horizon of a black hole, smashing into the singularity is inescapable.

In fact, I will show you that the radius of the innermost stable circular orbit, called the *ISCO*, is three times the Schwarzschild radius. Showing this takes some mathematical effort, but the result is important for ensuing discussion of accretion disks, quasars and the interpretation of data from gravitational wave observatories.

As an extra perk, we will be able to calculate our explorer Schwart's freefall plunge time from the event horizon to the central singularity... hidden from the view of outside observers, but still... ouch! Another fringe benefit is that you are going to learn why conserved quantities are conserved. Get your neurons humming and ready for Noether's theorem and Killing vectors.

14.1 Noether, Killing Vectors and Conservation Laws

If you are like me, you will have wondered what makes energy, linear momentum and angular momentum conserved quantities. In Sections 2.8 and 2.9 of Chapter 2, I showed you how to derive the relationship between energy and momentum from the invariant interval of Minkowski spacetime. Please look back for a quick refresh if you need to. You can see that energy and momentum are closely linked with the Minkowski metric.

When we say energy is conserved, we mean that as a quantity it is conserved in any given frame of reference (i.e. following a geodesic path). It is likely to change if you switch to a new reference frame. For example, one observer might see a rocket fly by with kinetic energy and momentum, but for an observer stationary relative to the rocket, it has neither. However, if you stick to the same reference frame (geodesic path), quantities such as energy, linear momentum and angular momentum are conserved.

You may have heard of Noether's theorem: *for every symmetry, there is a conserved quantity*. Emmy Noether was such an amazing woman that I have taken the liberty of including as Box 14.1 some material from my book *Quantum Untangling*. Some of you may know that the conservation of energy is related to a symmetry in physics over time, and the conservation of momentum is related to a symmetry over space. But what does this mean?

In mathematical language, sticking to the same reference frame means following a geodesic path through spacetime (in Minkowski spacetime, geodesics are straight unaccelerated paths). Conserved quantities do not change along the geodesics of whatever spacetime we are studying.

Jumping to the answer, the trick is to look for variables that the metric $[g_{\mu\nu}]$ does *not* depend on. For example, it might be that t does not appear in any of the components of $[g_{\mu\nu}]$. Whether we look at 11am or 2pm, or whenever, the metric will be the same. For each variable that does not appear in $[g_{\mu\nu}]$, the *lower index* version of the rate of change of that variable versus proper time is a conserved quantity. We label this u_x for any variable x. Obviously this

Untangling General Relativity: The Intuitive Self-Study Guide, First Edition. Simon Sherwood.
© 2026 John Wiley & Sons Ltd. Published 2026 by John Wiley & Sons Ltd.

quantity multiplied by rest mass m also is conserved. For example, the Minkowski metric does not depend on time t, so $m u_t$ is conserved. As the Minkowski metric is diagonal (no off-diagonal terms), we know: $m u_t = m g_{TT} u^t = m g_{TT} \frac{dt}{d\tau}$.

To keep things clear, I will use (t, x, y, z) units throughout this chapter. The Minkowski metric has the invariant interval in Equation 2.15. In this case, $g_{TT} = c^2$ and the conserved quantity associated with time is $mc^2 \frac{dt}{d\tau}$, which is the definition of *relativistic energy* E (see Equation 2.17). As shown in Equation 2.13, at non-relativistic speeds $mc^2 \frac{dt}{d\tau} \approx mc^2 + \frac{1}{2}mv^2$, which is rest mass energy plus classical kinetic energy. Thus, the conservation of energy is a direct consequence of the Minkowski metric not being time-dependent.

Similarly, the Minkowski metric does not depend on x so $m g_{XX} \frac{dx}{d\tau} = -m \frac{dx}{d\tau}$ is conserved, which is *relativistic momentum* p_x in the x direction (negative because of the metric signature I am using, as described in Box 3.1). Thus in classical physics, we have conservation of momentum in the x direction. The Minkowski metric also does not depend on y or z so there is also conservation of momentum in the y and z directions, unless there is some other external force.

Box 14.1 Symmetry and conservation: Emmy Noether (1882–1935)

Einstein: *Noether was the most significant creative mathematical genius thus far produced since the higher education of women began.*

Wiener: *Miss Noether is ... the greatest woman mathematician who has ever lived.*

Strong praise indeed. Noether's theorem proved that for every symmetry there is an associated conservation law, a conserved observable quantity. This was a key step in helping Einstein to sort out the maths of general relativity. It is also a central tenet of quantum mechanics. But consider the obstacles this genius faced as a woman...

- The challenge to study mathematics as one of only two female students in the whole of Erlangen University. The numbers are not surprising as the University's stated position was that mixed-sex education would *overthrow all academic order.*
- The challenge to get a position at Göttingen University in 1915 in spite of being a leading mathematician of her time. One faculty member's reaction was: *What will our soldiers think when they return to the university and find that they are required to learn at the feet of a woman?*
- The challenge to get paid. As a woman, her teaching went unpaid for the first eight years.

On top of this, she suffered Nazi persecution as a Jew and was summarily dismissed in 1933. She relocated to the USA, where sadly she died of an ovarian cyst at only 53 years. David Hilbert summed up the frustration of many physicists at her lack of advancement: *I do not see that the sex of the candidate is an argument against her admission... after all, the senate [university] is not a bath house.*

Written in spherical coordinates, the Minkowski metric depends on r and θ but does not depend on t or ϕ (see Equation 9.2 if you need a reminder). We have already addressed t, which is conservation of energy. Non-reliance on ϕ gives us *conservation of angular momentum*, which I label L. It is: $L = -m g_{\phi\phi} \frac{d\phi}{d\tau} = mr^2 \sin^2\theta \frac{d\phi}{d\tau}$. Do not be confused when you see this written elsewhere as $L = mr^2 \omega$. In that case, r is the distance from the axis of rotation that in spherical coordinates for a rotation in the ϕ direction, is $r \sin\theta$ (see Figure 9.1). And ω is the speed of rotation, which is almost exactly $\frac{d\phi}{d\tau}$ except at relativistic speeds.

Each of the variables t, x, y, z and ϕ, on which the Minkowski metric does *not* depend, gives what is called a *Killing vector*. The name suggests some dramatic vector operation slicing the metric to death. I am sorry to say that the origin is less exciting. It simply is named after the German mathematician Wilhelm Killing. For each Killing vector, there is a conservation law.

Turning to the Schwarzschild metric (see Box 13.2), you can see it does not depend on t and ϕ, which gives Killing vectors leading to conservation laws for energy and angular momentum, albeit subtly different from those of Minkowski spacetime. However, before we discuss that, I want to demonstrate why these Killing vectors lead to conservation laws.

14.2 Conserved Quantities Along Geodesics

In this book, I have often limited myself to diagonal metrics (no off-diagonal components). For simplicity, I will do the same again and show you that if a diagonal metric $[g_{\mu\nu}]$ does *not* depend on a variable (which I arbitrarily label x), then the quantity u_x is conserved along geodesic paths. The Schwarzschild metric is diagonal, so this is enough for now.

To demonstrate this, we are going to calculate the value of $\frac{du_x}{d\tau}$ and show the value is zero (meaning u_x is constant) for a geodesic pathway when the metric doesn't depend on x. As shown in Equation 14.1, the value of u_x for a diagonal metric is $g_{XX} u^x$, which we can differentiate by parts. The first term can be expressed using the multivariable chain rule for derivatives (see Box 14.2 if needed), summing across all values of **A**, a dummy variable, arriving at Equation 14.2.

$$\frac{du_x}{d\tau} = \frac{d(g_{XX} u^x)}{d\tau} = \frac{d(g_{XX})}{d\tau} u^x + g_{XX} \frac{du^x}{d\tau}, \tag{14.1}$$

$$= \frac{\partial g_{XX}}{\partial \mathbf{A}} \frac{d\mathbf{A}}{d\tau} u^x + g_{XX} \frac{du^x}{d\tau},$$

$$= \frac{\partial g_{XX}}{\partial \mathbf{A}} u^{\mathbf{A}} u^x + g_{XX} \frac{du^x}{d\tau}. \tag{14.2}$$

Box 14.2   Multivariable chain rule for derivatives

The derivative of a function can be expressed as the sum of derivatives across the coordinates. An example is shown below for function f and coordinates (t, x, y, z):

$$\frac{df}{d\tau} = \frac{\partial f}{\partial t} \frac{dt}{d\tau} + \frac{\partial f}{\partial x} \frac{dx}{d\tau} + \frac{\partial f}{\partial y} \frac{dy}{d\tau} + \frac{\partial f}{\partial z} \frac{dz}{d\tau}. \tag{14.3}$$

The next step is to calculate the value of $\frac{du^x}{d\tau}$ along a geodesic pathway, which is quantified by the geodesic equation from Box 5.2 in Section 5.2, shown below as Equation 14.4. As a reminder, this comes from setting the *covariant derivative* $\nabla_\tau u^x$ to zero and calculating the distortion of the geodesic paths in terms of $\frac{du^x}{d\tau}$. To evaluate the possible value of dummy variables **A** and **B**, you may need to refer back to the Christoffel symbol cheat sheet in Box 4.4. For a diagonal metric, any Christoffel symbol with three different indices is zero-value. Therefore, we know that the Christoffel symbol in Equation 14.4 has value only if either **A** or **B** is x, or **A** = **B**. This leads to Equation 14.5. As both dummy variables **A** and **B** run through all possible indices, the first two terms in this equation are equivalent and can be added, leaving only one dummy variable. Equation 14.6 applies the relevant formula from the cheat sheet for the first term.

$$\frac{du^x}{d\tau} = -\Gamma^X_{\mathbf{AB}} u^{\mathbf{A}} u^{\mathbf{B}}, \tag{14.4}$$

$$= -\Gamma^X_{\mathbf{A}X} u^{\mathbf{A}} u^x - \Gamma^X_{X\mathbf{B}} u^x u^{\mathbf{B}} - \Gamma^X_{\mathbf{AA}} u^{\mathbf{A}} u^{\mathbf{A}} = -2\,\Gamma^X_{\mathbf{A}X} u^{\mathbf{A}} u^x - \Gamma^X_{\mathbf{AA}} u^{\mathbf{A}} u^{\mathbf{A}}, \tag{14.5}$$

$$= -g^{XX} \frac{\partial g_{XX}}{\partial \mathbf{A}} u^{\mathbf{A}} u^x - \Gamma^X_{\mathbf{AA}} u^{\mathbf{A}} u^{\mathbf{A}}. \tag{14.6}$$

We then substitute this value into Equation 14.2. The first two terms in the result (Equation 14.7) cancel out. Applying the value of the Christoffel symbol to the last term gives the final result in Equation 14.8.

$$\frac{du_x}{d\tau} = \frac{\partial g_{XX}}{\partial \mathbf{A}} u^{\mathbf{A}} u^x + g_{XX} \frac{du^x}{d\tau},$$

$$= \frac{\partial g_{XX}}{\partial \mathbf{A}} u^{\mathbf{A}} u^x - g_{XX} g^{XX} \frac{\partial g_{XX}}{\partial \mathbf{A}} u^{\mathbf{A}} u^x - g_{XX} \Gamma^X_{\mathbf{AA}} u^{\mathbf{A}} u^{\mathbf{A}}, \qquad (14.7)$$

$$= -g_{XX} \Gamma^X_{\mathbf{AA}} u^{\mathbf{A}} u^{\mathbf{A}},$$

$$= g_{XX} \frac{1}{2 g_{XX}} \frac{\partial g_{\mathbf{AA}}}{\partial x} u^{\mathbf{A}} u^{\mathbf{A}},$$

$$\frac{du_x}{d\tau} = \frac{1}{2} \frac{\partial g_{\mathbf{AA}}}{\partial x} u^{\mathbf{A}} u^{\mathbf{A}}, \qquad \qquad \textit{for any diagonal metric.} \qquad (14.8)$$

This is the key result we want. $\mathbf{A}$ is a dummy variable, which means that $\frac{du_x}{d\tau}$ is the sum of terms, which all include a derivative of a component of the metric with respect to x. Therefore we have shown for any diagonal metric that if the metric $[g_{\mu\nu}]$ does not depend on x, the value of every term in the sum is zero.

More generally it can be shown that $\frac{du_x}{d\tau}$ is always zero along a geodesic pathway if $[g_{\mu\nu}]$ has no x dependence (shown later in Box 18.5). This is true even if the metric $[g_{\mu\nu}]$ has off-diagonal terms. This means that u_x always is a conserved quantity when $[g_{\mu\nu}]$ has no dependence on the x variable. To derive the conservation laws, we multiply by rest mass m allowing us to handle calculations dealing with multiple objects, such as two colliding billiard balls where total momentum is conserved.

For the Minkowski metric, the Killing vectors give conservation of energy (no time t dependence in the metric), conservation of linear momentum (no x, y or z dependence) and conservation of angular momentum (no ϕ dependence). You now can tell your friends that you know what drives these conservation laws. Emmy Noether and Wilhelm Killing would be proud of you.

14.3 Conserved Quantities of the Schwarzschild Metric

Based on what we have shown about Killing vectors, it is simple to construct the conserved quantities mu_t and mu_ϕ of the Schwarzschild metric. Equation 14.9 shows the Schwarzschild metric with conserved quantities below. On the left is the expression for mu_t, which I have labelled E_S. For comparison, the equivalent conserved quantities in Minkowski spacetime are also shown. At $r = \infty$, they are equivalent because at an infinite distance from the centre of the gravitational field, the Schwarzschild metric is flat and the metric is the same as Minkowski spacetime.

The conserved quantity mu_ϕ gives the same expression for conservation of angular momentum for the Schwarzschild metric as for Minkowski spacetime, so I use the same label L. However, differences in rotational behaviour will appear when we dig into more detail.

$$\left(1 - \frac{R_S}{r}\right) c^2 dt^2 - \frac{1}{\left(1 - \frac{R_S}{r}\right)} dr^2 - r^2 d\theta^2 - r^2 \sin^2\theta \, d\phi^2 = c^2 d\tau^2, \qquad (14.9)$$

Schwarzschild: $\quad m g_{TT} u^t = \left(1 - \frac{R_S}{r}\right) mc^2 \frac{dt}{d\tau} = E_S, \qquad m g_{\phi\phi} u^\phi = -mr^2 \sin^2\theta \frac{d\phi}{d\tau} = -L,$

Minkowski: $\quad m g_{TT} u^t = mc^2 \frac{dt}{d\tau} = E, \qquad m g_{\phi\phi} u^\phi = -mr^2 \sin^2\theta \frac{d\phi}{d\tau} = -L.$

I must give a word of warning about comparing the conserved quantities of Schwarzschild and Minkowski spacetime. The spacetimes have *different geodesics*. In Minkowski spacetime, a stationary object is following a geodesic. This means we can evaluate conserved quantities when an object is only moving through time, i.e. when only $\frac{dt}{d\tau} \neq 0$. In Schwarzschild spacetime, a stationary object is *not* following a geodesic because it is not in freefall. For example, standing still on the surface of the earth is not a Schwarzschild geodesic, because you are being accelerated away from freefall by the resisting force of the ground beneath you. Schwarzschild geodesics must involve radial motion, so $\frac{dr}{d\tau} \neq 0$, unless an object is in orbit when either $\frac{d\theta}{d\tau} \neq 0$ or $\frac{d\phi}{d\tau} \neq 0$.

14.4   Radial Plunge

Our aim is to examine the effect of the Schwarzschild conserved quantities E_S and L. However, I first want to review how things are handled classically using Newtonian mechanics and the Minkowski metric. For some of you this will be a bit tedious, but it will highlight the differences.

Let's start with a look at a radial freefall plunge in a gravitational field as described using Minkowski spacetime. This takes two steps. The first is to evaluate the properties of a radial path through Minkowski spacetime without any gravitational field. The second is to add in the gravitational effect using Newton's law.

As the motion is radial, $d\theta = d\phi = 0$, and the Minkowski invariant interval in spherical coordinates (such as shown in Figure 9.1) reduces to Equation 14.10. In the next lines, I have multiplied both sides by $\frac{m}{2}$, where m is the rest mass of the plunging object, and divided by $d\tau^2$. I have then rearranged the terms shifting some conserved quantities to the right-hand side. We arrive at Equation 14.11.

$$c^2 dt^2 - dr^2 = c^2 d\tau^2, \qquad \textit{Minkowski radial path}, \qquad (14.10)$$

$$\frac{mc^2}{2}\left(\frac{dt}{d\tau}\right)^2 - \frac{m}{2}\left(\frac{dr}{d\tau}\right)^2 = \frac{mc^2}{2},$$

$$\frac{m}{2}\left(\frac{dr}{d\tau}\right)^2 = \frac{mc^2}{2}\left(\frac{dt}{d\tau}\right)^2 - \frac{mc^2}{2} = \frac{E^2}{2mc^2} - \frac{mc^2}{2} = conserved. \qquad (14.11)$$

This puts the radial motion term in a form you will be familiar with $\frac{m}{2}\left(\frac{dr}{d\tau}\right)^2$ is kinetic energy (the object is only moving radially). All the terms on the right of Equation 14.11 are conserved quantities (E, m and c) in the sense that they do not vary along the path. This means that the object's kinetic energy (on the far left of the equation) is conserved if the path is *along a Minkowski geodesic*. This should not surprise you. Minkowski geodesics are straight-line acceleration-free paths.

To calculate what happens to the path in a gravitational field, we must make what you might call a *manual adjustment* to Minkowski spacetime. We adjust the conserved quantity by adding in Newton's expression for gravitational potential energy. This gives Equation 14.12, where M is the attracting mass. It says that the total of the kinetic energy of the object and its potential energy ($k.e + p.e$) is conserved. I know this is dull and obvious, but the punchline is coming.

$$\textit{Minkowski geodesic}: \qquad \frac{m}{2}\left(\frac{dr}{d\tau}\right)^2 = conserved,$$

$$\textit{Newtonian (gravity "added")}: \qquad \frac{m}{2}\left(\frac{dr}{d\tau}\right)^2 - \frac{GMm}{r} = conserved. \qquad (14.12)$$

We can perform exactly the same calculation using the Schwarzschild metric. On a radial plunge path, $d\theta = d\phi = 0$. If, as before, we multiply by $\frac{m}{2}$ and divide by $d\tau^2$, we get to Equation 14.13. We substitute with the conserved quantity E_S using the formula in Equation 14.9. We then rearrange, moving conserved quantities to the right, and substitute in the value of $R_S = \frac{2GM}{c^2}$. This gives Equation 14.14.

Take a look at Equation 14.14. All the quantities on the right (E_S, m and c) are conserved along a Schwarzschild geodesic. Therefore, the expression on the left is conserved. Furthermore, the radial plunge path in freefall *is* a Schwarzschild geodesic (a crucial difference from Minkowski spacetime). If you compare Equation 14.14 with Equation 14.12, you will see the conserved quantities are the same. However, we have not needed any manual adjustment to the Schwarzschild metric. The conservation law for what we classically call the total of kinetic energy plus gravitational potential energy ($k.e + p.e$) appears naturally from the metric.

$$\left(1 - \frac{R_S}{r}\right)c^2 \, dt^2 - \frac{1}{\left(1 - \frac{R_S}{r}\right)} dr^2 = c^2 \, d\tau^2, \qquad \textit{Schwarzschild radial plunge,}$$

$$\left(1 - \frac{R_S}{r}\right)\frac{mc^2}{2}\left(\frac{dt}{d\tau}\right)^2 - \frac{1}{\left(1 - \frac{R_S}{r}\right)}\frac{m}{2}\left(\frac{dr}{d\tau}\right)^2 = \frac{mc^2}{2},$$

$$\tag{14.13}$$

$$\frac{1}{\left(1 - \frac{R_S}{r}\right)}\frac{E_S^2}{2mc^2} - \frac{1}{\left(1 - \frac{R_S}{r}\right)}\frac{m}{2}\left(\frac{dr}{d\tau}\right)^2 = \frac{mc^2}{2}, \qquad \textit{note:}\, E_S = \left(1 - \frac{R_S}{r}\right)mc^2 \frac{dt}{d\tau},$$

$$\frac{E_S^2}{2mc^2} - \frac{m}{2}\left(\frac{dr}{d\tau}\right)^2 = \frac{mc^2}{2} - \frac{R_S mc^2}{2r},$$

$$\frac{m}{2}\left(\frac{dr}{d\tau}\right)^2 - \frac{R_S mc^2}{2r} = \frac{E_S^2}{2mc^2} - \frac{mc^2}{2} = \textit{conserved,}$$

$$\textit{Schwarzschild geodesic}: \qquad \frac{m}{2}\left(\frac{dr}{d\tau}\right)^2 - \frac{GMm}{r} = \textit{conserved.} \tag{14.14}$$

In addition to this showing you how the Schwarzschild metric incorporates gravitational effects, we can use the result to derive the equation of motion for the radial plunge path.

14.4.1 Plunge Time from Horizon to Singularity

Let's return to our intrepid explorer Schwart. Minky kicks him out of the rocket from her distant observation point, and watches his radial freefall into the black hole. We know Minky will never see Schwart reach the event horizon (Section 13.7). But how long does it take Schwart *on his clock* to travel from the event horizon of the black hole into the singularity? In this imaginary scenario that ignores his inevitable spaghettification near the singularity, does he have time for a nice cup of tea and some scones?

In his radial plunge, Schwart is in freefall following a Schwarzschild geodesic. At the start of his trip, Schwart is stationary far out from the black hole, so both terms in Equation 14.14 are zero. This gives Equation 14.15 for the trajectory. We integrate with respect to r and then substitute in the values $r = R_S$ and $r = 0$ to calculate the plunge time $\Delta\tau$ from the event horizon to the singularity. The result is Equation 14.16.

$$\frac{m}{2}\left(\frac{dr}{d\tau}\right)^2 = \frac{R_S mc^2}{2r}, \quad \Longrightarrow \quad \frac{d\tau}{dr} = \frac{\sqrt{r}}{c\sqrt{R_S}}, \tag{14.15}$$

$$\tau = \frac{2}{3c\sqrt{R_S}} r^{\frac{3}{2}} + C, \qquad \textit{where C is an integration constant,}$$

$$\Delta\tau = \left(\frac{2}{3c\sqrt{R_S}} R_S^{\frac{3}{2}} + C\right) - (0 + C) = \frac{2}{3c} R_S \quad \textit{seconds.} \tag{14.16}$$

The Schwarzschild radius R_S of the sun is about three kilometres (3×10^3 metres), which means Schwart has about 10^{-5} seconds travel time per solar mass as shown in Equation 14.17.

$$\textit{Per solar mass}: \qquad \frac{2(3 \times 10^3)}{3(3 \times 10^8)} \approx 10^{-5} \quad \textit{seconds.} \tag{14.17}$$

If Schwart falls into a small black hole of a few solar masses, he has less than a millisecond between event horizon and singularity. While Minky sees him slow to a stop as he approaches the event horizon, Schwart does not have time even to blink between crossing the event horizon and hitting the singularity. If he falls into Sagittarius A^*, the black hole at the centre of the Milky Way with four million solar masses, he has under a minute, which still isn't enough time for a decent brew. However, if he chooses M87 with its five billion solar masses, he has several hours. That leaves plenty of time for tea, scones and to write a few pages in his diary about the experience, although for obvious reasons this would be a rather futile task (and don't forget that this ignores those spaghetti-fying tidal forces).

14.5   Angular Momentum and Rotational Energy

The next step is to generalise the scenario beyond a pure radial plunge path. This will reveal a difference between Einstein's description of gravity and that of Newton. In addition to radial movement, we need to include rotational movement around the black hole. We take advantage of the spherical symmetry of the Schwarzschild solution to organise the scenario so that the rotational movement is in the ϕ direction with $\theta = \frac{\pi}{2}$, which means that $\sin \theta = 1$ (if this confuses you, there is an illustration of spherical coordinates back in Figure 9.1). I will call this *equatorial rotation* because it is around the equator of the sphere. This simplifies the distance to the axis of rotation from $r \sin \theta$ to r, and the expression for angular momentum becomes: $L = mr^2 \frac{d\phi}{d\tau}$, which makes for easy comparison with classically defined angular momentum: $mr^2\omega = mr^2 \frac{\partial \phi}{\partial t}$.

Again, we will start with Minkowski spacetime and then compare with the Schwarzschild metric. The Minkowski invariant interval including equatorial rotation is shown as Equation 14.18. Don't forget that the equatorial rotation means that $\sin \theta = 1$ and $d\theta = 0$. Going through the same steps as earlier (multiply by $\frac{m}{2}$, divide by $d\tau^2$ and shift conserved quantities to the right), we reach Equation 14.19, which now includes the term for rotational energy. In Equation 14.20 this is expressed using the conserved quantity, angular momentum L, based on the definition in Equation 14.9 (which is the same for both Minkowski and Schwarzschild spacetimes). Note that each term on the left varies with r, but the total is conserved.

$$\textit{"Equatorial" rotation}: \quad c^2 dt^2 - dr^2 - r^2 \, d\phi^2 = c^2 d\tau^2, \qquad \textit{Minkowski,} \tag{14.18}$$

$$\frac{mc^2}{2}\left(\frac{dt}{d\tau}\right)^2 - \frac{m}{2}\left(\frac{dr}{d\tau}\right)^2 - \frac{mr^2}{2}\left(\frac{d\phi}{d\tau}\right)^2 = \frac{mc^2}{2},$$

$$\frac{m}{2}\left(\frac{dr}{d\tau}\right)^2 + \frac{mr^2}{2}\left(\frac{d\phi}{d\tau}\right)^2 = \frac{E^2}{2mc^2} - \frac{mc^2}{2} = \textit{conserved,} \tag{14.19}$$

$$\textit{note: } L = mr^2\frac{d\phi}{d\tau}, \qquad \frac{m}{2}\left(\frac{dr}{d\tau}\right)^2 + \frac{L^2}{2mr^2} = \textit{conserved.} \tag{14.20}$$

As we are working with Minkowski spacetime, we make Newton's *manual adjustment* to Equation 14.21 in order to account for gravitational potential energy (see earlier in Equation 14.12). There are three terms: the radial kinetic energy, the gravitational potential energy and the rotational kinetic energy. It is often convenient to group the last two together and to treat them as a sort of combined potential energy (V_{eff}), which is called the *effective potential* as in Equation 14.22.

$$\textit{Minkowski + gravity}: \quad \frac{m}{2}\left(\frac{dr}{d\tau}\right)^2 + \frac{L^2}{2mr^2} - \frac{GMm}{r} = \textit{conserved,} \tag{14.21}$$

$$\frac{m}{2}\left(\frac{dr}{d\tau}\right)^2 + V_{eff} = \textit{conserved,}$$

$$\textit{where}: \quad V_{eff} = \frac{L^2}{2mr^2} - \frac{GMm}{r}. \tag{14.22}$$

Much of this will be familiar to readers with a strong background in physics. Now it is time for the fun stuff. We will go through the same exercise with the Schwarzschild metric. We go through exactly the same steps as in Equation 14.13, but with the rotational term included. The definition of angular momentum L in the Schwarzschild metric is the same as in the Minkowski metric (see left of Equation 14.20 and/or refer back to Equation 14.9, setting $\sin \theta = 1$).

$$Schwarzschild: \quad \left(1 - \frac{R_S}{r}\right)c^2\,dt^2 - \frac{1}{\left(1 - \frac{R_S}{r}\right)}dr^2 - r^2\,d\phi^2 = c^2\,d\tau^2, \quad \text{"Equatorial"},$$

$$\left(1 - \frac{R_S}{r}\right)\frac{mc^2}{2}\left(\frac{dt}{d\tau}\right)^2 - \frac{1}{\left(1 - \frac{R_S}{r}\right)}\frac{m}{2}\left(\frac{dr}{d\tau}\right)^2 - \frac{mr^2}{2}\left(\frac{d\phi}{d\tau}\right)^2 = \frac{mc^2}{2},$$

$$\frac{1}{\left(1 - \frac{R_S}{r}\right)}\frac{E_S^2}{2mc^2} - \frac{1}{\left(1 - \frac{R_S}{r}\right)}\frac{m}{2}\left(\frac{dr}{d\tau}\right)^2 - \frac{L^2}{2mr^2} = \frac{mc^2}{2},$$

$$\frac{E_S^2}{2mc^2} - \frac{m}{2}\left(\frac{dr}{d\tau}\right)^2 - \left(1 - \frac{R_S}{r}\right)\frac{L^2}{2mr^2} = \frac{mc^2}{2} - \frac{R_S\,mc^2}{2r}. \tag{14.23}$$

Note that an additional angular term appears in Equation 14.23 because we multiply through with the $\left(1 - \frac{R_S}{r}\right)$ factor. This creates an additional rotational energy term, which is $\left(-\frac{R_S}{r}\right)\frac{L^2}{2mr^2}$.

When the conserved quantities are moved to the right, there are now four terms on the left that vary with r, which means there is an additional term in the effective potential. Compare Equation 14.25, which is the effective potential of the Schwarzschild metric, with that of Minkowski spacetime in Equation 14.22.

$$Schwarzschild: \quad \frac{m}{2}\left(\frac{dr}{d\tau}\right)^2 - \frac{R_S\,mc^2}{2r} + \left(1 - \frac{R_S}{r}\right)\frac{L^2}{2mr^2} = \frac{E_S^2}{2mc^2} - \frac{mc^2}{2}, \tag{14.24}$$

$$\frac{m}{2}\left(\frac{dr}{d\tau}\right)^2 - \frac{GMm}{r} + \frac{L^2}{2mr^2} - \frac{L^2\,GM}{mc^2r^3} = conserved,$$

$$V_{eff} = \frac{L^2}{2mr^2} - \frac{L^2\,GM}{mc^2r^3} - \frac{GMm}{r}. \tag{14.25}$$

The effective potential is important because its derivative determines the radial acceleration of the object $\frac{d^2r}{d\tau^2}$, which can be written equivalently as $\frac{du^r}{d\tau}$. I described in the first few paragraphs of Subsection 10.1.1 the relationship in classical physics between the gravitational potential Φ and gravitational acceleration: $a^r = -\frac{\partial\Phi}{\partial r}$. It is worth flicking back and reading up to Equation 10.3 if you need a reminder. When we include the effect of angular movement, the net acceleration in the radial direction becomes $m\frac{du^r}{d\tau} = -\frac{\partial V_{eff}}{\partial r}$.

$$Minkowski + gravity: \quad V_{eff} = \frac{L^2}{2mr^2} - \frac{GMm}{r}, \tag{14.26}$$

$$Schwarzschild\ geodesic: \quad V_{eff} = \frac{L^2}{2mr^2} - \left(\frac{R_S}{r}\right)\frac{L^2}{2mr^2} - \frac{GMm}{r}. \tag{14.27}$$

14.6 A Few Words

Before I dive into the effect this has on orbits, let me simplify by explaining with words (*thank heavens* some readers may exclaim). Let's look first at the classical Newtonian picture. Imagine a rock being attracted towards a large mass. In Equation 14.26, V_{eff} has two terms. The last is the gravitational potential energy, which *decreases* (becomes more negative) as the rock approaches. This creates the attracting gravitational force. On the other hand, if the rock has angular momentum, then the first term, which is rotation-related, *increases* as the rock approaches, creating an apparent outward *centrifugal* force (an example is the outward pull you feel if you whirl around a rock tied to a piece of rope).

Look carefully at Equation 14.26. The rotational term is proportional to $\frac{1}{r^2}$ while the gravitational term is proportional to $\frac{1}{r}$. As the rock approaches, r gets smaller. At some stage, the rotational term will grow to fully offset the gravitational term, unless the rock has no angular momentum or has already hit the surface of the attracting mass. Where the inward gravitational force and apparent outward centrifugal force develop some sort of balance, stable circular orbits can form.

Now let's turn our attention to the Schwarzschild effective potential in Equation 14.27. There is an additional new rotational term. I have rewritten it in a way that makes it simple to see that it is $\frac{R_S}{r}$ in size, compared to the original rotational term. In the case of the earth orbiting the sun, its orbital radius is over 10^8 kilometres versus the sun's Schwarzschild radius of under 2 kilometres: $\frac{R_S}{r} \approx 10^{-8}$. The new rotational term seems inconsequential at this sort of level (although later we will discuss a tiny effect on Mercury's orbit). It is easy to understand how Newton missed it.

However, in the intense gravitational curvature near a black hole, this new term is significant. At the event horizon $r = R_S$, the new rotational term completely offsets the original. It does not matter how much angular momentum an object has, it will accelerate towards the singularity. Indeed, within the event horizon the new term is larger than the original, so angular momentum actually increases acceleration towards the singularity. Think back to Schwart crossing the event horizon in freefall. If he turns on his engine and accelerates in the ϕ direction to try to enter orbit, he gets to the singularity more quickly than he would with the engine off! You might joke that near the event horizon of a black hole, a swirling apple falls a whole lot faster than Newton thought.

14.7 Orbits and Trajectories

Let's discuss orbits. We are interested in the acceleration of an attracted object, which is the force per kilogram of mass ($F = ma$). The *acceleration* depends on the derivative of the *effective potential per kilogram of mass*, which I label U_{eff} and call the *specific effective potential* (watch out that some texts switch between V_{eff} and U_{eff} without warning).

Thus, we have radial acceleration: $\frac{d^2r}{d\tau^2} = -\frac{\partial U_{eff}}{\partial r}$ (see Box 14.3 for further explanation).

Box 14.3 Specific effective potential and radial acceleration

In order to provide an intuitive understanding, the starting point is conservation of energy as shown on the left of Equation 14.28. As an object moves radially, its total effective potential V_{eff} changes. For energy conservation, this must be balanced by a change in its radial kinetic energy. Dividing through by object mass m gives the right of Equation 14.28.

$$V_{eff} + \frac{1}{2}m(u^r)^2 = constant, \implies U_{eff} + \frac{1}{2}(u^r)^2 = constant. \tag{14.28}$$

We then differentiate the right of Equation 14.28 with respect to proper time τ, knowing the result must be zero (because it is a constant). For the second term, we use the relationship: $\frac{d}{d\tau} = \frac{d}{du^r}\frac{du^r}{d\tau} = \frac{d}{du^r}a^r$, where a^r is radial acceleration. As U_{eff} depends only on r, we can apply the chain rule giving the result on the right of Equation 14.29.

$$\frac{dU_{eff}}{d\tau} + u^r a^r = 0, \implies \frac{\partial U_{eff}}{\partial r}\frac{dr}{d\tau} + u^r a^r = 0. \tag{14.29}$$

Dividing through by u^r gives the result we are looking for. Radial acceleration is equal to the negative of the derivative with respect to r of the specific effective potential U_{eff}.

$$\frac{\partial U_{eff}}{\partial r}u^r + u^r a^r = 0, \implies a^r = -\frac{\partial U_{eff}}{\partial r}. \tag{14.30}$$

Up to now, I have worked with angular momentum L because it will be most familiar to readers. The underlying conserved angular quantity of both the Schwarzschild and the Minkowski metric is $g_{\phi\phi}u^\phi$, which I label ℓ. This is $\ell = r^2\sin^2\theta\frac{d\phi}{d\tau}$ (see Equation 14.9). Obviously $\ell = \frac{L}{m}$, so it is the attracted object's angular momentum divided by its mass, sometimes called its *specific angular momentum*. The U_{eff} specific effective potentials exclude the need for any reference to the attracted object's mass m, as shown in Equations 14.31 and 14.32 (M is the attracting mass). This simplifies things.

$$Minkowski + gravity: \quad U_{eff} = \frac{\ell^2}{2r^2} - \frac{GM}{r}, \tag{14.31}$$

$$Schwarzschild\ geodesic: \quad U_{eff} = \frac{\ell^2}{2r^2} - \frac{\ell^2 GM}{c^2 r^3} - \frac{GM}{r}. \tag{14.32}$$

14.7.1 Newton's Circular Orbits

The orbit of any planet in our solar system is an ellipse, a kind of off-shaped circle (we will discuss elliptical orbits later). For now, we will limit ourselves to the simplest orbit, which in theory is perfectly circular. This keeps the maths simple (well... as simple as possible). To calculate the classical circular Newtonian orbit, we need to work out the points where there is no radial acceleration: $\frac{d^2r}{d\tau^2} = 0$. This requires that based on Equation 14.31, $\frac{\partial U_{eff}}{\partial r} = 0$. If this is the case, then if we put an object in a circular orbit, meaning $\frac{dr}{d\tau} = 0$, then it will stay in orbit.

$$Minkowski + gravity: \qquad \frac{\partial U_{eff}}{\partial r} = -\frac{\ell^2}{r^3} + \frac{GM}{r^2} = 0, \qquad \Longrightarrow \qquad r = \frac{\ell^2}{GM}. \tag{14.33}$$

The calculation is shown in Equation 14.33. We take the derivative of U_{eff} using Equation 14.31 and set it to zero to find any maxima or minima. The result is the point of minimum effective potential where the gravitational and centrifugal terms are in balance. Under the classical Newtonian regime, this means that there is a stable orbit for any object, however small its value of ℓ (providing it is non-zero), if it can get close enough to the attracting mass without hitting its surface.

If an object with specific angular momentum ℓ is further than $r = \frac{\ell^2}{GM}$ from the attracting mass, it will experience a net acceleration towards the mass. If it gets closer than this, it will experience an accelerating force away from it.

14.7.2 Schwarzschild's Circular Orbits

We perform the same calculation using the definition of U_{eff} in Equation 14.32. This is a bit trickier because setting the derivative to zero leads to the quadratic Equation 14.34. To solve this, you must dig back into the dark recesses of your brain for the quadratic formula, or you can cheat by checking out Box 14.4. This gives the result in Equation 14.35. As a reality check, note that when $\frac{12\,G^2M^2}{\ell^2c^2}$ is very small, Equation 14.35 tends to $r = \frac{\ell^2}{GM}$ as the only non-zero solution, which matches the Newtonian result.

$$\frac{\partial U_{eff}}{\partial r} = -\frac{\ell^2}{r^3} + \frac{3\ell^2 GM}{c^2 r^4} + \frac{GM}{r^2} = 0, \qquad \Longrightarrow \qquad GMr^2 - \ell^2 r + \frac{3\ell^2 GM}{c^2} = 0, \tag{14.34}$$

$$r = \frac{\ell^2}{2GM}\left(1 \pm \sqrt{1 - \frac{12\,G^2M^2}{\ell^2c^2}}\right). \tag{14.35}$$

But cripes! The Schwarzschild solution in Equation 14.35 looks very different from the Newtonian version in Equation 14.33. The $\pm$ sign means there generally are two solutions: one minimum and one maximum. The minimum is similar to the stable Newtonian orbit. The maximum appears much closer to the attracting mass and is something new. What is more, there is not always any circular orbit solution at all. This seems weird. And don't forget that the Schwarzschild solution is not about black holes. It is *the* spherically symmetric vacuum solution, so it must be a decent model for our solar system.

Box 14.4   Quadratic solution

$$If: \quad ax^2 + bx + c = 0, \qquad \Longrightarrow \qquad x = \frac{-b \pm \sqrt{b^2 - 4ac}}{2a} = \frac{-b}{2a}\left(1 \pm \sqrt{1 - \frac{4ac}{b^2}}\right).$$

Before you get too flummoxed, let me put things in perspective with a couple of examples. In the case of the earth's orbit around the sun, the value in Equation 14.35 of the factor $\frac{12\,G^2M^2}{\ell^2c^2}$ is of the order of 10^{-7}. In the case of the moon's orbit around the earth, it is of the order of 10^{-10}. Their predicted stable Schwarzschild orbits are almost indistinguishable from those of Newton.

Figure 14.1a is a (reasonable) schematic illustration of the effective potential of the earth in the presence of the gravitational pull of the sun, taking into account the earth's orbital angular momentum. The vertical axis shows the effective potential U_{eff}. The horizontal axis is the distance from the sun measured in *tens of millions* of Schwarzschild radii. The potential is at a minimum about 50 million R_S away from the sun. This is about 150 million kilometres and is where the earth (shown in the figure as a blue ball) has a stable solar orbit. Its orbit is

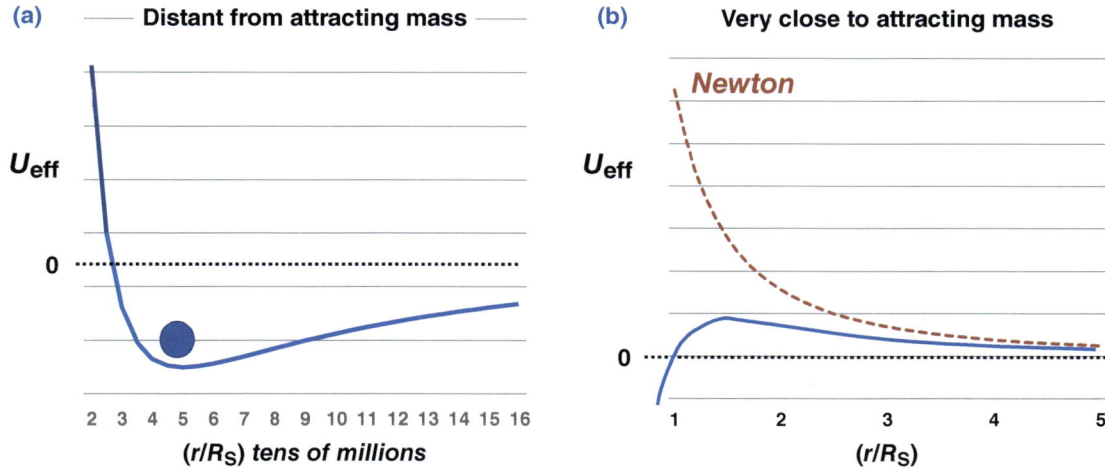

Figure 14.1 Schematic of earth's effective potential. The left (a) shows the calculation distant from the sun. The right (b) shows the theoretical result near to the sun's Schwarzschild radius.

actually somewhat elliptical, so you may want to think of the blue ball rolling slightly to and fro in the potential, rather than settled at one exact point.[1]

The single line shown for the U_{eff} potential holds for both the Schwarzschild and the Newtonian calculations. At 150 million kilometres from the sun, the difference between the two is only about one in 500,000. The two would be indistinguishable in the figure.

However, Equation 14.35 signals a major difference when objects get extremely close to the gravitational source. Figure 14.1b illustrates the effective potential U_{eff} were the earth able to get within a few Schwarzschild radii of a mass equivalent to the sun. In the case of the earth and sun, this is of course impossible. The radius of the sun is over 200,000 R_S. The approach of the earth would be limited by it crashing into the solar surface, but there is no comparable limit in the case of a black hole.

In this figure, the Schwarzschild specific effective potential U_{eff} is shown in blue. The (incorrect) Newtonian calculation is shown in orange (note that the scales for distance and U_{eff} are *not* the same in Figures 14.1a and 14.1b). Within a few R_S of the gravitational source, the results diverge. The Newtonian model suggests that the U_{eff} of any object with angular momentum rises rapidly and inexorably. This would create acceleration *away* from the singularity, offsetting the gravitational pull and creating stable orbits even inside the event horizon. In sharp contrast, the Schwarzschild metric tells us that the specific effective potential U_{eff} levels off and actually falls near the event horizon. Any object that ventures this close accelerates *towards* the singularity, whatever its angular momentum.

14.7.3 Innermost Stable Circular Orbit (ISCO)

One prediction of the Schwarzschild metric is that there is a limit to how close an object can approach a black hole and still enter a stable circular orbit. This is called the *innermost stable circular orbit* (ISCO). In Equation 14.35 there is only a real solution if the quantity in the square root is positive. This means that, for a stable circular orbit, ℓ^2 cannot be less than shown in Equation 14.36.

$$\frac{\ell^2 c^2}{G^2 M^2} = 12, \quad \implies \quad \ell^2 = \frac{12\,G^2 M^2}{c^2}, \qquad \textit{Minimum for circular orbit.} \qquad (14.36)$$

---

1 If you fancy creating some of these graphs, use the following relationship for Schwarzschild U_{eff}:

$$U_{eff} = \frac{c^2}{8}\left(\frac{K}{r_*^2} - \frac{K}{r_*^3} - \frac{4}{r_*}\right), \qquad where: \; r_* = \frac{r}{R_S}, \qquad K = \left(\frac{\ell c}{GM}\right)^2, \qquad U_{eff} \; units: \; \mathrm{m^2 s^{-2}}.$$

We substitute this value of ℓ^2 into Equation 14.35 as shown in Equation 14.37. This gives us the value of the ISCO radius as $3R_S$.

$$r = \frac{\ell^2}{2GM}(1) = \frac{12\,G^2M^2}{2\,GMc^2} = \frac{6\,GM}{c^2} = 3\,R_S. \tag{14.37}$$

The Innermost Stable Circular Orbit (ISCO) can be thought of as a boundary limiting the behaviour of objects that join the accretion disk around a black hole. Any object venturing closer than three times the Schwarzschild radius ($3R_S$) cannot hold circular orbit and will radially plunge into the black hole (this is called the *plunge region*). An interesting example is the coalescence of two black holes into one. Once they close in towards the ISCO, the pattern will change from a spiral towards each other into something more like a radial plunge.

Equation 14.36 also tells us that any object near a black hole with ℓ^2 less than that shown can never escape nor form a stable orbit. It must and eventually will freefall across the event horizon into the black hole. This freefall occurs as measured by a local clock. Don't forget that a distant observer will not see anything freefall across the event horizon because of gravitational time dilation (see Section 13.7).

I should note that these ISCO limits are affected by the spin of a black hole and, yes, all black holes are believed to spin. I will address this later, when we discuss the Kerr metric.

14.7.4 The Photon Sphere

One interesting quirk is the radius called the *photon sphere* where light orbits the black hole; at least in theory it does. This occurs at the point where the rotational element of the specific effective potential, which I will label U_{rot}, is at its maximum. This point can be calculated by taking the derivative of U_{rot}, which simply means taking the first two terms on the left of Equation 14.34 and setting the total to zero. The result is $\frac{3}{2}R_S$ as shown in Equation 14.38.

$$\frac{\partial U_{rot}}{\partial r} = -\frac{\ell^2}{r^3} + \frac{3\ell^2 GM}{c^2 r^4} = 0, \quad \Longrightarrow \quad r = \frac{3GM}{c^2} = \frac{3}{2}R_S, \quad U_{rot}\ maximum. \tag{14.38}$$

As an object freefalls inwards (before reaching the radius $r = \frac{3}{2}R_S$), the outward accelerating effect created by its orbital rotation will grow. Closer than $r = \frac{3}{2}R_S$ this orbital outward acceleration *decreases*. However high the object's angular momentum, any drift closer than $r = \frac{3}{2}R_S$ must generate net acceleration towards the singularity. It makes intuitive sense (at least to me) that this point of highest possible orbital outward acceleration should be associated with something moving at the maximum possible orbital speed, i.e. light. But let's be more rigorous.

Any path that light follows is called a *null geodesic* because for light $d\tau = 0$. We discussed this in Section 5.5 (see Equation 5.25). To evaluate the orbital path of a light beam, we orientate coordinates as usual so that it is what I have been calling an equatorial rotation. This means that it is in the ϕ direction with $\theta = \frac{\pi}{2}$ such that $\sin\theta = 1$ and $d\theta = 0$ (look back to the first paragraph of Section 14.5 if needed). If the light photons are in orbit around the black hole then $dr = 0$. In summary, $d\tau = 0$, $dr = 0$ and $d\theta = 0$. Only $d\phi$ and dt are non-zero, which reduces the Schwarzschild invariant interval to Equation 14.39.

$$\left(1 - \frac{R_S}{r}\right)c^2 dt^2 - \frac{1}{\left(1 - \frac{R_S}{r}\right)}dr^2 - r^2 d\theta^2 - r^2\sin^2\theta\,d\phi^2 = c^2 d\tau^2,$$

$$\left(1 - \frac{R_S}{r}\right)c^2 dt^2 - r^2\,d\phi^2 = 0, \quad \Longrightarrow \quad \left(\frac{d\phi}{dt}\right)^2 = \left(1 - \frac{R_S}{r}\right)\frac{c^2}{r^2}. \tag{14.39}$$

The next step is to calculate the radial acceleration of light using the geodesic equation. For light ($d\tau = 0$), we must work again with what is called an affine parameter: $\lambda = D\tau + E$, where D and E are non-zero constants. Using an affine parameter does not disturb the structure of the geodesic equation as was explained in Box 5.3 of Chapter 5.

The geodesic equation becomes Equation 14.40. Only $\frac{dt}{d\tau}$ and $\frac{d\phi}{d\tau}$ are non-zero, so we know that the only non-zero terms are when dummy variables **A** and **B** are either t or ϕ. Furthermore, Christoffel symbols with three different indices are zero, so the terms are only non-zero when **A** = **B**. This leads to the two terms on the right of Equation 14.40. You can calculate the first term using the Christoffel formula from Box 4.4, or you can cheat and pop in the value of Γ_{TT}^R, which we already derived in Equation 13.8. Don't forget to add in the c^2 term for g_{TT} because we are now working in coordinates of (t, r, θ, ϕ) rather than (ct, r, θ, ϕ). In reaching Equation 14.41, note that $g_{\phi\phi} = -r^2$ for this equatorial rotation.

$$\frac{d^2r}{d\lambda^2} = -\Gamma_{AB}^R \frac{d\mathbf{A}}{d\tau} \frac{d\mathbf{B}}{d\tau} = -\Gamma_{TT}^R \frac{dt}{d\lambda} \frac{dt}{d\lambda} - \Gamma_{\phi\phi}^R \frac{d\phi}{d\lambda} \frac{d\phi}{d\lambda}, \tag{14.40}$$

$$= -\frac{R_S c^2}{2r^2} \left(1 - \frac{R_S}{r}\right) \left(\frac{dt}{d\lambda}\right)^2 + \frac{1}{2} g^{RR} \frac{\partial g_{\phi\phi}}{\partial r} \left(\frac{d\phi}{d\lambda}\right)^2,$$

$$= -\frac{R_S c^2}{2r^2} \left(1 - \frac{R_S}{r}\right) \left(\frac{dt}{d\lambda}\right)^2 + r \left(1 - \frac{R_S}{r}\right) \left(\frac{d\phi}{d\lambda}\right)^2 = 0. \tag{14.41}$$

We are calculating an orbital path. This means that r is constant. There is no radial acceleration, so $\frac{d^2r}{d\lambda^2}$ is zero and therefore the expression in Equation 14.41 totals zero. This takes us to the result in Equation 14.42.

$$\frac{R_S c^2}{2r^2} \left(1 - \frac{R_S}{r}\right) \left(\frac{dt}{d\lambda}\right)^2 = r \left(1 - \frac{R_S}{r}\right) \left(\frac{d\phi}{d\lambda}\right)^2, \quad \implies \quad \left(\frac{d\phi}{dt}\right)^2 = \frac{R_S c^2}{2r^3}. \tag{14.42}$$

The grand finale is to equate the result in Equation 14.42 with our earlier result in Equation 14.39. With a bit of simple maths, this confirms that the radius of the photon sphere is indeed $\frac{3}{2}R_S$.

$$\frac{R_S c^2}{2r^3} = \left(1 - \frac{R_S}{r}\right) \frac{c^2}{r^2}, \quad \implies \quad \frac{R_S}{2} = r - R_S, \quad \implies \quad r = \frac{3}{2}R_S. \tag{14.43}$$

You may remember earlier, when describing light orbiting the black hole at the photon sphere, I added the words: *at least in theory it does*. This is because the photon sphere orbit is *unstable*. The slightest variation inwards will lead to the light spiralling down into the singularity. The slightest variation outwards will lead to it escaping the clutches of the black hole.

On the other hand, it does highlight the strange experience awaiting our astronaut Schwart as he heads towards the event horizon. If somehow he manages miraculously to avoid spaghettification, he would see light from all sides of the black hole bend around to him. Weird!

However, he might not have much fun if you add to this image the potential traffic jam of stuff in the accretion disk outside the event horizon. There would be material reaching the ISCO ($3R_S$) and spiralling in, along with radiation swirling around near the photon sphere $\left(\frac{3}{2}R_S\right)$. If a black hole encounters a region with enough stuff in it, just imagine the mess of swirling energetic particles and radiation smashing into each other. This brings us to the topic of quasars.

14.8 Quasars

The name quasar comes from *quasi-stellar object* because when they were first observed, nobody had any real clue what they were. And we are not talking ancient history. That was in the 1960s. One quasar can radiate energy thousands of times greater than that of a whole galaxy such as the Milky Way (which contains about 100 billion stars). Truly, they are humungous producers of radiation. Quasars are now known to be the result of the tremendous release of energy from gas and dust as it plunges down into black holes. Using the conservation laws of the Schwarzschild metric, we can start to make some calculations of this energy release.

The temptation is to use Equation 14.12, which balances kinetic energy with gravitational potential energy, but this is to get sucked (haha) back into the classical Newtonian view. Even ignoring the impact of spacetime curvature, anything falling into a black hole will move at relativistic speeds towards it. No, we must do better than that.

We want to know how much energy is released by an object falling into the accretion disk of a black hole. Therefore, our model is an object falling from infinity, moving through the accretion disk and ending up in a stable orbit at the ISCO. We can use the Schwarzschild conservation laws to compare the object's energy out at infinity with its energy in ISCO orbit.

Starting with Equation 14.24, we can do a quick reality check that the energy measure is correct when the object is stationary at $r = \infty$. All three terms on the left of the equation are zero (the object is not moving radially). As you can see in Equation 14.45, this means that the object has energy $E_S = mc^2$, which is rest mass energy (as expected).

$$\frac{m}{2}\left(\frac{dr}{d\tau}\right)^2 - \frac{R_S\, mc^2}{2r} + \left(1 - \frac{R_S}{r}\right)\frac{L^2}{2mr^2} = \frac{E_S^2}{2mc^2} - \frac{mc^2}{2}, \quad \textit{from Equation 14.24,} \tag{14.44}$$

$$\textit{At rest at } r = \infty : \qquad \frac{E_S^2}{2mc^2} - \frac{mc^2}{2} = 0, \quad \Longrightarrow \quad E_S = mc^2. \tag{14.45}$$

The next step is to calculate E_S in ISCO orbit. Equation 14.46 shows the value of L^2 in ISCO orbit using the result from Equation 14.36. We substitute this value for L^2 and the ISCO radius $r = 3\,R_S$ into Equation 14.44. As it is in orbit, $\frac{dr}{d\tau} = 0$. The calculation shows that in ISCO orbit, the object has about 0.94 mc^2 of energy.

$$L^2 = m^2\ell^2 = \frac{12\,G^2M^2m^2}{c^2} = 3\,R_S^2\,m^2c^2, \qquad R_S = \frac{2GM}{c^2}, \tag{14.46}$$

$$-\frac{R_S\, mc^2}{2r} + \left(1 - \frac{R_S}{r}\right)\frac{L^2}{2mr^2} = \frac{E_S^2}{2mc^2} - \frac{mc^2}{2}, \qquad \textit{from Equation 14.44,}$$

$$-\frac{R_S\, mc^2}{6\,R_S} + \left(1 - \frac{R_S}{3\,R_S}\right)\frac{3\,R_S^2\,m^2c^2}{18\,mR_S^2} + \frac{mc^2}{2} = \frac{E_S^2}{2mc^2}, \qquad \textit{substitute for L,}$$

$$mc^2\left(-\frac{1}{6} + \frac{1}{9} + \frac{1}{2}\right) = \frac{E_S^2}{2mc^2}, \qquad \textit{substitute for } R_S,$$

$$2m^2c^4\left(\frac{4}{9}\right) = E_S^2, \quad \Longrightarrow \quad E_S = \sqrt{\frac{8}{9}}\,mc^2 \approx 0.94\,mc^2.$$

The result is that in descending to ISCO orbit, the equivalent of about 6% of original rest mass energy is released. To put this in context, the nuclear fusion of hydrogen into helium releases energy equivalent to about 0.7% of original rest mass energy. The Schwarzschild black hole energy release is almost ten times that of fusion!

This is probably a conservative estimate. The ISCO of a spinning black hole is closer to the event horizon, and the energy release can reach over 40%. The spin of the black hole amplifies magnetic fields that result in powerful

Figure 14.2 Artist illustration of a quasar. *Source*: NASA, ESA, CSA, Joseph Olmsted (STScI) / Wikimedia Commons / Public Domain.

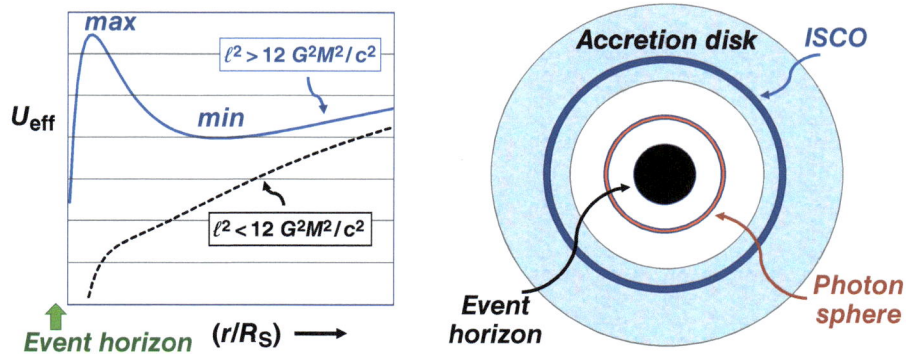

Figure 14.3 Illustrative schematic of Schwarzschild effective potential and black hole.

beams of intense electromagnetic radiation from either end of the spinning axis of the quasar black hole (see the artist illustration in Figure 14.2). Millions of quasars have been identified, providing further evidence of the intense spacetime curvature in black holes. We will discuss this further in Chapter 17 on spinning black holes.

14.9 Summary

In this chapter, we have looked at the implications of the conservation laws of the Schwarzschild metric. If the metric $[g_{\mu\nu}]$ does not depend on any particular variable x, it means that u_x is a conserved quantity along a geodesic. I demonstrated this for diagonal metrics (no off-diagonal terms). This leads to two conserved quantities for the Schwarzschild metric: E_S, that is the counterpart of energy E in the Minkowski metric, and L, that is the same as angular momentum in the Minkowski metric (see Equation 14.9).

I showed that we must manually add gravitational potential energy to the energy balance in the Minkowski metric, but it appears naturally in the Schwarzschild metric (Equation 14.14). While the maths of a radial plunge is similar, the conservation laws of the Schwarzschild metric lead to an extra term in the effective potential when rotational motion is factored in (Equation 14.25).

At large distances from the gravitational source, this extra term is small. Closer in, it creates major differences in the pattern of orbits and plunge paths. The left of Figure 14.3 illustrates the specific effective potential. If an object has less than a certain level of angular momentum, there is no stable circular orbit and it will radially plunge into the heart of the black hole (the dotted black line). Otherwise, its effective potential has a local minimum corresponding to a stable circular orbit, but the effective potential also has a maximum closer to the black hole after which the effective potential rapidly falls (the blue line in the figure).

The right of Figure 14.3 shows a schematic of a Schwarzschild black hole. The innermost stable circular orbit (ISCO) is at $r = 3R_S$. Inside this radius, the path of objects (for example, two black holes embracing each other) changes from a spiral orbit towards a radial plunge. Dr Andy Mummery, who helped edit this book, detected for the first time (in 2024) an example of this transition in the motion of gas in the accretion disk of a black hole.

I ended the chapter with a brief discussion of quasars. Based on the conservation laws of the Schwarzschild metric, an object releases the equivalent of about 6% of its rest mass energy when falling from an infinite distance down to the level of the accretion disk of a black hole (we modelled this by comparing the object at infinity with the ISCO orbit). This may be a conservative estimate because black holes rotate, which increases the energy release. With enough incoming material, it can create intense beams of outgoing electromagnetic radiation, sometimes more intense than the total electromagnetic radiation of all the stars in an entire galaxy.

Chapter 15

Revisiting Einstein's Success (Optional)

I have separated out two topics for this chapter. The first is the deflection of light in a weak gravitational field, such as around the sun. The second is the precession of Mercury. These two calculations were important in establishing Einstein's theory. As such, I have decided they have enough historic significance to justify inclusion in this book.

However, I must warn sensitive readers that the mathematics required is a little advanced and the topics themselves not very significant to understanding the mechanics of general relativity. You should feel free to skip this chapter if you want. While others read it, you can go out for a nice walk, have a bath or just lounge about. Relax... you deserve it.

15.1 The Deflection of Light and Gravitational Lensing

Einstein predicted that light would be deflected in the sun's gravitational field by twice as much as expected on the basis of Newton's theory. The results of Eddington's eclipse expedition were heralded as proof of Einstein's theory. This is discussed in Sections 5.5 and 5.6 of Chapter 5. I am not going to follow Einstein's calculation. Instead, I will step through a version using the Schwarzschild metric.

Fermat's principle, which Pierre de Fermat established in 1662 (the same Fermat who was famous for Fermat's Last Theorem), states that light follows the pathway that can be travelled in the least time. This is the equivalent of defining its geodesic pathway (Fermat got there many years earlier than everyone else). It takes longer for light to travel through materials, such as water or glass, because its path is effectively lengthened by the potential interaction with the material. If it helps, you can think of light bouncing to and fro along its path. This causes *refraction*, which is when the path of the light bends. If you want to fully understand why light travels in this fashion, I must refer you to Chapter 6 of *Quantum Untangling* on Feynman's path integral, which explains that the pathway is the result of combining all possible paths and analysing how they interfere with each other. This is too complicated a topic for me to address further in this book.

Getting back to our theme of light deflection, we need to calculate the geodesic pathway of light around a gravitational source. This pathway is the one that it takes the least time to travel. We can then assess what deflection the pathway involves. The set-up for the calculation is shown in Figure 15.1. Light travels from the star on the left, past an attracting mass (such as the sun), and on to the observer's eye on the right. If we call the total time taken from star to eye T_{tot} and the point of close approach to the attracting mass b, then the minimum pathway in terms of travel time is where $\frac{dT_{tot}}{db} = 0$. The closer the pathway to the attracting mass, the shorter the length of the pathway (shortening the travel time), but the bigger the effect of gravitational time dilation (lengthening the travel time).

We approximate the pathway with three straight lines. The first is the straight black line from the star to close to the attracting mass, which takes time T_1. As shown in the figure, you can rotate this straight-line path by deflection angle α to become the straight line to the observer marked as taking time T_3. This creates a gap that I show as a shorter green line marked as taking time T_2. Our goal is to add up the time of travel for the complete journey from star to observer: $T_1 + T_2 + T_3$.

We select a point (shown as a red dot in the figure) along the T_1 path of the light from the star. At this point, the travelling light is distance r away from the attracting mass. We approximate the path at that point as being in

Untangling General Relativity: The Intuitive Self-Study Guide, First Edition. Simon Sherwood.
© 2026 John Wiley & Sons Ltd. Published 2026 by John Wiley & Sons Ltd.

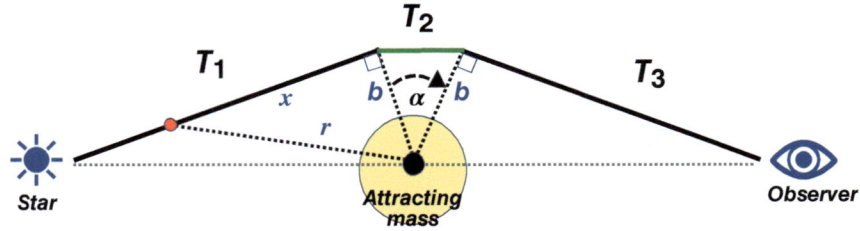

Figure 15.1 Schematic for calculation of the deflection of light.

the dr direction ($d\theta = d\phi = 0$). Light follows a null pathway ($d\tau = 0$), allowing us to calculate $\frac{dr}{dt}$ as shown in Equation 15.1.

$$0 = \left(1 - \frac{R_S}{r}\right)c^2 dt^2 - \frac{1}{\left(1 - \frac{R_S}{r}\right)}dr^2,$$

$$\left(\frac{dr}{dt}\right)^2 = c^2\left(1 - \frac{R_S}{r}\right)^2, \implies \frac{dr}{dt} = c\left(1 - \frac{R_S}{r}\right). \tag{15.1}$$

The challenge is to calculate time T_1. We assume the T_1 path starts from the star at $x = -\infty$ and ends at $x = 0$ close to the attracting mass. The time to travel dx along the path is the distance divided by the rate of progress from Equation 15.1. We integrate this from $x = -\infty$ to $x = 0$. Note in Figure 15.1 that by Pythagoras, $r = \sqrt{b^2 + x^2}$ where b is the closest distance to the gravitational source. For an object such as the sun, b must be much larger than R_S (the sun's radius is about 700,000 kilometres, whereas its Schwarzschild radius is about 3 kilometres). As a result $\frac{R_S}{r}$ is very small everywhere along the path, and we can use the Taylor approximation in Equation 15.2 to arrive at the expression for T_1 in Equation 15.3.

$$T_1 = \int_{-\infty}^{0} \frac{1}{c}\left(1 - \frac{R_S}{r}\right)^{-1} dx \approx \int_{-\infty}^{0} \frac{1}{c}\left(1 + \frac{R_S}{r}\right) dx, \tag{15.2}$$

$$= \int_{-\infty}^{0} \frac{1}{c}\left(1 + \frac{R_S}{\sqrt{b^2 + x^2}}\right) dx. \tag{15.3}$$

We now calculate how the time of travel T_1 varies with b (which depends on the path taken). We will then integrate this derivative over the path (i.e. over x). To do this we need to work out the derivative of $(b^2 + x^2)^{-\frac{1}{2}}$, which we do using differentiation by substitution as shown in Equation 15.4.

$$\text{Say}: \ f = b^2 + x^2, \quad \frac{df^{-\frac{1}{2}}}{db} = \frac{\partial f^{-\frac{1}{2}}}{\partial f}\frac{\partial f}{\partial b} = -\frac{1}{2}f^{-\frac{3}{2}}2b = -\frac{b}{(b^2 + x^2)^{-\frac{3}{2}}}. \tag{15.4}$$

I warned you at the start of the chapter that the maths is a bit tortuous. There is still more to come. Using the result from Equation 15.4 we derive Equation 15.5 for $\frac{dT_1}{db}$ by differentiating Equation 15.3 (the number 1 inside the brackets of the integral in Equation 15.3 is a constant and disappears in the derivative).

$$\frac{dT_1}{db} = \int_{-\infty}^{0} -\frac{R_S}{c}b(b^2 + x^2)^{-\frac{3}{2}} dx. \tag{15.5}$$

The integration is over x, which is over the course of a pathway. An important thing to note is that b does *not* change as you move along the pathway, i.e. the value of b doesn't change with x. We substitute variables with $k = \frac{x}{b}$, which means $x = bk$. As b is a constant for the pathway, this substitution means that $\frac{dk}{dx} = \frac{1}{b}$ and therefore $dx = (b)\,dk$. Given that b doesn't vary with x or k, we can move it in and out of the integral as we wish. This simplifies to Equation 15.6. The limits on the integral do not change because $k = 0$ when $x = 0$ and $k = -\infty$ when $x = -\infty$.

$$\frac{dT_1}{db} = \int_{-\infty}^{0} -\frac{R_S}{c} b(b^2 + b^2 k^2)^{-\frac{3}{2}} b \, dk,$$

$$= -\frac{R_S}{bc} \int_{-\infty}^{0} b^3 (b^2 + b^2 k^2)^{-\frac{3}{2}} \, dk = -\frac{R_S}{bc} \int_{-\infty}^{0} (1 + k^2)^{-\frac{3}{2}} \, dk. \tag{15.6}$$

My daughter Gabby is learning to rock climb and tells me there is often one tricky move per climb that really tests her. We have arrived at that point in our calculation. The solution to the integral is shown in Equation 15.7. It is actually quite well known, and you can find it on the Web (for anybody weird enough to want the gory details of how this works, please see Box 15.1). This gives us our final result for $\frac{dT_1}{db}$.

$$\frac{dT_1}{db} = -\frac{R_S}{bc} \int_{-\infty}^{0} (1 + k^2)^{-\frac{3}{2}} \, dk = -\frac{R_S}{bc} \left[\frac{k}{\sqrt{1 + k^2}} \right]_{-\infty}^{0} = -\frac{R_S}{bc} (0 + 1) = -\frac{R_S}{bc}. \tag{15.7}$$

Box 15.1 Integrate substituting with $k = \tan\theta$

$$k = \tan\theta, \implies \frac{dk}{d\theta} = \frac{1}{\cos^2\theta}, \qquad note\ that: \quad 1 + \tan^2\theta = (\cos^2\theta)^{-1},$$

$$\int (1 + k^2)^{-\frac{3}{2}} \, dk = \int \frac{(1 + \tan^2\theta)^{-\frac{3}{2}}}{\cos^2\theta} \, d\theta = \int \frac{(\cos^2\theta)^{\frac{3}{2}}}{\cos^2\theta} \, d\theta = \int \cos\theta \, d\theta = \sin\theta.$$

We then substitute back from θ to k, which gives the final result:

$$\sin\theta = \frac{\tan\theta}{\sqrt{1 + \tan^2\theta}} = \frac{k}{\sqrt{1 + k^2}}, \implies \int (1 + k^2)^{-\frac{3}{2}} \, dk = \frac{k}{\sqrt{1 + k^2}}.$$

We need the answer for the whole pathway. Path T_3 is symmetrical to T_1 so that is easy. The angle α is measured in radians and is small relative to b. Therefore, the segment T_2 is length αb (see Box 3.3 if needed) giving time of travel: $T_2 \approx \frac{\alpha b}{c}$. This leads to: $\frac{dT_2}{db} = \frac{\alpha}{c}$. We can now put the pieces together as shown in Equation 15.8 where T_{tot} is the total path time.

$$\frac{dT_{tot}}{db} = \frac{dT_1}{db} + \frac{dT_2}{db} + \frac{dT_3}{db} = \frac{\alpha}{c} - \frac{2R_S}{bc}. \tag{15.8}$$

This is the moment to apply Fermat's principle. The geodesic pathway of the light will be such that the time of passage is minimised. This is when $\frac{dT_{tot}}{db} = 0$ giving Equation 15.9 for α, which is the angle of deflection of the light pathway. This is twice the deflection predicted on the basis of Newtonian mechanics. Note again that this assumes b is much greater than the Schwarzschild radius R_S, as obviously is the case for light passing around the sun.

$$\frac{\alpha}{c} - \frac{2R_S}{bc} = 0, \implies \alpha = \frac{2R_S}{b} = \frac{4GM}{bc^2}. \qquad The\ final\ result! \tag{15.9}$$

I want to take a moment to reflect on Einstein's comment, which I quoted earlier: *according to the theory, half of this deflection is produced by the Newtonian field of attraction of the sun, and the other half by the geometrical modification [curvature] of space caused by the sun.* What happens if there is *only* time dilation and *no* spatial distortion? This would change the values of $\frac{dr}{dt}$ and T_1 to those in Equation 15.10. You should look back and compare with Equations 15.1 and 15.2 to make sure you understand why. If the approximation for T_1 confuses you, check out the Taylor expansion in Box 2.1.

$$\frac{dr}{dt} = c\left(1 - \frac{R_S}{r}\right)^{-\frac{1}{2}}, \implies T_1 \approx \int_{-\infty}^{0} \frac{1}{c}\left(1 + \frac{R_S}{2r}\right) dx,$$

$$\implies \alpha = \frac{R_S}{b} = \frac{2GM}{bc^2}, \qquad if\ no\ spatial\ distortion. \tag{15.10}$$

Figure 15.2 Hubble telescope images of gravitational lensing. *Source*: Courtesy of NASA.

Removing the spatial distortion runs through the calculation, approximately halving all the key values. This is what Einstein was getting at. The effect of time dilation gives a deflection for weak gravitational fields that closely matches Newton's result. The effect of the spatial distortion, due to the radial movement of the light in and out of the gravitational field, doubles the deflection, creating a detectable difference between the theories.

Let me summarise what we have done. We have balanced the gravitational time-lengthening effect of how close pathways come to the attracting mass (b) with the distance-lengthening effect of deviating from a straight line (αb). Our calculation gives the relationship between the two where the time of travel for light is minimised. The effect is called *gravitational lensing*. I should note that the approximations (such as the Taylor expansion in Equation 15.2) mean that our result is only accurate when $\frac{R_S}{r}$ is small. To apply it to the deflection of light around the sun, you simply set mass M as the solar mass and the closest distance b to be approximately the radius of the sun.

Back in 1919, Eddington could not have dreamt of the powerful telescopes we have today. These allow us to observe gravitational lensing at what you might call a galactic level. The curves in the image on the left of Figure 15.2 show the effect of gravitational lensing. This image taken by the Advanced Camera for Surveys (ACS) aboard NASA's Hubble Space Telescope is of a massive galaxy cluster known as Abell 1689. Even more impressive is the image on the right of the figure. Also by Hubble, this shows the distortion by a luminous red galaxy of a distant blue galaxy. The alignment is so precise that it creates almost a full circle image of the galaxy behind.

15.2 The Precession of Mercury

I have nothing against Mercury as a planet. I am sure it has its charm, but if it wasn't for Einstein's calculation, I suspect few university students would have heard about the precession of its perihelion.

Given the obscure nature of the topic, I better start at the beginning. Around 1610, a bright spark called Johannes Kepler (1571–1630) realised that the planets move around the sun in ellipses, not circles. Some years later, Isaac Newton (1643–1727) showed this to be consistent with his theory of gravitation. You might describe an ellipse as a squashed circle. As shown on the right of Figure 15.3, an ellipse has two focuses (or foci if you prefer). The sun occupies one focus.[1] The other is a mathematical construction rather than something physical. If you measure out from one focus to the ellipse and back to the other focus, the total length out and back is always the same (this is the dotted black line in the figure). The position in the elliptical path closest to the sun is called the *perihelion* and the most distant the *aphelion*.

The left of Figure 15.3 shows a schematic of Mercury orbiting the sun. In the simplest scenario of one planet orbiting the sun, Newton's laws predict that Mercury would sweep out an ellipse around the sun and return to exactly the same place in each orbit, i.e. Mercury's perihelion would not change (I will show you why). This is not what happens. As shown schematically in the figure, Mercury's orbit precesses, i.e. its perihelion moves. This precession is 575 arcseconds per century (ignoring what are called observer-planet distortions). The principal cause of this is the presence of other planets. However, if you use Newton's laws and account for all the anomalies

---

1 Technically, it is the centre of mass of the two-body system which occupies one focus. However, we can treat the sun as being at the focus because its mass is so much larger than that of Mercury.

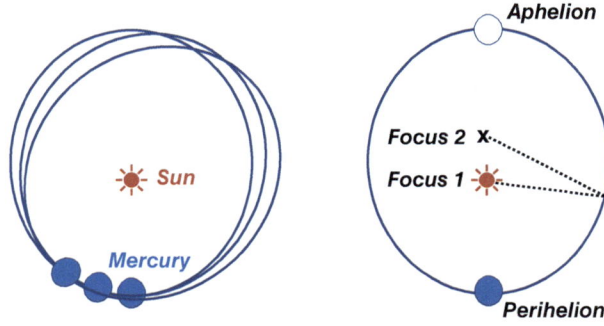

Figure 15.3 Schematic of Mercury precession and the structure of an ellipse.

(the other planets, the deformation of the sun due to its rotation, etc.), it still leaves 43 arcseconds per century unaccounted for.

It may seem picky to worry about such a small discrepancy, but it irked astronomers who could not come up with a good explanation. When Einstein used general relativity for the calculation, he found it precisely predicted the missing 43 arcseconds. You can get a sense of Einstein's elation from a couple of his quotes. He wrote *this is the most valuable discovery I have made in my life* and recalled later that *for several days I was beside myself with joyful excitement*. The difference between Einstein's and Newton's calculations comes down to the extra term in the effective potential of general relativity. To demonstrate the difference, I will use a number of simplifying approximations and shortcuts. Even so, the maths is not easy. Brace yourself!

15.2.1 Binet's Equation

The French physicist Jacques Binet (1786–1856) came up with a neat equation that saves legwork when working on orbital trajectories. If you substitute the radial variable r with $J = \frac{1}{r}$, the formula for Newtonian orbits simplifies to Equation 15.11. ℓ is the specific angular momentum we used in Chapter 14 (as described in Section 14.7). Note that I am using J for Binet's variable (J for Jacques—geddit?). Most texts use u but that might be confused with the effective potential.

$$\frac{d^2J}{d\phi^2} + J = \frac{GM}{\ell^2}, \qquad \textit{Binet's equation for Newtonian orbits.} \qquad (15.11)$$

This is an important equation so let me show you why it works. We need to calculate $\frac{d^2J}{d\phi^2}$ for $J = \frac{1}{r}$. As usual, we assume an equatorial rotation, which means $d\theta = 0$ and $\sin\theta = 1$. Therefore, the conserved quantity is $\ell = r^2\frac{d\phi}{d\tau}$. The first step is to derive $\frac{dJ}{d\phi}$ arriving at Equation 15.12.

$$\frac{dJ}{d\phi} = \frac{d}{d\tau}\left(\frac{1}{r}\right)\frac{d\tau}{d\phi} = \frac{r^2}{\ell}\frac{d}{d\tau}\left(\frac{1}{r}\right), \qquad \textit{note: } \ell = r^2\frac{d\phi}{d\tau}, \implies \frac{d\tau}{d\phi} = \frac{r^2}{\ell},$$

$$= \frac{r^2}{\ell}\frac{d}{dr}\left(\frac{1}{r}\right)\frac{dr}{d\tau} = \frac{r^2}{\ell}\left(-\frac{1}{r^2}\right)\frac{dr}{d\tau} = -\frac{1}{\ell}\frac{dr}{d\tau}. \qquad (15.12)$$

We differentiate again giving the expression in Equation 15.13 for $\frac{d^2r}{d\tau^2}$ in terms of $\frac{d^2J}{d\phi^2}$.

$$\frac{d^2J}{d\phi^2} = -\frac{1}{\ell}\frac{d}{d\phi}\left(\frac{dr}{d\tau}\right) = -\frac{1}{\ell}\frac{d}{d\tau}\left(\frac{dr}{d\tau}\right)\frac{d\tau}{d\phi} = -\frac{r^2}{\ell^2}\frac{d^2r}{d\tau^2} = -\frac{1}{\ell^2 J^2}\frac{d^2r}{d\tau^2},$$

$$\implies -\frac{d^2r}{d\tau^2} = \ell^2 J^2\frac{d^2J}{d\phi^2}. \qquad (15.13)$$

We know the radial acceleration $\frac{d^2r}{d\tau^2}$ is the derivative of the effective potential U_{eff}. The classical Newtonian formula is shown on the left in Equation 15.14 (the same as calculated on the left of Equation 14.33). This leads straight to Binet's equation.

$$-\frac{d^2r}{d\tau^2} = -\frac{\ell^2}{r^3} + \frac{GM}{r^2}, \quad \Longrightarrow \quad \ell^2 J^2 \frac{d^2J}{d\phi^2} = -\ell^2 J^3 + GMJ^2, \tag{15.14}$$

$$\Longrightarrow \quad \frac{d^2J}{d\phi^2} + J = \frac{GM}{\ell^2}, \qquad \textit{Binet's equation for Newtonian orbits}: J = \frac{1}{r}.$$

While each step in Binet's calculation is relatively simple, I am in awe that he managed to derive it. What is more, it makes things relatively painless because his equation is quite easy to solve. Equation 15.15 shows a general solution to Binet's equation. Below is the proof. A and B are constants. The solution is the formula for an ellipse with eccentricity e. You might describe e as a measure of how squished the ellipse is. The value of e is set between 0 and 1 with 0 being a circle and the ellipse becoming progressively more squished as the value of e rises (at the extreme value $e = 1$, the ellipse becomes a line).

$$\textit{If}: \frac{d^2J}{d\phi^2} + AJ = B, \quad \Longrightarrow \quad J = \frac{B}{A}\left(1 + e \cos \sqrt{A}\phi\right), \qquad \textit{General solution}, \tag{15.15}$$

$$\textit{Proof}: \quad \frac{dJ}{d\phi} = \frac{B}{\sqrt{A}}\left(-e \sin \sqrt{A}\phi\right), \qquad \frac{d^2J}{d\phi^2} = B\left(-e \cos \sqrt{A}\phi\right),$$

$$\Longrightarrow \quad \frac{d^2J}{d\phi^2} + AJ = B\left(-e \cos \sqrt{A}\phi\right) + A\frac{B}{A}\left(1 + e \cos \sqrt{A}\phi\right),$$

$$= B. \tag{15.16}$$

In the case of the Newtonian calculation, you can see that $A = 1$ and $B = \frac{GM}{\ell^2}$ by comparing Equations 15.14 and 15.15. Substituting back $J = \frac{1}{r}$ gives the solution for r shown in Equation 15.17.

$$J = \frac{GM}{\ell^2}(1 + e \cos \phi), \quad \Longrightarrow \quad r = \frac{\ell^2}{GM(1 + e \cos \phi)}, \qquad \textit{Newtonian orbit}. \tag{15.17}$$

We know at once that the Newtonian orbit does *not* precess because $\sqrt{A} = 1$ in Equation 15.15. Adding 2π to ϕ (a full orbit rotation) in Equation 15.17 brings you back to exactly the same expression for r because $\cos(\phi + 2\pi) = \cos \phi$. All points on the elliptical path remain the same distance from the focus orbit after orbit. It is now time to turn our attention to what happens when we use the effective potential of the Schwarzschild metric.

15.2.2 Binet with the Schwarzschild Metric

For general relativity, we must use the formula for the specific effective potential of the Schwarzschild metric. For convenience, this is shown again below (as in Equation 14.32 earlier). The key point is, of course, the extra term that appears.

$$\textit{Minkowski + gravity}: \quad U_{\textit{eff}} = \frac{\ell^2}{2r^2} - \frac{GM}{r},$$

$$\textit{Schwarzschild geodesic}: \quad U_{\textit{eff}} = \frac{\ell^2}{2r^2} - \frac{\ell^2 GM}{c^2 r^3} - \frac{GM}{r}. \tag{15.18}$$

We take the derivative of the specific effective potential and apply it as we did with the Newtonian version (see Equation 15.14). This gives Equation 15.19, which is the equivalent of Binet's equation for the Schwarzschild metric. It includes an extra term. That term is quadratic in J. Yuk!

$$-\frac{d^2r}{d\tau^2} = -\frac{\ell^2}{r^3} + \frac{3\ell^2 GM}{c^2 r^4} + \frac{GM}{r^2}, \quad \Longrightarrow \quad \ell^2 J^2 \frac{d^2J}{d\phi^2} = -\ell^2 J^3 + \frac{3\ell^2 GM}{c^2}J^4 + GMJ^2,$$

$$\Longrightarrow \quad \frac{d^2J}{d\phi^2} + J = \frac{GM}{\ell^2} + \frac{3\,GM}{c^2}J^2, \qquad \textit{Binet's equation; Schwarzschild}. \tag{15.19}$$

Some of you readers may be mathemagicians, but for me Equation 15.19 is the sort of ugly monster I fear may be sleeping under my bed at night. Our way forward will be to make approximations to simplify it into a more manageable form. Before starting, I must give a shout-out to Dr Michael Hall of Canberra whose 2022 paper introduced me to these simplifying tactics.

The key assumption is that the elliptical orbit is not heavily eccentric (i.e. it is near-circular, as is the case with Mercury), so the value of J never drifts too far from its average value, which I label J_0. This allows us to approximate J^2 in terms of J and J_0 as shown in Equation 15.20.

$$
\begin{aligned}
J^2 &= [J_0 + (J - J_0)]^2, \\
&= J_0^2 + 2J_0(J - J_0) + (J - J_0)^2, && \text{\textit{ignore small} } (J - J_0)^2 \text{ \textit{term},} \\
&\approx J_0^2 + 2J_0(J - J_0), && \implies \quad J^2 \approx 2J_0 J - J_0^2.
\end{aligned}
\tag{15.20}
$$

With this we can proceed to solve the Binet equation providing we acknowledge that the solution is approximate and is for near-circular orbits. Plugging the result of Equation 15.20 into Equation 15.19 gives Equation 15.21. Don't forget that J_0 is the average radius, so it is a constant.

$$
\frac{d^2 J}{d\phi^2} + J = \frac{GM}{\ell^2} + \frac{3\,GM}{c^2}(2J_0 J - J_0^2),
$$

$$
\implies \quad \frac{d^2 J}{d\phi^2} + \left(1 - \frac{6\,GMJ_0}{c^2}\right) J = \left(\frac{GM}{\ell^2} - \frac{3\,GMJ_0^2}{c^2}\right).
\tag{15.21}
$$

Comparing the result of Equation 15.21 with the original Binet's equation gives the value of constants A and B for the Schwarzschild metric, as shown in Equation 15.22.

$$
\frac{d^2 J}{d\phi^2} + AJ = B, \quad \text{where}: \quad A = \left(1 - \frac{6\,GMJ_0}{c^2}\right), \quad B = \left(\frac{GM}{\ell^2} - \frac{3\,GMJ_0^2}{c^2}\right).
\tag{15.22}
$$

All the hard work is done. We already have the general solution to this (Equation 15.15). As we are specifically searching for orbital precession, we are interested in the cosine term only.

$$
J = \frac{B}{A}\left(1 + e\cos\sqrt{A}\phi\right) = \frac{B}{A}\left(1 + e\cos\sqrt{\left(1 - \frac{6\,GMJ_0}{c^2}\right)}\phi\right).
\tag{15.23}
$$

The change in the cosine term means that the planet does not renew its path every 2π, but every $\sqrt{A}\,\phi = 2\pi$. If we label this angle of return $\phi_p = \dfrac{2\pi}{\sqrt{A}}$, the precession change with each orbit is $\Delta\phi_p = \phi_p - 2\pi$ as shown in Equation 15.24.

$$
\phi_p = \frac{2\pi}{\sqrt{A}} = \frac{2\pi}{\sqrt{\left(1 - \frac{6\,GMJ_0}{c^2}\right)}} \approx 2\pi\left(1 + \frac{3\,GM J_0}{c^2}\right), \quad \implies \quad \Delta\phi_p = \frac{6\pi\,GM J_0}{c^2}.
\tag{15.24}
$$

15.2.3 The Missing Precession

All that is left is to plug in values. The minimum and maximum values for the radius of Mercury's elliptical orbit give us J_0. This and the other values needed for the calculation are:

$$
\begin{aligned}
r_{min} &= 4.60 \times 10^{10}\,\text{m}, && \implies \quad J_{max} = 2.17 \times 10^{-11}\,\text{m}^{-1}, \\
r_{max} &= 6.98 \times 10^{10}\,\text{m}, && \implies \quad J_{min} = 1.43 \times 10^{-11}\,\text{m}^{-1}, && \implies \quad J_0 = 1.80 \times 10^{-11}\ \text{m}^{-1}, \\
G &= 6.67 \times 10^{-11}\,\text{m}^3\text{kg}^{-1}\text{s}^{-2}, && M = 1.99 \times 10^{30}\,\text{kg}, && c = 3.00 \times 10^8\ \text{ms}^{-1}.
\end{aligned}
$$

Feeding these values into the change in precession gives (2π radians $= 360°$):

$$
\frac{6\pi\,GM J_0}{c^2} \text{ radians \textit{per orbit}} = \frac{(3 \times 360)\,GM J_0}{c^2} \text{ degrees \textit{per orbit},}
$$

$$
= \frac{(3 \times 360)\,(6.67 \times 10^{-11})\,(1.99 \times 10^{30})\,(1.80 \times 10^{-11})}{(3.00 \times 10^8)^2} = 2.87 \times 10^{-5} \text{ degrees \textit{per orbit}.}
$$

The final (at last) step in the calculation is to convert this into arcseconds per century taking into account that Mercury orbits every 88 days (an arcsecond is 1/3,600 of a degree). This gives us the missing 43 arcseconds per century in Equation 15.25. I hope you feel the same thrill that Einstein did. I have added the $\approx$ sign to remind you that we have made a number of approximations in this calculation.

$$Per\ century:\quad \frac{100 \times 365}{88}(3{,}600)(2.87 \times 10^{-5}) \approx 43 \quad \text{arcseconds } per\ century. \tag{15.25}$$

Some of you may be surprised by the accuracy of the result. However using the numbers above, the value of the $2J_0J$ term that affected A in Equation 15.21 (thereby creating precession) is almost 50× greater than the value of $(J - J_0)^2$ that we ignored in the approximation of Equation 15.20. In any case, the important point is to see that the extra term in the effective potential U_{eff} of the Schwarzschild metric generates precession.

15.3 The Aftermath

With apologies to those readers holding ice packs to their foreheads after all the maths in this chapter, the deflection of light and the precession of Mercury were important in building widespread belief in the theory of general relativity. Einstein was thrilled at his explanation of the mysterious anomaly in the precession of Mercury, which he described as *the most valuable discovery I have made in my life*. It was followed by Eddington's (somewhat dubious) confirmation of Einstein's prediction regarding the deflection of light by the sun. Yeah! Well done Albert! You are now a superstar!

However, the Schwarzschild metric still appears to have a snag. The usual coordinates do not allow us to plot the transit of anything across the event horizon. Can we patch together coordinates that cover both outside the event horizon and inside it? This is the focus of Chapter 16.

Chapter 16

Schwarzschild: Other Coordinates

16.1 Introduction for Dummies

Some of you may find the title of this section patronising, but it is not. I am a self-confessed dummy, which is why I wrote this book. I get it that things are not always easy *to get*. So, at the risk of boring the brainiacs, let me wade through what for them will be some dull background. In this chapter, we will be working with unusual coordinates. The important lesson is that changing coordinates has no effect on the underlying reality of what is happening. A change of coordinates is simply a different way of expressing the same thing, like explaining something in Italian or Spanish instead of English: whether you call it beer, *birra* or *cerveza* doesn't change what it is.

Figure 16.1 is designed to illustrate this. Imagine we climb a hillside that slopes upwards to the north. We trek north-east from hut *a* to *b* to *c* as shown on the left of the figure. The numbered lines are contours showing altitude in metres. The hillside has an upward slope in the northerly direction, but different parts slope upwards at different rates, so the contour lines twist and turn.

The left of the figure expresses the trek with north and east coordinates. However, we can also express it with altitude and distance east by using the contour lines as a coordinate. Every point along the trek can be expressed in coordinates (north, east) or in coordinates (altitude, east) as shown on the right of the figure.[1] While the paths on the left and right of the figure look very different, I hope it is obvious that the coordinate change (switching from the measure of north to that of altitude) makes no difference to the trek. Try to keep this in mind when we start switching coordinates for the Schwarzschild metric.

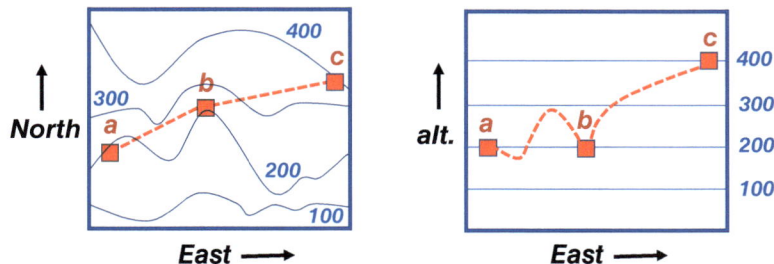

Figure 16.1 Schematic contour map to illustrate a coordinate change.

Box 16.1 Natural units

This chapter is more conceptual than numerical. Having factors of c in the equations just confuses things, so I have set $c = 1$. Where necessary, I will label clearly, such as with: $(t, r, \theta, \phi \, [c = 1])$. Where I use logarithms, (log) indicates the *natural* logarithm, which you may know as (ln).

---

1 This assumes the slope has an upward pitch in the northerly direction, so each point is uniquely defined by east and altitude coordinates.

Untangling General Relativity: The Intuitive Self-Study Guide, First Edition. Simon Sherwood.
© 2026 John Wiley & Sons Ltd. Published 2026 by John Wiley & Sons Ltd.

In this chapter, we will discuss two different coordinate systems: Eddington–Finkelstein (EF) coordinates and Kruskal–Szekeres (KS) coordinates. The bad news is that the underlying maths is a little laborious. Therefore, I have done my best to explain the logic and what the coordinate systems reveal, while separately providing optional mathematical detail for those who want it.

16.2 Eddington–Finkelstein Coordinates

Let's admit there is a big problem with our Schwarzschild coordinates. For good reason, we have used the coordinates familiar to a distant observer sitting in spacetime, which has a metric similar to that of flat Minkowski spacetime. This is familiar and comfortable… but very messy on this occasion. Consider the left of Figure 16.2 labelled *Schwarzschild Coordinates*. It is a spacetime diagram. The usual time coordinate t is shown on the y-axis. This is time as measured by an observer infinitely far away from the black hole, and I am using natural units so $c = 1$ (see Box 16.1). If we consider the two space explorers I introduced in Section 13.3, time t is as measured on Minky's clock far from the black hole while she watches Schwart freefall into it. Radial coordinate r is shown on the x-axis. This is the reduced circumference (see Section 13.9). The event horizon of the black hole $r = R_S$ is marked with a dotted black line and the singularity is $r = 0$.

To the right of the event horizon in the *Schwarzschild Coordinates* figure, there are arrowed red lines bending inwards and upwards. These are radial ingoing beams of light. The blue lines are radially outgoing light beams, and the green rings are light cones (check back to Box 13.7 if you need a reminder about light cones). Let's start with the bits that make sense. In flat Minkowski spacetime, light always moves diagonally (see Box 13.7). In the Schwarzschild diagram, when r grows large, the ingoing and outgoing light beams cross approximately as diagonals and the light cones are similar to those of Minkowski spacetime. That is no surprise. If r is large, you are far from the attracting mass, so it resembles flat spacetime. To the left of the event horizon, the light cones all point to the singularity. Again, this makes sense as nothing, not even light, can escape the black hole.

Now let's discuss the trickier parts of the figure. Using the coordinates of a distant observer for the Schwarzschild metric, the distortion of time (gravitational time dilation) along with the distortion of space (dr) mean that the observer will never observe anything reach the event horizon, not even light. From Minky's distant perspective, the ingoing light's progress (red line) towards the event horizon slows and takes an infinite amount of time to reach it. This is why the ingoing red line light rays bend upwards in the figure as time passes with less and less inward progress towards $r = R_S$.

The result is a coordinate singularity at the event horizon $r = R_S$. According to Minky's measure of time, the ingoing light never reaches the event horizon. Time and eternity tick by. On the other hand, you can see weird stuff happening on the left of the event horizon. The incoming light rays (marked again with arrowed red lines) head backwards in time in theory. Everything is a mess. Her time no longer makes sense, which is not surprising given that there is no possible link back to her. Yet somehow these ingoing light rays hit the $r = 0$ singularity on the left.

An obvious solution (obvious now but perhaps not back in the early days) is to use different coordinates; something to track paths through spacetime while avoiding the fatal flaw of infinite time dilation. The EF solution is to use the ingoing light beams as coordinates. Think of the ingoing light beams in the diagram on the left of Figure 16.2 as being like the contour lines on the left of Figure 16.1. Using those contour lines, we created the diagram on the right of Figure 16.1. Let's do something similar, using light beams as a coordinate for the Schwarzschild metric.

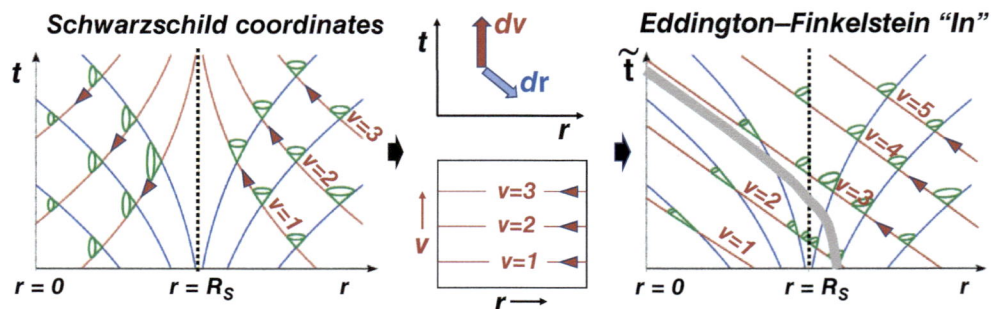

Figure 16.2 Transition to Eddington–Finkelstein (EF) coordinates. *Source*: Ray d'Inverno (1995) / Oxford University Press/ Public Domain.

16.3   Intuitive EF

I am going to label the new ingoing light beam coordinate v (often called *advanced time*), because everybody does. Please note that this v label has nothing to do with velocity. It takes the place of Minky's time t coordinate in the Schwarzschild metric. The aim is to measure paths through spacetime using the new v coordinate alongside the existing r coordinate. This is illustrated schematically in the lower central grid of Figure 16.2.

In order to use the ingoing light beam as a coordinate, we want v to be *constant* for the beam of light as it travels into the black hole. The simplest way to understand v is as a measure of the *emission time* of the light beam when it started its journey from far away. Thus, the bottom central graphic shows the $v = 1$ emitted light beam as a flat line. Above it, later in emission time, is the ingoing $v = 2$ beam and so forth. As each light beam travels inwards, the value of r reduces, but there is no change to the original emission time (obviously) so v is constant.

Effectively, we take the $v = 1$, $v = 2$ and $v = 3$ emission-time red ingoing light lines that bend upwards and downwards when measured in Schwarzschild coordinates (the left of Figure 16.2) and flatten them out. Conceptually, this is similar to the way we switched the altitude contours to be a coordinate in Figure 16.1. A path through spacetime can be labelled with Schwarzschild coordinates (t, r, θ, ϕ) or the new coordinates (v, r, θ, ϕ). The advantage of the latter is that, while the t coordinate of any path heads to $t = \infty$ at the event horizon, the v coordinate does not.

Can this work practically? Let's think about Minky and Schwart. Minky hovers stationary relative to the black hole and kicks her boyfriend Schwart out of the airlock. Can we track Schwart's path using the new v coordinate instead of the old t coordinate? Yes! Minky sends a light beam (or sequence of beams) into the black hole, coded with the emission time according to her clock (for example, using radio messages or coding with shifts in frequency). At any stage of his journey, Schwart can measure the incoming light from Minky and know the emission time of the light reaching him. This gives him a value of v that is specific for his position in spacetime.

If this puzzles you, it may help to look back at the illustrative spacetime diagram on the right of Figure 13.3. Schwart's path into the black hole is the vertical red line. The dotted red lines show light signals travelling from Schwart (S) to Minky (M). Think instead of light signals from Minky (M) to Schwart (S), which would be dotted blue lines moving diagonally up and left. As Schwart moves towards the black hole, he will receive light signals which are emitted later and later by Minky. The crucial point is that *nothing changes* when Schwart crosses the event horizon (the solid black line). Schwart will continue to receive light signals with increasing emission time. The new coordinate system does not have a singularity at the event horizon. Yippee!

In summary, we use the emission time of the ingoing light beams to give us the EF coordinates $(v, r, \theta, \phi\,[c = 1])$. In order to make the EF spacetime diagrams easier to compare with others, the norm is to display the time axis as $\tilde{t} = v - r$ as shown on the right of Figure 16.2. I find it easiest to think of it just as a *display change* so that the ingoing light beams run at 45° diagonally rather than flat. The result is a spacetime diagram on which we can track geodesics. For example, the thick grey line is the geodesic of an object released from stationary relative to the black hole, so it could be Schwart's path in freefall through the event horizon and on towards the singularity (or, from his perspective, as the black hole hurtles towards him). If you want a little history on Finkelstein and EF coordinates, check out Box 16.2.

We will discuss the EF spacetime diagram further, but first I offer a couple of optional sections digging into the maths that sits behind the coordinate change. Don't worry if this is beyond you. You don't need the detailed maths to understand the logic for the EF coordinates.

Box 16.2   David Finkelstein (1929–2016)

David Finkelstein was one of the first to suggest that what we now call a black hole can exist at the centre of the Schwarzschild metric. At a lecture in 1957, he proposed that topological changes might address the coordinate singularity at the event horizon. The EF coordinate system was formally published by Roger Penrose but, in a generous gesture rarely seen among competing researchers, Penrose passed credit to Finkelstein whose lecture he had attended. Finkelstein spent the rest of his career trying to develop a quantum theory of gravity. The following quote gives a measure of the difficulty involved:

When I began my own research, I took it for granted that it had three stages: I would first find a theory in which I could at least potentially believe, then compute its consequences, test it against experimental data, and return to stage 1 for an improved version. After about forty years, I could not help noticing that I was still in the first stage.

16.3.1 EF Maths Step 1 (Optional)

We want the new coordinate v to be constant (which means $dv = 0$) for a radial ingoing light beam. For a radial light beam we know $d\tau = 0$ because it is a null path (Section 5.5) and also $d\theta = d\phi = 0$ (radial path). The Schwarzschild invariant interval (Box 13.2) reduces to Equation 16.1 leading to the relationship between dt and dr shown in Equation 16.2. The $\pm$ sign appears because the radial light ray can be ingoing or outgoing. For an *ingoing* light ray it is a plus sign, because as time passes the light gets closer. Thus dt increases as dr decreases, and you must *add* the changes to reach the zero result. By design, $dv = 0$ for the radial light beam giving us the formulas in Equation 16.3. This allows us to express dt^2 for ingoing light beams in terms of dv and dr in Equation 16.4.

$$0 = \left(1 - \frac{R_S}{r}\right) dt^2 - \frac{1}{\left(1 - \frac{R_S}{r}\right)} dr^2, \qquad (t, r, \theta, \phi \,[c = 1]), \qquad (16.1)$$

$$\implies \quad dt = \pm \frac{dr}{\left(1 - \frac{R_S}{r}\right)}, \quad \implies \quad dt \pm \frac{dr}{\left(1 - \frac{R_S}{r}\right)} = 0, \quad \text{radial light beam,} \qquad (16.2)$$

$$dv = dt + \frac{dr}{\left(1 - \frac{R_S}{r}\right)} \quad \text{ingoing light,} \qquad du = dt - \frac{dr}{\left(1 - \frac{R_S}{r}\right)} \quad \text{outgoing light,} \qquad (16.3)$$

$$\text{Ingoing}: \quad dt^2 = \left(dv - \frac{dr}{\left(1 - \frac{R_S}{r}\right)}\right)^2 = dv^2 - \frac{2\,dv\,dr}{\left(1 - \frac{R_S}{r}\right)} + \frac{dr^2}{\left(1 - \frac{R_S}{r}\right)^2}. \qquad (16.4)$$

Substituting Equation 16.4 into the full Schwarzschild invariant interval allows us to switch the dt coordinate to dv. This gives Equation 16.5, which is the Schwarzschild metric using the v coordinate, where v is a constant for *ingoing* light beams. If you do the same calculation using a different coordinate, which I will call u for *outgoing* light beams, the formula on the right of Equation 16.3 gives a similar result except that it is $+2\,du\,dr$ instead of $-2\,dv\,dr$.

$$d\tau^2 = \left(1 - \frac{R_S}{r}\right) dt^2 - \frac{1}{\left(1 - \frac{R_S}{r}\right)} dr^2 - r^2 d\theta^2 - r^2 \sin^2\theta \, d\phi^2, \qquad (t, r, \theta, \phi \,[c = 1]),$$

$$= \left(1 - \frac{R_S}{r}\right) dv^2 - 2\,dv\,dr + \frac{1}{\left(1 - \frac{R_S}{r}\right)} dr^2 - \frac{1}{\left(1 - \frac{R_S}{r}\right)} dr^2 - r^2 d\theta^2 - r^2 \sin^2\theta \, d\phi^2, \qquad (16.5)$$

$$= \left(1 - \frac{R_S}{r}\right) dv^2 - 2\,dv\,dr - r^2 d\theta^2 - r^2 \sin^2\theta \, d\phi^2, \qquad (v, r, \theta, \phi \,[c = 1]) \; "IN".$$

I think it is easier to see what is going on when it is displayed in metric form as in Equation 16.6. For the v coordinate, which is constant for radially *ingoing* light, the value of $g_{VR} (= g_{RV})$ is negative: -1 (when we work later with the equivalent u coordinate for outgoing light rays this becomes $+1$).

$$[g_{\mu\nu}] = \begin{bmatrix} \left(1 - \frac{R_S}{r}\right) & -1 & 0 & 0 \\ -1 & 0 & 0 & 0 \\ 0 & 0 & -r^2 & 0 \\ 0 & 0 & 0 & -r^2 \sin^2\theta \end{bmatrix}, \quad \text{Schwarzschild EF } (v, r, \theta, \phi \,[c = 1]). \qquad (16.6)$$

If you are a normal sort of person, your head will be spinning. I mean what just happened? Off-diagonal components have appeared in the metric, and the dr^2 term has completely disappeared. Do not panic! All will be explained. I need you to stretch your mind all the way back to Figure 3.1 in Section 3.3, or you can just flick over it again. The appearance of off-diagonal terms means that the v and r coordinates are not orthogonal. A move in the v direction (dv) is not orthogonal to a move in the r direction (dr).

Look back at the two central schematics in Figure 16.2. The bottom one shows the v and r coordinates on separate axes. The ingoing light rays are flat lines because they move across r with constant value of v. The top schematic shows dv and dr in terms of t and r. Suppose that our dear friend Schwart is falling into the black hole and decides to move dv in the v coordinate direction *without* changing his r position. He turns on his rocket engine to hover, maintaining his distance relative to the black hole. As time passes, the v emission-time of the light beams arriving from Minky grows. By sitting stationary and letting time pass, Schwart can move dv in the v coordinate direction independent of any change in r. The dv and dt basis vectors are aligned.

What if Schwart decides to move dr *without* changing the value of v? Maintaining a constant value of v means moving along a light path. A shift of dr with no change in v is a null light path, so dr^2 in the metric has zero length measured using the invariant interval. This is why g_{RR} is zero. Note also that to change dr independent of any change to v, involves a light path, which is a movement in space (changing the r value) *and* in time (changing the old coordinate t value). This means that the dv and dr basis vectors are not orthogonal. Hence, there are off-diagonal terms in the metric.

16.3.2 EF Maths Step 2: Along Comes a Tortoise (Optional)

I have shown you the relationship between dv, dt and dr. You probably want to know what this means in terms of v, t and r. Sorry, but the answer is not simple. The ingoing version of v is shown in Equation 16.7, where you will see a new label r_* that is called the *tortoise coordinate* (note that log is the natural logarithm). I am not going to step through the derivative calculation (dull, dull, dull), which you can find online, but the value of r_* is defined such that the derivative $\dfrac{dr_*}{dr}$ gives the answer in Equation 16.8 (note again that the symbol $|x|$ means the absolute value of x, whether positive or negative).

$$v = t + r_*, \quad \text{where} \quad r_* = r + R_S \log\left|\frac{r}{R_S} - 1\right|, \qquad r_* \text{ is the tortoise coordinate,} \tag{16.7}$$

$$\implies \quad \frac{dr_*}{dr} = 1 + R_S\left(\frac{1}{r - R_S}\right) = \frac{1}{1 - \frac{R_S}{r}}. \tag{16.8}$$

Why? Because by design we want the ingoing coordinate v to be constant for an ingoing light beam, so $\dfrac{dv}{dr} = 0$ for the light. We know the value of $\dfrac{dt}{dr}$ for an ingoing light beam from Equation 16.2. The value of r_* is set in order for its derivative to offset this as shown in Equation 16.9.

$$\frac{dv}{dr} = \frac{d(t + r_*)}{dr} = \frac{dt}{dr} + \frac{dr_*}{dr} = -\frac{1}{1 - \frac{R_S}{r}} + \frac{1}{1 - \frac{R_S}{r}} = 0, \qquad \frac{dt}{dr} \text{ is for ingoing light.} \tag{16.9}$$

The name *tortoise coordinate* for r_* comes from this offsetting role. As you approach the event horizon ($r = R_S$), the time coordinate t blows up due to gravitational time dilation as shown on the left of Figure 16.2. The change in the value of the tortoise coordinate r_* is the opposite. Rather than blowing up, it slows down... like a tortoise... sort of... um....

16.4 Crossing the Black Hole Event Horizon

Leaving the maths to one side, let's review the spacetime diagram of a black hole using EF ingoing v coordinates. This is shown again on the left of Figure 16.3. The ingoing radial light beams are shown as red-straight lines. We use the $\tilde{t} = v - r$ axis so that they are 45° diagonal, which gives them a sort of familiar feel. Using the ingoing v coordinate straightens the ingoing beams, but the outgoing beams, shown as blue lines, are curved. They are useful in showing the light cones (green).

The ingoing (red) radial beams are heading for the singularity from the start of their journey and continue until they hit it. An outward-going light beam (for example, the light from a torch pointing away from the black hole) will escape providing it is emitted outside the event horizon. You can see that the blue outgoing lines beyond the event horizon travel away from it. Inside the event horizon, there is no escape. The light cones show that even outgoing light beams finish up at the singularity.

The grey line illustrates the path of an object into the black hole. As it crosses the event horizon, its progress can be measured against the ingoing light coordinate, which is the emission time from some distance outside the black hole. There is no discontinuity at the event horizon in this emission time, nor in the transit time of the light.

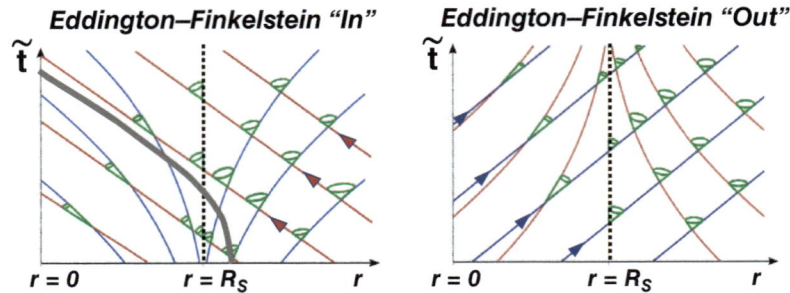

Figure 16.3 EF ingoing and outgoing. *Source*: Ray d'Inverno (1995) / Oxford University Press/ Public Domain.

The EF coordinates were important historically. As early as 1939, Oppenheimer and Snyder had proposed a model of how a dying star, out of fusion fuel, could implode under gravitational pressure into a black hole (the Oppenheimer-Snyder model). However, sceptics argued that this collapse is impossible because of a singularity at the event horizon. EF coordinates proved that there is no physical barrier to light or matter crossing the event horizon. There is no singularity there. It is just that, because of gravitational time dilation, we distant observers will never observe objects cross the event horizon or a star collapse inside it. With no physical singularity to point at, sceptics became believers. The scientific community began to accept the physical existence of black holes.

16.5 White Holes

The EF coordinates reveal something else. There is another legitimate solution to the Schwarzschild metric that we can demonstrate with *outgoing* light beams as coordinates (instead of ingoing ones). This is called the *white hole* solution. Think of it as the opposite of a black hole. Let me start with a warning. Arguably a white hole, albeit a legitimate solution in general relativity, is a hypothetical mathematical creation.

The intuitive rationale for the white hole solution is that the Schwarzschild metric is static. As such, the solution is *time-independent* (see Section 12.1). This means that you can reverse the flow of time and still have a solution that is valid for the Schwarzschild metric. If you reverse time, any object falling into a black hole would be seen as an object ejected by it. Ingoing attraction becomes outgoing repulsion. As a black hole is a legitimate solution of the Schwarzschild metric, then so is a white hole.

The outgoing EF spacetime diagram is shown on the right of Figure 16.3. Again, the display has been shifted, this time using ($\tilde{t} = u + r$), so that the outgoing light beams are at 45°. If you examine the light cones, you will see that everything, including light, is ejected out of the event horizon, and nothing outside the event horizon can get in.

However, the black hole described by the Schwarzschild metric is unchanging and therefore *eternal* in the sense that it has always existed. Once you inject time dependence into the metric, such as the formation of a black hole as a result of stellar collapse, there is no longer an obvious white hole solution.

Do white holes really exist? Without evidence, we must be sceptical. The formation of a new white hole would require a drop in entropy, which is generally considered a no-no (there is more detail on entropy later in Chapter 28, Box 28.4). On the other hand, there are obvious mathematical similarities with the Big Bang. Might that have been a white hole? Perhaps a new universe bursts into existence with every new black hole created in our universe (Lee Smolin has suggested this and he is one of the fathers of Loop Quantum Gravity, so no dummy). Physicists love playing with such ideas... and let's not forget that many were doubtful about the reality of black holes, which we now know to exist.

Optional maths: Those readers who absorbed the maths of the EF coordinates should note that the outgoing solution recognises the validity of the negative solution in the ± square root of Equation 16.2. The value of $\frac{dt}{dr}$ for outgoing light beams is the opposite of that of ingoing light beams. As shown in Equation 16.10 (the outgoing equivalent of Equation 16.8), the outgoing coordinate, which we label u, is ($u = t - r_*$) as compared to the ingoing coordinate ($v = t + r_*$).

$$\frac{du}{dr} = \frac{d(t - r_*)}{dr} = \frac{dt}{dr} - \frac{dr_*}{dr} = +\frac{1}{1 - \frac{R_S}{r}} - \frac{1}{1 - \frac{R_S}{r}} = 0, \qquad \frac{dt}{dr} \textit{ is for outgoing light.} \qquad (16.10)$$

The result is two independent EF coordinate systems for the Schwarzschild metric. The first is (v, r, θ, ϕ) and is based on coordinate v being constant for any particular *ingoing* light beam. The second is (u, r, θ, ϕ) and is based on coordinate u being constant for any particular *outgoing* light beam.

$$\begin{aligned} \textit{Ingoing } (v, r, \theta, \phi) \quad &v = t + r_*, \;\; v = C, \;\; \textit{constant for ingoing light beam,} \\ \textit{Outgoing } (u, r, \theta, \phi) \quad &u = t - r_*, \;\; u = C, \textit{ constant for outgoing light beam.} \end{aligned} \qquad (16.11)$$

16.6 Kruskal–Szekeres Coordinates

The EF coordinates don't give us the full picture of the Schwarzschild metric. The ingoing (v, r, θ, ϕ) coordinates show us the black hole solution. The outgoing (u, r, θ, ϕ) coordinates show us the white hole solution. But both of these solutions are part of the same Schwarzschild metric. We use Kruskal–Szekeres (KS) coordinates to reveal both solutions and how they are linked. Most of the black hole diagrams that you will see are based on KS coordinates, including Penrose–Carter diagrams, which we will discuss later.

To see the EF black hole solution, we penetrate through the event horizon coordinate singularity by replacing the old t coordinate with v, which has constant value for any ingoing light beam. We get the EF white hole solution by replacing t with u, which has constant value for any outgoing light beam. With KS we replace both old coordinates t and r with versions of v and u. The logic is simple. Both t and r have coordinate singularities at the event horizon (check back to Equation 13.17 if needed), so let's replace both for a full view of the Schwarzschild metric.

Rather than dive right in, I want to set the stage using flat Minkowski spacetime. The diagrams in Figure 16.4 illustrate how v and u coordinates work in Minkowski spacetime. The spacetime diagrams are in natural units, so the speed of light is one and the light rays are 45° diagonal. The one on the left shows the direction of increase in v and u. You can compare with the central diagram, which shows ingoing light rays travelling up and left, each with constant value of v; and outgoing light rays travelling up and right, each with constant value of u. The diagram on the right of the figure takes the black spacetime track from the central diagram and expresses it in terms of v and u. And do not forget: *switching coordinates makes no difference to the actual track*. It is just a different way of measuring the same thing.

Before discussing what the KS coordinates tell us, I offer again a more detailed mathematical description for those readers who are feeling brave. If you want to skip the maths, feel free to jump to Section 16.7.

16.6.1 KS Maths (Optional)

Let's see how the Schwarzschild metric looks using the v and u radial light coordinates, starting from the values we derived for dv and du in Equation 16.3, shown again as Equation 16.12. Using these values in Equation 16.13, we combine dv and du in a way that allows us to replace both the dt and dr coordinates in the Schwarzschild metric.

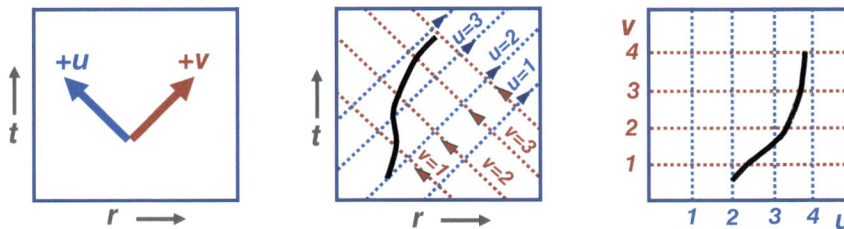

Figure 16.4 Illustrating KS coordinates using Minkowski spacetime. The solid black line is a spacetime track shown first in (t, r) coordinates and then in (v, u) coordinates.

$$dv = dt + \frac{dr}{\left(1 - \frac{R_S}{r}\right)} \quad \textit{ingoing light,} \qquad du = dt - \frac{dr}{\left(1 - \frac{R_S}{r}\right)} \quad \textit{outgoing light,} \qquad (16.12)$$

$$\left(1 - \frac{R_S}{r}\right) dv\, du = \left(1 - \frac{R_S}{r}\right)\left(dt^2 - \frac{dr^2}{\left(1 - \frac{R_S}{r}\right)^2}\right) = \left(1 - \frac{R_S}{r}\right) dt^2 - \frac{dr^2}{\left(1 - \frac{R_S}{r}\right)}. \qquad (16.13)$$

This allows us to restate the Schwarzschild invariant interval from its original form in Equation 16.14 into the one using radial light coordinates in Equation 16.15. The result has no dt or dr terms. This is important because in the Schwarzschild metric g_{TT} and g_{RR} contain coordinate singularities at the event horizon.

$$d\tau^2 = \left(1 - \frac{R_S}{r}\right) dt^2 - \frac{1}{\left(1 - \frac{R_S}{r}\right)} dr^2 - r^2 d\theta^2 - r^2 \sin^2\theta\, d\phi^2, \quad (t, r, \theta, \phi\, [c = 1]), \qquad (16.14)$$

$$d\tau^2 = \left(1 - \frac{R_S}{r}\right) dv\, du - r^2 d\theta^2 - r^2 \sin^2\theta\, d\phi^2, \qquad (v, u, \theta, \phi\, [c = 1]). \qquad (16.15)$$

Note also that Equation 16.15 has no dv^2 or du^2 terms, so g_{vv} and g_{uu} are both zero. This is because increasing v without any change to u is a move directly along the path of an outgoing light beam. Similarly, increasing u without changing v is along the path of an ingoing light beam. Both are null light paths and result in no movement in terms of the invariant interval $d\tau$.

Our work is not yet complete. We need to organise things so that we don't get any new coordinate singularities in the g_{vu} ($= g_{uv}$) component of the metric. This takes a little mathematical legwork because v and u have an inbuilt incompatibility. In terms of coordinate u, an outgoing beam has value $u = C$ where C is a constant. The right of Equation 16.16 shows this value using coordinate v. The problem is that $(\log 0)$ is undefined, so the value v of the outgoing beam has a coordinate singularity at the event horizon ($r = R_S$). This is why the outgoing rays head off to infinity at the event horizon in the lefthand diagram of Figure 16.3. And the same is true if you plot an ingoing ray against the u coordinate.

$$v = t + r_*, u = t - r_*, \implies v = u + 2r_*,$$
$$\textit{Outgoing ray}: \quad u = C, \implies v = C + 2\left(r + R_S \log\left|\frac{r}{R_S} - 1\right|\right). \qquad (16.16)$$

The way to avoid this is disarmingly obvious once you see it. We create an exponential form of v and u. Taking the exponential of a log cancels it out removing the singularity. To perform this wonderful sleight of hand, we create the KS coordinates in Equation 16.17. I will show you in a moment how they are set such that the log part of Equation 16.16 disappears. These KS coordinates V and U are purely an exponential *re-scaling* of the EF coordinates v and u, as $2R_S$ is a constant. The values of v and u are constant for ingoing and outgoing light beams, respectively, so the same is true of the values for the light beams in coordinates V and U.

$$V = \exp\left(\frac{v}{2R_S}\right), \qquad\qquad U = -\exp\left(-\frac{u}{2R_S}\right). \qquad (16.17)$$

Note that $\exp(x)$ is the same as e^x, but a bit clearer to read. Let's plot the outgoing light ray using KS coordinate V. We start with the outgoing ray expressed in terms of v (Equation 16.16 written out in full). We then express this in terms of KS coordinate V and simplify to give Equation 16.18. Using V, the log disappears and the formula for an outgoing ray is what mathematicians describe as well-behaved (I wish I had seen as much from my kids when they were teenagers). At the event horizon, the V value of the outgoing light beam is zero.

$$\textit{Outgoing ray (v value)}: = C + 2r + 2R_S \log\left|\frac{r}{R_S} - 1\right|, \qquad \textit{New coordinate}: \quad V = \exp\left(\frac{v}{2R_S}\right),$$

$$\textit{Outgoing ray (V value)}: = \exp\left(\frac{C + 2r}{2R_S}\right) \exp\left(\frac{2R_S}{2R_S} \log\left|\frac{r}{R_S} - 1\right|\right), \qquad (16.18)$$

$$= \left|\frac{r}{R_S} - 1\right| \exp\left(\frac{C + 2r}{2R_S}\right).$$

The same calculation for an ingoing beam using KS coordinate U is shown as Equation 16.19. You can see how the form of U is precisely designed to remove the logarithm. As a result, for an ingoing light beam $U = 0$ at the event horizon, and there is no coordinate singularity.

$$\textit{Ingoing ray (u value)} : = C - 2r - 2R_S \log\left|\frac{r}{R_S} - 1\right|, \qquad \textit{Coordinate}: \quad U = -\exp\left(-\frac{u}{2R_S}\right),$$

$$\textit{Ingoing ray (U value)} : = -\exp\left(\frac{C - 2r}{-2R_S}\right)\exp\left(\frac{-2R_S}{-2R_S}\log\left|\frac{r}{R_S} - 1\right|\right), \tag{16.19}$$

$$= -\left|\frac{r}{R_S} - 1\right|\exp\left(\frac{2r - C}{2R_S}\right).$$

I have shown you all this so that you understand what sits behind KS coordinates. The concept is more important than the mathematical detail. In summary, the KS V and U coordinates are purely an exponentially re-scaled version of the EF v and u coordinates. As g_{vv} and g_{uu} are zero in the metric, we know that g_{VV} and g_{UU} are zero. The exponential re-scaling removes the coordinate singularities at the event horizon, so the off-diagonal terms in the metric are also well-behaved. Removing all the coordinate singularities allows us to view the black hole and white hole solutions of the Schwarzschild metric together. For those interested, Equation 16.20 is the final KS metric.

$$[g_{\mu\nu}] = \begin{bmatrix} 0 & \frac{2R_S^3}{r}e^{\frac{-r}{R_S}} & 0 & 0 \\ \frac{2R_S^3}{r}e^{\frac{-r}{R_S}} & 0 & 0 & 0 \\ 0 & 0 & -r^2 & 0 \\ 0 & 0 & 0 & -r^2\sin^2\theta \end{bmatrix}, \quad \textit{Schwarz. KS } (V, U, \theta, \phi\, [c = 1]). \tag{16.20}$$

For display purposes, we make the null pathways diagonal using a similar trick to the $\tilde{t}$ axis for EF. We create two new axis variables T and X where $T = \frac{V + U}{2}$ and $X = \frac{V - U}{2}$, or conversely $V = T + X$ and $U = T - X$. The overall effect is that both ingoing and outgoing light beams are at 45° in the KS spacetime diagram.

16.7 The KS Big Picture Schwarzschild Diagram

That is enough maths. It is time to pop the champagne corks! We can now look at the Schwarzschild metric with KS coordinates to see what is revealed. Let's start with the KS spacetime diagram on the left of Figure 16.5. T is the sort-of time-like axis and, as usual, points upwards. X is the sort-of space-like axis. The light beams move exactly as you would see in a flat Minkowski spacetime diagram. The ingoing light beam (red) at 45° up to the left and the outgoing light beam (blue) at 45° up to the right. Wherever light moves in the diagram, it is always on a 45° diagonal.

Of course, the Schwarzschild metric is not flat. It has curvature. Something has to give for the light beams to remain diagonal in the KS Schwarzschild spacetime diagram. If you track a path with constant value of the old t time or r radial coordinates, it is distorted and does not read straight across from the new T time-like axis or straight up from the new X space-like axis.

Continuing to focus on the left diagram of Figure 16.5, look carefully at the outgoing (blue) light path. This will help you get your bearings. Using the old (t, r) coordinates the light is emitted at $t = 0$ at $r = 1.4R_S$ (the r scale is shown in units of R_S). The old t time coordinate is now shown as straight radial lines out from the centre of the diagram. The old r coordinate is a sequence of curves (actually hyperbola). The (blue) outgoing light ray starts from $(t, r) = (0, 1.4\,R_S)$ goes through $(t, r) = (0.6, 1.6\,R_S)$ and heads on outwards.

Turn your attention now to the red ingoing light beam also emitted from $(t, r) = (0, 1.4\,R_S)$. As it heads into the event horizon ($r = 1R_S$), the radial t coordinate time lines get closer and closer. At the event horizon $t = \infty$. This is the effect of gravitational time dilation from the perspective of an observer outside of the event horizon who will never see the light reach the event horizon. However, the ingoing light beam does cross the event horizon and reach the singularity. Note that the event horizon is a straight diagonal line corresponding to $T = X$, which is $U = 0$. The singularity at $r = 0$ is an extended hyperbola.

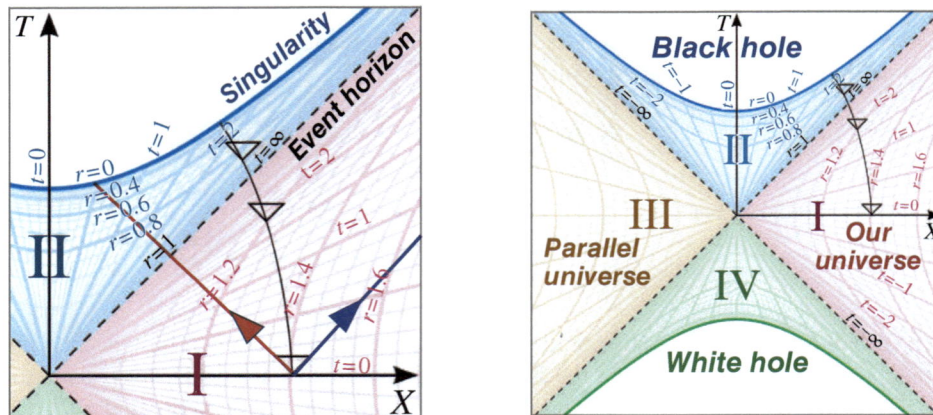

Figure 16.5 KS spacetime diagrams (adapted from Dr. Greg, CC Share-Alike 3.0 Unported). The left diagram shows an ingoing light beam (in red), an outgoing light beam (in blue) and the geodesic path of an object falling across the event horizon into the black hole (in black). The right diagram is zoomed out to show the whole Schwarzschild metric.

Between the example outgoing and ingoing light paths is a black line that is a geodesic. It is the track of a massive object falling from stationary into the black hole (the same as the geodesic grey track in the EF diagram on the left of Figure 16.3, but shown in KS coordinates). As we saw with the ingoing EF coordinates, the object from its own perspective continues unobstructed into the black hole. Once the event horizon has been crossed (entering the blue zone), there is no escape from the singularity. But don't forget that an outside observer never experiences this as is discussed in depth in Section 13.7 (*Time: For Minky the Clock Stops*). For more details on how KS coordinates help us to see what is going on, see Box 16.3.

Box 16.3 Minky and Schwart's excellent adventure

Referring to the left diagram of Figure 16.5, imagine that Minky kicks Schwart out of the airlock at $(t, r) = (0, 1.4\,R_S)$. This is the start of the black track geodesic in the diagram. Let me show you how KS coordinates illustrate Minky and Schwart's experience.

Minky hovers so she follows the $r = 1.4R_S$ path, curving up to the right in the diagram with increasing t. Schwart in freefall follows the black track geodesic. For Minky to see Schwart cross the event horizon, an outgoing light beam must travel from him to her. This is at 45° up and right, so it follows the dotted black line. This intersects with Minky's $r = 1.4R_S$ path at $t = \infty$ on Minky's clock. Therefore, she never sees him cross the event horizon. From Minky's perspective, she sees Schwart's clock progressively slow (getting ever closer to stopping) as he approaches the event horizon.

What does Schwart see? An ingoing light beam must arrive at Schwart from Minky. If you plot back an upward line at 45° (like the red line) arriving where Schwart crosses the event horizon, he sees Minky just after $t = 1$. While Minky sees Schwart at the event horizon for eternity (albeit redshifted and fading to nothing), Schwart does *not* see any of Minky's future.

Perhaps the biggest difference with KS versus EF coordinates is that we can zoom out to get a bigger picture of the Schwarzschild metric as shown on the right of Figure 16.5. The values of V and U are extended outwards until they hit a singularity (or with our diagram, leave the page and head off to T or X at infinity). For obvious reasons this is called the *maximally extended Schwarzschild solution*.

Zone I (red) is our universe outside the event horizon of the black hole. Zone II (blue) is inside it. You can still see poor old Schwart's black track path from Zone I into oblivion in Zone II. The KS diagram also shows the time-reversed solution, Zone IV (green), which is inside the event horizon of the white hole. Let's compare the white hole with the black hole. For the black hole, any light beam inside the event horizon (blue) travels at 45° diagonally upwards and will inevitably hit the singularity (blue hyperbola line). Nothing inside the event horizon can ever get out. For the white hole, any light beam inside the event horizon (green) also travels upwards at 45° diagonally and will inevitably exit the event horizon. And nothing outside the white hole event horizon can ever get in.

If one uses EF coordinates (instead of KS), the EF ingoing diagram on the left of Figure 16.3 is the spacetime diagram of our universe and the black hole. It is the equivalent of KS Zones I and II. The EF outgoing diagram on the right of Figure 16.3 is the spacetime diagram of our universe and the white hole. It is the equivalent of KS Zones I and IV.

The weird thing is that, based on pure mathematics, the extended solution includes a *parallel universe*, shown as Zone III (orange). Theoretical physicists who get excited by white holes go positively loopy-loo about parallel universes! In theory, there is an instant in time when the two universes of Zone I and III might meet... think of the Flamm paraboloid of one (Figure 13.5) connecting with the Flamm paraboloid of the other creating a wormhole between them with the radius of the event horizon (also called an *Einstein–Rosen bridge*). Even in theory, a Schwarzschild wormhole is not open long enough for anything, even light, to traverse it.

Let me offer a strong mental health warning. The white hole and parallel universe (Zones IV and III) in the Schwarzschild solution rest on its being *time-independent*. For any solution forwards in time, there is a legitimate solution backwards in time. However, black holes are not unchanging time-independent objects. And for any Schwarzschild black hole that is not *eternal*, there is no Zone IV or Zone III. But ho-hum, it is a legitimate solution to the Einstein field equations and the idea of parallel universes is too much fun to ignore, especially when you start toying with the notion of wormholes connecting them. We will return to the topic later.

16.8 Penrose–Carter Diagrams

I should take a moment to mention *Penrose–Carter* diagrams, which I think of as another way to *display* coordinates rather than a separate coordinate system (albeit that the distinction is a bit arbitrary). Penrose–Carter diagrams allow you to see the whole of space and time, even with additional universes, in one image. This is achieved with *conformal mapping*.

Let me try to explain with a simple example. Suppose we want to display a universe that is an infinite number of light years across, in a diagram only 2 centimetres wide. The universe runs from $-\infty$ to $+\infty$ light years. The diagram runs from -1 centimetre to $+1$ centimetre. Equation 16.21 shows you an example of how you could scale this from the origin 0 out to 1 centimetre. Move $\frac{1}{2}$ centimetre out and label that 1 light year distance out from the origin. Another $\frac{1}{4}$ centimetre out is 2 light years from the origin. Another $\frac{1}{8}$ centimetre out is 3 light years from the origin and so forth. Every light year out is scaled down by another half. Adding up the length in centimetres, you can see that the total from the origin all the way out to infinite light years is 1 centimetre in total, because: $\frac{1}{2} + \frac{1}{4} + \frac{1}{8} + \frac{1}{16}... = 1.$

$$\begin{bmatrix} Centimetres \\ Light\ years \end{bmatrix} \begin{bmatrix} 0 & +\frac{1}{2} & +\frac{1}{4} & +\frac{1}{8} & +\frac{1}{16} & +\frac{1}{32} & ... & +\frac{1}{1,024} & ... & +\frac{1}{2^{\infty}} \\ 0 & 1 & 2 & 3 & 4 & 5 & ... & 10 & ... & \infty \end{bmatrix} \tag{16.21}$$

Do the same in the negative direction to give the full horizontal axis running through the origin from $-\infty$ to $+\infty$ light years distance in two centimetres. Now run a vertical axis at the origin, built with the same rules. This is for time and runs from $-\infty$ to $+\infty$ years in two centimetres vertically up and down through the origin. For spacetime diagrams, we like light to travel at 45° diagonally so the coordinates of distance (light years) and time (years) must match 1:1 everywhere on the spacetime diagram. If you go to the far right of the horizontal axis, an infinite distance in light years is displayed on an infinitely small scale. The same must apply to time, so at the far right and left, all the infinity of time becomes a point. Similarly, at the top and bottom, where time is scaled infinitely small on the vertical scale, the entire length of the universe scales down to a point.

The actual maths behind the Penrose–Carter diagram maps $x = -\infty$ to $+\infty$ using $\tan^{-1}x$, across to values from $-\frac{\pi}{2}$ to $+\frac{\pi}{2}$, but the concept is similar. This gives the diamond shape of a Penrose–Carter diagram covering even infinite time and space in a universe, such as on the left of Figure 16.6. Light rays move diagonally. The curved lines are lines of constant distance and time. Extending this structure to the Schwarzschild metric allows us to show the entirety of Zones I and III in one diagram (compare the right of Figure 16.6 with the right of Figure 16.5).

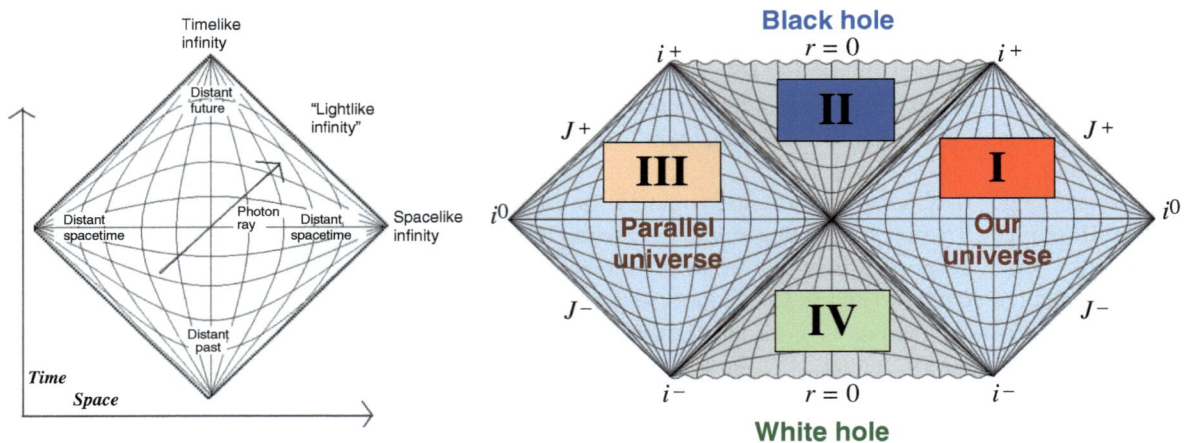

Figure 16.6 Penrose–Carter diagrams. On the left is the universe (Eric B, public domain) and on the right, the extended Schwarzschild Solution (Ray d'Inverno (1995) / Oxford University Press/ Public Domain).

I am not sure whether this is very illuminating with the Schwarzschild metric, but it will become helpful when we examine the Kerr metric for rotating black holes.

16.9 Schwarzschild Metric: Final Thoughts

This chapter ends our examination of the Schwarzschild metric, so I will take a moment to summarise. The start of the story in Chapter 12 was to find a spherically symmetrical vacuum solution to the EFEs. I used its curvature footprint to show you how the requirements that the metric be both Ricci-0 and spherically symmetrical lead to only one possible pattern of Riemann curvature (Section 12.2). As a result, the following relationships must be true: $g_{RR} = -\frac{1}{g_{TT}}$ and $g_{TT} = 1 - \frac{C}{r}$ (where C is a constant).

The fact that there is only one such solution means that we cannot ignore the implications of the Schwarzschild metric for extreme gravitational fields. This led to a discussion in Chapter 13 of black holes, which we now know sit at the heart of most galaxies. Sagittarius A^* in our Milky Way galaxy weighs in at about four million solar masses, while the much larger M87 black hole in the constellation of Virgo is estimated to be over five billion solar masses.

The Schwarzschild metric reveals two singularities in a black hole. The first is the singularity at $r = 0$ where spacetime curvature becomes infinite. The second is at the event horizon R_S where g_{TT} becomes zero and g_{RR} becomes infinite. However, the curvature footprint indicates no singularity in the underlying Riemann curvature at the event horizon. We confirmed later with EF coordinates that indeed this is a coordinate singularity.

This coordinate singularity means that observers outside will never see an object cross the event horizon (it will redshift and gradually fade from view). However, from the perspective of the object itself, it is business as usual. It will cross the event horizon unaffected by gravitational time dilation... providing we ignore the inevitable spaghettification that awaits any object falling into a black hole due to tidal forces.

In Chapter 14, we developed the physics surrounding black holes by looking at conserved quantities. I introduced you to the work of Emmy Noether and Wilhelm Killing, who showed that symmetries in the metric lead to conservation laws. Specifically, if the metric $[g_{\mu\nu}]$ is independent of a variable x, then u_x is conserved along any geodesic pathway.

The symmetries of the Schwarzschild metric lead to a slightly different definition from that of Newton for effective potential in a gravitational field. An additional term appears. Although negligible at large distances from the attracting mass, this additional term leads to different behaviour near the black hole such as an innermost stable circular orbit (ISCO) and a photon sphere where light can orbit the black hole (albeit in an unstable orbit).

The Schwarzschild conservation laws show that an object descending to the ISCO can release energy equivalent to over 6% of its rest mass energy. This is far greater than the energy output of nuclear fusion and can generate immense jets of radiation from active black holes (quasars).

In Chapter 15, I took you on a short trip down memory lane, revisiting a couple of calculations that were important in establishing the validity of Einstein's theory. I used the Schwarzschild metric to show that the deflection of light in a gravitational field is twice that expected based on Newtonian mechanics. I also stepped through some of the mathematics behind the missing 43 arcseconds per century of precession in the orbit of Mercury.

In this final chapter on the Schwarzschild metric we looked at some coordinates that allow us to peer inside the event horizon. We started with EF coordinates. The EF trick is to give ingoing and outgoing light beams a set value (I find it helpful to think of this as a measure related to their emission time). This creates a basis for new coordinates. The ingoing EF coordinates show how light and objects can cross the event horizon. The outgoing coordinates reveal another possible solution from the Schwarzschild metric, a time-reversed version: a white hole that ejects material from the event horizon as opposed to the black hole that drags material in.

The limitation of EF coordinates is that when you use the ingoing EF coordinates, outgoing light beams hit a coordinate singularity at the event horizon and, equivalently, when you use outgoing EF coordinates, the ingoing light beams hit a coordinate singularity. This is solved with KS coordinates, which are related to the EF ingoing and outgoing light beam coordinates, and are used simultaneously to replace both the time and radial coordinates.

The result is what you might call the full picture of the Schwarzschild metric: black hole, white hole and even an extra parallel universe. While this is exciting and interesting (assuming you like such things), there is good reason to argue that these are just mathematical anomalies rather than anything physical. On the other hand, there have been many occasions when things thought to be mathematical anomalies have turned out to exist. Let's not forget that many doubted the existence of black holes. So, perhaps we should keep a sceptical but open mind.

Having built up this picture of the Schwarzschild metric, let me now burst the bubble somewhat. The Schwarzschild metric is an excellent model for much, but is a bit unrealistic for an actual black hole. This is because black holes spin. The chance of material crashing into a black hole with no net angular momentum is negligible. And the material being so tightly compressed, any object arriving with angular momentum is going to make the black hole spin very fast. It is time to discuss spinning black holes and the Kerr metric.

Chapter 17

Kerr Metric: An Intuitive Introduction

Box 17.1 The most shattering experience

It was Subrahmanyan Chandrasekhar (1910–1995) who showed that a star of more than a few solar masses can become a black hole when it runs out of fusion fuel (see Box 13.5). His reaction to Kerr's discovery conveys the importance of the Kerr metric:

In my entire scientific life, extending over forty-five years, the most shattering experience has been the realisation that an exact solution of Einstein's equations of general relativity, discovered by the New Zealand mathematician, Roy Kerr, provides the absolutely exact representation of untold numbers of massive black holes that populate the universe.

The Kerr metric describes the gravitational field in a vacuum around a spinning mass. In most day-to-day scenarios, the attracting mass is not spinning fast enough to affect things significantly (for example, the sun and the earth), so the Schwarzschild metric works just fine. Black holes are an exception. The material they pull in is likely to have considerable angular momentum. That angular momentum is conserved and finally absorbed into the black hole. If you maintain angular momentum while decreasing the radius of an object, the spin rate increases. The usual example is an ice-skater who pulls in his or her arms to perform a dramatic spin. A black hole with no significant angular momentum would be strange.

There is not much you can say about a black hole because you cannot observe or identify the stuff inside it. A black hole has a certain mass, a certain angular momentum and a certain charge. The last is relatively unimportant because there is no reason to expect a typical black hole to carry much charge. That leaves its mass and angular momentum as the only key defining features. The physicist John Archibald Wheeler famously described this as: *black holes have no hair*. The Kerr metric quantifies the gravitational field around a black hole based on its mass and angular momentum… and that is exactly what we need. This means it describes virtually all black holes (see Chandrasekhar's reaction in Box 17.1).

Therefore, it may strike you as odd that the general reaction to Kerr's discovery was a lack of interest. He presented his metric at a symposium in 1962. Reputedly, his talk didn't even make it into the final conference summary notes. And Kerr himself is quoted as not realising the full importance of his work. My guess (and this really is a guess) is that the muted reaction was due to a lack of intuition. The coordinates Kerr used are confusing and don't tell a story. If you want to be scared, take a look at Equation 17.1, which is Kerr's metric written in his original coordinates. Note that the two lines of maths form one long equation. If you work through his original coordinates $(du, dr, d\theta, d\phi)$, you will see in the invariant interval alongside du^2, dr^2, $d\theta^2$ and $d\phi^2$, there are a whole bunch of off-diagonal terms: $du\,dr$, $du\,d\phi$, and $dr\,d\phi$. Clearly, his u coordinate is far from simple. I mean, what?

$$ds^2 = -\left(1 - \frac{2mr}{r^2 + a^2\cos^2\theta}\right)(du + a\sin^2\theta\,d\phi)^2 + 2(du + a\sin^2\theta\,d\phi)(dr + a\sin^2\theta\,d\phi)\dots$$

$$\dots + (r^2 + a^2\cos^2\theta)(d\theta^2 + \sin^2\theta\,d\phi^2), \qquad \textit{Kerr metric, original coordinates.}$$

(17.1)

Untangling General Relativity: The Intuitive Self-Study Guide, First Edition. Simon Sherwood.
© 2026 John Wiley & Sons Ltd. Published 2026 by John Wiley & Sons Ltd.

Let me stress that I am not criticising Roy Kerr. The guy is a genius to have come up with this metric (although Box 17.3 reveals an amusing occasion when he did less well). Remember that this was almost 50 years after Einstein's work and nobody else had managed to find a rotational vacuum solution to the EFEs. Good on yer, Roy!

My point is that the Kerr metric can be intimidating, but I will not let that happen to you, my precious readers. By the end of this chapter, I hope to have convinced you that the basic structure of the Kerr metric is intuitive and unsurprising. That is my goal. It does mean taking things a bit more slowly than in other texts. But all good things come to those who wait.

The most important constraint on the Kerr metric is that it must be *Ricci-0*. However, I am not going to be able to give you that derivation. Using again a quote from Chandrasekhar, the Kerr metric *simply does not allow a presentation that can be followed in detail with modest effort... and, on occasion, may require as many as ten, twenty, or even fifty pages.* If someone like Chandrasekhar says something like that, you'd better believe it.

The best we can do is sleuth our way towards an explanation. For this, I am going to employ the skills of Rational Rachel, an imaginary detective named after my wife who has an extraordinary ability to find the things I lose... spectacles, keys, cats and even children. Rational Rachel will help us explain the mysterious Kerr metric by building a detailed profile of it.

17.1 The Kerr Metric Using BL Coordinates

Box 17.2 The Kerr metric in BL coordinates

$$d\tau^2 = \left(1 - \frac{2GM\,\mathbf{r}}{p^2}\right) dt^2 - \frac{p^2}{\mathbf{r}^2 + a^2 - 2GM\,\mathbf{r}}\, d\mathbf{r}^2 - p^2\, d\theta^2 \dots$$

$$\dots - \left(\mathbf{r}^2 + a^2 + a^2 \sin^2\theta \, \frac{2GM\,\mathbf{r}}{p^2}\right) \sin^2\theta \, d\phi^2 + 2a \sin^2\theta \, \frac{2GM\,\mathbf{r}}{p^2}\, dt\, d\phi,$$

$$a = \frac{J}{M} \equiv \frac{Angular\ Momentum}{Mass}, \qquad p^2 = \mathbf{r}^2 + a^2 \cos^2\theta, \qquad (c = 1).$$

Box 17.2 shows the Kerr metric in what are called *Boyer–Lindquist* (BL) coordinates. The two lines of maths form one long equation. Some readers may protest that this doesn't look any simpler than Kerr's original version, but it is. Let me point out the salient features of this beauty.

Look carefully at the first four terms in the Kerr BL invariant interval. Yuk! At first, they look messy. I have labelled what is called the BL radial coordinate **r** in bold (BoLd for Boyer–Lindquist, geddit?). We will discuss it in detail later, but let me explain the underlying difference between it and the spherical coordinate *r*. While the spherical coordinate *r* reflects spherical symmetry, the BL radial coordinate **r** reflects the symmetry of an *oblate spheroid*, which is like a squashed sphere.

Although the first four terms of the Kerr metric look complicated in BL coordinates, the use of the BL radial coordinate **r** simplifies the overall structure of the metric by reducing the number of off-diagonal terms from three in Kerr's original metric (Equation 17.1) down to one. If you analyse the coordinate components $(dt, d\mathbf{r}, d\theta, d\phi)$ in the invariant interval, you can see the usual dt^2, $d\mathbf{r}^2$, $d\theta^2$ and $d\phi^2$ terms plus one (and only one) off-diagonal term: $dt\, d\phi$. Take a careful look at Box 17.2 to make sure you can spot it. It is the fifth and last term in the invariant interval.

My first mission is to show why at least one off-diagonal term appears in the Kerr metric. My second mission is to explain the use of the symmetry of an oblate spheroid. Along the way, Rational Rachel will help us build a more detailed profile of the metric.

Box 17.3 Young Roy underperforms

In 1950, Roy Kerr (1934–) sat his university entrance exam in mathematics. It consisted of two papers each worth 300 points. Overall, Kerr scored a mediocre 298. What happened? He turned up in the afternoon for the second paper. The exam had been that morning. More than impressed by Kerr's score on the first paper, the university accepted him anyway. So sleeping in can work out.

17.2 Why Angular Momentum Matters

The off-diagonal term is shown separately as Equation 17.2. You might wonder why there is this off-diagonal term in the Kerr invariant interval.

$$+ 2a \sin^2 \theta \, \frac{2GM \, \mathbf{r}}{p^2} \, dt \, d\phi, \qquad \textit{Off-diagonal term, Kerr metric (BL coordinates)},$$

$$a = \frac{J}{M} \equiv \frac{\textit{Angular Momentum}}{\textit{Mass}}, \qquad p^2 = \mathbf{r}^2 + a^2 \cos^2 \theta, \qquad (c = 1).$$

(17.2)

Let's start with a simple question. The Kerr metric describes a rotating attracting mass, i.e. one with angular momentum. Why would its angular momentum make a difference to the spacetime metric surrounding it? Of course, rotational energy adds to effective rest mass energy and very slightly increases the inward radial gravitational pull, but beyond that, why would it matter? The answer lies in the EFEs shown in spherical coordinates in Equation 17.3.

$$R_{\mu\nu} - \frac{1}{2} R g_{\mu\nu} = k \, T_{\mu\nu},$$

Spacetime curvature ⟺ Energy-momentum,

$$
\begin{bmatrix}
R_{TT} & R_{TR} & R_{T\theta} & R_{T\phi} \\
R_{RT} & R_{RR} & R_{R\theta} & R_{R\phi} \\
R_{\theta T} & R_{\theta R} & R_{\theta\theta} & R_{\theta\phi} \\
R_{\phi T} & R_{\phi R} & R_{\phi\theta} & R_{\phi\phi}
\end{bmatrix}
- \frac{1}{2} R
\begin{bmatrix}
g_{TT} & g_{TR} & g_{T\theta} & g_{T\phi} \\
g_{RT} & g_{RR} & g_{R\theta} & g_{R\phi} \\
g_{\theta T} & g_{\theta R} & g_{\theta\theta} & g_{\theta\phi} \\
g_{\phi T} & g_{\phi R} & g_{\phi\theta} & g_{\phi\phi}
\end{bmatrix}
= k
\begin{bmatrix}
T_{TT} & T_{TR} & T_{T\theta} & T_{T\phi} \\
T_{RT} & T_{RR} & T_{R\theta} & T_{R\phi} \\
T_{\theta T} & T_{\theta R} & T_{\theta\theta} & T_{\theta\phi} \\
T_{\phi T} & T_{\phi R} & T_{\phi\theta} & T_{\phi\phi}
\end{bmatrix}.
$$

(17.3)

Our scenario is an attracting mass rotating in the $d\phi$ direction. As a reminder, take a look at Figure 17.1. This is the same as Figure 9.1 but in natural units ($c = 1$). The figure also shows the metric of flat spacetime in spherical coordinates, which will be useful for reference later. The left includes arrows showing the direction of rotation of the central attracting mass (around what you might call the equator). Whatever the direction of rotation, we can always position our spherical coordinates to match this schematic. Let's think about the spacetime metric *inside* the rotating mass. At this stage, we will avoid the extreme black hole scenario with its tricky singularity. Think instead perhaps of our sun, which basically is a rotating sphere of compressed hydrogen gas, or think of a rapidly spinning high-density neutron star.

As usual, I begin with some comments on symmetry. Our scenario is rotating. It is not static, but it is stationary. This means that the metric may change with time reversal. If we switch from $+dt$ to $-dt$, the direction of rotation would reverse. We cannot eliminate the possibility of off-diagonal terms in the metric in the way we did in our analysis of the Schwarzschild metric (Section 12.1). However, the symmetries in the scenario do tell us that the

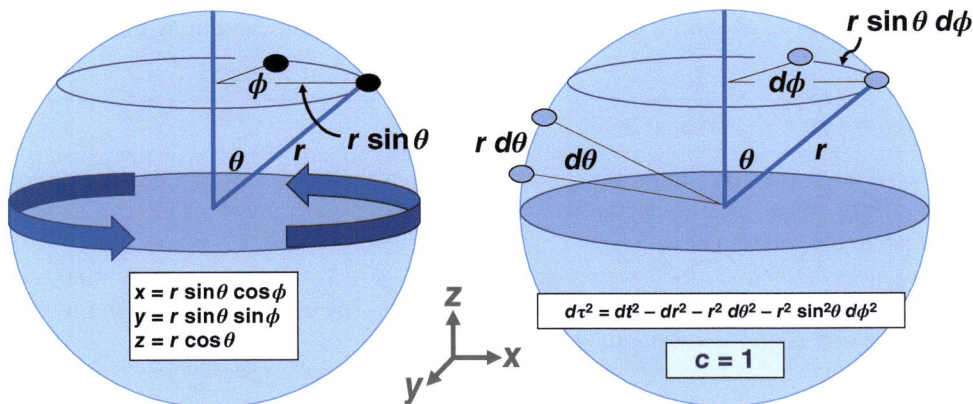

Figure 17.1 Spherical coordinates. The invariant interval shown is flat Minkowski spacetime.

metric should *not* be dependent on time t or angle ϕ. Neither t nor ϕ should appear in any of the components of $[g_{\mu\nu}]$. If we set t to 11am or 3pm, the metric should not change. Similarly, the metric should be the same whether we measure it at angle $\phi = 0$ or $\phi = \pi$. The rotation is in the $d\phi$ direction. There is circular symmetry around the equator.

Returning to the EFEs in Equation 17.3, our interest lies in $R_{T\phi}$ ($= R_{\phi T}$) and $g_{T\phi}$ ($= g_{\phi T}$) of the *internal* spacetime metric of the rotating star, so I have separated out that particular EFE as Equation 17.4. I need you to cast your mind back to Chapter 8 on the energy-momentum tensor $[T_{\mu\nu}]$. You may want to check back to the summary in Section 8.8, including Figure 8.2. This will remind you that T_{TT} contains the rest mass energy and in Cartesian coordinates T_{TX} (which is the same as T_{XT}) is momentum in the x direction, T_{TY} momentum in the y direction and so forth. In spherical coordinates T_{TR} reflects momentum in the radial direction, $T_{T\theta}$ angular momentum in the θ direction and $T_{T\phi}$ angular momentum in the ϕ direction. The important point is that the angular momentum of the rotation shown in Figure 17.1 around the z axis creates a $T_{T\phi}$ term in the energy-momentum tensor: $T_{T\phi} \neq 0$.

$$R_{T\phi} - \frac{1}{2} R g_{T\phi} = k T_{T\phi} \neq 0, \qquad \textit{Internal metric: angular momentum in } d\phi \textit{ direction.} \qquad (17.4)$$

This suggests that the $g_{T\phi}$ term of the spacetime metric inside the star is *not zero*. We know for a fact that there *must be* a non-zero off-diagonal term in the metric. If $g_{T\phi}$ were zero, then the Ricci tensor component $R_{T\phi}$ would have to be non-zero for the total of Equation 17.4 to be non-zero. However, mixed-index components of $R_{\mu\nu}$ such as $R_{T\phi}$ are always zero if the spacetime metric has no off-diagonal terms *unless* the metric depends on both the variables (the proof is back in Box 9.3). And we know from symmetry that this is not the case; the metric contains neither t nor ϕ terms.

What is the significance of an off-diagonal $g_{T\phi}$ term in the *internal* metric of the star? It means that the t and ϕ coordinates are no longer orthogonal inside the star. A move along the time t axis also moves in the ϕ direction and vice versa (check back to Section 3.3 and the right of Figure 3.1 for a simple 2-D example). At first this may strike you as odd, but remember that the presence of energy-momentum affects the spacetime volume between infinitely close parallel geodesics (Section 6.4). If the energy-momentum is moving, it makes intuitive sense that this movement distorts the spacetime metric.

Some smarty-pants readers may question what this has to do with the Kerr metric, which covers the surrounding vacuum, not the rotating star's interior. My answer is that the distortion between t and ϕ from the $g_{T\phi}$ term will be most significant at the star surface, where the angular motion is strongest. If it were suddenly to disappear beyond the star surface, that would represent a weird discontinuity in the curvature of spacetime at the boundary (not good, as discussed in Section 10.2). To allow a smooth transition, you would expect the Kerr metric to reflect the distortions of the spacetime inside the rotating mass.

In summary, the internal metric of the rotating mass must have an off-diagonal term, so the metric of the surrounding vacuum should also have an off-diagonal term. This appears as a non-zero $g_{T\phi}$ term in the Kerr metric (Box 17.4).

How can you extend the logic to black holes? It is far from obvious where any boundary is! I just imagine the spacetime volume containing the rotating mass becoming smaller and smaller until it reduces to a singularity. What that singularity is, we do not know, but the distortion of the energy-momentum within it remains.

Box 17.4 The Kerr metric's off-diagonal term

The Kerr metric has a non-zero off-diagonal $g_{T\phi}$ ($= g_{T\phi}$) component. This means that the time t and angular ϕ axes are not fully orthogonal. It results in *frame dragging*: objects in the vicinity of the rotating mass experience some acceleration in the rotating direction.

This arises because the angular momentum appears as $T_{T\phi} \neq 0$ in the energy-momentum tensor of the rotating mass, leading to $g_{T\phi} \neq 0$ in its internal spacetime metric. The off-diagonal term in the Kerr metric is the boundary effect of this on the surrounding vacuum.

The classical Newtonian view of gravity is based on mutual attraction between masses. This creates a force of attraction between their centres of mass. There is no reason to think that angular momentum would play a significant role. Einstein's theory is very different. The presence of energy-momentum distorts the spacetime containing it. If it is rotating rapidly, it alters the pattern of distortion. One consequence that we can spot immediately in the EFEs is the emergence of an off-diagonal $g_{T\phi}$ ($= g_{\phi T}$) term in the metric. This leads to what is called *frame-dragging*. Nearby objects are dragged around in the direction of rotation (discussed later in Section 18.5).

It is time for our sleuth Rational Rachel to start to build her Kerr profile. She points to the following things we know about the $dt\,d\phi$ term in the Kerr metric. It must be a function of angular momentum J (no rotation, no off-diagonal term) and **r** (the effect will decay with distance from the mass). It should be zero at the poles (no rotation) and grow to a maximum at the equator so it might well be a function of $\sin\theta$. It involves dt, so it is affected by gravitational time dilation, which depends on GM. It should *not* depend on t or ϕ because of the symmetries discussed.

Rational Rachel's profile for $dt\,d\phi$ is that it should depend on $(J, \mathbf{r}, \sin\theta, G, M)$ but not on (t, ϕ). Hopefully, if you compare back with the Kerr off-diagonal term in Equation 17.2, the structure will start to make sense.

17.3 An Oblate Spheroid

My second mission is to throw some light on the other major distortion created by the rotation of the attracting mass. The title I have given this section is one of my favourites, guaranteed to frighten the hardiest reader. As mentioned earlier, the symmetry of the Kerr metric is no longer spherical like that of the Schwarzschild metric but resembles that of an *oblate spheroid*. Think of a sphere squashed down at the poles and bulging at the equator. One result of the change in symmetry is that the event horizon of the Kerr metric has this oblate spheroid (squashed sphere) shape.

The different symmetry causes havoc to the Kerr metric unless you use coordinates that respect it. If only I could think of an analogy, a familiar surface with this symmetry. Oh yes... the earth! Our dear home planet is not actually a sphere. It bulges slightly at the equator due to its rotation. The earth's radius from the centre to the pole is 6,357 kilometres as opposed to 6,378 out to the equator. This may seem small, but think of the mess we'd be in if we treated the earth as spherically symmetrical and applied spherical coordinates.

A typical jet flies at an altitude of about 10 kilometres. Imagine an air traffic controller applying this in spherical coordinates. A flight plan leaving the north pole and proceeding at an earth-radius of 6,367 kilometres is going to get ugly near the equator. Ouch! The obvious thing to do is to create a coordinate that measures things relative to the surface of our oblate spheroidal home... we call this coordinate height above sea level. I am fortunate enough currently to be writing this book at Whistler ski resort. The top of the resort is 7,000 feet above sea level. From a quick look on the Web, I know that water will boil there at 93 degrees centigrade, so it will be hard to make a decent cup of tea. Try working that out with spherical coordinates.

We can use something similar to sea level in our choice of coordinates for the Kerr metric. We use the BL radial coordinate as illustrated in Figure 17.2. On the left are the usual spherical coordinates. Any point with $r = 1$ sits on the surface of the blue sphere. At $r = 2$ it is the surface of the orange sphere, and for $r = 3$ the green sphere. The point $(1, \frac{\pi}{2}, 0)$ is shown. It is one unit of length out from the centre and angle $\theta = \frac{\pi}{2}$ down from the pole. This should look familiar.

On the right is the equivalent diagram using the BL radial coordinate, which I have labelled in the figure as r_{BL} to distinguish it clearly from spherical coordinate r. A point with $r_{BL} = 1$ sits on the *blue oblate spheroid*. Increasing the value of the r_{BL} coordinate takes you to the surface of the orange and then the green oblate spheroids. The BL radial coordinate is this sequence of layer after layer like those Russian Matryoshka dolls. If you look at the point labelled $(r_{BL}, \theta, \phi) = (3, 0, 0)$, it is in the same spatial location as for spherical coordinates $(r, \theta, \phi) = (3, 0, 0)$. However, the location of $(r_{BL}, \theta, \phi) = (1, \frac{\pi}{2}, 0)$ is different from $(r, \theta, \phi) = (1, \frac{\pi}{2}, 0)$. To create the oblate spheroidal symmetry, we have sort of stretched out the x and y axes. If the radius to the pole is r, then the radius to the equator is $\sqrt{r^2 + a^2}$ where a is a constant. Each layer of oblate spheroidal symmetry becomes progressively less distorted (squashed) until at an infinite distance out ($r = \infty$), it is spherical.

To manage the oblate spheroidal symmetry of the Kerr metric most easily (see Box 17.5), we must cope with the BL radial coordinate. Writing r_{BL} is a bore so, as mentioned earlier, I label the BL radial coordinate as **r** in bold (except in the figures where I show it as r_{BL} for clarity).

Using the BL radial coordinate **r** is not pain-free. The downside is that it messes with the way the metric appears. For example, a move of $d\theta$ near the poles is further in terms of the invariant interval than it is for

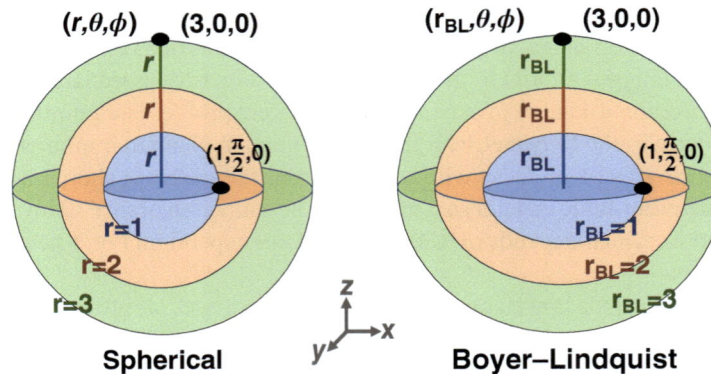

Figure 17.2 Comparison of spherical and BL coordinates. The BL radial coordinate is shown on the right, labelled as r_{BL} for clarity.

spherical coordinates. With the BL radial coordinate, it is a move around the perimeter of an ellipse (rather than around a circle as it is in the case of spherical coordinates). Again, the surface of the earth serves as an analogy. The equivalent of coordinate θ is latitude. Each degree of latitude is an angular degree (measured from the centre of the earth) further away from the north pole. The distance between lines of latitude one degree apart varies. It is 110.6 kilometres at the equator, but 111.7 kilometres at the poles.

Box 17.5 The Kerr metric's oblate spheroidal symmetry

Let me try to show you why this symmetry is not a complete surprise. I should warn that serious physicists may cough and splutter at my lack of rigour.

We know that angular momentum affects the internal spacetime metric of the rotating mass (Section 17.2), and must expect some similar effect on the surrounding vacuum to avoid a discontinuity in curvature at the boundary. This distortion should grow from near the poles (where the effect of angular momentum is felt less) up to a maximum at the equator (where the effect of angular motion is felt most).

In the case of the Schwarzschild metric, the vacuum curvature manifests spatially as an increase in radial separation dr between spherical shells around the attracting mass (Section 13.9). A similar sort of effect from the extra angular momentum in a rotating mass could lead to a change in this radial separation that gradually increases from pole to the equator. The result might well have a symmetry resembling an oblate spheroid.

17.4 BL Radial Coordinate Mathematics (Optional)

To understand the structure of the Kerr metric, I believe some (not too challenging) BL maths helps. I have labelled this section optional because it may be more detailed than some readers want. If you are one of them, then feel free to skip to Section 17.5.

Figure 17.3 shows an oblate spheroid on the right. Mathematicians may be interested to know that this is a type of ellipsoid. The equator (the x/y plane) is a *circle* with radius L. Any plane cutting through the poles (for example, the x/z or y/z planes) is an *ellipse* with long radial axis length L and short radial axis length $\mathbf{r}$ (this is the BL radial coordinate shown as r_{BL} in the figure). The ellipse is shown in detail on the left of the figure. I introduced ellipses in Section 15.2 on the precession of Mercury's orbit, but some of you may have skipped that chapter, so let me start from scratch.

An ellipse has two focus points. The distance of each focus from the centre is called the *focal length* and is labelled a in the figure. An important feature of an ellipse is that the distance in a straight line from one focus to the perimeter and then back to the other focus is always the same. If you calculate this along the long radial axis, it is $(L + a)$ from the left focus out to the right perimeter and $(L - a)$ back to the right focus, for a total of $2L$. Using

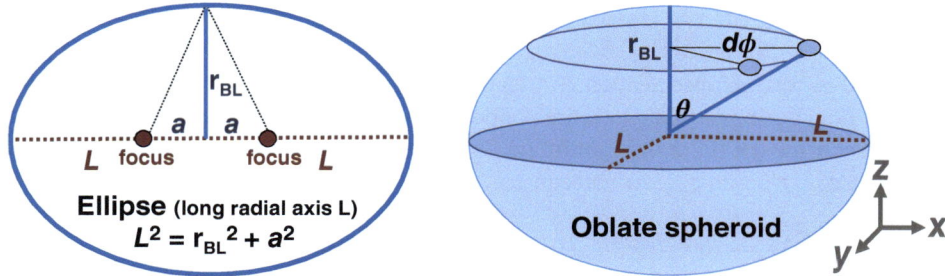

Figure 17.3 The mathematics of BL coordinates. In the figure, the BL radial coordinate is labelled as r_{BL} for clarity.

Pythagoras, the same calculation along the short radial axis (the dotted black line) is $2\sqrt{\mathbf{r}^2 + a^2}$. The two paths must be the same length so $L = \sqrt{\mathbf{r}^2 + a^2}$.

As the short radial axis is $\mathbf{r}$ and the long radial axis is $\sqrt{\mathbf{r}^2 + a^2}$, we must scale up the x and y axes by $\frac{\sqrt{\mathbf{r}^2 + a^2}}{\mathbf{r}}$ to give the required BL oblate spheroid symmetry. As you would expect, the level of scaling depends on the focal length a. In the extreme that $a = 0$, no scaling is required because the oblate spheroid is a sphere. For any doubters who question whether scaling up x and y like this (while leaving z alone) accurately produces an oblate spheroid, there is a proof in Box 17.6.

What happens to the metric when we switch to the new BL radial coordinate $\mathbf{r}$? For spherical coordinates, $g_{\phi\phi} = r^2\sin^2\theta$. This is because a shift of $d\phi$ is a move of length $r\sin\theta\,d\phi$ as shown on the right of Figure 17.1. You multiply the angular change $d\phi$ by the distance out from the axis of $r\sin\theta$. This distance out is in the x/y plane, so it increases with the $\frac{\sqrt{\mathbf{r}^2 + a^2}}{\mathbf{r}}$ scaling factor to become $(\sqrt{\mathbf{r}^2 + a^2})\sin\theta$. Therefore, for $g_{\phi\phi}\,d\phi^2$ using the BL radial coordinate, $g_{\phi\phi} = -(\mathbf{r}^2 + a^2)\sin^2\theta$. The minus sign appears because of the $< + - - - >$ signature of the metric.

The calculation of $g_{\theta\theta}$ also is easy. You can see from the right of Figure 17.3 that a shift of $d\theta$ is a move around the perimeter of the ellipse. The formula for calculating the length of a section of the perimeter of an ellipse is well known. It is shown in Equation 17.5 in terms of $\mathbf{r}$ (the short radial axis) and a (the focal length). The integration is the sum of tiny $d\theta$ shifts, so squaring the integrand (the thing integrated) leads directly to $g_{\theta\theta} = -(\mathbf{r}^2 + a^2\cos^2\theta)$.

$$\begin{aligned} Perimeter &= \int \sqrt{(\mathbf{r}^2 + a^2)\cos^2\theta + \mathbf{r}^2\sin^2\theta}\ \ d\theta, \qquad Note: L^2 = \mathbf{r}^2 + a^2, \\ &= \int \sqrt{\mathbf{r}^2 + a^2\cos^2\theta}\ \ d\theta, \implies \quad g_{\theta\theta} = -(\mathbf{r}^2 + a^2\cos^2\theta). \end{aligned} \tag{17.5}$$

We also need to calculate $g_{\mathbf{RR}}$ for the BL radial coordinate. This is the distance moving from one oblate sphere surface out to the next. It depends on θ and is a bit more complicated. Keen mathematicians may want to calculate this (I suggest using the formula for the radius of an ellipse). The answer is $g_{\mathbf{RR}} = -\left(\dfrac{\mathbf{r}^2 + a^2\cos^2\theta}{\mathbf{r}^2 + a^2}\right)$.

Box 17.6 Showing the BL radial coordinate creates an oblate spheroid (Optional)

This box is for any super-mathematical readers who want to dig fully into the argument. We take the coordinates for a sphere ($x^2 + y^2 + z^2 = r^2$) and create new BL coordinates XYZ using BL scaling for X and Y. Labelling $r^2 + a^2 = K^2$ for clarity gives Equation 17.6. Using the relationship for a sphere, we arrive at Equation 17.7, which mathematicians will recognise as the formula for an ellipsoid. As $r < k$ it is an oblate spheroid.

$$X^2 = \frac{r^2 + a^2}{r^2}x^2 = \frac{K^2 x^2}{r^2}, \qquad Y^2 = \frac{r^2 + a^2}{r^2}y^2 = \frac{K^2 y^2}{r^2}, \qquad Z^2 = z^2, \tag{17.6}$$

$$\frac{x^2}{r^2} + \frac{y^2}{r^2} + \frac{z^2}{r^2} = 1 = \left(\frac{X}{K}\right)^2 + \left(\frac{Y}{K}\right)^2 + \left(\frac{Z}{r}\right)^2, \quad Formula\ for\ oblate\ spheroid. \tag{17.7}$$

17.5 Minkowski Spacetime Using the BL Radial Coordinate

Welcome back to readers who skipped Section 17.4. In order for you to understand the structure of the Kerr metric, I want to show you how boring old flat spacetime appears using the BL radial coordinate. Equation 17.8 shows Minkowski spacetime using the BL radial coordinate. Those who read the optional maths section will note I have simply substituted in $g_{\mathbf{RR}}, g_{\theta\theta}$ and $g_{\phi\phi}$. As you can see, this significantly alters the look of the Minkowski metric!

$$d\tau^2 = dt^2 - \frac{\mathbf{r}^2 + a^2 \cos^2\theta}{\mathbf{r}^2 + a^2}\, d\mathbf{r}^2 - (\mathbf{r}^2 + a^2 \cos^2\theta)\, d\theta^2 - (\mathbf{r}^2 + a^2)\, \sin^2\theta\, d\phi^2. \tag{17.8}$$

Equation 17.8 looks befuddling, but it is still plain vanilla flat Minkowski spacetime. Changing the coordinates does not change the spacetime. And all we have done is change one, and only one, coordinate from r to $\mathbf{r}$, where the new $\mathbf{r}$ has the same value at all places on the surface of an oblate sphere (to switch back to regular spherical coordinates, simply set $a = 0$). Typically, the expression is simplified visually by labelling $p^2 = \mathbf{r}^2 + a^2 \cos^2\theta$ giving the form in Equation 17.9.

$$d\tau^2 = dt^2 - \frac{p^2}{\mathbf{r}^2 + a^2}\, d\mathbf{r}^2 - p^2\, d\theta^2 - (\mathbf{r}^2 + a^2)\, \sin^2\theta\, d\phi^2. \tag{17.9}$$

We will be using the BL radial coordinate to examine the Kerr metric. When you look at the Kerr metric, you will recognise much of what we have been discussing. Don't let these changes in the metric frighten you. They just cater for its oblate spheroidal symmetry.

Before moving on, we need Rational Rachel to help us build a profile of a, the focal length of the squashed ellipse as shown on the left of Figure 17.3 (in the Kerr metric a determines what is called the *spin parameter,* as will be discussed later). In layman's language, we want to know what determines how squashed the oblate spheroidal symmetry is. We would expect the primary driver to be the amount of angular momentum J (it is rotation that leads to the ellipsoidal distortion). It also should depend on mass M (a large slowly rotating mass is different from a small rapidly rotating mass). Rachel concludes that a will be affected by J and M. If you check back to the Kerr metric in Box 17.2, you will see that this is the case and that $a = \dfrac{J}{M}$.

17.6 The Other BL Coordinates

So far, we have altered only the radial coordinate $\mathbf{r}$. We also have to toy with t and ϕ to switch to the full BL coordinate system. This may seem a big deal, but it is not. All the important calculations in the next chapter give results dependent only on $\mathbf{r}$ (which hopefully you now understand) and θ, the usual spherical angular coordinate. The reason we alter t and ϕ is to isolate the effect of the off-diagonal term to $g_{T\phi}$ ($= g_{\phi T}$). If we do not do this, the effect of the $g_{T\phi}$ term leaks elsewhere in the metric and we end up with additional confusing off-diagonal entries, as was the case with Kerr's original coordinate system discussed back in Equation 17.1.

For anybody strange enough to want the gory details, the t and ϕ BL coordinate changes are shown as Equation 17.10.

$$dt_{BL} = dt - \frac{2GM\mathbf{r}}{\mathbf{r}^2 + a^2 - 2GM\mathbf{r}}\, d\mathbf{r}, \qquad d\phi_{BL} = d\phi + \frac{a}{\mathbf{r}^2 + a^2 - 2GM\mathbf{r}}\, d\mathbf{r}. \tag{17.10}$$

17.7 Summary

In order to avoid your constantly having to refer back, I have shown again the Kerr metric in full BL coordinates in Box 17.7. Please note that the two lines of maths form one long equation. The bold form of $\mathbf{r}$ is to remind you that this is the BL radial coordinate, not the usual spherical coordinate r. It is the distance (in flat spacetime) along the shorter radial axis that runs from the centre to the pole of the relevant oblate spheroid shell. Note that most texts and lecture series set $G = 1$ so $2GM$ is shown simply as $2M$. I have left G in for clarity and as a gesture of respect to Isaac Newton, who formulated G in 1680.

The Kerr metric is an *exact* solution to the EFEs. It has been shown to be the unique vacuum solution for the spacetime surrounding a rotating (uncharged) black hole. Its key feature is that it is Ricci-0 ($[R_{\mu\nu}] = 0$). Unfortunately, for a mere mortal like me, a full derivation of the Kerr metric is unfathomable.

Box 17.7 The Kerr metric in BL coordinates

$$d\tau^2 = \left(1 - \frac{2GM\,\mathbf{r}}{p^2}\right) dt^2 - \frac{p^2}{\mathbf{r}^2 + a^2 - 2GM\,\mathbf{r}} dr^2 - p^2\,d\theta^2 \ldots$$

$$\ldots - \left(\mathbf{r}^2 + a^2 + a^2 \sin^2\theta\, \frac{2GM\,\mathbf{r}}{p^2}\right) \sin^2\theta\, d\phi^2 + 2a \sin^2\theta\, \frac{2GM\,\mathbf{r}}{p^2}\, dt\, d\phi,$$

$$a = \frac{J}{M} \equiv \frac{Angular\ Momentum}{Mass}, \qquad p^2 = \mathbf{r}^2 + a^2 \cos^2\theta, \qquad (c = 1).$$

The presence of the rotating mass leads to two major differences in the Kerr metric compared with the Schwarzschild metric: the emergence of a new non-zero off-diagonal component $dt\,d\phi$ in the invariant interval plus the spacetime metric has oblate spheroidal symmetry (rather than the spherical symmetry of the Schwarzschild metric).

The non-zero off-diagonal $dt\,d\phi$ term in the invariant interval causes frame-dragging, as will be discussed later. The value of $g_{T\phi}$ depends as you would expect on the amount of angular momentum (via the values of a and p^2) and the distance out from the rotating mass (via the values of $\mathbf{r}$ and p^2). It is proportional to $(\sin^2\theta)$, which makes sense as this varies from zero at the poles (no angular motion, no off-diagonal term) to one at the equator (maximum angular momentum, maximum value of the off-diagonal term). It also is affected by time dilation because it contains the dt term, so GM appears. The dependence on these variables is to be expected but, as a Scot might say, the off-diagonal term is a complicated wee beastie.

The key variable driving the oblate spheroidal symmetry is focal length a. In the language of the Kerr metric, a defines what is called the spin parameter. It depends on angular momentum and mass: $a = \frac{J}{M}$. The larger a is, the more squashed the oblate spheroidal symmetry (if needed, check back to Figure 17.3). In Box 17.5, I attempted to give you an intuitive explanation of why this symmetry makes sense.

If there is no angular momentum ($a = 0$), the BL radial coordinate reverts to being the usual spherical coordinate ($\mathbf{r} = r$) and ($p = r$). In addition, the off-diagonal term disappears and the mass term $\frac{2GM\,\mathbf{r}}{p^2}$ becomes the more familiar $\frac{2GM}{r}$. The result is the Schwarzschild metric as shown in Equation 17.11. This has to be the case because this is the metric for the vacuum surrounding a spherical non-rotating mass.

$$d\tau^2 = \left(1 - \frac{2GM\,r}{r^2}\right) dt^2 - \frac{r^2}{r^2 - 2GM\,r} dr^2 - r^2\,d\theta^2 - r^2 \sin^2\theta\, d\phi^2, \qquad Kerr\,(a=0),$$

$$= \left(1 - \frac{2GM}{r}\right) dt^2 - \frac{1}{1 - \frac{2GM}{r}} dr^2 - r^2\,d\theta^2 - r^2 \sin^2\theta\, d\phi^2, \quad Schwarzschild.$$

(17.11)

If you analyse the Kerr metric as you get further and further from the rotating mass, $\mathbf{r}$ tends to infinity, and all the BL coordinates $(t, \mathbf{r}, \theta, \phi)$ become the same as the usual spherical coordinates (note that θ is the usual spherical coordinate anyway). At $\mathbf{r} \to \infty$, the Kerr metric is the same as that of flat Minkowski spacetime.

In the absence of a derivation, I hope I have gone some way to getting you comfortable with the Kerr metric. In the next chapter we will examine the implications of the Kerr metric for conditions surrounding a rotating black hole. You will see that things are much more complicated than with the good old non-rotating Schwarzschild black hole. There is even a legitimate (but not necessarily physically realistic) mathematical argument that it might be possible to pass through the singularity of a Kerr black hole and emerge out into another universe.

It appeals to me that black holes still hold many mysteries. In fact, we don't even know who first came up with the name *black hole* (see Box 17.8).

Box 17.8 Who first called them black holes?

John Archibald Wheeler (1911–2008) is generally thought to have invented the term black hole. But the truth is, we will never know. The term was coined by an unidentified person in a conference audience, as Wheeler himself acknowledged:

In the fall of 1967, [I was invited] to a conference. ... In my talk, I argued that we should consider the possibility that the centre of a pulsar is a gravitationally completely collapsed object. I remarked that one couldn't keep saying 'gravitationally completely collapsed object' over and over. One needed a shorter descriptive phrase. 'How about black hole?' asked someone in the audience. I had been searching for the right term for months, mulling it over in bed, in the bathtub, in my car, whenever I had quiet moments. Suddenly, this name seemed exactly right. When I gave a more formal Sigma Xi-Phi Beta Kappa lecture ... on December 29, 1967, I used the term, and then included it in the written version of the lecture published in the spring of 1968.

Chapter 18

Kerr Black Holes

Box 18.1 The Kerr metric in BL coordinates

$$d\tau^2 = \left(1 - \frac{2GM\,\mathbf{r}}{p^2}\right) dt^2 - \frac{p^2}{\mathbf{r}^2 + a^2 - 2GM\,\mathbf{r}}\, d\mathbf{r}^2 - p^2\, d\theta^2 ...$$

$$... - \left(\mathbf{r}^2 + a^2 + a^2 \sin^2\theta \, \frac{2GM\,\mathbf{r}}{p^2}\right) \sin^2\theta\, d\phi^2 + 2a \sin^2\theta \, \frac{2GM\,\mathbf{r}}{p^2}\, dt\, d\phi,$$

$$a = \frac{J}{M} \equiv \frac{Angular\ Momentum}{Mass}, \qquad p^2 = \mathbf{r}^2 + a^2 \cos^2\theta, \qquad (c = 1),$$

$$[g_{\mu\nu}] = \begin{bmatrix} 1 - \dfrac{2GM\,\mathbf{r}}{p^2} & 0 & 0 & a \sin^2\theta \, \dfrac{2GM\,\mathbf{r}}{p^2} \\ 0 & -\dfrac{p^2}{\mathbf{r}^2 + a^2 - 2GM\,\mathbf{r}} & 0 & 0 \\ 0 & 0 & -p^2 & 0 \\ a \sin^2\theta \, \dfrac{2GM\,\mathbf{r}}{p^2} & 0 & 0 & -\left(\mathbf{r}^2 + a^2 + a^2 \sin^2\theta \, \dfrac{2GM\,\mathbf{r}}{p^2}\right)\sin^2\theta \end{bmatrix}.$$

18.1 The Outer Event Horizon

Box 18.1 shows the Kerr metric. Our first assignment is to find the event horizon(s) of the Kerr metric, i.e. the equivalent of the Schwarzschild event horizon R_S. Your instinct may be to look at the scenario $g_{TT} = 0$ because the dt^2 term becomes zero. This worked when we studied the Schwarzschild metric. It revealed a coordinate singularity where time stops from the perspective of a distant observer (see Section 13.7). However, this approach doesn't work for the Kerr metric because the non-zero off-diagonal $dt\,d\phi$ term in the invariant interval means that time will still be observed to pass if the object is moving in the ϕ direction (we will address this later).

$$Kerr\ Metric : g_{RR} = -\frac{p^2}{\mathbf{r}^2 + a^2 - 2GM\,\mathbf{r}}, \qquad Schwarzschild : g_{RR} = -\frac{1}{\left(1 - \frac{2GM}{r}\right)}. \tag{18.1}$$

There is no such ambiguity with singularities in the value of g_{RR} because there is no off-diagonal $d\mathbf{r}$ term, so we will start with that. Equation 18.1 compares the Kerr and Schwarzschild terms ($c = 1$). In the case of Schwarzschild, there is a singularity at $r = 2GM = R_S$ when the denominator is zero and $g_{RR} = \infty$. The equivalent for the Kerr metric is: $(\mathbf{r}^2 + a^2 - 2GM\,\mathbf{r}) = 0$. This is solved using the quadratic formula (Box 14.4) to give Equation 18.2. Note that if there is no rotation, then the BL $\mathbf{r}$ coordinate becomes r and $a = 0$. In this case, the only non-zero solution is $r = 2GM$, which is, of course, the Schwarzschild event horizon.

Untangling General Relativity: The Intuitive Self-Study Guide, First Edition. Simon Sherwood.
© 2026 John Wiley & Sons Ltd. Published 2026 by John Wiley & Sons Ltd.

$$\mathbf{r}^2 - 2GM\mathbf{r} + a^2 = 0, \quad \Longrightarrow \quad \mathbf{r} = \frac{2GM}{2}\left(1 \pm \sqrt{1 - \frac{4a^2}{4G^2M^2}}\right),$$

Kerr event horizons : $\qquad \mathbf{r} = GM \pm \sqrt{(GM)^2 - a^2}.$ (18.2)

Equation 18.2 looks innocent enough, but highlights significant differences between the Kerr and Schwarzschild metrics. The $\pm$ in the equation means there are *two* solutions, called the *outer event horizon* and *inner event horizon*. And there are no solutions if $a^2 > (GM)^2$, so instead of two event horizons, there are *no* event horizons at all. Oops... more on this later.

The *outer* horizon is the Kerr equivalent of the Schwarzschild radius. Based on M, which is the *energy-equivalent mass* of the black hole (we must include its rotational energy) and its angular momentum (which determines a), we can calculate the value $\mathbf{r}$ of its outer horizon. Don't forget that $\mathbf{r}$ is the BL radial coordinate, so the event horizon is the surface of an oblate spheroid, not a sphere (Section 17.3). This is the surface of the central blue spheroid in Figure 18.1, which illustrates an *extremal* black hole (which means it is rotating at maximum speed, as I will explain later). Just as with the Schwarzschild event horizon, once an object crosses the Kerr outer event horizon there is no coming back. Not even light can escape back out.

The outer event horizon marks the point inside which $\mathbf{r}^2 - 2GM\mathbf{r} + a^2 < 0$. This means that g_{RR} goes from negative to positive (p^2 is always positive). Therefore, $d\mathbf{r}^2$ goes from negative to positive in terms of the invariant interval. All paths become time-like and lead inwards from the horizon. The same happens when crossing the Schwarzschild radius of a non-rotating black hole (see Section 13.10) and the result is similar. Outside observers will not see any object ever cross the Kerr outer event horizon because of the coordinate singularity. However, in the frame of reference of an object (or astronaut), there is no theoretical obstacle to crossing it. Of course, any astronaut will almost certainly already have been ripped apart by tidal forces due to the variation in inward radial acceleration and rotational frame-dragging of different parts of the body (we will discuss frame-dragging in a moment).

In summary, the outer event horizon of a Kerr black hole has much in common with the Schwarzschild horizon of a non-rotating black hole:

- The coordinate singularity means observers outside will never see anything cross it.
- In their own frame of reference, objects can cross the horizon.
- Once across the outer horizon, the radial coordinate becomes time-like and all paths point inwards, so there is no way to escape back out.

Notable differences are that the outer horizon of a Kerr black hole is an oblate spheroid, not spherical, and is closer in than the Schwarzschild equivalent (see Figure 18.1, which shows the Schwarzschild equivalent as a

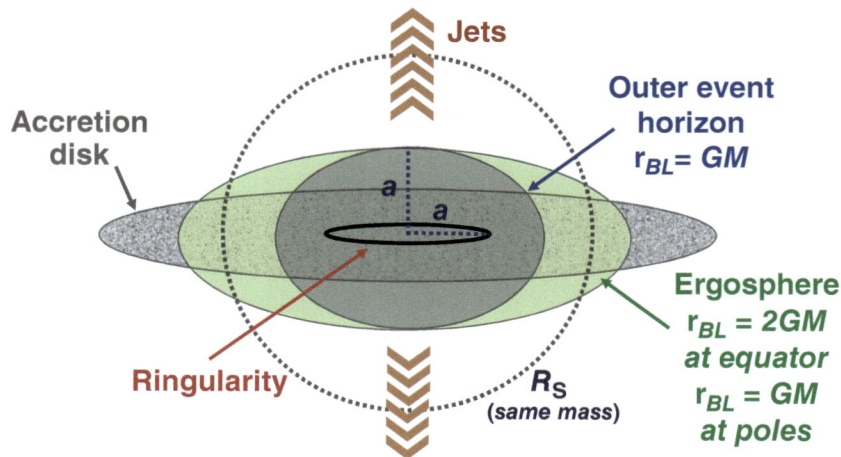

Figure 18.1 Schematic of an extremal Kerr black hole ($a \approx GM$). The Schwarzschild event horizon for an equivalent non-rotating black hole is shown ($R_S = 2GM$ in spherical coordinates).

dotted line). Indeed, you can see from Equation 18.2 that the greater the angular momentum of the black hole (i.e. the greater the value of *a*) the tighter in the Kerr event horizons.

18.2   The Inner Event Horizon

The inner event horizon is the other possible solution for the Kerr event horizon. It is found by applying the minus sign of the ± in Equation 18.2. For example, if the value of *a* is small relative to *GM*, then the outer horizon will be near **r** = 2*GM* and the inner horizon near **r** = 0. On the other hand, if the value of *a* is almost *GM* (as is believed to be the case with many black holes), then the outer and inner horizons are close together in the region of **r** ≈ *GM*.

What is this inner horizon? Most think of the inner horizon as a mathematical anomaly that is irrelevant. After all, we cannot see inside the outer event horizon. In the words of physicist and author Matt Visser: *almost all relativists would agree that it is safe to discard the entire region inside the inner horizon.* On the other hand, the metric changes significantly, so I want to mention it even if your professors curse me and tell you to ignore it.

The change in the metric crossing the *outer* horizon is that moves along the spatial radial coordinate dr^2 become positive (time-like) in terms of the invariant interval. Progress inwards from the outer horizon is unavoidable. Game over, you might think, but there is a subtle twist to the tale. In the case of the Schwarzschild metric, dr^2 remains time-like right down to *r* = 0, so all paths lead to the singularity at *r* = 0 (ignoring possible quantum effects and obstacles). In the case of the Kerr metric, dr^2 becomes positive (time-like) upon crossing the outer event horizon but *switches back to negative* (space-like) upon crossing the inner event horizon.

What does this mean? We don't really know. We can say that the path of an object crossing the outer horizon is drawn inexorably towards the inner horizon. However, once across the inner horizon, there is a legitimate question as to whether all paths lead to the singularity.

18.3   Underlying Riemann Curvature

We can examine the underlying Riemann curvature in the Kerr metric by comparing its curvature footprint with that of the Schwarzschild metric. The calculations are horrific. They are complicated (a lot!) by the off-diagonal term in the Kerr metric. No longer can we calculate things like $g^{TT}R_{T\phi\phi T} = R^T{}_{\phi\phi T}$, but must sum over all coordinates to catch the effect of any off-diagonal terms. This is written as $g^{T\mu}R_{\mu\phi\phi T} = R^T{}_{\phi\phi T}$, where the repeated μ term is summed over (see Einstein summation in Box 3.5).

To make matters worse, none of my shortcut cheat sheets for Christoffel symbols and Ricci tensor components work. To calculate a Christoffel symbol, you have to go back to the detailed definition in Box 4.3 and be sure not to miss the extra entries created by the non-zero off-diagonal metric term. Aarrgghh! Thankfully, people much smarter than your author have done the work. I offer special thanks to Matt Visser, whom I quoted earlier. His results form the basis of much in this section. If you are up for some heavier maths, you can refer back to his paper, *The Kerr Spacetime: A Brief Introduction.*

Equation 18.4 shows the curvature footprint for the Kerr metric. To facilitate the calculation, the coordinates are switched to what is called an orthonormal rational polynomial basis. This eliminates any trigonometric functions in the metric making computer-based computations easier. I have labelled these coordinates (0, 1, 2, 3). You will have to refer to Visser's paper if you want the full details,[1] but the choice of coordinates is not critical to the result because the Riemann tensor measures true curvature. It is not misled by any curvature that is coordinate-generated (see discussion in Section 7.2).

Compare the curvature footprint of the Kerr metric in Equation 18.4 with that of the Schwarzschild metric in Equation 18.3 (this is the footprint in Equation 12.22, which we derived in Chapter 12). All the entries add across to zero, as must be the case for Ricci-0 spacetime ($[R_{\mu\nu}] = 0$). You can see that the pattern is identical. Using these orthonormal BL coordinates with their oblate spheroidal symmetry matches up the Kerr metric curvature of the θ and ϕ coordinates in the same way that spherical symmetry does for the Schwarzschild metric. This θ and ϕ symmetry leads to only one possible Ricci-0 pattern of curvature as we showed in Section 12.2, so the similarity between the Kerr and Schwarzschild curvature footprints is no coincidence.

---
1 *The Kerr Spacetime: A Brief Introduction*, Matt Visser: *https://arxiv.org/pdf/0706.0622* (as of February 2025).

$$\text{Curvature Footprint of Schwarzschild Metric} \qquad S = \frac{GM}{c^2 r^3} \qquad (18.3)$$

$$
\begin{aligned}
g^{TT}R_{TT} &= g^{TT}R^T_{TTT} + g^{TT}R^R_{TRT} + g^{TT}R^\theta_{T\theta T} + g^{TT}R^\phi_{T\phi T} = & 0 & & -2S & & +S & & +S & & = 0,\\
g^{RR}R_{RR} &= g^{RR}R^T_{RTR} + g^{RR}R^R_{RRR} + g^{RR}R^\theta_{R\theta R} + g^{RR}R^\phi_{R\phi R} = & -2S & & +0 & & +S & & +S & & = 0,\\
g^{\theta\theta}R_{\theta\theta} &= g^{\theta\theta}R^T_{\theta T\theta} + g^{\theta\theta}R^R_{\theta R\theta} + g^{\theta\theta}R^\theta_{\theta\theta\theta} + g^{\theta\theta}R^\phi_{\theta\phi\theta} = & S & & +S & & +0 & & -2S & & = 0,\\
g^{\phi\phi}R_{\phi\phi} &= g^{\phi\phi}R^T_{\phi T\phi} + g^{\phi\phi}R^R_{\phi R\phi} + g^{\phi\phi}R^\theta_{\phi\theta\phi} + g^{\phi\phi}R^\phi_{\phi\phi\phi} = & S & & +S & & -2S & & +0 & & = 0.
\end{aligned}
$$

$$\text{Comparative Footprint of Kerr Metric} \qquad K = \frac{GM\mathbf{r}\left(\mathbf{r}^2 - 3a^2\cos^2\theta\right)}{\left(\mathbf{r}^2 + a^2\cos^2\theta\right)^3} \qquad (18.4)$$

$$
\begin{aligned}
g^{00}R_{00} &= g^{00}R^0_{000} + g^{00}R^1_{010} + g^{00}R^2_{020} + g^{00}R^3_{030} = & 0 & & -2K & & +K & & +K & & = 0,\\
g^{11}R_{11} &= g^{11}R^0_{101} + g^{11}R^1_{111} + g^{11}R^2_{121} + g^{11}R^3_{131} = & -2K & & +0 & & +K & & +K & & = 0,\\
g^{22}R_{22} &= g^{22}R^0_{202} + g^{22}R^1_{212} + g^{22}R^2_{222} + g^{22}R^3_{232} = & K & & +K & & +0 & & -2K & & = 0,\\
g^{33}R_{33} &= g^{33}R^0_{303} + g^{33}R^1_{313} + g^{33}R^2_{323} + g^{33}R^3_{333} = & K & & +K & & -2K & & +0 & & = 0.
\end{aligned}
$$

The important thing is the complicated-looking underlying curvature term on the right of Equation 18.4. At the Kerr event horizons there is *no* underlying curvature singularity. If you want to check this, substitute into the value of K in Equation 18.4, the two Kerr event horizon values from Equation 18.2. As with the Schwarzschild horizon, the outer and inner event horizons of the Kerr metric are *coordinate singularities*. From the perspective of an astronaut crossing the horizon, there is no sudden change in curvature.

Note that when there is no rotation in the Kerr metric ($a = 0$), the BL coordinate $\mathbf{r}$ becomes the same as the spherical coordinate r and the curvature in Equation 18.4 matches the Schwarzschild metric, as you would expect.

The *only real curvature singularity* in the Kerr metric is at $(\mathbf{r}^2 + a^2\cos^2\theta) = 0$ or, using the p label, when $p^2 = 0$. This is when the curvatures in Equation 18.4 blow up to infinity.[2] The comparable value in the Schwarzschild metric is at $r = 0$.

18.4 The Kerr Singularity (Ringularity)

Let's see what we can learn about the Kerr curvature singularity at $(\mathbf{r}^2 + a^2\cos^2\theta) = 0$. As a is a fixed positive number for the black hole and both $\mathbf{r}^2$ and $\cos^2\theta$ must be positive, the singularity can occur only when $\mathbf{r} = 0$ and $\theta = \frac{\pi}{2}$ (because $\cos\frac{\pi}{2} = 0$). The angle $\theta = \frac{\pi}{2}$ is at the equator. Switching back to the usual spherical radial coordinate r, the value at the equator is: $r = \sqrt{\mathbf{r}^2 + a^2}$ (check back to Section 17.4 if needed), which for BL radial coordinate $\mathbf{r} = 0$ gives in spherical coordinates $r = a$. This means that the singularity of the Kerr metric is an equatorial ring with radius a, the focal length. This is generally called the *ringularity* of the Kerr metric and is illustrated in Figure 18.1.

This along with the mystery surrounding the inner event horizon where the radial coordinate $d\mathbf{r}^2$ becomes negative again in terms of the invariant interval (see Section 18.2), excites physicists to wonder whether it might be possible to pass through the Kerr black hole missing the singularity... to sneak through the centre of the ringularity. Could it be a wormhole and, if so, to where?

A related weird-sounding question is whether there is something beyond the point marked by a spherical coordinate $r = 0$ (the centre of the black hole). In the case of the Schwarzschild metric, there is a singularity at $r = 0$ such that space and time become undefined. This does not happen with the Kerr metric. Back in Section 16.7, I introduced you to the maximally extended Schwarzschild solution, which follows coordinates from $-\infty$ to $+\infty$ or until they hit a singularity (the result was summarised on the right of Figure 16.5 and includes the black hole

---

2 For completeness, note that the curvature singularity in the other non-zero Riemann components is also at $p^2 = 0$.

and a white hole). For the Kerr metric with no singularity at $r = 0$, the same argument leads out to $r = -\infty$ and a sequence of black and white holes. We will discuss this further in Section 18.10 with my usual caveat that it may be more math-magical than physical.

18.5 Frame-Dragging

It now is time to have a look at the off-diagonal term in the metric that is making mathematical calculations so much more difficult: $g_{T\phi} = g_{\phi T} \neq 0$. Let me remind you that the off-diagonal term means that the time t dimension and the ϕ dimension are *not orthogonal*. In the Kerr metric, moving through time dt carries with it some movement $d\phi$ in the direction of rotation.

There is a simple 2-D example back in Figure 3.1 of a metric that is not orthogonal. Let's imagine the effect of a non-zero g_{TX} term added to the Minkowski metric of flat spacetime. You are in a rocket with the engines off, moving in the y direction, minding your own business, happily following a geodesic. You would find that you start moving in the x direction simply because you are travelling through time and each bit of dt brings with it some dx.

Let's get specific with the Kerr metric. Suppose you are bored with this book and jettison it from your rocket out in space far from a rotating black hole ($\mathbf{r} \approx \infty$). I scream insults at you for mistreating my masterpiece, but then we settle down and watch how the book moves as it accelerates towards the rotating black hole. What happens?

The Kerr metric doesn't depend on t or ϕ, which means that u_t and u_ϕ are conserved quantities. This is an important point. If you have any doubts, refer back to Section 14.1 on conservation laws. As the Kerr metric is not diagonal, Box 18.5 has the relevant conservation law proof if your maths is up to it. Initially the book is stationary at $\mathbf{r} = \infty$, so $u^t = \frac{dt}{d\tau} = 1$ (no time dilation) and $u^\phi = \frac{d\phi}{d\tau} = 0$ (no angular velocity). We also know at $\mathbf{r} = \infty$ that the Kerr metric is the same as flat Minkowski spacetime (see Section 17.7), so initially $g_{T\phi} = 0$. This gives the initial value $u_\phi = 0$ at $\mathbf{r} = \infty$ as shown in Equation 18.5. This is conserved, so along the whole geodesic $u_\phi = 0$. Before moving on, any reader struggling with the formula structures in Equation 18.5 should check back to Section 3.7, which describes how the maths works for metrics with off-diagonal terms.

$$\text{Initial value at } \mathbf{r} = \infty : \quad u_\phi = g_{\phi\phi}u^\phi + g_{T\phi}u^t = (0) + (0)(1) = 0, \tag{18.5}$$

$$\text{ZAMO}: \quad u_\phi = 0, \quad \implies \quad g_{\phi\phi}u^\phi = -g_{T\phi}u^t, \quad \implies \quad -\frac{g_{T\phi}}{g_{\phi\phi}} = \frac{u^\phi}{u^t} = \frac{d\phi}{dt}. \tag{18.6}$$

We know the value $u_\phi = 0$ is conserved along the Kerr geodesic. Using the expression for u_ϕ on the right of Equation 18.5 gives Equation 18.6, which is the angular velocity of my abandoned book in terms of $g_{T\phi}$ and $g_{\phi\phi}$. The relationship works only for an object like this starting with zero angular velocity, so I have given it the label ZAMO (zero angular momentum object) for clarity. All that remains is to feed in the values of $g_{T\phi}$ and $g_{\phi\phi}$ from the Kerr metric invariant interval in Box 18.1. Don't forget that $g_{T\phi}$ is *half* the $d\phi\, dt$ term in the invariant interval because this is split $g_{T\phi} = g_{\phi T}$. The result is Equation 18.7. As shown, for values of $\mathbf{r}$ much larger than a, this varies inversely with $\mathbf{r}^3$.

$$\begin{aligned}
\frac{d\phi}{dt} &= -\frac{g_{T\phi}}{g_{\phi\phi}} = -\frac{\sin^2\theta}{p^2}(2aGM\mathbf{r})\frac{p^2}{\sin^2\theta}\left(\frac{-1}{(\mathbf{r}^2 + a^2)p^2 + 2a^2GM\mathbf{r}\sin^2\theta}\right), \\
&= \frac{2aGM\mathbf{r}}{(\mathbf{r}^2 + a^2)(\mathbf{r}^2 + a^2\cos^2\theta) + 2a^2GM\mathbf{r}\sin^2\theta}, \\
&\approx \frac{2aGM}{\mathbf{r}^3} \approx \frac{2GJ}{\mathbf{r}^3}, \quad \text{for } \mathbf{r} >> a, \quad (c = 1).
\end{aligned} \tag{18.7}$$

As we watch my beautiful book descend towards the black hole, we will see it pick up angular speed in the direction of rotation (of the black hole). This is the effect of the spacetime around the Kerr black hole being sort-of dragged with its rotation. It is called frame-dragging or the *Lens–Thirring effect*. One word of warning about the expression frame-dragging. Nothing is being dragged. The object will not feel any accelerating force. It is simply that the freefall geodesic includes this angular motion. When we think of the Kerr metric we tend to think of black holes, but this frame-dragging formula is applicable to any rotating mass (see Box 18.2).

Box 18.2 Measuring frame-dragging

In 2004, NASA launched Gravity Probe B to a height 400 miles above the earth. Its mission was to use four on-board gyroscopes to measure the frame-dragging effect of the earth's rotation. To put the challenge in context, NASA estimated the likely effect as only 42 milliarcseconds of angle per year, which I am told is the equivalent angle of the width of a human hair viewed from a quarter of a mile away. Although the effect is miniscule, NASA's report confirmed the frame-dragging effect to within 20% of that predicted.

Frame-dragging also has been tested to a claimed accuracy of 5–10% based on data from the two NASA LAGEOS satellites (Laser Geometric Environmental Observation Survey) some 3,700 miles above earth, and the Italian Space Agency's 2012 LARES 1 satellite. More recently, in November 2023, preliminary results from the LARES 2 satellite launched in 2022 reportedly showed *complete agreement with the predictions of Einstein's gravitational theory.*

In addition, there is growing observational evidence of frame-dragging such as that from the rotation of pulsar PSR J1141-6545 (a neutron star), which is over 10,000 light years away in orbit around a white dwarf star. Analysing the variation in the pulsar is believed to have revealed the impact of frame-dragging on its orbit.

18.6 The Ergosphere

We have not yet addressed what happens when $g_{TT} = 0$ in the Kerr metric. In the Schwarzschild metric, $g_{TT} = 0$ at the event horizon. An outside observer never sees any object reach the point where $g_{TT} = 0$ because, from the observer's perspective, its time slows towards a stop (Section 13.7). The Kerr metric is different because of the presence of the $dt\,d\phi$ term in the invariant interval.

As you can see from Box 18.1, $g_{TT} = 0$ for the Kerr metric when $p^2 = 2GM\mathbf{r}$. This is the start of what is called the *ergosphere*. At this point, a distant observer still can see time pass because of the non-zero $dt\,d\phi$ term, *providing $d\phi$* is non-zero. A distant observer can see an object enter the ergosphere, but it must move in the direction of rotation of the black hole.

In layman's language, nothing can stand still. Within the ergosphere *everything* rotates with the black hole, even light. However, while everything must rotate with it, this does not mean nothing can escape its clutches. The ergosphere sits above the event horizon. With enough acceleration, an object can still escape the black hole.

Let's do some maths. We can calculate the path of a light beam moving in the ϕ direction at the edge of the ergosphere ($g_{TT} = 0$). This simplifies the maths because $dr = d\theta = 0$, and for light, proper time $d\tau = 0$. Remembering $g_{TT} = 0$ at the edge of the ergosphere, we can generate an expression for $\frac{d\phi}{dt}$ as shown on the right of Equation 18.8.

$$g_{TT}dt^2 + 2g_{T\phi}dt\,d\phi + g_{\phi\phi}d\phi^2 = 0, \implies g_{\phi\phi}\left(\frac{d\phi}{dt}\right)^2 + 2g_{T\phi}\left(\frac{d\phi}{dt}\right) = 0, \quad At\ ergosurface. \tag{18.8}$$

There are two solutions to Equation 18.8 for $\frac{d\phi}{dt}$ as we would expect for light, which can move in the $+\phi$ or $-\phi$ direction. These are shown in Equation 18.9 (to save you time, note that the amount calculated on the right is twice that in Equation 18.7). For completeness, I have shown the result on the right simplified with $p^2 = 2GM\mathbf{r}$ for the ergosurface.

$$\frac{d\phi}{dt} = 0, \quad or \quad \frac{d\phi}{dt} = -2\frac{g_{T\phi}}{g_{\phi\phi}} = \frac{4aGM\mathbf{r}}{(\mathbf{r}^2 + a^2)p^2 + 2a^2GM\mathbf{r}\sin^2\theta}, \quad At\ ergosurface,$$
$$= \frac{2a}{\mathbf{r}^2 + a^2 + a^2\sin^2\theta}. \tag{18.9}$$

At the ergosurface, if the light moves with the rotation of the black hole, you get the result on the right of Equation 18.9. But here is the punchline: if the light travels directly against the rotation of the black hole, then the only remaining solution is $\frac{d\phi}{dt} = 0$. The result shows that even moving at the speed of light against the rotation, there is no net progress versus the frame-dragging effect of the black hole's rotation. Nothing travels faster than the speed of light, so inside the ergosphere nothing can hover or fall directly in. Everything, including light, must rotate in the same direction as the black hole.

Where is the start of the ergosphere, which is called the *outer ergosurface*? As noted earlier, $p^2 = 2GM\mathbf{r}$ when g_{TT} is zero, which means: $\mathbf{r}^2 - 2GM\mathbf{r} + a^2\cos^2\theta = 0$. We solve this quadratic equation in $\mathbf{r}$ with the traditional formula (Box 14.4) giving Equation 18.10. At the poles $\cos\theta = 1$, so this matches the event horizons (compare with Equation 18.2). At the equator $\cos\theta = 0$ and the outer surface of the ergosphere extends beyond the Kerr event horizon out to $\mathbf{r} = 2GM$. This is illustrated in green in Figure 18.1.

$$\mathbf{r} = GM \pm \sqrt{\left(GM\right)^2 - a^2\cos^2\theta}, \qquad \textit{Ergosurface.} \tag{18.10}$$

Some distressed readers will write in if I do not mention there is a second solution to Equation 18.10. The minus option of the $\pm$ is called the *inner ergosurface*. At the poles, this matches the inner event horizon. At the equator, the inner ergosurface is at $\mathbf{r} = 0$, which in spherical coordinates is $r = a$, the Kerr ringularity. My instinct is to dismiss this as a mathematical anomaly of interest to others but not to me. Perhaps this is why I am an author rather than a Nobel Laureate.

18.7 Penrose, Blandford–Znajek and Quasars (Revisited)

What happens when you combine the humungous energy release from objects falling into the accretion disk (Section 14.8) with the enormous rotational power of frame-dragging in the ergosphere? The answer is that it generates crazy levels of power. Let's revisit the topic of quasars adding in what we have learned from the Kerr metric.

The presence of frame-dragging and the ergosphere makes it possible, at least in theory, to use a rotating black hole as a power source, although this is not something I would recommend you to try at home! The methodology was first unearthed by Roger Penrose and is named in his honour the *Penrose Process*. It is illustrated on the left of Figure 18.2. Your rocket descends into the ergosphere picking up huge rotational speed from frame-dragging. It releases or fires an object down into the black hole (the dotted red line) that allows it enough outward momentum to escape. Your rocket exits with the extra rotational energy picked up in the ergosphere. The offset is that the rotation of the black hole slightly slows. Hey presto, you have extracted energy from the black hole.

The *Blandford–Znajek Process* builds on the idea as an explanation of quasars. To quote Roger Blandford, one of the originators, Penrose showed that black holes *are not one-way membranes, as it were; you can extract the spin energy. We showed a way of doing that with electromagnetic fields.* In Section 14.8 of Chapter 14 we used the conserved quantities of the Schwarzschild metric to demonstrate the energy release from particles and objects descending to the accretion disk of a Schwarzschild black hole. This will ionise much of the material in the accretion disk, creating a plasma with strong magnetic fields.

The frame-dragging rotation of the black hole is believed to twist the magnetic field lines as illustrated on the right of Figure 18.2. These twisted magnetic field lines accelerate a stream of particles (probably electrons, positrons, protons and nuclei) out from both poles of the rotating black hole. There is still much debate about the details of the mechanism, but observations from the Event Horizon Telescope (especially of the black hole M87) reveal stripes of polarised light that suggest strong coherent magnetic fields do indeed play a role in creating quasar jets.[3]

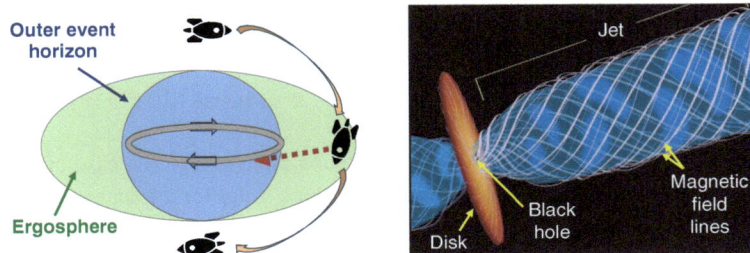

Figure 18.2 Energy output from a black hole. The left illustrates the Penrose process. The right illustrates the intense magnetic fields and resulting jets from a quasar (courtesy of NASA).

---

3 For more detail and images, see: https://eventhorizontelescope.org/blog/astronomers-image-magnetic-fields-edge-m87s-black-hole (as of February 2025).

The power of these jets is remarkable and worth mentioning again. Researchers on M87, the super-massive black hole of more than five billion solar masses in the constellation of Virgo (there is a picture earlier in Figure 13.1) estimate the power of its jets at about 10^{36} watts, which is 10^{10} times the output of the sun. Other quasar jets have been estimated at power levels in excess of 10^{40} watts, which is thousands of times the power output of our entire Milky Way galaxy.

18.8 Extremal Black Holes and Cosmic Censorship

Let me drag your brains back to the discussion of the Kerr event horizons in Section 18.1 culminating with Equation 18.2 shown again as Equation 18.11. I raised the problem that if $a^2 > (GM)^2$ then there are no solutions and therefore no event horizons... oops! The magnitude can be written as $|a| > GM$ noting that the value of a can be positive or negative depending on the direction of rotation.

$$\text{Kerr event horizons} \qquad \mathbf{r} = GM \pm \sqrt{(GM)^2 - a^2}. \tag{18.11}$$

If $|a| > GM$ then an outside observer can see all the way down to the Kerr singularity (ringularity). This poses a major concern for general relativity theorists. A super-smart friend, Matt von Hippel, who helped me both with this book and my last, once told me that *nature should not have sharp edges*. His point was (and is) that singularities are ugly and difficult. Taken at face value, a curvature singularity such as that at $r = 0$ in the Schwarzschild metric or at $p^2 = 0$ in the Kerr metric is a place where spacetime curvature becomes infinite. Space and time cease to exist. Infinite curvature, infinite energy-density... anything and everything can happen. It is a point where and when all physical laws break down. Indeed, *where* and *when* no longer have any meaning.

Roger Penrose received the 2020 Nobel Prize for his paper (of January 1965!) arguing that all black holes contain a singularity. Without going into the mathematical detail that is beyond me, the nub of his argument is that the event horizon of a black hole is what he calls a *trapped surface*. All light cones are angled along or into the surface towards a central point. You might want to look back at the right of Figure 16.2 to see this. Light, for which proper time does not pass, must end up somewhere. Once inside the event horizon, it cannot escape to infinity, so it must end on a singularity.

The one saving grace of the black hole singularity is that it seems to be hidden behind the event horizon. From the perspective of outside observers (like us), it sits outside our time horizon and cannot be observed. But returning to the contents of Equation 18.11, if $|a| > GM$ then there is no event horizon and the singularity exists, observable in our time.

Roger Penrose's solution was to conjure up the *cosmic censorship conjecture*, which forbids any such *naked singularity*. His conjecture states as a law that any singularity must be hidden by an event horizon, effectively removing it from our timeline. This gives a limit of $|\frac{a}{GM}| = 1$. This is called the *Kerr parameter* or *spin parameter* (often labelled a_*), a convenient rule of thumb as it is a dimensionless quantity as shown in Equation 18.12. A black hole with angular momentum close to $a_* = 1$ is called *extremal*. Observational evidence suggests that the majority of black holes have $a_* > 0.9$.

$$|a| < GM, \qquad |\frac{a}{GM}| = |\frac{J}{GM^2}| < 1, \qquad \textit{cosmic censorship} : \textit{max angular momentum},$$

$$\textit{Units (c included)} : \qquad a_* = |\frac{Jc}{GM^2}| \equiv \frac{(\text{kg m}^2\,\text{s}^{-1})(\text{m s}^{-1})}{(\text{m}^3\,\text{kg}^{-1}\,\text{s}^{-2})(\text{kg}^2)} \equiv \textit{dimensionless Kerr parameter}. \tag{18.12}$$

Penrose's conjecture is not plucked out of thin air. There is substance to it in the case of the Kerr metric. Matter falls in from the accretion disk adding angular momentum, so black holes tend towards extremal. However, as the angular momentum grows, frame dragging throws off objects with initial rotation matching the black hole. Penrose believes that once the black hole becomes extremal, any object falling in that would add angular momentum ends up in a scattering orbit. For those interested, there is a mathematical demonstration related to this at the end of this chapter in Section 18.12.

In short, based on the cosmic conjecture, the only objects that can fall into an extremal black hole are those that don't increase its net angular momentum. This provides a seductive logic for the cosmic censorship conjecture, but far from proof. Naked singularities remain a hot topic, with opinions varying from Penrose's position (there are no naked singularities) to scepticism (cosmic censorship is conjured up from nothing), to questioning the existence of singularities in black holes at all (Roy Kerr's position), to wondering if worrying about singularities is a complete waste of time and effort (actually, this is a fair question, see Box 18.3).

Box 18.3 Singularities: searching for a single answer; take your pick

Say no to naked: Penrose's conjecture is quite widely accepted. It fits the Kerr metric and observational data. Plus, he is a Nobel Laureate and one of the smartest humans on the planet. Do you really want to argue with him?

Go naked, don't be shy: Many theoretical scenarios appear to result in naked singularities (there is a nice layman summary in *Scientific American*, May 21, 2013). Are we going to believe a rather-too-convenient conjecture without proof?

What singularity?: Roy Kerr, the genius who discovered the Kerr metric, believes Penrose is wrong. He believes stuff chaotically thrashes around within the inner event horizon without forming a singularity (*Do Black Holes Have Singularities*, November 2023).

Studying singularities is a waste of time: Quantum Electrodynamics (QED) faced a similar problem with the singularity of point-like electrons. The solution (Effective Field Theory) was to accept that current theories cannot handle these extremes. In the case of GR, clearly the very small will be influenced by quantum effects when other as yet unknown rules apply.

Singularities might hold the answer: The Large Hadron Collider (LHC) flings particles around a 27-kilometre loop. This is dwarfed by quasars, which are the largest particle accelerators in the universe with jets extending out over thousands of light years. Might studying related GR singularities give us a hint as to how to build a quantum theory of gravity? That prospect for physicists is like nectar to bees.

18.9 Conserved Quantities and Contorted Orbits

I can imagine all you readers licking your lips with anticipation of some neat analysis of Kerr orbits. Sorry, but it is a mess. Let me explain why. The Kerr metric has this sort of oblate spheroidal symmetry, so bang goes basic symmetry except for orbits exactly around the equator. Add to that the frame-dragging effect, which varies with the angular momentum of the black hole and the location of the orbit (travelling over the pole is completely different from travelling over the equator). It is like trying to plot the course of a yacht across the ocean but with its velocity changing because the sea level goes up and down, added to which there is a varying crosswind. Aarrgghh!

To calculate orbits and paths, we rely on conserved quantities. As discussed earlier, the Kerr metric is not dependent on t or ϕ, which means that u_t and u_ϕ are conserved quantities along geodesics. The cross-term in the metric complicates things (for example, check out Box 18.4 to see how it affects the structure of four-vectors). Equation 18.13 shows the conserved quantity u_t for the Kerr metric compared with that of the Schwarzschild metric. It uses the BL oblate spheroidal coordinate $\mathbf{r}$ and is also dependent on a and θ.

$$Kerr: \quad u_t = g_{TT}u^t + g_{T\phi}u^\phi = \left(1 - \frac{2GM\,\mathbf{r}}{p^2}\right)u^t + 2a\sin^2\theta\,\frac{2GM\,\mathbf{r}}{p^2}u^\phi,$$

$$Schwarzschild: \quad u_t = g_{TT}u^t = \left(1 - \frac{2GM}{r}\right)u^t.$$

$$(18.13)$$

Table 18.1 Kerr ISCO and photon spheres for an extremal black hole.

Kerr	Prograde	Retrograde
ISCO	$1\,GM$	$9\,GM$
Photon sphere	$1\,GM$	$4\,GM$

The Kerr and Schwarzschild versions of u_ϕ are compared in Equation 18.14. Again, you can see that the Kerr definition is bafflingly complicated.

$$Kerr: \quad u_\phi = g_{\phi\phi}u^\phi + g_{T\phi}u^t = -\left(\mathbf{r}^2 + a^2 + a^2\sin^2\theta\,\frac{2GM\,\mathbf{r}}{p^2}\right)\sin^2\theta u^\phi + 2a\sin^2\theta\,\frac{2GM\,\mathbf{r}}{p^2}u^t,$$

$$Schwarzschild: \quad u_\phi = g_{\phi\phi}u^\phi = -r^2\sin^2\theta u^\phi. \tag{18.14}$$

This being said, we can simplify by considering an extremal black hole ($|a| = GM$) and looking at purely equatorial orbits ($\sin\theta = 1$) to give the results in Table 18.1. The Kerr equatorial Innermost Stable Circular Orbits (ISCOs) are radius GM out for a prograde orbit (orbiting in the same direction as the rotation of the black hole) and $9\,GM$ out for retrograde orbit (the opposite direction). Note that the Schwarzschild ISCO is at $6\,GM$ (Subsection 14.7.3). This makes intuitive sense because the Kerr frame-dragging effectively gives the prograde orbit an extra angular push, so the stable orbit is closer in; in fact, it is at the event horizon for an extremal black hole. In contrast, the object in retrograde orbit is moving against the frame-dragging. The Kerr photon spheres are also shown in the table. In comparison, the Schwarzschild photon sphere is at $3\,GM$.

Much of the work on Kerr orbits is computer generated. The true picture is more dynamic than the simple scenarios suggest. Orbits can shift to and fro across the equatorial plane, producing a torus-like pattern as illustrated on the left of Figure 18.3, while the right shows how general geodesic paths can trace out an intricate contorted web around the black hole.

18.10 Maximal Extension of the Kerr Metric

The final feature of the Kerr metric that I want to address is its maximal extension. Mathematicians pierced through the coordinate singularities of the Schwarzschild metric with the Kruskal–Szekeres (KS) coordinates to visualise its maximal extension (as described in Section 16.7). This extension includes a white hole. For convenience, I have shown this in the form of its Penrose–Carter diagram on the right of Figure 18.4 (this is from Figure 16.6 earlier). The white hole becomes part of the solution only if the black hole is *eternal*, rather than one formed from the remnants of a supernova.

On the left of Figure 18.4 is a schematic of the equivalent exercise for a Kerr black hole. As you can see, things get a lot more interesting. The maximal extension follows all possible coordinates from $-\infty$ to $+\infty$ unless they hit a singularity. With the Schwarzschild metric (right of figure), all pathways in the black hole lead to the singularity shown as a wavy horizontal line... any object that travels into the Schwarzschild black hole hits the singularity... game over.

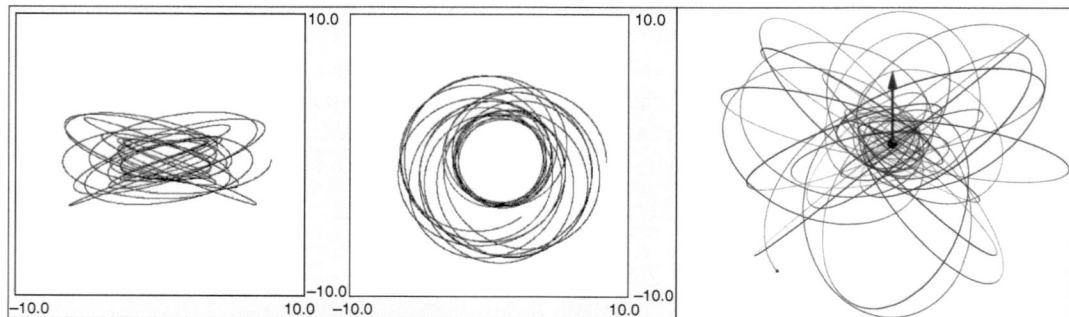

Figure 18.3 Illustrations of Kerr orbits and trajectories (left: B wlk; right: Maarten van de Meent / Wikimedia Commons / CC BY 4.0).

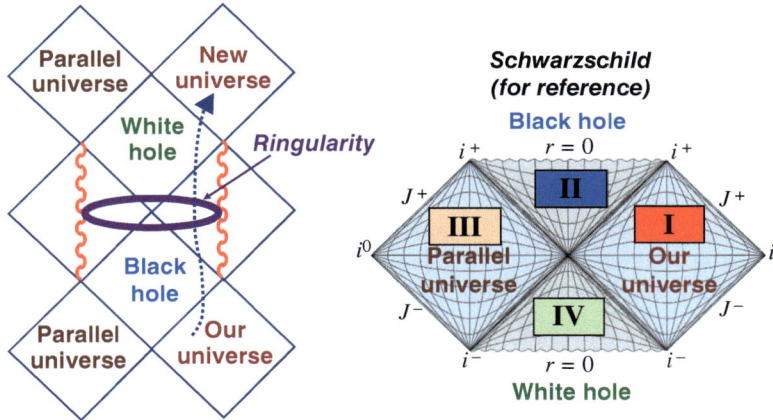

Figure 18.4 Penrose–Carter diagram of maximally extended Kerr metric (left) and for comparison the extended Schwarzschild Solution (right, *Source*: Osanshouo / Wikimedia Commons / CC BY 4.0.).

Box 18.4 Off-diagonal terms in the metric

In Section 3.7, I showed how to raise and lower the indices of metrics with off-diagonal terms (which I called wonky). Let me use the Kerr metric as an example and show you that, based on Einstein summation (Box 3.5), $u_\mu u^\mu = 1$, if you use natural units ($c = 1$) and metric signature $< + - - - >$. Using signature $< - + + + >$ it is $u_\mu u^\mu = -1$. This is another way to express the invariant interval. For the Kerr metric, using the approach back in Equation 3.25, note the extra terms in u_t and u_ϕ because of the off-diagonal metric term. Starting in Equation 18.15 with the generalised invariant interval including the $dt d\phi$ off-diagonal term, we divide both sides by $d\tau^2$ to give Equation 18.16. With a little reorganising we arrive at $u_\mu u^\mu = 1$.

$$u_\mu = g_{\mu\nu}u^\nu \; : \quad u_t = g_{TT}u^t + g_{T\phi}u^\phi, \quad u_\mathbf{r} = g_{RR}u^\mathbf{r}, \quad u_\theta = g_{\theta\theta}u^\theta \quad u_\phi = g_{\phi\phi}u^\phi + g_{T\phi}u^t,$$

$$d\tau^2 = g_{TT}dt^2 + g_{RR}d\mathbf{r}^2 + g_{\theta\theta}d\theta^2 + g_{\phi\phi}d\phi^2 + 2g_{T\phi}dt d\phi, \tag{18.15}$$

$$1 = g_{TT}(u^t)^2 + g_{RR}(u^\mathbf{r})^2 + g_{\theta\theta}(u^\theta)^2 + g_{\phi\phi}(u^\phi)^2 + 2g_{T\phi}u^t u^\phi, \tag{18.16}$$

$$= (g_{TT}u^t + g_{T\phi}u^\phi)u^t + g_{RR}(u^\mathbf{r})^2 + g_{\theta\theta}(u^\theta)^2 + (g_{\phi\phi}u^\phi + g_{T\phi}u^t)u^\phi,$$

$$= u_t u^t + u_\mathbf{r}u^\mathbf{r} + u_\theta u^\theta + u_\phi u^\phi = u_\mu u^\mu.$$

With the Kerr metric there is the mathematically possible pathway *through the ringularity*. The pathway through is shown in the Penrose–Carter diagram by a gap between the edges of the singularity (shown as red wavy lines). I have included an image of the ringularity (this is for clarity and would not appear in a formal Penrose–Carter diagram). If you follow the mathematics, this pathway leads out through a white hole into another universe such as shown by the dotted line. In that new universe, the object could travel into another Kerr black hole, pass through its ringularity and on to yet another universe and so on ad infinitum!

This sort of hypothetical pathway is referred to as a *wormhole* or *Einstein-Rosen bridge*. The idea of jumping through wormholes to another universe (or to another time or place in our universe) is much beloved of science-fiction writers. While wormholes are legitimate solutions to the Einstein Field Equations (EFEs) and therefore of interest to theoretical physicists, there is no evidence that they really exist, and there is every reason to be sceptical about their presence at the heart of Kerr black holes.

For one thing, the Kerr metric is only an approximation. It is not time-dependent, so it describes an idealised black hole unchanging in time. For another, the interior solution of the metric (inside the inner event horizon) is poorly understood and likely to be unstable especially with stuff flying around the ringularity, let alone passing through it. Even if we ignore all of this, we must admit to having no idea what happens at the heart of black holes.

We know quantum mechanics must play a leading role but, with no theory of quantum gravity, we can only guess blindly what that role is.

18.11   Summary

I opened this chapter with an analysis of the *outer event horizon* of the Kerr metric. This is the point at which the radial coordinate switches from negative to positive in the invariant interval equation, effectively becoming time-like. The outer event horizon of a Kerr black hole is oblate spheroidal in shape. The higher the angular momentum of the black hole, the tighter in the outer event horizon. For an extremal black hole ($|a| = GM$) it is at $r = GM$. It is the closest equivalent to the Schwarzschild event horizon with similar features:

- Once crossed, an object cannot escape back out.
- An observer outside it will never see any object cross it.

There is a less understood *inner event horizon*, which many physicists dismiss as a mathematical anomaly. At this point, the radial coordinate switches back to negative value in the invariant interval equation, so effectively it switches back from time-like to space-like, raising the mathematical possibility that within it not all paths lead to the singularity.

I introduced you to the underlying Riemann curvature using the curvature footprint for the Kerr metric showing that the outer and inner event horizons are coordinate singularities (as is the case with the Schwarzschild equivalent). The only real curvature singularity of the Kerr metric is at $p^2 = 0$, which is when $\mathbf{r} = 0$ and $\theta = \frac{\pi}{2}$. In normal spherical coordinates, this is an equatorial ring at $r = a$ and is often called the Kerr *ringularity*.

Jumping to the final topic in this chapter, the pathway freedom inside the inner event horizon and the theoretical route through the ringularity lead to an intriguing *maximal extension* of the Kerr metric with pathways running into the black hole, out through a white hole and on to alternative universes. As I have warned many times, this may be more math-magical than physical.

Another interesting feature of the Kerr metric is *frame dragging*. I showed you that an object approaching the Kerr black hole with zero angular momentum (ZAMO) will pick up angular rotational velocity in the direction of its spin, dragged around by the curvature of spacetime surrounding the black hole. This reaches an extreme at the *ergosphere* where nothing, not even light, can resist the dragging effect.

This makes it possible in theory to extract energy from the spinning black hole by a rocket dipping into the ergosphere, picking up rotational speed, and releasing an object into the black hole to allow the rocket to escape back out: the *Penrose Process*. The *Blandford-Znajek Process* builds on this idea, adding the presence of strong twisted magnetic fields that accelerate particles out from the poles of the black hole; a potential explanation for the enormously powerful emissions from quasars.

If the angular momentum of a black hole were to grow beyond the extremal level $|a| = GM$, then in theory the event horizon would disappear and the ringularity would become observable. This led us to a discussion of Roger Penrose's conjecture of *cosmic censorship*, that no black hole can have $|a| > GM$. There is much controversy surrounding this, but Penrose believes that any object approaching the black hole with angular momentum that would break cosmic censorship, would scatter to infinity rather than descend into it. There is an optional mathematical illustration related to this in Section 18.12 at the end of this chapter.

We looked at the Kerr metric in terms of conservation laws. The conserved quantities are much more complicated than for the Schwarzschild metric. Except for simple scenarios such as equatorial orbits, computer analysis is needed and reveals contorted almost chaotic patterns of orbits as you can see in the examples in Figure 18.3.

Below is the added bonus for any readers who want to delve a little deeper. In addition to the promised proof of conservation laws (Box 18.5), there is the optional section illustrating how cosmic censorship ties in with the maths of the Kerr metric. Or if you want to skip all of this, please feel free to jump on to Chapter 19.

Box 18.5 Conservation laws: metrics with off-diagonal terms Optional

In Section 14.2 I showed that if a diagonal metric $[g_{\mu\nu}]$ is not dependent on x, then u_x is conserved along geodesics. We can extend this to metrics including off-diagonal terms. This calculation uses Einstein summation (see Box 3.5). Using natural units ($c = 1$), the starting point is the general four-vector relationship: $u_\mu u^\mu = 1$ (as shown in Box 18.4). This is a constant, which gives us Equation 18.17.

$$\frac{d}{d\tau}(u_\mu u^\mu) = 0, \implies u^\mu \frac{du_\mu}{d\tau} + u_\mu \frac{du^\mu}{d\tau} = 0. \tag{18.17}$$

In Equation 18.18 the geodesic equation (Box 5.2) is substituted in for $\frac{du^\mu}{d\tau}$. As μ, **A** and **B** are all dummy variables, we can switch μ and **A**.

$$u^\mu \frac{du_\mu}{d\tau} - u_\mu \Gamma^\mu_{AB} u^A u^B = 0, \implies u^\mu \frac{du_\mu}{d\tau} - u_A \Gamma^A_{\mu B} u^\mu u^B = 0, \tag{18.18}$$

$$\implies u^\mu \left(\frac{du_\mu}{d\tau} - \Gamma^A_{\mu B} u_A u^B \right) = 0. \tag{18.19}$$

This gives Equation 18.19. As u^μ need not be zero for a variable, we conclude that the contents inside the brackets must be zero for each *individual* variable (it takes a little more maths to cover off all options, but this turns out to be true). Starting from this, I have substituted μ with x to emphasise that it is now a single variable, not a dummy variable. In Equation 18.20, we substitute for the Christoffel symbol using the full formula from Box 4.3 to cover the possibility of off-diagonal terms in the metric.

$$0 = \frac{du_x}{d\tau} - \Gamma^A_{XB} u_A u^B = \frac{du_x}{d\tau} - \frac{1}{2} g^{DA} u_A u^B \left(\frac{\partial g_{XD}}{\partial B} + \frac{\partial g_{BD}}{\partial x} - \frac{\partial g_{XB}}{\partial D} \right), \tag{18.20}$$

$$= \frac{du_x}{d\tau} - \frac{1}{2} u^D u^B \left(\frac{\partial g_{XD}}{\partial B} + \frac{\partial g_{BD}}{\partial x} - \frac{\partial g_{XB}}{\partial D} \right), \tag{18.21}$$

$$= \frac{du_x}{d\tau} - \frac{1}{2} u^D u^B \frac{\partial g_{BD}}{\partial x} = 0. \tag{18.22}$$

In Equation 18.21, the action of g^{DA} changes u_A to u^D. If you look at Equation 18.21, you will see that dummy indices **B** and **D** are interchangeable, so the first and third terms inside the brackets are the same and cancel each other out. This leads to Equation 18.22, which is what we need for the proof.

If $\frac{\partial g_{BD}}{\partial x} = 0$ then it follows that $\frac{du_x}{d\tau} = 0$. In words, if the metric does not depend on variable x, then based on the geodesic equation, we have shown that u_x is conserved along geodesics.

18.12 Cosmic Censorship and the Kerr Metric (Optional)

This somewhat lengthy optional section gives an illustration of how Penrose's cosmic censorship conjecture fits with the Kerr metric. The maths is challenging as it involves three calculations, although each is a useful exercise in its own right. First, we derive the equatorial effective potential of the Kerr metric. Next, we consider an object falling into an extremal black hole (we will use $a = GM$ for the black hole). The object has the maximum possible angular momentum without breaking the cosmic censorship conjecture (I call it *MAMO* for *Maximum Angular Momentum Object*). Finally, we use the Kerr effective potential to track the *MAMO*'s descent. We add a little more angular momentum (which would break cosmic censorship) and show that such an object doesn't cross the event horizon, but scatters off from the black hole. In contrast, objects with a little less angular momentum cross the event horizon and become part of the black hole.

18.12.1 Effective Potential of the Kerr Metric (Equatorial)

The derivation of the effective potential of the Schwarzschild metric is described back in Section 14.5. We are going to do a similar calculation for the Kerr metric. To simplify the maths, we consider a descent in the equatorial plane, so at all stages $\sin\theta = 1$, $\cos\theta = 0$ and $d\theta = 0$. I label the Kerr conserved quantities $\epsilon = u_t$, which means $(\epsilon = g_{TT}u^t + g_{T\phi}u^\phi)$ and $\ell = -u_\phi$, which means $(\ell = -g_{\phi\phi}u^\phi - g_{T\phi}u^t)$ from Equations 18.13 and 18.14. In words, ϵ is the conserved energy per kilogram (or mass equivalent as $c = 1$) along a geodesic and ℓ is the conserved angular momentum per kilogram. We can express u^t and u^ϕ in terms of these conserved quantities. To derive the answers shown as Equation 18.23, add together the two equations above each.

$$g_{\phi\phi}\epsilon = g_{TT}g_{\phi\phi}u^t + g_{T\phi}g_{\phi\phi}u^\phi, \qquad\qquad g_{T\phi}\epsilon = g_{TT}g_{T\phi}u^t + g_{T\phi}g_{T\phi}u^\phi,$$

$$g_{T\phi}\ell = -g_{T\phi}g_{T\phi}u^t - g_{T\phi}g_{\phi\phi}u^\phi, \qquad\qquad g_{TT}\ell = -g_{TT}g_{T\phi}u^t - g_{TT}g_{\phi\phi}u^\phi,$$

$$\Rightarrow \quad u^t = \frac{g_{\phi\phi}\epsilon + g_{T\phi}\ell}{g_{TT}g_{\phi\phi} - g_{T\phi}^2}, \qquad\qquad u^\phi = \frac{-g_{T\phi}\epsilon - g_{TT}\ell}{g_{TT}g_{\phi\phi} - g_{T\phi}^2}. \tag{18.23}$$

We feed these definitions along with $\epsilon = u_t$ and $\ell = -u_\phi$ into the invariant interval equation $u_\mu u^\mu = 1$ (Box 18.4 has detail on this formula). It is an equatorial descent so $u^\theta = 0$, $\sin\theta = 1$ and $p^2 = r^2$ because $\cos\theta = 0$. The maths is not complicated and is helped by the fact that the numerator in Equation 18.24 works out simply to r^2, but it is still a bit of a drag to arrive at last at Equation 18.25.

$$1 = u_\mu u^\mu = u_t u^t + u_r u^r + u_\phi u^\phi = \epsilon\left(\frac{g_{\phi\phi}\epsilon + g_{T\phi}\ell}{g_{TT}g_{\phi\phi} - g_{T\phi}^2}\right) + g_{rr}u^r u^r - \ell\left(\frac{-g_{T\phi}\epsilon - g_{TT}\ell}{g_{TT}g_{\phi\phi} - g_{T\phi}^2}\right),$$

$$\left(\frac{dr}{d\tau}\right)^2 = u^r u^r = \frac{1}{g_{rr}} - \frac{(g_{\phi\phi}\epsilon^2 + g_{TT}\ell^2 + 2g_{T\phi}\epsilon\ell)}{g_{rr}(g_{TT}g_{\phi\phi} - g_{T\phi}^2)}, \tag{18.24}$$

$$= \frac{1}{g_{rr}} - \frac{(g_{\phi\phi}\epsilon^2 + g_{TT}\ell^2 + 2g_{T\phi}\epsilon\ell)}{r^2},$$

$$= \frac{1}{r^2}\left[(-r^2 - a^2 + 2GMr) + \left(r^2 + a^2 + \frac{2a^2 GM}{r}\right)\epsilon^2 - \left(1 - \frac{2GM}{r}\right)\ell^2 - \frac{4aGM}{r}\epsilon\ell\right],$$

$$\left(\frac{dr}{d\tau}\right)^2 = \epsilon^2 - 1 + \frac{2GM}{r} + \frac{a^2(\epsilon^2 - 1) - \ell^2}{r^2} + \frac{2GM}{r^3}(\ell - a\epsilon)^2. \tag{18.25}$$

We have in Equation 18.25 an expression for radial velocity $\frac{dr}{d\tau}$ along an equatorial geodesic in terms of r and the conserved quantities. All that remains is to put it in the familiar form giving U_{eff} the effective potential (per kilogram of mass) for the Kerr metric along an equatorial geodesic. The result is Equation 18.26. You can compare with U_{eff} of the Schwarzschild metric in Equation 14.32. If you set $a = 0$ in Equation 18.26, you will see the two match ($c = 1$ in the Kerr version).

$$\frac{\epsilon^2 - 1}{2} = \frac{1}{2}\left(\frac{dr}{d\tau}\right)^2 - \frac{GM}{r} + \frac{\ell^2}{2r^2} - \frac{a^2(\epsilon^2 - 1)}{2r^2} - \frac{GM}{r^3}(\ell - a\epsilon)^2, \quad Kerr\ (c = 1),$$

$$U_{eff} = \frac{\ell^2}{2r^2} - \frac{a^2(\epsilon^2 - 1)}{2r^2} - \frac{GM}{r^3}(\ell - a\epsilon)^2 - \frac{GM}{r}. \tag{18.26}$$

18.12.2 Characterising the MAMO

Our aim is to model the path into an extremal black hole ($a = GM$) of an object that would break cosmic censorship, i.e. lead to ($a > GM$). First we need to determine the angular momentum of the *MAMO* (maximum angular momentum object) that the black hole can absorb *without* this happening. The *MAMO* has mass m, so it has angular momentum $m\ell$ and energy-equivalent mass of $m\epsilon$. Before the object arrives, the Kerr parameter of the extremal black hole is $a_* = 1 = \frac{a}{GM} = \frac{J}{GM^2}$, so the black hole has angular momentum $J = GM^2$. After the object arrives, let's say it has angular momentum J' and mass M'. For the black hole to remain extremal, $\dfrac{J'}{G(M')^2} = 1$. Adding in the angular momentum and mass-energy of the *MAMO* gives the result in Equation 18.27, noting that the *MAMO* mass is small: $m << M$ (which must be the case as the Kerr metric would not apply anyway to two large masses). The result shows that the *MAMO*'s angular momentum is: $\ell = 2GM\epsilon$. Anything greater than this would break cosmic censorship if it entered the extremal black hole, because its addition would lead to $a > GM$.

$$Extremal: \quad 1 = \frac{J'}{G(M')^2} = \frac{GM^2 + m\ell}{G(M + m\epsilon)^2} \approx \frac{GM^2 + m\ell}{GM^2 + 2GMm\epsilon}, \quad \Rightarrow \quad \ell = 2GM\epsilon. \tag{18.27}$$

18.12.3 Tracking the MAMO

We want to know what happens to the *MAMO* at the outer event horizon, which for an extremal black hole is at $\mathbf{r} = GM$ (you can see this from Equation 18.11 setting $|a| = GM$). Does the *MAMO* cross it and enter the black hole? We can work out the *MAMO*'s velocity by plugging in to Equation 18.26 the values $\ell = 2GM\epsilon$, $a = GM$ and $\mathbf{r} = GM$ as shown in Equation 18.28. The answer is that the *MAMO*'s radial velocity is zero at the event horizon.

$$\frac{\epsilon^2 - 1}{2} = \frac{1}{2}\left(\frac{d\mathbf{r}}{d\tau}\right)^2 + \frac{(2GM\epsilon)^2}{2(GM)^2} - \frac{(GM)^2(\epsilon^2 - 1)}{2(GM)^2} - \frac{GM}{(GM)^3}(2GM\epsilon - GM\epsilon)^2 - \frac{GM}{GM},$$

$$= \frac{1}{2}\left(\frac{d\mathbf{r}}{d\tau}\right)^2 + 2\epsilon^2 - \frac{\epsilon^2}{2} + \frac{1}{2} - \epsilon^2 - 1, \tag{18.28}$$

$$= \frac{1}{2}\left(\frac{d\mathbf{r}}{d\tau}\right)^2 + \frac{\epsilon^2 - 1}{2}, \quad \Longrightarrow \quad \frac{d\mathbf{r}}{d\tau} = 0, \quad \textit{at event horizon.}$$

If we want to know what happens next, we have to look at the radial acceleration, which is the differential of the effective potential in Equation 18.26: $\dfrac{d^2\mathbf{r}}{d\tau^2} = \dfrac{dU_{eff}}{d\mathbf{r}}$ (check back to Section 14.5 if needed). As calculated in Equation 18.29, the *MAMO*'s radial acceleration at the event horizon is also zero. The *MAMO* at the event horizon has zero radial velocity and zero radial acceleration!

$$\frac{dU_{eff}}{d\mathbf{r}} = -\frac{\ell^2}{\mathbf{r}^3} + \frac{a^2(\epsilon^2 - 1)}{\mathbf{r}^3} + \frac{3GM}{\mathbf{r}^4}(\ell - a\epsilon)^2 + \frac{GM}{\mathbf{r}^2}, \quad [\ell = 2GM\epsilon, \quad \mathbf{r} = a = GM],$$

$$= \frac{1}{GM}\left(-4\epsilon^2 + \epsilon^2 - 1 + 3\epsilon^2 + 1\right), \quad \Longrightarrow \quad \frac{dU_{eff}}{d\mathbf{r}} = \frac{d^2\mathbf{r}}{d\tau^2} = 0, \quad \textit{at event horizon.} \tag{18.29}$$

The *MAMO* arrives at the event horizon orbit, but neither crosses it nor scatters off. When I stumbled across this, it gave me a black-cat-cosmic-censorship-creepy-maths sort of feeling. The next step is to consider an object *MAMO+* with slightly higher angular momentum ($\ell = 2GM\epsilon + \Delta$) than the *MAMO*. If absorbed by the black hole, it would break cosmic censorship. We compare this with an object *MAMO−* with slightly lower angular momentum ($\ell = 2GM\epsilon - \Delta$) than the maximum. Δ is very small, so we can ignore any Δ^2 terms.

Let's first look at how the $\pm\Delta$ changes the radial velocity at the event horizon. It alters two terms in the Kerr effective potential in Equation 18.26. The two affected terms are shown below. The change to the first (ignoring Δ^2 terms) is shown in Equation 18.30 and increases the value of that term in the calculation in Equation 18.28 by $\dfrac{2\Delta\epsilon}{GM}$ at the event horizon. The change to the second term is equal and opposite at the event horizon as shown in Equation 18.31. This means that the radial velocity at the event horizon for *MAMO+* and *MAMO−* is still zero: $\dfrac{d\mathbf{r}}{d\tau} = 0$.

$$\frac{\ell^2}{2\mathbf{r}^2}, \quad \Longrightarrow \quad \frac{(2GM\epsilon \pm \Delta)^2}{2(GM)^2} \approx 2\epsilon^2 \pm \left(\frac{2\Delta\epsilon}{GM}\right), \quad \textit{ignoring } \Delta^2 \textit{ terms,} \tag{18.30}$$

$$-\frac{GM}{\mathbf{r}^3}(\ell - a\epsilon)^2, \quad \Longrightarrow \quad -\frac{GM}{(GM)^3}(2GM\epsilon \pm \Delta - GM\epsilon)^2 \approx -\epsilon^2 \pm \left(-\frac{2\Delta\epsilon}{GM}\right). \tag{18.31}$$

Now for the great reveal! Let's look at how the change in *MAMO* angular momentum affects the *radial acceleration* at the event horizon. The radial acceleration is $\dfrac{d^2\mathbf{r}}{d\tau^2} = \dfrac{dU_{eff}}{d\mathbf{r}}$. The derivatives of the same two affected terms are shown in Equations 18.32 and 18.33 along with the ± effects of the change in the *MAMO* angular momentum. The effects *do not cancel*.

$$-\frac{\ell^2}{\mathbf{r}^3}, \quad \Longrightarrow \quad -\frac{(2GM\epsilon \pm \Delta)^2}{(GM)^3} \approx -\frac{4\epsilon^2}{GM} \pm \left(\frac{-4\Delta\epsilon}{(GM)^2}\right), \qquad \textit{ignoring } \Delta^2, \tag{18.32}$$

$$\frac{3GM}{\mathbf{r}^4}(\ell - a\epsilon)^2, \quad \Longrightarrow \quad \frac{3GM}{(GM)^4}(2GM\epsilon \pm \Delta - GM\epsilon)^2 \approx \frac{3\epsilon^2}{GM} \pm \left(\frac{+6\Delta\epsilon}{(GM)^2}\right). \tag{18.33}$$

We can compare the behaviour of *MAMO+* and *MAMO−* at the black hole event horizon. As shown in Equation 18.34, the *MAMO+*, which poses a threat to cosmic censorship, has no radial velocity at the event horizon but has positive radial acceleration. Therefore, the *MAMO+ accelerates away from* the black hole and scatters off. It does *not* cross the event horizon. In sharp contrast, Equation 18.35 shows that *MAMO−*, which does not threaten cosmic censorship, has negative radial acceleration, so it *accelerates towards* the black hole, *crossing* the event horizon and entering the black hole.

$$MAMO+ : \quad \frac{d\mathbf{r}}{d\tau} = 0, \quad \frac{d^2\mathbf{r}}{d\tau^2} = -\frac{4\Delta\epsilon}{(GM)^2} + \frac{6\Delta\epsilon}{(GM)^2} = +\frac{2\Delta\epsilon}{(GM)^2}, \tag{18.34}$$

$$MAMO- : \quad \frac{d\mathbf{r}}{d\tau} = 0, \quad \frac{d^2\mathbf{r}}{d\tau^2} = +\frac{4\Delta\epsilon}{(GM)^2} - \frac{6\Delta\epsilon}{(GM)^2} = -\frac{2\Delta\epsilon}{(GM)^2}. \tag{18.35}$$

The effective potential of the Kerr metric means in this example that objects that would break cosmic censorship by entering the black hole do not do so. They scatter off. While this is not proof of cosmic censorship, it certainly suggests that Penrose's conjecture has some decent mathematical backing and is a neat example of how nature does not appear to like overspinning black holes.

Chapter 19

Gravitational Waves

Box 19.1 Detection is unthinkable

The German physicist Gustav Mie (1868–1957) considered how gravity and electromagnetism might compare. In his words: *Any gravitational radiation emitted… is so extraordinarily weak that it is unthinkable ever to detect it… if one could ever prove the existence of gravitational waves, the processes responsible for their generation would probably be much more curious and interesting than even the waves themselves.*

Since 2015, we have observatories that can detect gravitational waves (GW). This is an exciting step forward for astronomy. Up to now, our observational research has been based entirely on electromagnetic (EM) radiation. The upside of EM radiation is that it is easier to detect because it strongly interacts with charged particles. The downside is that it is affected by these interactions distorting the original EM radiation en route. In contrast, GW radiation is much harder to detect (see Box 19.1), and therefore is less distorted in its journey from source to earth. GW observatories such as the Laser Interferometer Gravitational-Wave Observatories (LIGO) open up a whole new way of studying the universe.

In this chapter, I will step you through a somewhat simplified mathematical analysis of how GW radiation fits with the Einstein Field Equations (EFEs). To achieve this, I focus the search on transverse waves for reasons I will explain. The rest of the chapter covers the discoveries of LIGO and how to interpret the GW oscillations detected, and looks to the future, examining other GW observational facilities and research.

19.1 Einstein's Flip-Flop

The existence of GW radiation was far from obvious. Our starting point is a letter in 1916 (now lost) from Einstein to Schwarzschild reportedly stating: *There are no gravitational waves analogous to light waves* (flop!). A couple of months later Einstein concluded in a paper that GW radiation does exist (flip!), although he got the details wrong. Two years later in 1918, he acknowledged that his work was *marred by a regrettable error in calculation* (flop!) but then came up with the correct formula (flip!) and it is this paper that is often quoted as Einstein's prediction of GW.

Einstein's 1918 analysis used a simplified version of the EFEs. In 1936, he revisited it, applying the full EFEs and concluded that GW radiation does not exist (flop!). It turned out that this conclusion was based on problems with coordinate singularities and did not signal an underlying issue; but Einstein maintained a sceptical stance for the rest of his life. In the late 1950s, shortly after Einstein died in 1955, further work concluded that GW radiation does indeed fit with the EFEs. It took another 60 years for GW radiation to be detected.

I tell you all of this because I want you to understand that GW radiation is not an obvious prediction of the EFEs. Hopefully, this will explain why I will be making simplifying assumptions to help steer through the maths in a manageable way.

Untangling General Relativity: The Intuitive Self-Study Guide, First Edition. Simon Sherwood.
© 2026 John Wiley & Sons Ltd. Published 2026 by John Wiley & Sons Ltd.

19.2 The Maths of GW Radiation

In this section, we are going to examine the existence of GW radiation as a *transverse* wave moving at *light speed*. I use the term transverse to mean that the wave oscillates perpendicular to the wave's direction of advance (like an EM wave). This is in contrast to a longitudinal wave, which oscillates in the direction of its advance (like a sound wave or compression wave). To justify these assumptions for GW, I am going to lean on some concepts from quantum mechanics along with a bit of special relativity.

19.2.1 Massless Graviton

To those of you unfamiliar with quantum mechanics, the following argument may be hard to grasp. However, there is no hiding from the fact that physics is quantum at its core. The EFEs relate the energy-momentum tensor $[T_{\mu\nu}]$ to spacetime curvature in the metric $[g_{\mu\nu}]$. We *know* for a fact that on the right side of the EFEs, energy-momentum is quantised. If the EFEs are correct, then the left side equals the right side, so the spacetime curvature must also be quantised.

A key feature of quantisation is wave-particle duality. Excitations of quantum fields come in lumps, which we call particles. The same must be expected of GW radiation. We call the related particle the *graviton*. Those of you who have studied quantum field theory (QFT) will know that we expect the graviton to be a massless boson (force-carrier) that is spin-2 (it is an excitation of a rank-2 tensor field).

The reason we expect the graviton to be massless is that the gravitational force operates over such vast distances. Interactions in QFT involve the transfer of energy-momentum through the field, creating attraction or repulsion. The transfer requires the field to be excited, which means borrowing energy from the vacuum. This is allowed by Heisenberg's uncertainty principle (HUP), which requires that nothing can have exactly zero energy and zero momentum. Physicists call these ghost excitations *virtual particles*.

If the field's particles are massive, each excitation requires a lot of borrowed energy. The HUP tells us that the more energy required, the less time it can be borrowed, so the excitations must be short-lived. The result is a short-range force. This is the case with the *weak force* with its (very) massive bosons. It operates within the atomic nucleus and has a range of only 10^{-17} metres.

How does this compare with the range of the gravitational force? Consider, for example, the size of our Milky Way galaxy. It alone is over 10^{20} metres across. The seemingly infinite range of the gravitational force is a strong reason to expect that the graviton, like the photon, is massless.[1]

19.2.2 Transverse Wave

Massless particles move at light speed, so our assumption means that we are looking for GW radiation moving at light speed, which points to a transverse wave. Let me explain. Basically, there are two types of waves: *transverse* and *longitudinal*. With a transverse wave, the variation is perpendicular to the direction of motion as the wave moves along. An example is EM radiation. With a longitudinal wave, the variation is in the direction of motion. An example is a compression wave such as a sound wave.

It may help you to think of a Slinky (the wire coil toy). If you have a friend hold one end while you shake the other from side to side, you get a transverse wave moving along the slinky. Alternatively, you can push and pull the slinky in and out generating a longitudinal compression wave that flows up it to your friend. If this distinction confuses you, please take a moment to check out on the Web the difference between transverse and longitudinal waves to be sure you understand.

---

1 Technical note: this is not proof that the graviton is massless. It might have a very tiny mass, but the most obvious intuitive expectation is that it is massless.

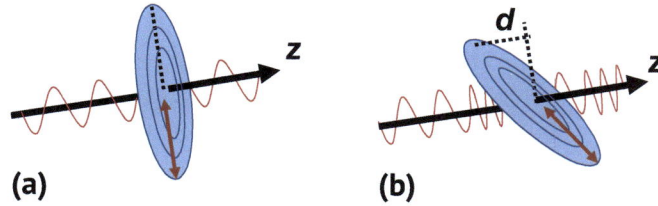

Figure 19.1 The variation of a transverse wave is perpendicular to the direction of motion as shown in a. Longitudinal waves involve variation that is not perpendicular to the direction of motion as shown in b. However, at light speed this plane doesn't exist because $d = 0$.

At light speed, longitudinal waves do not fit sensibly within the requirements of special relativity. Figure 19.1 illustrates the challenge. On the left, Figure 19.1a shows a pure *transverse* wave moving along in the z direction. As it moves along, all the variation in the wave is in the plane perpendicular to the direction of motion as shown. That works fine even at light speed. In Figure 19.1b, I have introduced some *longitudinal* variation in the wave along the z direction of motion. This creates variation in a plane that is not perpendicular to the motion. In the figure it is illustrated in an off-perpendicular plane. However, at light speed length contraction is infinite so $d = 0$ (check back to Section 2.5 if needed), which makes it hard to come up with a reasonable consistent structure. A longitudinal wave at light speed would require variation in a plane that does not exist geometrically for any observer. Oops!

As an aside for those familiar with quantum mechanics and spin, this is why massless particles have only two distinct spin states. At light speed, their spin must be perpendicular with the direction of motion, leaving only two options: clockwise and anticlockwise (or some superposition of the two). It should come as no surprise that the massless graviton does indeed have only two independent states (or polarisations), as will be revealed in a moment.

19.2.3 Light Speed GW Radiation in the EFEs

I hope to have convinced you that GW radiation makes most sense as a transverse wave moving at light speed. This allows us to narrow down the mathematical search. Our other assumption, which Einstein himself used in his 1918 paper, is that we are analysing GW radiation far from its source, so the perturbation of the metric is very small. Put simply, the metric containing the GW radiation is only very slightly different from that of flat Minkowski spacetime.

Taking all these assumptions together leads us to the metric in Equation 19.1. Note that I have included the c speed of light factor as this will prove useful later. The scenario is that we are analysing GW radiation moving in the z direction a long way from the source. All variations in the metric are in the xy plane, perpendicular to the direction of motion. The variation from the Minkowski metric is shown as h_{XX}, h_{YY} and h_{XY}. The values of h_{XX}, h_{YY} and h_{XY} will vary with time t and along the direction of motion z. Our mission is to discover the nature of these variations in Ricci-0 spacetime ($[R_{\mu\nu}] = 0$) i.e. what type of GW radiation can travel through the vacuum.

$$\text{GW analysis metric } (t,x,y,z): \quad [g_{\mu\nu}] = \begin{bmatrix} c^2 & 0 & 0 & 0 \\ 0 & -1 - h_{XX} & -h_{XY} & 0 \\ 0 & -h_{XY} & -1 - h_{YY} & 0 \\ 0 & 0 & 0 & -1 \end{bmatrix}. \tag{19.1}$$

As $[R_{\mu\nu}] = 0$, we know $R_{XY} = 0$. The first step in our analysis is to show what this requires in terms of the h variations. We know $R_{XY} = R^T_{XTY} + R^X_{XXY} + R^Y_{XYY} + R^Z_{XZY}$ (check back in Equation 7.11), but the second and third terms are zero by symmetry (bottom left of Box 7.3). This leaves: $R_{XY} = R^T_{XTY} + R^Z_{XZY} = 0$. To avoid having to look back, the formulas for calculating the components of the Riemann tensor and the Christoffel symbols are shown again in Box 19.2.

Box 19.2 Formula reminder (dummy indices shown in bold)

$$R^A_{BCD} = \partial_C \Gamma^A_{BD} - \partial_D \Gamma^A_{BC} + \Gamma^A_{CE} \Gamma^E_{BD} - \Gamma^A_{DE} \Gamma^E_{BC}, \qquad \textit{Riemann Tensor,} \qquad (19.2)$$

$$\Gamma^A_{BC} = \frac{1}{2} g^{AD} \left(\frac{\partial g_{DC}}{\partial B} + \frac{\partial g_{DB}}{\partial C} - \frac{\partial g_{BC}}{\partial D} \right), \qquad \textit{Christoffel Symbol.} \qquad (19.3)$$

We cannot use my Christoffel or Ricci cheat sheets because of the off-diagonal terms in the metric. However, we can simplify by ignoring the last two terms in the Riemann formula of Equation 19.2. This is because we assume that the variations h and their derivatives $\frac{\partial h}{\partial t}$ and $\frac{\partial h}{\partial z}$ are very small. The value of each Christoffel symbol depends on the derivatives and so is tiny. Multiply two together (as in the two last terms of Equation 19.2) and you have tiny multiplied by tiny... which we dismiss as negligible. This brings us to the calculation.

$$R_{XY} = R^T_{XTY} + R^Z_{XZY} \approx \partial_t \Gamma^T_{XY} - \partial_y \Gamma^T_{XT} + \partial_z \Gamma^Z_{XY} - \partial_y \Gamma^Z_{XZ}, \qquad (19.4)$$

$$\approx \partial_t \Gamma^T_{XY} + \partial_z \Gamma^Z_{XY}, \qquad \textit{metric only depends on t and z,}$$

$$= \partial_t \left[\frac{1}{2} g^{TT} \left(\frac{\partial g_{TY}}{\partial x} + \frac{\partial g_{TX}}{\partial y} - \frac{\partial g_{XY}}{\partial t} \right) \right] + \partial_z \left[\frac{1}{2} g^{ZZ} \left(\frac{\partial g_{ZY}}{\partial x} + \frac{\partial g_{ZX}}{\partial y} - \frac{\partial g_{XY}}{\partial z} \right) \right],$$

$$= \partial_t \left[\frac{1}{2} \frac{1}{c^2} \left(0 + 0 + \frac{\partial h_{XY}}{\partial t} \right) \right] + \partial_z \left[\frac{1}{2} (-1) \left(0 + 0 + \frac{\partial h_{XY}}{\partial z} \right) \right],$$

$$= \frac{1}{2c^2} \frac{\partial^2 h_{XY}}{\partial t^2} - \frac{1}{2} \frac{\partial^2 h_{XY}}{\partial z^2} \approx 0, \qquad \Longrightarrow \qquad \frac{1}{c^2} \frac{\partial^2 h_{XY}}{\partial t^2} - \frac{\partial^2 h_{XY}}{\partial z^2} \approx 0. \qquad (19.5)$$

Let me step through this for clarity. Equation 19.4 is the definition based on Equation 19.2 ignoring the multiple Christoffel $\Gamma\Gamma$ terms. Any ∂x and ∂y terms are zero because the metric doesn't depend on x or y. I have then substituted using the Christoffel definition from Equation 19.3, noting that the value of the dummy index **D** is limited. For example, it must be t in the first case because g^{TT} has value but $g^{TX} = g^{TY} = g^{TZ} = 0$. It is then simple to insert values noting that $-\frac{\partial g_{XY}}{\partial t} = \frac{\partial h_{XY}}{\partial t}$ and $-\frac{\partial g_{XY}}{\partial z} = \frac{\partial h_{XY}}{\partial z}$.

Setting the value of R_{XY} to zero, as we require for GW radiation in a vacuum, gives Equation 19.5, which leads to the equation of a transverse wave moving at light speed in the z direction. For those of you less familiar with this wave equation, there is a refresher in Box 19.3.

Box 19.3 Refresher on sinusoidal wave equation

$\Psi = A \sin \left(\frac{2\pi}{\lambda} \right) x$ is a stationary sinusoidal wave of amplitude A and wavelength λ.

$\Psi = A \sin \frac{2\pi}{\lambda}(x - vt)$ is the same wave travelling in the x direction at velocity v.

$$\frac{\partial \Psi}{\partial x} = \frac{2\pi}{\lambda} A \cos \frac{2\pi}{\lambda}(x - vt), \qquad \frac{\partial^2 \Psi}{\partial x^2} = -\left(\frac{2\pi}{\lambda} \right)^2 A \sin \frac{2\pi}{\lambda}(x - vt) = -\left(\frac{2\pi}{\lambda} \right)^2 \Psi,$$

$$\frac{\partial \Psi}{\partial t} = -v \frac{2\pi}{\lambda} A \cos \frac{2\pi}{\lambda}(x - vt), \qquad \frac{\partial^2 \Psi}{\partial t^2} = -v^2 \left(\frac{2\pi}{\lambda} \right)^2 A \sin \frac{2\pi}{\lambda}(x - vt) = -v^2 \left(\frac{2\pi}{\lambda} \right)^2 \Psi,$$

$$\Longrightarrow \frac{1}{v^2} \frac{\partial^2 \Psi}{\partial t^2} - \frac{\partial^2 \Psi}{\partial x^2} = 0, \qquad \textit{wave moving at velocity v in x direction.}$$

So far we only know about h_{XY}, which based on metric symmetry is the same as h_{YX}. We know nothing about h_{XX} and h_{YY}. We can uncover their secrets by calculating R_{XX} and R_{YY} and setting them to zero as required for a vacuum solution. The calculation of R_{XX} looks a bit more complicated as only R^X_{XXX} is zero by symmetry.

However, there are still only two non-zero terms because the metric depends on only t and z. The result shows, h_{XX} also matches the equation of a wave at light speed. By symmetry, the same is true for h_{YY} based on $R_{YY} = 0$.

$$R_{XX} = R^T_{XTX} + R^Y_{XYX} + R^Z_{XZX},$$

$$\approx \partial_t \, \Gamma^T_{XX} - \partial_x \, \Gamma^T_{XT} + \partial_y \, \Gamma^Y_{XX} - \partial_x \, \Gamma^Y_{XY} + \partial_z \, \Gamma^Z_{XX} - \partial_x \, \Gamma^Z_{XZ},$$

$$= \partial_t \, \Gamma^T_{XX} + \partial_z \, \Gamma^Z_{XX},$$

$$= \partial_t \left[\frac{1}{2} g^{TT} \left(\frac{\partial g_{TX}}{\partial x} + \frac{\partial g_{XT}}{\partial y} - \frac{\partial g_{XX}}{\partial t} \right) \right] + \partial_z \left[\frac{1}{2} g^{ZZ} \left(\frac{\partial g_{ZX}}{\partial x} + \frac{\partial g_{XZ}}{\partial x} - \frac{\partial g_{XX}}{\partial z} \right) \right], \quad (19.6)$$

$$= \partial_t \left[\frac{1}{2} \frac{1}{c^2} \left(0 + 0 + \frac{\partial h_{XX}}{\partial t} \right) \right] + \partial_z \left[\frac{1}{2} (-1) \left(0 + 0 + \frac{\partial h_{XX}}{\partial z} \right) \right],$$

$$= \frac{1}{2c^2} \frac{\partial^2 h_{XX}}{\partial t^2} - \frac{1}{2} \frac{\partial^2 h_{XX}}{\partial z^2} \approx 0, \quad \Longrightarrow \quad \frac{1}{c^2} \frac{\partial^2 h_{XX}}{\partial t^2} - \frac{\partial^2 h_{XX}}{\partial z^2} \approx 0,$$

$$\Longrightarrow \quad \frac{1}{c^2} \frac{\partial^2 h_{YY}}{\partial t^2} - \frac{\partial^2 h_{YY}}{\partial z^2} \approx 0, \quad \textit{by symmetry.}$$

Therefore, the perturbations h_{XX}, h_{YY} and h_{XY} all appear to match the formula for a transverse wave moving at the speed of light in the z direction. You will often see this wave relationship written as $\Box h = 0$, using what is called the *d'Alembert operator*. Equation 19.7 shows its meaning using Cartesian coordinates in Minkowski space-time. The wave equations at the end of the calculations in Equations 19.5 and 19.6 give the same result but with the Cartesian coordinates oriented so the gravitational wave travels in the z direction.

$$\Box h = 0, \quad \Longrightarrow \quad \frac{1}{c^2} \frac{\partial^2 h}{\partial t^2} - \frac{\partial^2 h}{\partial x^2} - \frac{\partial^2 h}{\partial y^2} - \frac{\partial^2 h}{\partial z^2} = 0, \quad \textit{d'Alembert Operator.} \quad (19.7)$$

19.2.4 Two Distinct Polarisations

We can show that two of the perturbations are related: $h_{XX} = -h_{YY}$ by using the calculation of R_{TZ}, which must also be zero value. Equation 19.8 sets out the start of the calculation. We can dismiss the terms that are derivatives of x and y because the metric depends on only t and z. This reduces the calculation to two Christoffel terms. Using the Christoffel definition from Equation 19.3 leads to Equation 19.9. While g^{XX} and g^{YY} are about -1, we know that g^{XY} (which is h^{XY}) is very small because if a matrix has small off-diagonal terms, then its inverse also has small off-diagonal terms (see Box 3.7 on the inverse of a matrix for a 2-D example). As our calculation is approximate, we ignore the g^{XY} term. The result is Equation 19.10. The value of all Ricci components must be zero. We know from the result of Equation 19.6 that h_{XX} and h_{YY} vary with t and z. Therefore, we can conclude that $h_{XX} = -h_{YY}$.

$$R_{TZ} = R^X_{TXZ} + R^Y_{TYZ} \approx \partial_x \Gamma^X_{TZ} - \partial_z \Gamma^X_{TX} + \partial_y \Gamma^Y_{TZ} - \partial_z \Gamma^Y_{TY}, \quad (19.8)$$

$$= -\partial_z \left(\Gamma^X_{TX} + \Gamma^Y_{TY} \right),$$

$$= -\partial_z \left(\frac{g^{XX}}{2} \frac{\partial g_{XX}}{\partial t} + \frac{g^{YY}}{2} \frac{\partial g_{YY}}{\partial t} + g^{XY} \frac{\partial g_{XY}}{\partial t} \right), \quad (19.9)$$

$$\approx -\partial_z \left(\frac{1}{2} \frac{\partial h_{XX}}{\partial t} + \frac{1}{2} \frac{\partial h_{YY}}{\partial t} \right),$$

$$= -\frac{1}{2} \partial_z \partial_t \left(h_{XX} + h_{YY} \right) = 0, \quad \Longrightarrow \quad h_{XX} = -h_{YY}. \quad (19.10)$$

In summary, we have used an approximate calculation of the perturbed Minkowski metric to mimic how a weak gravitational wave might appear at a distance. It reveals that the perturbations are similar to transverse waves moving at light speed. The perturbation has two independent forms. The first is the perturbation of $(h_{XY} = h_{YX})$. The second is the perturbation of $(h_{XX} = -h_{YY})$. These are the two polarisations of the GW radiation.

Before moving on, let me offer a word of warning. The analysis we have done involves a lot of simplification. To fully prove that GW radiation is consistent with the EFEs takes much more effort. Don't forget that from the date of Einstein's original 1918 paper, which also was based on approximations, it took physicists about 40 years to prove the details.

19.3 Tell Me More About Gravitational Waves

You might think that the spherically symmetrical collapse of a star into a black hole would produce GW, but it does not. Consider the change such a collapse would have on the Schwarzschild or Kerr metrics. The collapse does not change the mass nor the angular momentum. Therefore, if the collapse is spherically symmetrical, there is no change to the spacetime metric of the surrounding vacuum. No change... no GW radiation.

However, perfect spherical symmetry is an ideal and there is plenty going on in astrophysics that generates GW radiation. The most relevant example is the orbit of compact binaries. Figure 19.2a is an artist's illustration of the GW radiation (ripples) flowing out from two orbiting masses (the black dots). For GW radiation to be detectable, you need a lot of mass in a very tight orbit. The sources we can detect are binary systems of black holes and neutron stars. In the case of black holes, there is an immense release of GW radiation when they get close enough to merge in a rapid plunge (of which more later).

GW radiation spreads out through space at light speed, distorting the spatial dimensions perpendicular to the wave's direction of travel. Figure 19.2b is a schematic of the distortion (vastly exaggerated) to the x and y spatial dimensions showing two underlying polarisations. On the left is the $h_{XX} = -h_{YY}$ perturbation. The result is a slight stretch in the x direction (lighter blue) followed by a slight stretch in the y direction (darker blue). On the right is the h_{XY} perturbation. GW radiation can be any combination of the two polarisations.

The observation of GW radiation requires super-sensitive detectors. The first GW radiation detected by the two LIGO observatories involved a change to the spatial dimensions by a factor of only 10^{-21}. Figure 19.2c shows the basic set-up. Two arms, each about four kilometres long, are set at right angles. Laser light is shone along each arm using a beam-splitter. The beams are reflected back by mirrors at the end of each arm. The beams are recombined. Any change in the length of either arm results in a detectable change to the interference pattern in the recombined beam. This sounds simple, but the requirements in terms of accuracy create huge engineering headaches. At its most sensitive, LIGO can detect a change 1/10,000th the width of a proton.

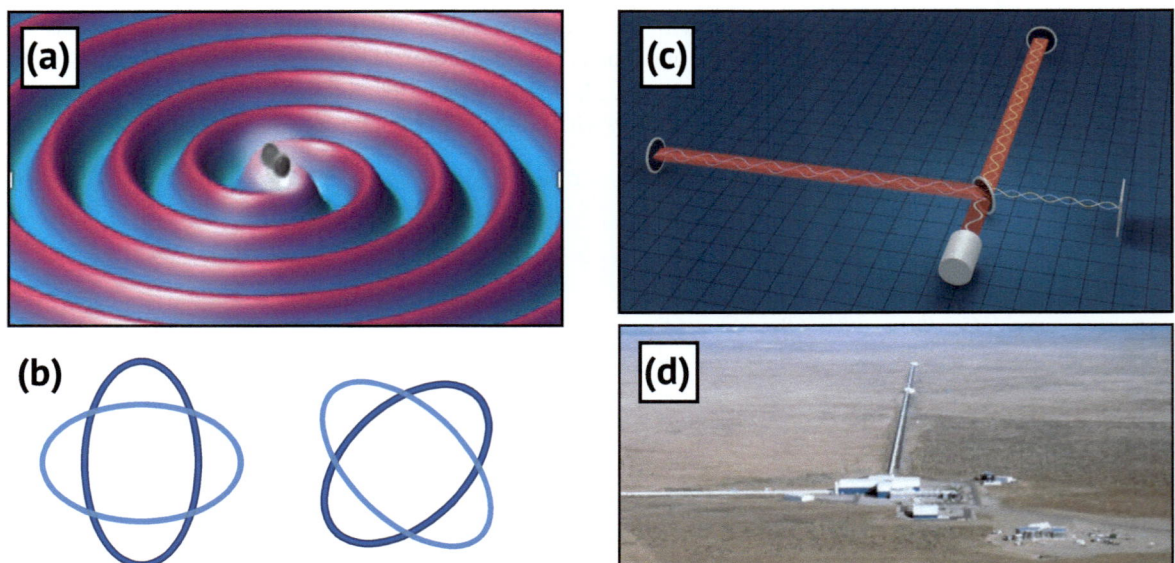

Figure 19.2 (a) Artist illustration of gravitational wave production, *Source:* MoocSummers / Wikimedia Commons / CC BY 4.0, (b) Gravitational wave polarisation, (c) Diagram of LIGO interferometer, T. Pyle, Caltech/MIT/LIGO lab CC share-alike 4.0, (d) Aerial view of LIGO Hanford, *Source:* Caltech/MIT/LIGO Laboratory / Wikimedia Commons / Public Domain.

Figure 19.2d is an aerial photograph of the LIGO observatory at Hanford in Washington state on the northwest coast of the USA. There is also LIGO Livingston about 3,000 kilometres away in the state of Louisiana on the southeast coast. The distance between the two LIGO observatories allows them to compare distortions created by incoming GW and eliminate local distortions such as seismic events.

19.4 Chirp GW150914: A Case Study

On 14 September 2015, the LIGO observatories made the first ever detection of GW radiation from the merger of two black holes weighing in at 36 and 29 solar masses. Since then, LIGO has detected many gravitational wave events. It is clear that the observation of GW radiation is going to play a major role in the future of astronomy, so let me try to give you a flavour of how the data is analysed. After all, it seems reasonable to question how the LIGO research team concluded from a wiggle on a graph that there were two black holes of 36 and 29 solar masses!

Let me use as a case study the original LIGO detection event GW150914 (Gravitational Wave, year 2015, month 09, day 14). The top left of Figure 19.3 shows the original gravitational wave data from LIGO Livingston (blue) and LIGO Hanford (red). The signals (called *chirps*) have been shifted and inverted to reflect the time delay between the observatories and their orientation to the waves. On the bottom left of the figure is the reconstructed wavelet (combining the observations) and a blue line labelled *numerical relativity*, which is a fancy label for a series of computer-based simulations of the generation of GW radiation that go through various scenarios to find a match with the observed data.

The right of Figure 19.3 shows a model based on numerical relativity, which should help you understand what is happening. The entire chirp lasted less than half a second. At time 0.30, the two black holes are orbiting each other. They release gravitational wave energy and inspiral towards each other. Once they get close enough to be in the vicinity of the Innermost Stable Circular Orbit (ISCO) they descend in a radial plunge. The mathematics of this is discussed back in Section 14.7.3. In Section 14.8, I showed you that objects descending to the ISCO release energy equivalent to 6% of their rest mass for a Schwarzschild black hole, and more if it is a rotating black hole. A useful rule of thumb is that the energy release in the final plunge is equivalent to about 10% of the rest mass of the smaller of the black holes. All this energy is released in a few milliseconds.

Following events as labelled on the right of Figure 19.3, the inspiral becomes a radial plunge at the ISCO. The result is that the black holes merge. There is a final smaller release of GW radiation as the new merged black hole settles down in a process called *ring-down*.

Let's make a *very rough* estimate of the size of the black holes involved. Imagine for a moment two equivalent black holes A and B in binary orbit. A full orbit goes $(A - B)$ to $(B - A)$ and finally back to the original positions

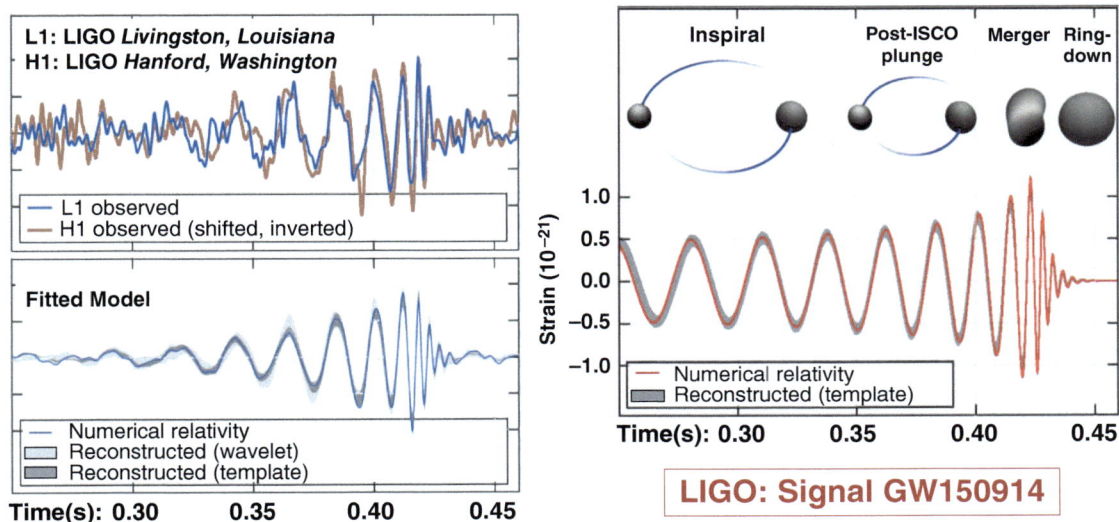

Figure 19.3 LIGO GW150914, courtesy of B.P. Abbott et al., LIGO CC 3.0 Unported.

$(A-B)$. It turns out (trust me on this) that $(A-B)$ to $(B-A)$ is one gravitational wave, with a second equivalent wave coming from $(B-A)$ to $(A-B)$, so the frequency of the gravitational wave produced is twice the orbital frequency. I must emphasise that this is a gross simplification, which risks making astrophysicists wince, but hopefully will give you a flavour of the analysis.

LIGO reported chirp GW150914 as increasing from about 30 Hz to 200 Hz (Hz: Hertz is frequency per second). Note that the effect of gravitational time dilation around the ISCO is less than you might expect (at least it is less than I expected). If you feed $r = 3R_S$ into Equation 13.5, it gives $d\tau \approx 0.8\, dt$ (you should use the square root rather than the Taylor approximation). Anyway, let's assume for illustration that at the moment before the radial plunge, the orbital frequency of the two black holes was somewhere around 100 Hz or 100 orbits per second. Modelling leads to an estimate that the black holes were moving at about half the speed of light. This allows us to make an approximate initial estimate of the mass of the black holes involved. The first step is Equation 19.11, which gives the perimeter of the tightest orbit (the moment of radial plunge): an orbit in one hundredth of a second at half light speed. This gives the orbital radius (the r coordinate measure) in Equation 19.12, which is the ISCO (when the plunge occurs). We then use the Schwarzschild formula for the ISCO calculated back in Equation 14.37: $r = \frac{6GM}{c^2}$. This gives an estimate of mass equivalent to about 30 solar masses.

$$\text{Orbit ISCO perimeter} \quad \frac{0.5c}{100} = \frac{1.5 \times 10^8}{100} = 1.5 \times 10^6 \text{ metres,} \tag{19.11}$$

$$\text{Orbit ISCO radius} \quad \frac{1.5 \times 10^6}{2\pi} = 2.4 \times 10^5 \text{ metres,} \tag{19.12}$$

$$\text{Mass } M = \frac{R_{ISCO}\, c^2}{6G}, \quad \frac{2.4 \times 10^5 (3 \times 10^8)^2}{6(6.7 \times 10^{-11})} = 5.4 \times 10^{31} \text{ kilograms,} \tag{19.13}$$

$$= \frac{5.4 \times 10^{31}}{2 \times 10^{30}} \approx 30 \, solar \, masses.$$

Of course, LIGO's analysis is way more sophisticated than this. Using the first guesstimate, you calculate the likely energy release in the plunge and compare this with the amplitude of the gravitational wave chirp for an idea of how far away the merger occurred. In turn, this allows adjustment also for change in wavelength of the GW en route due to the expansion of the universe. And changes in the frequency of the chirp encode data on the relative masses of the black holes, their spin rates, and their orbital speed. It is a highly iterative process gradually homing in on a signal match. However, I hope the crude illustration above gives you a feel of the process.

LIGO announced the confirmed detection of the GW of chirp GW150914 in 2016. The 2017 Nobel Prize in physics was awarded to Rainer Weiss, Barry Barish and Kip Thorne in recognition of the achievement.

19.5 Chirp GW170817

The success of LIGO has awakened wider interest in gravitational wave observatories. In addition to the two LIGO facilities there are Virgo (Italy), GEO600 (Germany) and TAM300 and KAGRA (Japan). There are also space-based observatories planned for the 2030s (LISA and DECIGO). One benefit of multiple observatories is that it allows physicists to better pinpoint the directional origin of the GW radiation. This brings us to chirp GW170817, the first observation of GW radiation that has been confirmed with matching observations of EM waves.

On 17 August 2017, the two LIGO observatories picked up a 100-second chirp. The calculated mass from the frequency was too small for black holes so it indicated the merger of two neutron stars (see Section 13.5 for a discussion of the mass limit for a black hole). The chirp was not seen at Virgo, which indicated that it was at a blind-spot angle to the Virgo observatory. Combined with the LIGO data, this narrowed down the location of its origin in the sky.

This allowed other observatories to focus in on the location with positive sightings reported in gamma ray, optical, infrared and X-ray observatories. It provided the final confirmation and verification of gravitational wave observations. In the words of the famous physicist Stephen Hawking (1942–2018): *It is a genuine milestone. It is the first ever detection of a gravitational wave source with an electromagnetic counterpart.*

19.6 Pulsar Timing Arrays

Pulsars are magnetised spinning neutron stars that emit EM radiation from their poles. This creates a pulsing flash as the beam swings round and round. The rate of spin is stable, so the pulse rate is regular and predictable. Pulsars are spread out through the sky thousands of light years apart. For the past 15 years, data has been collected on 68 pulsars spread throughout our Milky Way galaxy in an effort to use variations in the signals to detect GW radiation. The results of this work were published in June 2023.

The research compares variation in the arrival time of pulsar pulses. The presence of long wavelength GW radiation changes the distances between pulsars and earth in a correlated fashion, altering the arrival time of pulses. This allows for the detection of GW radiation with very long wavelengths (light years) that could never be picked up at an earth-based observatory.

Box 19.4 Excerpts from NANOGrav press release, June 2023

Astrophysicists using large radio telescopes... have found evidence for gravitational waves that oscillate with periods of years to decades... While earlier results from NANOGrav uncovered an enigmatic timing signal common to all the pulsars they observed, it was too faint to reveal its origin. The 15-year data release demonstrates that the signal is consistent with slowly undulating gravitational waves passing through our galaxy... After years of work, NANOGrav is opening an entirely new window on the gravitational-wave universe.

The North American Nanohertz Observatory for Gravitational Waves (NANOGrav), a collaboration of almost 200 American and Canadian physicists who monitor pulsars, has reported a background hum of GW radiation. There is supporting evidence from telescopes in Europe, China, India and Australia. What is this hum? It might be the merger of super-massive black holes. But it also could be the result of events early in the formation of the universe... a window further into the past than is possible with EM radiation observations (see Box 19.4).

The findings of these pulsar timing arrays are creating a lot of excitement. It is early days and I suspect there is a lot more to come. This is one occasion when I fear this book will quickly be out of date, so you may want to look up NANOGrav on the Web to get the latest. At time of writing in February 2025, their news website is: *https://nanograv.org/news-events/news*.

19.7 A Note on Hawking Radiation

Let me start by highlighting that Hawking radiation is *not* a form of GW radiation. It has nothing to do with GW. It is very weak radiation from a black hole. It is almost exclusively low-energy EM radiation (photons), although in more extreme scenarios it can theoretically include particle emissions.

Hawking radiation is a *quantum* phenomenon caused by the constraints imposed on underlying quantum fields by the presence of the black hole event horizon. Readers with no background in quantum theory are going to find Hawking radiation bizarre and confusing. Even Stephen Hawking himself found it hard to describe in layman's language. Any of you who have read his book *A Brief History of Time* will have come across his explanation of Hawking radiation as the result of the spontaneous formation of particle–antiparticle pairs at the event horizon, one of which crosses into the black hole while the other escapes... creating radiation.

This description bears little resemblance to the maths in his original paper and is arguably downright misleading. It suggests that Hawking radiation occurs only at the event horizon (wrong), involves real particles with negative energy crossing into the black hole (wrong) and results in the creation of high-energy particles (wrong).

Let me try to give an intuitive explanation to those of you who know some quantum physics. QFT describes particles (photons, electrons, quarks etc.) as excitations of quantum fields. If you spatially constrain the quantum field, you limit the possible excitation. Think of a vibrating guitar string. Shorten the string (by sticking your finger on the fret) and the vibration pattern changes. The result is a higher note.

Compare this with what happens to an electron when you spatially constrain it near the positively charged nucleus of an atom. The QFT field excitation, which we call an electron, must fit within the spatial constraint. Longer wavelengths are not possible. The tighter the constraint, the shorter the wavelength of the excitation and the higher the energy. Removing the longer wavelength options means that the constrained electron has higher energy and its energy levels are quantised.

One of the weirder features of quantum theory is that the development of QFT fields, and therefore the behaviour of particles, is calculated by taking into account all the ways the field could develop. Each possible development pathway is weighted according to its *action*, which depends on the *Lagrangian* of the field. The action of each possible outcome affects it. This leads to an interaction between pathways similar to that between waves in general. Some possible outcomes constructively interfere and others destructively interfere. It turns out that the overall development of the quantum field tends to resemble the path of least action, but the underlying truth is that it is a weighted amalgam of all the possibilities.

Any of you who have read my book *Quantum Untangling* should dredge your memory for the chapter on Feynman's Path Integral, which uses the example of electrons travelling from *A* to a screen at *B*. If we put a barrier between *A* and *B* with two tiny slits (the *Double-slit* experiment), it creates an interference pattern on the screen. The point of arrival of each electron depends on both available pathways. Open a third slit and the interference pattern changes to reflect the extra development pathway. The physicist Richard Feynman chased this argument all the way down the following rabbit hole. What if we add more and more slits until there are an infinite number? That is the same as no barrier. Therefore, the path of any particle is really a weighted amalgam of *all* development pathways. He then showed how these pathways add up such that they all cancel out except those closest to the straight-line path from *A* to *B*. We perceive the particle as going in a straight line from *A* to *B* but in reality it is affected by all possible pathways... anywhere in space, anywhere in time. What a mind-blowing illusion.

Stephen Hawking knew all this. He wondered what happens to Feynman's Path Integral calculation in the presence of an event horizon. Development pathways into the event horizon are blocked because there is no way back out. This must change the cancelling balance with other pathways away from the event horizon, creating an asymmetry. He calculated what this means for the vacuum near to an event horizon. We know from QFT that the vacuum is not a void but has a minimal excitation level. If there is an event horizon for an observer, that observer will see the vacuum vibrations of the quantum field constrained by it. Hawking did a long (very complicated) calculation and showed that to this observer, the vacuum appears slightly higher energy. The imbalance in the available pathways leads to low-energy radiation... Hawking radiation.

Note carefully that Hawking radiation is *observer-dependent*. It is a result of the presence of an event horizon. The event horizon of a black hole is a *coordinate singularity* (see Section 13.6). For us observing from afar there is an event horizon, so we observe the Hawking radiation. The same is true of an astronaut stationary near to the event horizon (with rockets at full blast to maintain her position). However, things are different for an astronaut in freefall into the black hole. For this observer, there is no event horizon. There is no constraint on the development pathways. There is no Hawking radiation.

How does all this square with Hawking's description in his book of particle–antiparticle formation at the event horizon? Hawking tried to simplify by picturing the loss of available development pathways as negative energy particles crossing the event horizon into the black hole, while their positive energy counterparts escape to infinity. In fact, the emissions are low-energy photons throughout the region from the event horizon up to 10–20 times the Schwarzschild radius out. Hawking was a genius, but on this occasion his imagery does not stack up.[2]

19.7.1 Taking the Temperature of a Black Hole

Hawking's herculean mathematical calculation is well beyond our scope. However, I want to share a short-cut calculation of the relationship between the temperature of a black hole (based only on its Hawking radiation) and its size. This largely is an excuse to introduce *dimensional analysis*. If this is new to you, take a look at Box 19.5.

---

2 For more on this, check out Forbes article: https://www.forbes.com/sites/startswithabang/2020/07/09/yes-stephen-hawking-lied-to-us-all-about-how-black-holes-decay/ (as of February 2025).

Box 19.5 Dimensional analysis

Dimensional analysis is a fancy term for matching up the dimensions (or units) in an equation. Imagine that we are looking for the relationship between energy E and mass m and the speed of light c. The dimensions of a quantity are shown with the symbol $[x]$. The dimensions of energy are: $[E] = \frac{ML^2}{t^2}$ where M is mass, L is length and t is time. For the dimensions to match, the relationship between E, m and c must be of the form: $E \propto mc^2$. It could not, for example, be $E \propto mc^3$. If you calculate a relationship and the dimensions do not match, it is a sure sign that you have made a mistake.

$$[E] = \frac{ML^2}{t^2}, \qquad\qquad [m][c]^2 = M\left(\frac{L}{t}\right)^2 = \frac{ML^2}{t^2}.$$

I will work with a Schwarzschild black hole (no charge, no rotation) for simplicity. Things are further simplified by the no-hair theorem: the characteristics of a Schwarzschild black hole and its event horizon are captured in full by the Schwarzschild radius R_S. Therefore, the temperature of the Schwarzschild black hole must be a function of R_S, but what function? We ask ourselves what universal constants could make sense in this relationship. R_S captures the essence of the event horizon. This distorts the pathways resulting in a change in energy of the vacuum and consequent radiation. One example of a relationship between radiation and temperature is Planck's law of black-body radiation, which involves the Boltzmann constant k_B, the speed of light c and the Planck constant h (shown here reduced as $\frac{h}{2\pi}$, which is the usual form for quantum physics).

We can use dimensional analysis to deduce how these factors might combine in the relationship between the temperature of the black hole and R_S. First we create a relationship with unknown powers of each labelled arbitrarily α, β, γ and δ and lay out the dimensions of each factor as shown in Equation 19.14 where I have used T^o as the symbol for degrees of temperature, L is length (as is the dimension of R_S), M is mass and t is time. In Equation 19.15 the dimensions are shown in terms of the unknown variables. We know that the dimensions must match for the relationship to make sense. For the temperature units to match, $\alpha = -1$. For the mass units to match, $\alpha + \gamma = 0$, which means $\gamma = 1$. Applying these to the units of time, $-2\alpha - \beta - \gamma = 0$ gives $\beta = 1$. Feeding these values into the units of length gives $\delta = -1$.

$$Temp \propto [k_B]^\alpha [c]^\beta [\hbar]^\gamma [R_S]^\delta, \quad [k_B] = \frac{ML^2}{T^o\, t^2}, \quad [c] = \frac{L}{t}, \quad [\hbar] = \frac{ML^2}{t}, \quad [R_S] = L, \tag{19.14}$$

$$T^o \propto \left[\frac{ML^2}{T^o\, t^2}\right]^\alpha \left[\frac{L}{t}\right]^\beta \left[\frac{ML^2}{t}\right]^\gamma [L]^\delta \propto [T^o]^{-\alpha}[M]^{\alpha+\gamma}[t]^{-2\alpha-\beta-\gamma}[L]^{2\alpha+\beta+2\gamma+\delta}, \tag{19.15}$$

$$\implies \quad \alpha = -1, \quad \beta = 1, \quad \gamma = 1, \quad \delta = -1.$$

This simple piece of dimensional analysis suggests the relationship on the left of Equation 19.16. The actual result as calculated by Stephen Hawking is shown on the right. It is surprising how far you can get with dimensional analysis. If you want the result in terms of the mass of the black hole, simply convert with: $R_S = \frac{2GM}{c^2}$.

$$Temp \propto \frac{\hbar c}{R_S\, k_B}, \qquad\qquad Actual\ result: \quad Temp = \frac{\hbar c}{4\pi\, R_S\, k_B} = \frac{\hbar c^3}{8\pi\, k_B\, GM}. \tag{19.16}$$

The temperature due to Hawking radiation is *very* small for black holes that might form from a supernova. The smallest such black hole is about three solar masses, which has R_S of about ten kilometres. Based on Equation 19.16, this gives a Hawking radiation temperature of 2×10^{-8} Kelvin. Typically, this would be completely overwhelmed by emissions from the accretion disk. Theoretically the black hole could shrink, but the universe is bathed in what is called the cosmic microwave background radiation (CMB), which has a spectrum of radiation equivalent to a temperature of 2.7 Kelvin. We will discuss the CMB in detail later, but the point here is that the energy of this background radiation far offsets the energy loss of Hawking radiation for this sort of black hole, even if it has no accretion disk feeding it.

However, note that the Hawking radiation *increases* as the mass of the black hole decreases. In theory, a very small black hole will radiate strongly and finally disappear in a massive release of energy. Nothing of this sort has ever been detected. Furthermore, with no established theory of quantum gravity, Hawking's maths is a work-around. This being said, several different approaches to combining the maths of GR and quantum physics give similar results. I think it is fair to say that the existence of Hawking radiation is generally accepted as a scientific truth albeit the radiation may never be directly measurable.

19.8 Summary

The primary topic of this chapter is GW radiation. I used a simplified model of how GW radiation might appear at a great distance by considering a small perturbation of the Minkowski metric. The result is a perturbation that bears much in common with a transverse wave moving at light speed. It has two possible polarisations: ($h_{XY} = h_{YX}$) and ($h_{XX} = -h_{YY}$). All this fits well with quantum theory's predictions for a graviton particle. Although this simple picture encouraged physicists, it took many years to establish that GW radiation is consistent with the full details of the EFEs. Even Einstein flip-flopped on this.

We then discussed gravitational wave observatories, particularly the detection of GW radiation by LIGO. As an example, I explained how the first detected chirp GW150914 can be analysed based on the relationship between the radius of the black hole ISCO and the frequency of the GW radiation just before the radial plunge. Chirp GW170817 was another important milestone as EM radiation observatories were able to confirm LIGO's GW detection by measuring EM activity in the same region of the sky.

The observation of GW radiation offers a whole new tool for research (see Hawking's comment in Box 19.6). In addition to LIGO, several further observatories are either on stream or in development. Added to this is the prospect of observations from Pulsar Timing Arrays that can detect much longer wavelength GW radiation and may offer a tantalising glimpse into the distant past, revealing information about a period outside the range of EM radiation observatories.

Box 19.6 Hawking on GW observatories

Stephen Hawking pointed out the importance of GW observations:
Gravitational wave observations let us test general relativity in situations where a gravitational field is strong and highly dynamical. Some people think that general relativity needs modifying in order to avoid introducing dark energy and dark matter. Gravitational waves are a new way to search for a signature of possible modifications of general relativity. A new observational window on the universe typically leads to surprises that cannot yet be foreseen. We are still rubbing our eyes, or rather ears, as we have just woken up to the sound of gravitational waves.

I finished the chapter with a note on Hawking radiation. Stephen Hawking's analysis combines the maths of general relativity with that of quantum theory. Although we have no quantum theory of gravity, he was able to piece together enough maths to develop a picture of how the presence of a black hole event horizon would alter the quantum fields in the surrounding vacuum. The behaviour of a quantum field is the result of a complicated interaction between all the available development pathways (i.e. of every possibility). These pathways constructively and destructively interfere. The event horizon constrains some pathways and thus alters the behaviour of the field. Hawking showed that this results in a faint radiation from black holes. The smaller the black hole, the greater the radiation.

In Chapter 20, I will review the material in this module on *Vacuum Curvature*. And then we will move on to the module about *Cosmology*.

Chapter 20

Module Summary: Vacuum Curvature

In the first module in this book, *The Essentials*, we studied the complicated structure of the Einstein Field Equations (EFEs) and quantified the relationship between the Ricci tensor $[R_{\mu\nu}]$, the Ricci scalar R, the spacetime metric $[g_{\mu\nu}]$, the energy-momentum tensor $[T_{\mu\nu}]$, some constants and Einstein's fudge factor, the cosmological constant Λ.

After all that work, this module on *Vacuum Curvature* has been based on a much simpler relationship: $[R_{\mu\nu}] = 0$. We can make this simplification because if there is no energy-momentum ($[T_{\mu\nu}] = 0$) and we ignore the cosmological constant, which is negligible except at vast cosmological scales, then $[R_{\mu\nu}] = 0$ (as shown in Section 9.5.1).

20.1 Schwarzschild Metric

In Chapter 12, I introduced you to the Schwarzschild metric, which is the metric of the vacuum surrounding a time-independent (i.e. unchanging) un-charged stationary non-rotating spherically symmetric mass. This was discovered within a few months of the publication of the EFEs, much to Einstein's surprise.

I showed you how the combination of spherical symmetry with $[R_{\mu\nu}] = 0$, leads to only one possible pattern of curvature. Using this pattern and making a few calculations took us to one unique solution: the Schwarzschild metric.

The unique nature of the Schwarzschild metric makes it a particularly powerful tool. As the only available solution, it must apply to the spacetime metric of an extreme gravitational field as well as to weak fields. In Chapter 13, we studied the implications of this, especially for black holes.

We looked at the event horizon, $R_S = \frac{2GM}{c^2}$, of a Schwarzschild black hole. This is the point where nothing, not even light, can escape. It would require an infinite amount of acceleration to hover at the event horizon.

Initially, physicists were sceptical of the physical existence of black holes. However, Chandrasekhar showed that supernovae above a certain size collapse to neutron stars. And there is a clear limit to the possible size of a neutron star beyond which it forms a black hole (Section 13.5). We now know there are black holes at the centre of most galaxies, including our own Milky Way with its black hole Sagittarius A^* of about four million solar masses (see Box 20.1 for more background on black holes).

Box 20.1 Black hole facts and figures (as of February 2025)

The *first detection* of a black hole was in 1964 due to the strong emissions of X-rays from its accretion disk. The source was identified as a bright blue star orbiting what appeared to be a strange dark object. This is the black hole Cygnus X-1 located in the constellation of Cygnus in our Milky Way. We don't know how many black holes there are in the Milky Way, but current guesstimates are as many as 100 million.

The *first image* of a black hole was of M87 in 2019 by the Event Horizon Telescope, shown on the left of Figure 13.1.

As of February 2025, the *nearest* known black hole to earth is Gaia BH1, about 1,500 light years away. The *furthest* (i.e. most distant) detected is in galaxy QSO J0313-1806, around 13 billion light years away. The *most massive* black hole so far detected is TON 618 estimated at 66 billion solar masses, while the *least massive* so far detected is only 3.8 solar masses.

Untangling General Relativity: The Intuitive Self-Study Guide, First Edition. Simon Sherwood.
© 2026 John Wiley & Sons Ltd. Published 2026 by John Wiley & Sons Ltd.

Weird things happen at the event horizon. I compared the perspectives of a distant observer (Minky) with that of an astronaut freefalling into the black hole (Schwart). From Minky's perspective, Schwart's time slows exponentially and in theory stops altogether at the horizon (Section 13.7). Also, near to the horizon radial distances are distorted (Section 13.9). As a result, Minky never sees Schwart cross the event horizon. The perspective of Schwart is very different. The event horizon is a coordinate singularity not a curvature singularity. For Schwart, there is no sudden change at the event horizon. He can cross it blissfully unaware of his impending doom... until he is spaghettified by tidal forces.

Chapter 14 explored the conservation laws associated with the Schwarzschild metric. The metric doesn't depend on t or ϕ, which means the quantities u_t and u_ϕ are conserved along geodesics. I showed you how the conservation of u_t leads to the same definition of gravitational potential energy for a radial plunge as was calculated by Newton, but the conservation of u_ϕ adds an additional angular momentum term to the effective potential.

The change is negligible at large distances from the attracting mass but radically alters orbital behaviour near the event horizon. There is an Innermost Stable Circular Orbit (ISCO) at $3R_S$. When gas in the accretion disk crosses the ISCO, its motion changes to a radial plunge down into the black hole. This is one of the factors that allows LIGO to deduce the relationship between the gravitational waves detected and the size of the black holes involved.

Chapter 15 was a trip down memory lane, revisiting a couple of calculations that proved important in building the credibility of Einstein's theory. The first was his prediction that the deflection of light by the sun's gravitational field is twice that predicted based on Newtonian mechanics. The second was the unexplained 43 arcseconds of precession of Mercury's orbit. I stepped through the calculations using the Schwarzschild metric.

In Chapter 16, I introduced some coordinate systems that allow us to peer inside the Schwarzschild event horizon. Eddington–Finkelstein (EF) coordinates use the emission times of incoming or outgoing light beams to replace one of the coordinates. The emission time does not change in crossing the event horizon, eliminating the coordinate singularity. The EF *in* coordinates show that objects, from their perspective, cross the event horizon and enter the black hole. The EF *out* coordinates reveal a different part of the Schwarzschild metric. As the metric is time-independent, it is equally valid with time running forward or backward. Instead of the black hole attracting everything inwards, the maths points to the option of a white hole repelling everything from it.

The combined black-hole-white-hole picture can be shown with Kruskal–Szekeres (KS) coordinates that replace both the original Schwarzschild time and radial coordinates with those related to ingoing and outgoing light beams. This leads to the maximally extended Schwarzschild metric with a black hole and white hole linked together in perpetuity along with a parallel universe. I warned (and now warn again) that this may be more mathematics than physical reality. Indeed, the white hole is only part of the Schwarzschild solution if the black hole is eternal (i.e. has always existed).

20.2 Kerr Metric

The Kerr metric is the equivalent of the Schwarzschild metric but for the vacuum around a *rotating* mass. I should add that there are adjusted versions of the Schwarzschild and Kerr metrics if the attracting mass is *electrically charged*. This gives the four different metrics in Table 20.1. However, for black holes, the most important is the Kerr metric because all black holes are expected to rotate and there is no reason to expect them to have a significant net electric charge.

The Kerr metric is somewhat intimidating at first sight, so I devoted the whole of Chapter 17 to an intuitive introduction. I explained how the presence of angular momentum in the energy-momentum tensor of the rotating mass leads to an off-diagonal term in its internal spacetime metric. This in turn leads to the off-diagonal term $g_{T\phi}$ in the Kerr metric of the surrounding vacuum.

Table 20.1 Vacuum solutions: Schwarzschild, Kerr and associated electrically charged versions.

Metrics	Static	Rotating
No charge	*Schwarzschild*	*Kerr*
Electrically charged	*Reissner–Nordström*	*Kerr–Newman*

I tried to give you an intuitive rationale for why the symmetry of the metric switches from spherical (as in the Schwarzschild metric) to an oblate spheroidal symmetry. I also went into mathematical detail of how we use Boyer–Lindquist (BL) coordinates to reflect this symmetry and the changes this creates in the metric.

This launched us into the discussion of Kerr black holes in Chapter 18. The Kerr black hole has an outer event horizon, which is somewhat similar to the event horizon of the Schwarzschild metric in that distant observers will not see anything cross it and the radial coordinate switches from space-like to time-like (its value switches from negative to positive in terms of the invariant interval).

The Kerr outer event horizon radius is smaller than the Schwarzschild equivalent. The greater the angular momentum J, the smaller the radius until when the spin is such that $|a| > GM$ there is no horizon at all. This creates a paradox. If a black hole can have $|a| > GM$, then its singularity would be observable from afar. But how can a naked singularity morph smoothly into well-behaved spacetime if there is no event horizon to hide the ugly mathematical discontinuity? Roger Penrose's solution is the cosmic censorship conjecture, which simply states that $|a|$ must always be less than GM. This is unproven and controversial. I offered an optional mathematical section at the end of the chapter to highlight some of the mathematical logic behind Penrose's position (Section 18.12).

The underlying Riemann curvature of the Kerr metric shows that the Kerr event horizon is a coordinate singularity (as is true of the Schwarzschild event horizon). The only curvature singularity of the Kerr metric is at $(\mathbf{r}^2 + a^2\cos^2\theta) = 0$. This equation reveals that the Kerr singularity is a *ringularity*; a ring oriented equatorially. Using spherical coordinates, this is at radius $r = a$.

The interplay of the Kerr event horizons and ringularity permits some intriguing wiggle room: could an object fall into a Kerr black hole and miss the singularity? At the outer event horizon the radial coordinate switches from space-like to time-like (negative to positive invariant interval value). Anything crossing the outer event horizon is led inexorably inwards. Nothing can escape back out. However, there is also an inner event horizon inside the outer one. When an object crosses this inner event horizon, the mathematics indicates that the radial coordinate reverts to being space-like (negative value in terms of the invariant interval). Many physicists dismiss this as a mathematical anomaly. However, might it mean that objects falling into the Kerr black hole are not constrained to hit the ringularity? Might they orbit the ringularity or even pass through the centre of it?

If true (with a big *if*), then the maximal mathematical extension of the Kerr metric raises the theoretical possibility that the ringularity could act as a wormhole out through a white hole and on to alternate universes. But don't get too excited. First, this is highly speculative and second, if Penrose's conjecture is correct, we will never be able to see behind the event horizon.

Another interesting feature of the Kerr metric is frame-dragging and the ergosphere. The presence of the non-zero off-diagonal $g_{T\phi}$ $(= g_{\phi T})$ component of the metric means that the time t coordinate and the ϕ coordinate are no longer orthogonal. I used the example of a ZAMO (zero angular momentum object) approaching the black hole on a direct radial descent. The off-diagonal term means that the ZAMO will pick up angular velocity as it freefalls along a geodesic towards the black hole. This is frame-dragging. At longer ranges, it falls off with $\mathbf{r}^3$ so it becomes hard to detect. However, it reaches an extreme at what is called the ergosphere within which nothing, not even light, can resist the rotating effect.

Penrose (him again) showed how, in theory, frame-dragging might allow rotational energy to be extracted from a rotating black hole (the Penrose process). Blandford and Znajek built on this idea proposing that this frame-dragging energy source, combined with the energy release of objects descending into the black hole, fuel particle emissions from the poles of the black hole. These are accelerated by twisted magnetic field lines, and the result is the intense radiation of quasars.

20.3 Gravitational Waves and Hawking Radiation

This brings us to Chapter 19 on gravitational waves. It took 40 years from Einstein's GR publication for physicists to determine mathematically that gravitational waves are consistent with the full EFEs. I showed you that transverse GW radiation at the speed of light fits neatly with a highly simplified GR metric (faintly perturbed Minkowski spacetime). What pops out of the model is light-speed GW radiation with two distinct polarisations. This is exactly what is expected from quantum theory.

The first detection of gravitational waves created much excitement (see Box 20.2). In Section 19.4, I tried to give you a flavour of how GW radiation data can be interpreted, particularly the importance of the change in frequency as the trajectory switches from orbital to a radial plunge.

LIGO and other observatories offer a new way of looking at the universe. GW radiation is much less interactive than EM radiation. While the EM radiation is affected by much en route, the GW radiation can be observed pretty much in original form. In addition to the higher frequency GW radiation detectable by GW observatories on earth, there is now the exciting potential to observe lower frequency GW radiation using pulsar timing arrays. GW radiation should teach us a lot in the next few years.

Box 20.2 LIGO's detection: quotes

Ladies and gentlemen, we have detected gravitational waves. We did it!
David Reitze, Caltech physicist and LIGO lab director

It will give us ears to the universe where before we've only had eyes.
Karsten Danzmann, LIGO collaboration member

The universe has spoken and we have understood.
David Blair, astrophysicist and LIGO collaboration member

I keep telling people I'd love to be able to see Einstein's face right now!
Rainer Weiss, inventor of the interferometric gravitational wave detector and LIGO co-founder

---

I also touched on Hawking radiation. This is not GW radiation. It is predominantly electromagnetic (photons). You might describe it as a side effect of the black hole event horizon removing some of the possible ways that quantum fields can develop. To some extent, the quantum fields in the surrounding vacuum are constrained. Hawking demonstrated mathematically that this results in a very faint radiation from black holes. As an exercise in dimensional analysis, I showed what this means in terms of temperature. Unless the black hole is very small (smaller than any discovered to date), this is only a few billionths of a Kelvin.

20.4 Module Memory Jogger

Below is a memory-jogger of some of the key topics in this module. And Box 20.3 provides some further resources.

- *Schwarzschild metric:* $\begin{bmatrix} 1 - \dfrac{2GM}{c^2 r} & 0 & 0 & 0 \\ 0 & -(1 - \dfrac{2GM}{c^2 r})^{-1} & 0 & 0 \\ 0 & 0 & -r^2 & 0 \\ 0 & 0 & 0 & -r^2 \sin^2\theta \end{bmatrix}$, (ct, r, θ, ϕ).

- *The Schwarzschild metric is the only (electrically neutral) static spherically symmetrical vacuum solution, so we must take seriously what it tells us about intense gravitational fields.*
- *Schwarzschild radius (event horizon):* $R_S = \dfrac{2GM}{c^2}$.
- *From the perspective of a distant observer, nothing crosses the event horizon because time stops and it takes an infinite amount of progress to move radially $g_{TT} = 0$, $g_{RR} = \infty$.*
- *From the perspective of an observer in freefall, there is no singularity at the event horizon. She crosses the event horizon and falls into the black hole.*
- *Schwarzschild metric is not dependent on t or ϕ, so u_t and u_ϕ are conserved along geodesic paths.*
- *Schwarzschild effective potential per kilogram:* $U_{eff} = \dfrac{\ell^2}{2r^2} - \dfrac{\ell^2\,GM}{c^2 r^3} - \dfrac{GM}{r}$.
- *Schwarzschild ISCO:* $3R_S$, *Photon sphere:* $\dfrac{3}{2}R_S$.
- *Light deflects twice as much as predicted by classical theory:* $\alpha = \dfrac{2\,R_S}{b} = \dfrac{4GM}{bc^2}$.

- Mercury's missing 43 arcseconds of precession explained: $\Delta\phi_p = \dfrac{6\pi\,GM\,J_0}{c^2}$.
- Using EF and KS coordinates to remove the event horizon coordinate singularity reveals that for an eternal black hole, the Schwarzschild metric theoretically could include an associated white hole and alternate universe.
- Penrose-Carter diagrams can be used to display the entirety of space and time.

- Kerr $(c = 1)$:

$$\begin{bmatrix} 1 - \dfrac{2GM\,\mathbf{r}}{p^2} & 0 & 0 & a\sin^2\theta\,\dfrac{2GM\,\mathbf{r}}{p^2} \\[2ex] 0 & -\dfrac{p^2}{\mathbf{r}^2 + a^2 - 2GM\,\mathbf{r}} & 0 & 0 \\[2ex] 0 & 0 & -p^2 & 0 \\[2ex] a\sin^2\theta\,\dfrac{2GM\,\mathbf{r}}{p^2} & 0 & 0 & -\left(\mathbf{r}^2 + a^2 + a^2\sin^2\theta\,\dfrac{2GM\,\mathbf{r}}{p^2}\right)\sin^2\theta \end{bmatrix}.$$

- Kerr metric has oblate spheroidal symmetry and therefore is expressed above using BL coordinates.
- Note: $a = \dfrac{J}{M} \equiv \dfrac{Angular\ Momentum}{Mass}$, $p^2 = \mathbf{r}^2 + a^2\cos^2\theta$, $(c = 1)$.
- The interior spacetime metric of a rotating mass must include a non-zero off-diagonal term, so the non-zero $g_{T\phi}$ term in the Kerr metric is expected based on boundary conditions.
- Frame dragging on zero angular momentum object (ZAMO): $\dfrac{d\phi}{dt} \approx \dfrac{2aGM}{\mathbf{r}^3}$ for $\mathbf{r} \gg a$.
- Ergosphere inside which everything must rotate: $\mathbf{r} = GM \pm \sqrt{(GM)^2 - a^2\cos^2\theta}$.
- Penrose process: rotational energy can be extracted in theory from the rotating black hole.
- Kerr event horizons: $\mathbf{r} = GM \pm \sqrt{(GM)^2 - a^2}$.
- At the outer horizon, the radial coordinate becomes time-like and nothing escapes (similar to Schwarzschild event horizon).
- No horizon exists if $|a| > GM$. Penrose conjectures this is impossible (cosmic censorship).
- Kerr singularity is a ring (ringularity): $\mathbf{r} = 0$, and $\theta = \dfrac{\pi}{2}$, $r = a$ in spherical coordinates.
- The meaning of the inner event horizon is not clear. The radial coordinate switches back to space-like. Might it be possible for something to travel through the centre of the ringularity?
- As a result, the maximally extended Kerr metric includes wormhole pathways through the Kerr black hole to white holes and multiple parallel universes.
- GW radiation: transverse waves at light speed with two polarisations: $\dfrac{1}{c^2}\dfrac{\partial^2 h}{\partial t^2} - \dfrac{\partial^2 h}{\partial z^2} = 0$.
- First gravitational wave detection: the merger of two black holes, Chirp GW150914.
- GW radiation provides a new view of the universe: LIGO, Virgo, further observatories planned, plus data from NANOGrav and pulsar timing arrays.
- Hawking radiation for the black holes we know (3 solar masses or higher) is equivalent to only a few billionths of a Kelvin, but the smaller the black hole the higher the radiation.
- Hawking radiation temperature equivalent: $\dfrac{\hbar c}{4\pi R_S k_B} = \dfrac{\hbar c^3}{8\pi k_B GM}$.

Box 20.3 Module 2: further resources

A great classic is Kip Thorne's book, *Black Holes and Time Warps: Einstein's Outrageous Legacy*. It avoids the mathematics, but still has lots of insight.

PBS offers some short videos that are well worth a look:
https://www.youtube.com/@pbsspacetime/videos
(as of February 2025).

Module III

Cosmology

Chapter 21

The Friedmann–Robertson–Walker Metric

Box 21.1 The mystery of the night sky

Sherlock Holmes and Dr Watson go on a camping trip, set up their tent and fall asleep. Some hours later, Holmes wakes his faithful friend. *Watson, look up at the sky and tell me what you see.* Watson replies, *I see millions of stars.* Holmes asks, *What does that tell you?* Watson ponders for a minute. *Astronomically speaking, it tells me that there are millions of galaxies and potentially billions of stars and planets. Astrologically, it tells me that Saturn is in Leo. Time wise, it appears to be approximately a quarter past three. Meteorologically, it seems we will have a beautiful day tomorrow. What does it tell you?* Holmes is silent for a moment, then speaks. *Watson, you idiot, someone has stolen our tent*!

In the module on *Vacuum Curvature,* we looked at the spacetime metric surrounding a single mass. It is time to expand our horizon. Look up at the night sky (see Box 21.1 for a laugh). You don't need a telescope to see the Milky Way, which contains about 100 billion (10^{11}) stars plus possibly 100 million black holes. And the Milky Way is just one of about a trillion (10^{12}) galaxies in the observable universe (some estimates put the number even higher). My wife Rachel describes this as *mind-boggling-oggling* in size.

Throughout this module I will set the speed of light to $c = 1$, unless otherwise indicated.

21.1 The Cosmological Principle

In this module on *Cosmology,* we will explore what the Einstein field equations (EFEs) tell us about the development of the wider universe. This presents a different challenge from that of the vacuum metrics. For one thing, the universe clearly is *not* a vacuum, so the metric is *not* Ricci-0. For another, we are interested in the development of the universe, so the metric must be *time-dependent.* That wipes out a lot of the useful symmetries that helped us determine the Schwarzschild and Kerr metrics. In order to make progress with the peskily-complicated EFEs, cosmologists have to simplify things with broad new assumptions about the symmetry of the universe. Let's get started.

The simplifying symmetries used for cosmology are called *the cosmological principle* and can be summarised as:

- The universe is *homogeneous*: it is the same at every location.
- The universe is *isotropic*: it is the same in every direction.

This deserves some explanation. When you look up at the sky, things appear neither homogeneous nor isotropic. There is the Milky Way in one direction, which rather upsets the notion of isotropy. Nor is our solar system homogeneous. There are some big round lumps like the earth and moon, to say nothing of the even bigger lump called the sun. Conditions on the surface of the sun and earth are not similar, not at all. The patch of universe near to us is *not* the same everywhere *nor* in every direction. What are the cosmologists thinking?

It all comes down to scale. In cosmology, we zoom out to encompass the whole universe. At this range, our solar system and its irregularities become insignificant. The whole Milky Way galaxy is like a speck of dust alongside a

Untangling General Relativity: The Intuitive Self-Study Guide, First Edition. Simon Sherwood.
© 2026 John Wiley & Sons Ltd. Published 2026 by John Wiley & Sons Ltd.

trillion other such specks. Cosmologists argue that at this range the universe today is like a homogeneous isotropic dust field, with each speck of dust being a galaxy. Thinking back to the early universe, it was a hot plasma of interacting radiation and particles, which cosmologists view as a homogeneous fluid. Indeed, later we will use the *perfect fluid* model for the energy-momentum tensor (Section 8.7) in analysing its development.

21.2 The Hubble Parameter

Let me quickly summarise the discovery of the expansion of the universe. In 1929, the work of Edwin Hubble and others revealed that the universe is expanding. It all started with Henrietta Leavitt (Box 21.2) who found a direct relationship between the period of dimming of a certain type of star (Cepheid variable) and its luminosity (its actual brightness close-up). This created what is called a *standard candle*. Astronomers could observe the periodic dimming of a Cepheid variable in a galaxy, determine its brightness based on Leavitt's work and then calculate the distance to the galaxy by comparing it with the Cepheid star's observed brightness from earth.

Box 21.2 Henrietta Swan Leavitt (1868–1921)

Henrietta Leavitt's ground-breaking research began in 1892 when she was a volunteer assistant at the Harvard Observatory. She had studied mathematics and astronomy at the Harvard Annex (now Radcliffe College). At the time, Harvard University didn't accept female students. Her task was to examine the change in brightness of Cepheid variable stars. She studied over 1,700 of these stars and noticed a relationship between their brightness and the time between peaks of brightness.

Plagued by ill-health, she took several breaks from her work and finally published her paper 16 years later in 1908 writing: *It is worthy of notice that the brighter variables have the longer periods.* Cepheid variables now could be used as a standard candle. Famously, this allowed Edwin Hubble to compare distance away with rate of recession (determined by redshift) and announce in 1929 that the universe is expanding. Prior to this in 1924, Leavitt's Cepheids allowed Andromeda to be the first galaxy identified as outside the Milky Way.

Tragically, Leavitt never learned the importance of her work. In 1925, the Swedish mathematician Mittag-Leffler attempted to nominate her for the 1926 Nobel Physics Prize. She would have been the first female nominee since Madame Curie won it in 1903, but she died of stomach cancer in 1921, only 53 years old.

It was Harlow Shapley, the new Director of the Harvard Observatory, who informed Mittag-Leffler of her death. Having used Leavitt's discovery in papers of his own, he attempted to claim the credit for her work, in the hope that he would be nominated in her place! Pah!

Hubble compared the calculated distance away of galaxies with the redshift in some of their electromagnetic line emissions. For example, hydrogen has a specific emission pattern (the Lyman, Balmer and Paschen series). The redshift is caused by the Doppler effect and reveals how fast a galaxy is moving away from earth. Although Hubble's standard candle was out by a factor of seven, the pattern was clear. On average, the further the galaxy from earth, the faster it is receding.

The relationship between the two is called the *Hubble parameter* or H (I prefer this to calling it the Hubble constant because it changes over time). By modern estimates, the current Hubble parameter H is about 70 km s^{-1} per megaparsec (Mpc), i.e. two galaxies separated by a megaparsec move away from each other at 70 km s^{-1}. There are various different estimates. For consistency, I will use the value 70 as the basis for calculations throughout this book. A megaparsec is a traditional astronomical measure of distance equivalent to 3.26 million light years. Historical it may be, but sometimes I wonder if there is a committee of physicists who work out the most contorted possible measures in order to confuse students. A more intuitive description is that at this current rate, spatial distances grow by about 7% every billion years. Another way to describe the same thing is that a metre of space grows by about 10^{-10} metres per annum equivalent to the diameter of an atom. This expansion does not affect us or the solar system because it is far outweighed by greater forces, as I will explain in more detail later.

If we label the distance of a galaxy away from us as a, then H is the velocity of separation ($\dot{a}$) divided by distance (a). You might think of it as a measure of the doubling rate of the size of the universe. In addition to providing observational data on the current value of $\left(\frac{\dot{a}}{a}\right)$, Hubble woke the scientific community to the expansion of the universe and the need to explore its relationship with the curvature of spacetime.

$$Current\ Hubble\ parameter:\quad H = \frac{\dot{a}}{a} \approx 70\,\mathrm{km\,s}^{-1}\,\mathrm{Mpc}^{-1} \approx 0.07\ per\ billion\ years. \tag{21.1}$$

21.3 The Expanding Universe: A Newtonian View

Let me start from a Newtonian perspective. I warn in advance that this is the *wrong* way to describe things. However, it is a good launching point because it is familiar. Dear old Isaac knew, based on his law of gravity, that the universe could not be static. Consider two nearby stars. By the law of gravity, the two attract each other. They accelerate towards each other. How does the distance between them change? That depends on their relative velocity. But there is no scenario that results in the stars remaining static relative to each other.

Let me illustrate with the example of you throwing this book up into the air (please don't). Give the book a casual toss into the air and you'd better watch your head, because it is going to come back down. However, throw it up so fast that its velocity *v* exceeds the *escape velocity* from the earth's surface, and it will escape the clutches of the earth's gravitational field.

- *v < escape velocity*: Book returns to earth with a crunch.
- *v = escape velocity*: Book continues to separate from the earth, but the speed of separation slows (theoretically at infinite distance apart, the book finally would be stationary relative to the earth, although it will never actually reach this point).
- *v > escape velocity*: Book continues to separate at speed from the earth in perpetuity.

There is no scenario that leaves the book stationary relative to the earth, hovering static above you. This is because the book must always accelerate towards the earth. By the same argument, there is no stable scenario that leaves the galaxies of the universe static relative to each other.

Newton realised that his law of gravity doesn't naturally accommodate a static universe, but he was religious and chose to ignore the possibility that it is expanding or contracting. He concluded it is held static by a *voluntary agent* (God). In any case, it would have taken a brave soul to conclude anything else and risk religious disapproval... especially after what happened to Galileo 40 years earlier (see Box 21.3).

Box 21.3 Galileo and the Inquisition

In 1632, Galileo Galilei (1564–1642) published the *Dialogue Concerning the Two Chief World Systems* comparing the heliocentric theory of Copernicus with the prevalent Ptolemaic view (earth at the centre of the universe). Several of his books had already been censored. He was ordered to Rome by the Inquisition and declared guilty of *vehement suspicion of heresy*. He then was forced to renounce his assertion that the earth moves around the sun, reciting the following:

I have been judged vehemently suspect of heresy, that is, of having held and believed that the sun is in the centre of the universe and immovable, and that the earth is not at the centre of same, and that it does move. Wishing however, to remove from the minds of your Eminences and all faithful Christians this vehement suspicion reasonably conceived against me, I abjure with a sincere heart and unfeigned faith, I curse and detest the said errors and heresies, and generally all and every error, heresy, and sect contrary to the Holy Catholic Church.

At the end of this Galileo is reputed to have whispered, *Eppur si muove* (and yet it moves). He was put under house arrest and held until his death 10 years later.

21.4 General Relativity View: A Co-Moving Frame

You will have heard of the Big Bang. The universe is believed to have originated from a denser high energy state (perhaps a singularity). Had Newton known of it, I suspect he would have imagined a massive explosion that flung matter out at high speed through space. Cosmologists have a different perspective on what happened in the Big

Bang. Based on general relativity (GR), nothing was flung out through space, nor is there any braking force. What we call gravitational acceleration is the galaxies quietly following spacetime geodesics. Think of the galaxies as moving through time but stationary relative to the surrounding space. Galaxies do have some local motion of their own called *peculiar motion*, but at large scale this is overwhelmed by the expansion of the universe. At this scale, it is *space* itself that is expanding. The galaxies are just carried along for the ride.

We can simulate the expansion of the universe with what is called the Friedmann–Robertson–Walker (FRW) metric. Let's start with a simplified FRW metric: space is flat (later, we will examine scenarios where the space has positive and negative curvature). Imagine a section of the universe with galaxies distributed throughout like specks of dust. We create a three-dimensional (3-D) frame or grid. This is illustrated in two dimensions in Figure 21.1 and is called a *co-moving frame*. The distance between grid lines is distance D, which increases over time. If we say dx is a small measure of length (such as a metre), then the distance between adjacent galaxies is $D = a(t)\,dx$, where a is a dimensionless scale factor. As time passes, a increases and the grid expands carrying along the galaxies (red dots).

We reflect this in the metric by having the spatial dimensions scale with time. If we use Cartesian coordinates, then dx, dy and dz scale up with $a(t)$, which means that we set $g_{XX} = -a^2$, where a is a function of time t. The scaling must be the same in all three spatial dimensions to conform to the cosmological principle, so $g_{YY} = -a^2$ and $g_{ZZ} = -a^2$. This gives the invariant interval and spacetime metric shown in Equation 21.2. As mentioned earlier, please remember that I am using natural units: $c = 1$.

$$d\tau^2 = dt^2 - (a\,dx)^2 - (a\,dy)^2 - (a\,dz)^2,$$

Simplified FRW metric, flat space $(t, x, y, z; c = 1)$: $\qquad [g_{\mu\nu}] = \begin{bmatrix} 1 & 0 & 0 & 0 \\ 0 & -a^2 & 0 & 0 \\ 0 & 0 & -a^2 & 0 \\ 0 & 0 & 0 & -a^2 \end{bmatrix}.$ $\qquad$ (21.2)

The time dependence in the metric means that the overall *spacetime* is *not* Ricci-0. We can examine the curvature in the *spatial* dimensions by choosing a particular moment t in time. In the case of the metric in Equation 21.2, you can see at any given time that $g_{XX} = g_{YY} = g_{ZZ}$, which means the space (as distinct from the spacetime) is Euclidean flat. It remains isotropic (the same in every direction) because the spatial dimensions all expand at the same rate.

By the end of this module, you will have learned that the universe viewed at large scale is approximately spatially flat. Sadly, we cannot leap to that assumption at this stage. We know spatial curvature can and does exist at least locally within the universe. For example, the Schwarzschild metric has spatial curvature. It is this that doubles the deflection of light (see Section 15.1). Therefore, we have to build into the metric the possibility that the universe as a whole has spatial curvature. This complicates the metric and requires the use of spherical coordinates. Sorry!

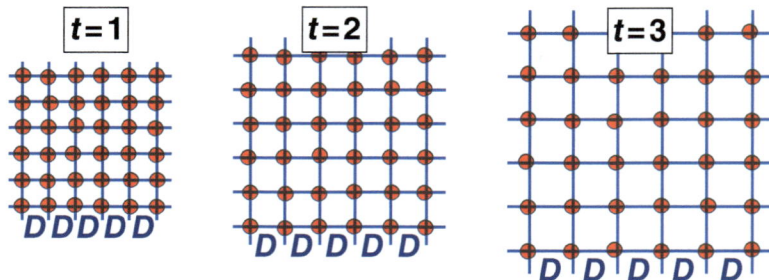

Figure 21.1 Co-moving frame: galaxies (red) co-move with the underlying expansion of space.

21.5   Introduction to 3-D Spatial Curvature

It is important to distinguish the curvature of four-dimensional (*4-D*) *spacetime* from the curvature of *3-D space*. The universe contains energy-momentum so it has *4-D spacetime curvature*. However. it may or may not have *3-D spatial curvature*. We discussed spatial curvature back in Section 3.4 with the two-dimensional (2-D) example of bugs living on the surface of a sphere compared with life on a flat surface. As a reminder, Figure 21.2 shows the three possibilities illustrated in 2-D: positive curvature (like the surface of a sphere), zero curvature (flat) and negative curvature (like the surface of a saddle).

All of this is shown in 2-D, but you need to imagine the curvature applying in 3-D between the three spatial dimensions. In flat space, adjacent parallel geodesics remain parallel and the angles of a triangle add up to the familiar total of π radians (180 degrees). With positive spatial curvature, adjacent parallel geodesics get progressively closer and the angles of a triangle add up to more than π radians. With negative spatial curvature, adjacent parallel geodesics get progressively further apart and the angles of a triangle add up to less than π radians. There is more detail back in Section 7.1 if you need it.

It is impossible to draw curved 3-D space, but let me try to explain what it is. Mathematically, we construct curved 3-D space by embedding it in 4-D space. Consider the 2-D surface of the sphere on the left of Figure 21.2. The bugs happily crawl around in their 2-D world, but the curvature changes things. If they travel in a straight line, they end up back where they started, having gone around the sphere. Imagine now that there is a 4-D sphere. Its surface would be 3-D. This surface is spatially 3-D like our universe. However, if you travel far enough in a straight line through the 3-D space, you end up back where you started because of the positive curvature.

How might we know if our universe has positive curvature? If the curvature were very (very) strong, we would look out and see our own behinds. A test on a 2-D surface could be to measure the sum of the angles in a triangle, which would add up to more than π radians (180°) if there is positive curvature. Alternatively, we could measure the circumference of a circle, which would be less than $2\pi r$. For a 3-D measure of curvature, we could check if the surface area of a sphere is exactly $4\pi r^2$ or its volume exactly $\frac{4}{3}\pi r^3$ (as discussed back in Section 6.6).

A more subtle test is to compare at an instant in time the number of galaxies at different distances (but bear in mind that looking out far across space means looking back in time). If the universe is homogeneous, it has the same density of galaxies everywhere. If it is spatially flat, you would expect to see more galaxies the further you look out because your view encompasses more space. This is not necessarily the case if the universe has positive spatial curvature. The easiest way to visualise this is to think of the bugs on the surface of the sphere in Figure 21.2. Imagine the surface embedded uniformly with 2-D galaxies. The bug sits at the north pole and measures the number of galaxies at different distances. The first measurement might be the first circle down from the pole in the figure. The bug looks further out, reaching the equator. The circle is bigger, so at that range the bug sees more galaxies as you would expect. However, beyond the equator the circle gets smaller, which means it contains fewer galaxies. Looking out at that distance and beyond, the bug will see the density of galaxies decrease.

A different way of thinking about the same thing is to consider the effect of curvature on how large an object appears at distance. Figure 21.3 shows an exaggerated illustration of how the angle subtended by an object changes with spatial curvature. In the centre is flat space with straight-line geodesic light paths (shown in red). The left shows positive spatial curvature where the geodesics followed by the light rays draw together. The right shows negative curvature with the geodesics separating. As you can see, a distant object appears bigger with positive

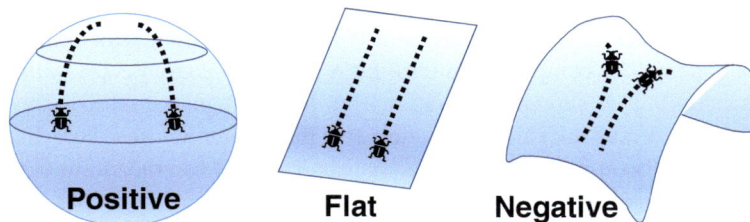

Figure 21.2 Parallel paths draw together with positive spatial curvature. If negative, they separate.

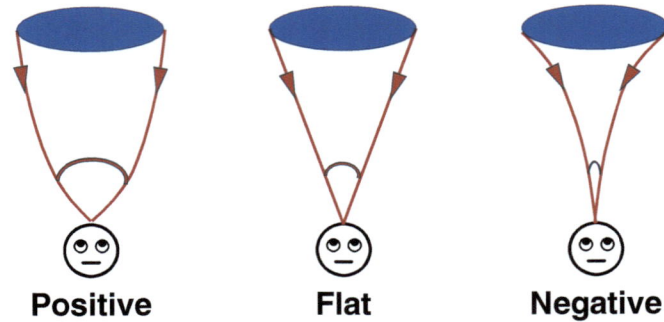

Figure 21.3 Objects at a distance appear larger if there is positive spatial curvature.

spatial curvature. Therefore, if you observe a certain angle of space, the apparent density of galaxies will be lower if there is positive spatial curvature and higher if there is negative spatial curvature.

I hope it is obvious that every spot and whatever direction you move on the 2-D surface of a 3-D sphere is equivalent. The same is true of the 3-D surface of a 4-D sphere, which means it is homogeneous and isotropic. Yippee! That is what we need to conform to the cosmological principle. This also holds for the negatively curved equivalent (trust me on this). Therefore, we can have homogeneous and isotropic 3-D space that is flat, positively curved or negatively curved. These are the only three homogeneous and isotropic possibilities unless you extend to some more exotic topology with a hole such as a universe shaped like a doughnut. We will not bother ourselves with that here.

With this introduction, I will now show you the FRW metric that allows us to model potential spatial curvature in one super-clever metric. From there, we will derive the *Friedmann* equations in Chapter 22, which relate the expansion of the universe with spatial curvature and the amount and type of energy present (matter, radiation, vacuum energy).

21.6 Spatial Curvature in the FRW Metric (Optional)

Introducing *spatial* curvature into the metric complicates things. I have labelled this section optional, because it requires some more advanced tools, including the use of hyperbolic sines and cosines. I will step through things slowly, but please have a nice cup of tea and get your brain in top gear for what is to come. And if you find it too much, just skip to the description of the FRW metric in Section 21.7.

To accommodate spatial curvature, we must work in spherical coordinates. Equation 21.3 shows the invariant interval for *flat* space (compare with the spatial components of the metric in Equation 9.2). Our focus is only on the spatial components of the metric, so I ignore time dt and use ds for the invariant interval measured in terms of spatial distance.

$$\text{Flat space}: \ ds^2 = dr^2 + r^2 \, d\theta^2 + r^2 \sin^2\theta \, d\phi^2, \quad (r, \theta, \phi). \tag{21.3}$$

We need the equivalent metrics for curved 3-D space. Let's start with positive curvature. Back in Chapter 3, I used Figure 3.3 to show you why the *2-D surface* of a unit sphere (called a 2-sphere) has the metric of Equation 3.17. If we label the radius of the sphere as R (not to be confused with coordinate r), the metric is Equation 21.4 for the 2-sphere. The comparable metric for a positively curved *3-D volume* of space (called a 3-sphere) using coordinates (ω, θ, ϕ) is shown as Equation 21.5. I am not going to derive the 3-sphere metric. It is a standard formula. Note that versus the earlier Equation 3.17 in Chapter 3, I have switched θ to ω and ϕ to θ to make the comparison clearer. Also, don't forget that the value of R depends on the radius of the sphere (in the non-existent illustrative extra dimension), so it is determined by the curvature. In the metric, it too is a dimensionless scale factor. It is *not* a coordinate.

$$\text{2-sphere}: \ ds^2 = R^2 \left(d\omega^2 + \sin^2\omega \, d\theta^2 \right), \qquad\qquad (\omega, \theta), \tag{21.4}$$

$$\text{3-sphere}: \ ds^2 = R^2 \left(d\omega^2 + \sin^2\omega \, d\theta^2 + \sin^2\omega \sin^2\theta \, d\phi^2 \right), \qquad (\omega, \theta, \phi). \tag{21.5}$$

Now for the super-clever FRW trick. We make a coordinate change setting $r = \sin \omega$. This means we can substitute r^2 for the $\sin^2 \omega$ terms and that $d\omega^2 = \frac{dr^2}{1-r^2}$, as shown in Equation 21.6.

$$r = \sin \omega, \quad \Longrightarrow \quad \frac{dr}{d\omega} = \cos \omega, \quad \Longrightarrow \quad d\omega^2 = \frac{dr^2}{\cos^2 \omega} = \frac{dr^2}{1 - \sin^2 \omega} = \frac{dr^2}{1 - r^2}. \tag{21.6}$$

This gives the metric for a 3-sphere in Equation 21.7. It has a similar form to that of flat space in Equation 21.3. Note that the r coordinate is not the same as for flat space. In the case of flat space, it is the radial distance out to a point from the origin. At the risk of providing unnecessary detail, in the case of the positively curved 3-sphere, it is the sine of the circular angle around the sphere from the origin to the point. Importantly, because $dr = \cos \omega \, d\omega$, a move in the dr coordinate direction is in the same direction as $d\omega$, so we know that the new r coordinate remains orthogonal to θ and ϕ. This ensures that no off-diagonal terms suddenly appear in the metric.

$$\textit{3-sphere}: \ ds^2 = R^2 \left(\frac{1}{1-r^2} dr^2 + r^2 \, d\theta^2 + r^2 \sin^2 \theta \, d\phi^2 \right), \quad (r, \theta, \phi). \tag{21.7}$$

We can do something similar to generate the metric for negatively curved space. This is called a hyperbolic 3-space. The metric is shown as Equation 21.8. It looks like the 3-sphere but uses the hyperbolic sinh. I can imagine many readers wincing. Relax! The metric is a standard formula and you will not need to be familiar with the intricacies of hyperbolic trigonometry to understand what follows. We do a coordinate change setting $r = \sinh \omega$. This means we can substitute r^2 for the $\sinh^2 \omega$ terms and that $d\omega^2 = \frac{dr^2}{1+r^2}$. Note that $\cosh^2 \theta - \sinh^2 \theta = 1$. This gives Equation 21.9. Again, the coordinate r has a different meaning, but remains orthogonal to θ and ϕ. It is the sine of the hyperbolic angle from the origin around to the point.

$$\textit{Hyperbolic 3-space}: \ ds^2 = R^2 \left(d\omega^2 + \sinh^2 \omega \, d\theta^2 + \sinh^2 \omega \sin^2 \theta \, d\phi^2 \right), \quad (\omega, \theta, \phi). \tag{21.8}$$

$$r = \sinh \omega, \quad \Longrightarrow \quad \frac{dr}{d\omega} = \cosh \omega, \quad \Longrightarrow \quad d\omega^2 = \frac{dr^2}{\cosh^2 \omega} = \frac{dr^2}{1 + \sinh^2 \omega} = \frac{dr^2}{1 + r^2},$$

$$\textit{Hyperbolic 3-space}: \ ds^2 = R^2 \left(\frac{1}{1+r^2} dr^2 + r^2 \, d\theta^2 + r^2 \sin^2 \theta \, d\phi^2 \right), \quad (r, \theta, \phi). \tag{21.9}$$

These cunning coordinate changes allow us to combine flat, positive and negative spatial curvature into one spatial metric as shown in the invariant interval in Equation 21.10. It includes a new variable, which I label bold **K** to avoid any confusion with the k constant in the EFEs. Set **K** = +1 and the metric matches that of the 3-sphere in Equation 21.7. Set **K** = 0 and the metric matches the flat space in Equation 21.3. Set **K** = −1 and the metric matches that of hyperbolic 3-space in Equation 21.9. Therefore, **K** is +1 for positive curvature, 0 for flat space and −1 for negative curvature.

$$ds^2 = R^2 \left(\frac{1}{1 - \mathbf{K}r^2} dr^2 + r^2 \, d\theta^2 + r^2 \sin^2 \theta \, d\phi^2 \right), \quad (r, \theta, \phi),$$

$$\mathbf{K} = +1 \quad \textit{positive curvature}, \qquad \mathbf{K} = 0 \quad \textit{flat}, \qquad \mathbf{K} = -1 \quad \textit{negative curvature}. \tag{21.10}$$

The last steps are to add in the time dimension t in order to have a spacetime metric and to give it a time-dependent dimensionless spatial scale factor generally labelled $a(t)$. Both a and R are dimensionless scale factors. The former is the time-dependent scale factor. The latter depends on the amount of spatial curvature. The R^2 term in Equation 21.10 is absorbed into a^2. Voila... we have the FRW invariant interval and metric shown in Box 21.5. As recompense for the head-cracking maths in this section, I offer the joke Box 21.4, which you should by now understand doesn't work if there is positive spatial curvature.

Box 21.4 Just for laughs

Parallel lines have so much in common... *it is a shame they will never meet.*

21.7 The FRW Metric

Congratulations to readers who understood optional Section 21.6. Any readers who got lost shouldn't worry. You don't have to know the details in order to use the FRW metric. Let me summarise how it works. There are three possible forms of spatial curvature that are consistent with the cosmological principle, i.e. the resulting space is homogeneous and isotropic. These can be ranked as positively curved space, flat space (no curvature) and negatively curved space.

The FRW metric combines all three into one clever metric by introducing a new curvature variable **K** that is value **K** $= +1$ for positive spatial curvature (in 2-D this would be the surface of a sphere), **K** $= 0$ for zero spatial curvature (in 2-D a flat surface) and **K** $= -1$ for negative spatial curvature (in 2-D the surface of a saddle). The resulting FRW metric is shown in Box 21.5.

We now can look at the implications of spatial curvature simply by toggling the value of **K** between $+1, 0$ and -1. I should point out that many texts on cosmology focus on invariant distance ds rather than invariant time $d\tau$ and use the $< -+++>$ convention where the spatial components of $[g_{\mu\nu}]$ are positive (this is called the East-coast convention). I have chosen to stick with $d\tau$ and $< +--->$ because it matches earlier modules and my previous book *Quantum Untangling*. This has absolutely no effect on the final result of the calculations.

The r coordinate is not the same in the three scenarios, but this is not important. The coordinates r, θ and ϕ are orthogonal in all three cases, as you can see from the absence of any off-diagonal terms in the metric. All three spacetimes are homogeneous and isotropic. This means that a move in any spatial direction has the same effect. You get the same result whatever spatial coordinates you use, providing they are orthogonal to each other.

Note that r, θ and ϕ do *not* vary over time, because all the time variation in the FRW metric is encapsulated in the scale factor a.

Box 21.5 FRW metric

The spatial scale factor a varies with time t. The FRW metric:
- has *positive spatial* curvature when **K** $= +1$
- is *spatially flat* when **K** $= 0$
- has *negative spatial* curvature when **K** $= -1$

$$d\tau^2 = -ds^2 = dt^2 - a^2 \left(\frac{1}{1 - \mathbf{K}r^2}\, dr^2 + r^2\, d\theta^2 + r^2 \sin^2\theta\, d\phi^2 \right), \qquad (t, r, \theta, \phi;\ c = 1),$$

$$\textit{FRW metric } (t, r, \theta, \phi;\ c = 1): \quad [g_{\mu\nu}] = \begin{bmatrix} 1 & 0 & 0 & 0 \\ 0 & -\dfrac{a^2}{1 - \mathbf{K}r^2} & 0 & 0 \\ 0 & 0 & -a^2 r^2 & 0 \\ 0 & 0 & 0 & -a^2 r^2 \sin^2\theta \end{bmatrix}.$$

21.8 Ricci Curvature and the Cosmological Principle

In Chapter 22, we are going to calculate the Ricci tensor of the FRW metric. This will lead us to the Friedmann equations. First, I want to highlight the curvature symmetries linked to the cosmological principle. This is going to save us a lot of calculating. Again we use what I call the *curvature footprint*, which shows the Riemann components of the Ricci tensor. As a reminder, Equation 21.11 shows the Ricci scalar and breakdown of one of its components (check back to Sections 7.3 and 7.4 if needed).

$$R = g^{00}R_{00} + g^{11}R_{11} + g^{22}R_{22} + g^{33}R_{33}, \quad \textit{Ricci scalar,}$$

$$g^{00}R_{00} = g^{00}\left(R^0_{000} + R^1_{010} + R^2_{020} + R^3_{030} \right), \quad \textit{Example: Time Component.}$$

(21.11)

Our starting point is the curvature footprint in Equation 21.12, which shows the symmetries that exist for *any diagonal* spacetime metric. The coordinates are shown as $(0, 1, 2, 3)$ where 0 is the time coordinate and $(1, 2, 3)$ are the spatial coordinates. It is based on the symmetries inherent in the Riemann tensor components that add together to give the Ricci tensor. For example, $g^{00}R^1_{010} = g^{11}R^0_{101}$. This is described in full in Subsection 9.3.1. Equation 21.12 is the same as Equation 9.4 earlier if you want to refer back.

Underlying Symmetries for any 4-D Diagonal Metric (21.12)

$$
\begin{aligned}
g^{00}R_{00} &= g^{00}R^0_{000} + g^{00}R^1_{010} + g^{00}R^2_{020} + g^{00}R^3_{030} = 0 \;+a \;+b \;+c, \\
g^{11}R_{11} &= g^{11}R^0_{101} + g^{11}R^1_{111} + g^{11}R^2_{121} + g^{11}R^3_{131} = a \;+0 \;+d \;+e, \\
g^{22}R_{22} &= g^{22}R^0_{202} + g^{22}R^1_{212} + g^{22}R^2_{222} + g^{22}R^3_{232} = b \;+d \;+0 \;+f, \\
g^{33}R_{33} &= g^{33}R^0_{303} + g^{33}R^1_{313} + g^{33}R^2_{323} + g^{33}R^3_{333} = c \;+e \;+f \;+0.
\end{aligned}
$$

We can build on this with the symmetries created by the cosmological principle. The principle states that the universe is homogeneous and isotropic. This means that all three spatial dimensions are equivalent, providing they are orthogonal, which they are in the FRW metric.

I introduced you to the Riemann tensor many chapters ago, so let me remind you that it uses parallel transport to measure the underlying curvature and is not fooled by weird or wonky coordinates (Section 7.2). Therefore, based on the cosmological principle, Riemann curvature must be equal in all spatial directions if measured on the same scale (i.e. an identical amount of invariant interval).

As a result, we know $R^1_{010} = R^2_{020} = R^3_{030}$ (which is $a = b = c$). In each case, the spatial coordinates $(1, 2, 3)$ have one upper and one lower index, which scales them to be equivalent with respect to the invariant interval. Similarly, $R^2_{121} = R^3_{131}$ $(d = e)$ and $R^1_{313} = R^2_{323}$ $(e = f)$.

Note that we used this logic to derive the Schwarzschild metric. In that case, we used spherical symmetry to match up the Riemann curvature in the θ and ϕ directions (refer back to Box 12.2 for more detail if needed). In the case of the FRW metric, the cosmological principle allows us to match up the Riemann curvature of all three spatial directions.

As a result, there are only two independent values in the curvature footprint if the metric conforms to the cosmological principle. This is shown in Equation 21.13. I have labelled these b and d to avoid confusion with the a scaling factor in the FRW metric and with the speed of light.

Curvature footprint of any diagonal metric conforming to cosmological principle (21.13)

$$
\begin{aligned}
g^{00}R_{00} &= g^{00}R^0_{000} + g^{00}R^1_{010} + g^{00}R^2_{020} + g^{00}R^3_{030} = 0 \;+b \;+b \;+b = 3b, \\
g^{11}R_{11} &= g^{11}R^0_{101} + g^{11}R^1_{111} + g^{11}R^2_{121} + g^{11}R^3_{131} = b \;+0 \;+d \;+d = b + 2d, \\
g^{22}R_{22} &= g^{22}R^0_{202} + g^{22}R^1_{212} + g^{22}R^2_{222} + g^{22}R^3_{232} = b \;+d \;+0 \;+d = b + 2d, \\
g^{33}R_{33} &= g^{33}R^0_{303} + g^{33}R^1_{313} + g^{33}R^2_{323} + g^{33}R^3_{333} = b \;+d \;+d \;+0 = b + 2d.
\end{aligned}
$$

This symmetry is true for *any* and *all* diagonal metrics conforming to the cosmological principle, whether the spacetime is flat or has curvature... whatever the coordinates... Cartesian, spherical... it is always true. Also note that it means $g^{11}R_{11} = g^{22}R_{22} = g^{33}R_{33}$, and the value of the Ricci scalar is as shown in Equation 21.14.

$$
R = g^{00}R_{00} + g^{11}R_{11} + g^{22}R_{22} + g^{33}R_{33} = 6b + 6d. \tag{21.14}
$$

21.9 Summary

The cosmological principle assumes that the universe is *homogeneous* (the same at every location) and *isotropic* (the same in every direction). Obviously, this is not true locally; the sun is different from the moon. The assumption is that at huge scale, variations disappear and the universe today is like a homogeneous dust field with each galaxy a speck of dust, while the early universe, a plasma of interacting radiation and particles, is treated mathematically like a perfect fluid.

Throw this book into the air at less than escape velocity and it will fall back down. If above escape velocity, it will have enough energy to break free of the earth's gravitational field and sail into the distance. But at whatever speed you throw it, the book will never rest static above your head.

Similarly, neither Isaac Newton's nor Einstein's theories of gravity permit a stable static universe. Newton believed the universe to be held static by what he called a voluntary agent (even Newton made mistakes—see Box 21.6). And Einstein's effort to stabilise things with the cosmological constant Λ was doomed to fail. Setting Λ at a level to offset the effective gravitational acceleration between two galaxies might sound good, but the slightest peculiar motion or difference in distance between galaxies would upset the balance. Any equilibrium would be temporary and unstable.

In any case, the concept of a static universe was consigned to the dustbin once Hubble revealed his findings. The further a galaxy is from us, the faster it tends to be receding. The relationship between the two is called the *Hubble parameter H*, which is $\frac{\dot{a}}{a}$ in the co-moving frame.

If you visualise this expansion model flowing back in time, you conclude that the universe started from a much denser state... the *Big Bang*. An obvious question is how it will end. Are the galaxies receding from each other at faster than escape velocity? Will it end in a *big crunch,* or will the expansion continue for ever, leading to *heat death* as the galaxies grow ever further apart?

To analyse things we build a co-moving framework where the spatial dimensions expand over time and the galaxies are carried along for the ride. The simplest metric is for flat space (Equation 21.2), but sadly this is not enough. We have to cater for the possibility that the universe has spatial curvature. Illustrating with 2-D surfaces, positive spatial curvature is like the surface of a sphere, and negative spatial curvature is like the surface of a saddle (Figure 21.2). If there is significant spatial curvature, it would be detectable because positive spatial curvature makes distant objects appear larger and negative spatial curvature makes them appear smaller (Figure 21.3).

We use the FRW metric to cover the possibility of positively and negatively curved space. The trick is to construct homogeneous, isotropic positively curved space using the surface of a 3-sphere embedded in 4-D space. We do the same for negative spatial curvature by embedding what is called a hyperbolic 3-space in 4-D space. The mathematics of these coordinate changes is a bit gruesome, so I labelled it optional. The result is the FRW spacetime metric in Box 21.5, where we can toggle the value of **K** from +1 to 0 to −1 for metrics that have respectively positive, zero (flat) and negative curvature.

The coordinate *r* is not the same in the three FRW scenarios, but this doesn't matter because in each scenario the coordinates remain orthogonal to each other and the space is homogeneous and isotropic. I showed you that this also creates symmetries within the Riemann components of the Ricci tensor (Equation 21.13). These symmetries will assist us in Chapter 22 when we calculate the Ricci curvature of the FRW metric.

Combining the result of that calculation with the EFEs, we will derive the famous Friedmann equations. These are the tools that cosmologists use to study how the development of the universe is affected by the presence of various types of energy and by any spatial curvature.

Box 21.6 Newton's bubble bursts (1720)

As well as being the most famous scientist of his age, Isaac Newton was heavily involved in the financial affairs of Britain as Master of the Mint. Did this combination of scientific genius and knowledgeable financier make him a peerless investor? No!

In the early 1700s, the British government set up the South Sea Company, giving it a monopoly on much trade with the Americas and redeeming a chunk of government debt for shares in the company. Newton invested and then sold his initial position in the first half of 1720. As the shares traded at about £100 before 1720 and he would have sold at £200–300, he must have made a tidy profit.

This was the famous South Sea Bubble when wild speculation led the company's share price to leap five-fold for a few months before collapsing back down. Unfortunately for Newton, he jumped right back in, buying more shares at the height of the bubble. It is estimated that he lost the current equivalent of £20–40 million ($25–50 million) in the ensuing debacle.

In spite of the crash, the South Sea Company continued to trade and a few years later was managing over 70% of the British national debt. In fact, any investors who bought its stock in 1719 and resisted buying or selling in the short 1720 bubble did just fine.

And what of Newton? The good news is that he rebuilt his financial position and died a wealthy man, although legend has it that he forbade anyone from ever again mentioning in his presence the name *South Sea Company*.

Chapter 22

The Friedmann Equations

Box 22.1   Simon's Ricci cheat sheet for 4-D diagonal metrics

The eight-term formula below helps in calculating the Ricci tensor of any diagonal spacetime metric (no off-diagonal components). The dimensions are labelled arbitrarily (A, B, C, D).

$$R^A{}_{BAB} = -\frac{\partial}{\partial A}\left(\frac{g^{AA}}{2}\frac{\partial g_{BB}}{\partial A}\right) - \frac{\partial}{\partial B}\left(\frac{g^{AA}}{2}\frac{\partial g_{AA}}{\partial B}\right) - \frac{g^{AA}g^{AA}}{4}\frac{\partial g_{AA}}{\partial A}\frac{\partial g_{BB}}{\partial A} + \frac{g^{AA}g^{BB}}{4}\frac{\partial g_{AA}}{\partial B}\frac{\partial g_{BB}}{\partial B}$$

$$\cdots - \frac{g^{AA}g^{AA}}{4}\left(\frac{\partial g_{AA}}{\partial B}\right)^2 + \frac{g^{AA}g^{BB}}{4}\left(\frac{\partial g_{BB}}{\partial A}\right)^2 - \frac{g^{AA}g^{CC}}{4}\frac{\partial g_{AA}}{\partial C}\frac{\partial g_{BB}}{\partial C} - \frac{g^{AA}g^{DD}}{4}\frac{\partial g_{AA}}{\partial D}\frac{\partial g_{BB}}{\partial D}.$$

$$R^A{}_{BAB} = [1] + [2] + [3] + [4] + [5] + [6] + [7] + [8], \qquad \text{terms numbered for easy reference.}$$

22.1   FRW Metric: Ricci Calculation

We are going to calculate the Ricci tensor for the FRW metric. The result, combined with the Einstein Field Equations (EFEs), will lead to the Friedmann equations. The maths is a bit tedious, but I will step through it slowly. If you find it too much, just skip on to Section 22.2 and the Friedmann equations. For those who are willing to follow the maths, I have good news. The Friedmann equations sit at the heart of cosmology, so it is worth it. And this is the last Ricci calculation in the book—yippee! After this, the maths gets easier. For convenience, Equation 22.1 shows again the FRW metric (as in Box 21.5).

$$\text{FRW metric } (t, r, \theta, \phi;\ c = 1): \qquad [g_{\mu\nu}] = \begin{bmatrix} 1 & 0 & 0 & 0 \\ 0 & -\dfrac{a^2}{1 - \mathbf{K}r^2} & 0 & 0 \\ 0 & 0 & -a^2 r^2 & 0 \\ 0 & 0 & 0 & -a^2 r^2 \sin^2\theta \end{bmatrix}. \tag{22.1}$$

Equation 22.2 shows the curvature footprint for any diagonal metric conforming to the cosmological principle (as explained back in Equation 21.13). This is true of the FRW metric and our target is to calculate the values of b and d for it. If you have any doubts about these symmetries, please refer back to Section 21.8.

Curvature footprint of any diagonal metric conforming to cosmological principle (22.2)

$$g^{00}R_{00} = g^{00}R^0{}_{000} + g^{00}R^1{}_{010} + g^{00}R^2{}_{020} + g^{00}R^3{}_{030} = \quad 0 \quad +b \quad +b \quad +b = \quad 3b,$$

$$g^{11}R_{11} = g^{11}R^0{}_{101} + g^{11}R^1{}_{111} + g^{11}R^2{}_{121} + g^{11}R^3{}_{131} = \quad b \quad +0 \quad +d \quad +d = \quad b + 2d,$$

$$g^{22}R_{22} = g^{22}R^0{}_{202} + g^{22}R^1{}_{212} + g^{22}R^2{}_{222} + g^{22}R^3{}_{232} = \quad b \quad +d \quad +0 \quad +d = \quad b + 2d,$$

$$g^{33}R_{33} = g^{33}R^0{}_{303} + g^{33}R^1{}_{313} + g^{33}R^2{}_{323} + g^{33}R^3{}_{333} = \quad b \quad +d \quad +d \quad +0 = \quad b + 2d,$$

Ricci scalar $\qquad R = g^{00}R_{00} + g^{11}R_{11} + g^{22}R_{22} + g^{33}R_{33} = 6b + 6d.$

Untangling General Relativity: The Intuitive Self-Study Guide, First Edition. Simon Sherwood.
© 2026 John Wiley & Sons Ltd. Published 2026 by John Wiley & Sons Ltd.

22.1.1 Calculation Step 1

We will start with the value of b for the FRW metric. You can see in the curvature footprint that we could choose any of the six b entries to work with. When confronted with this sort of calculation, it is worth spending time experimenting to find the easiest calculation. We will use $b = g^{00}R^2{}_{020}$. For the FRW metric in spherical coordinates, this is $g^{TT}R^{\theta}{}_{T\theta T}$, which is simply $R^{\theta}{}_{T\theta T}$ because $g^{TT} = 1$.

Referring to *Simon's Ricci cheat sheet* of Box 22.1 (introduced back in Section 12.3) we set in the cheat sheet: $A = \theta$, $B = t$, $C = r$ and $D = \phi$. As $g_{TT} = 1$, any term involving a derivative of g_{TT} is zero, which wipes out all the terms except [2] and [5]. This gives us Equation 22.3.

The calculation is shown step by step. Don't forget that a is the scale factor. Any increase in a means that distant galaxies are getting further apart. $\dot{a}$ is $\frac{\partial a}{\partial t}$, which is the rate of growth of the scale factor. The higher the value of $\dot{a}$, the higher the velocity of separation of distant galaxies. $\ddot{a}$ is $\frac{\partial^2 a}{\partial t^2}$, which is the acceleration of the scale factor. If $\ddot{a}$ is negative, the velocity of separation of distant galaxies falls over time. If $\ddot{a}$ is positive, it increases.

The result is Equation 22.4. The values of r, θ and ϕ do *not* vary with time because all the time variation is encapsulated in the scale factor a. Note from the result that the value of b is not affected by the spatial curvature factor $\mathbf{K}$.

$$R^{\theta}{}_{T\theta T} = -\frac{\partial}{\partial t}\left(\frac{g^{\theta\theta}}{2}\frac{\partial g_{\theta\theta}}{\partial t}\right) - \frac{g^{\theta\theta}g^{\theta\theta}}{4}\left(\frac{\partial g_{\theta\theta}}{\partial t}\right)^2, \tag{22.3}$$

$$= -\frac{\partial}{\partial t}\left(\frac{-1}{2a^2r^2}(-2\dot{a}ar^2)\right) - \frac{1}{4a^4r^4}(4\dot{a}^2a^2r^4),$$

$$= -\frac{\partial}{\partial t}\left(\frac{\dot{a}}{a}\right) - \frac{\dot{a}^2}{a^2}, \qquad note: \frac{\partial}{\partial t}\left(\frac{1}{a}\right) = \frac{\partial}{\partial a}\left(\frac{1}{a}\right)\frac{\partial a}{\partial t} = -\frac{\dot{a}}{a^2},$$

$$= -\frac{1}{a}\frac{\partial\dot{a}}{\partial t} - \dot{a}\frac{\partial}{\partial t}\left(\frac{1}{a}\right) - \frac{\dot{a}^2}{a^2} = -\frac{\ddot{a}}{a} + \frac{\dot{a}^2}{a^2} - \frac{\dot{a}^2}{a^2} = -\frac{\ddot{a}}{a},$$

$$\implies b = g^{TT}R^{\theta}{}_{T\theta T} = -\frac{\ddot{a}}{a}. \tag{22.4}$$

22.1.2 Calculation Step 2

The calculation of d is more convoluted. We could choose for the calculation any of the six d entries in the curvature footprint. In my opinion, the easiest is $g^{22}R^3{}_{323}$, which means we need $R^{\phi}{}_{\theta\phi\theta}$. Therefore, we set in the cheat sheet: $A = \phi$, $B = \theta$, $C = t$ and $D = r$.

The metric is not dependent on ϕ so any terms including $\frac{\partial}{\partial\phi}$ are zero, which eliminates terms [1], [3] and [6]. In addition, $\frac{\partial g_{\theta\theta}}{\partial\theta} = 0$ eliminating term [4]. This leaves terms [2], [5] [7] and [8] as potentially non-zero.

If you want to work through the calculation, the following trigonometric identities should help: $\frac{\partial(\sin^2\theta)}{\partial\theta} = 2\sin\theta\cos\theta$ and $\frac{\partial}{\partial\theta}\left(\frac{\cos\theta}{\sin\theta}\right) = -\frac{1}{\sin^2\theta}$. If part of the calculation looks familiar, it is because several steps appeared in Equation 12.15 when we worked with the Schwarzschild metric.

$$R^{\phi}{}_{\theta\phi\theta} = -\frac{\partial}{\partial\theta}\left(\frac{g^{\phi\phi}}{2}\frac{\partial g_{\phi\phi}}{\partial\theta}\right) - \frac{g^{\phi\phi}g^{\phi\phi}}{4}\left(\frac{\partial g_{\phi\phi}}{\partial\theta}\right)^2 - \frac{g^{\phi\phi}g^{TT}}{4}\frac{\partial g_{\phi\phi}}{\partial t}\frac{\partial g_{\theta\theta}}{\partial t} - \frac{g^{\phi\phi}g^{RR}}{4}\frac{\partial g_{\phi\phi}}{\partial r}\frac{\partial g_{\theta\theta}}{\partial r},$$

$$= -\frac{\partial}{\partial\theta}\left(\frac{2a^2r^2\sin\theta\cos\theta}{2a^2r^2\sin^2\theta}\right) - \frac{4a^4r^4\sin^2\theta\cos^2\theta}{4a^4r^4\sin^4\theta} + \frac{4\dot{a}^2a^2r^4\sin^2\theta}{4a^2r^2\sin^2\theta} - \frac{4a^4r^2\sin^2\theta}{4a^4r^2\sin^2\theta}(1 - \mathbf{K}r^2),$$

$$= -\frac{\partial}{\partial\theta}\left(\frac{\cos\theta}{\sin\theta}\right) - \frac{\cos^2\theta}{\sin^2\theta} + \dot{a}^2r^2 - 1 + \mathbf{K}r^2,$$

$$= \frac{1}{\sin^2\theta} - \frac{\cos^2\theta}{\sin^2\theta} + \dot{a}^2r^2 - 1 + \mathbf{K}r^2,$$

$$= \frac{\sin^2\theta}{\sin^2\theta} + \dot{a}^2r^2 - 1 + \mathbf{K}r^2 = \dot{a}^2r^2 + \mathbf{K}r^2,$$

$$\implies d = g^{\theta\theta}R^{\phi}{}_{\theta\phi\theta} = -\frac{1}{a^2r^2}(\dot{a}^2r^2 + \mathbf{K}r^2) = -\frac{\dot{a}^2}{a^2} - \frac{\mathbf{K}}{a^2}. \tag{22.5}$$

It is an exhausting slog, but most of the terms come from the use of spherical coordinates and cancel out. Equation 22.5 gives the value of d. You can see that it *does* depend on $\mathbf{K}$. It is affected by spatial curvature in the metric.

The Ricci curvature of the FRW metric does not depend on r, θ or ϕ. The cosmological principle states that the universe is the same in all locations (homogeneous) and all directions (isotropic), so the curvature cannot depend on the choice of spatial coordinates. The coordinate r is not the same in the three FRW scenarios ($\mathbf{K} = +1, 0, -1$), but this cannot matter so long as r, θ and ϕ are orthogonal (which they are).

22.1.3 Ricci Tensor Components of FRW Metric

We can use the value of b from Equation 22.4 and d from Equation 22.5 to fill in the curvature footprint of the FRW metric. To keep things general, I use the broader coordinate labels $(0, 1, 2, 3)$ where, as usual, 0 is time and 1, 2 and 3 are the spatial coordinates. The result for the FRW metric is shown as Equation 22.6. It is true whatever orthogonal spatial coordinates are used.

Curvature footprint of FRW metric (22.6)

$$
\begin{aligned}
g^{00}R_{00} &= 0 & -\frac{\ddot{a}}{a} & & -\frac{\ddot{a}}{a} & & -\frac{\ddot{a}}{a} & = -3\frac{\ddot{a}}{a}, \\
g^{11}R_{11} &= -\frac{\ddot{a}}{a} & +0 & & \left(-\frac{\dot{a}^2}{a^2} - \frac{\mathbf{K}}{a^2}\right) & & \left(-\frac{\dot{a}^2}{a^2} - \frac{\mathbf{K}}{a^2}\right) & = -\frac{\ddot{a}}{a} - \frac{2\dot{a}^2}{a^2} - \frac{2\mathbf{K}}{a^2}, \\
g^{22}R_{22} &= -\frac{\ddot{a}}{a} & \left(-\frac{\dot{a}^2}{a^2} - \frac{\mathbf{K}}{a^2}\right) & & +0 & & \left(-\frac{\dot{a}^2}{a^2} - \frac{\mathbf{K}}{a^2}\right) & = -\frac{\ddot{a}}{a} - \frac{2\dot{a}^2}{a^2} - \frac{2\mathbf{K}}{a^2}, \\
g^{33}R_{33} &= -\frac{\ddot{a}}{a} & \left(-\frac{\dot{a}^2}{a^2} - \frac{\mathbf{K}}{a^2}\right) & & \left(-\frac{\dot{a}^2}{a^2} - \frac{\mathbf{K}}{a^2}\right) & & +0 & = -\frac{\ddot{a}}{a} - \frac{2\dot{a}^2}{a^2} - \frac{2\mathbf{K}}{a^2}.
\end{aligned}
$$

22.2 Deriving the Friedmann Equations

The Friedmann equations combine the Ricci tensor results with the EFEs to quantify the relationship between the key values of the curvature of the FRW metric, b and d, with the presence of energy-momentum in the universe. The energy-momentum tensor $[T_{\mu\nu}]$ (also known as the stress-energy tensor) of the universe is modelled as that of a *perfect fluid*. We discussed this back in Section 8.7. It allows us to incorporate energy-density ρ and internal pressure P. I have labelled energy-density simply as ρ because, using $c = 1$, there is no distinction between energy and mass density anyway. Internal pressure P will prove particularly significant during the early stages of the universe, when it was a dense plasma of radiation and particles.

The left of Equation 22.7 shows the energy-momentum tensor of a stationary perfect fluid using the coordinates of the Minkowski metric (this is Equation 8.14 in Section 8.7). The right side shows the same energy-momentum tensor in the coordinates of the FRW metric. Switching to the FRW coordinates is straightforward because all tensors transform between coordinate systems in the same way. Therefore, we change the energy-momentum tensor in the same way that transforms the Minkowski metric (Equation 3.4) into the FRW metric (Equation 22.1).

Perfect fluid : Minkowski   *Perfect fluid : FRW metric* $(c = 1)$

$$
[T_{\mu\nu}] = \begin{bmatrix} \rho & 0 & 0 & 0 \\ 0 & P & 0 & 0 \\ 0 & 0 & P & 0 \\ 0 & 0 & 0 & P \end{bmatrix}, \quad \begin{bmatrix} \rho & 0 & 0 & 0 \\ 0 & \frac{a^2}{1 - \mathbf{K}r^2}P & 0 & 0 \\ 0 & 0 & a^2r^2P & 0 \\ 0 & 0 & 0 & a^2r^2\sin^2\theta P \end{bmatrix}.
\tag{22.7}
$$

Note that in using this we are taking a specific frame of reference called the *cosmic rest frame*, which I will explain in more detail in a moment in Section 22.3. It is the rest frame relative to the expansion of the universe. As a result, all the entries in the energy-momentum tensor are diagonal (no off-diagonal terms). Equation 22.8 shows a couple of calculations for the perfect fluid in FRW coordinates that we will need to derive the Friedmann equations. Similar calculations for the other components show: $g^{11}T_{11} = g^{22}T_{22} = g^{33}T_{33} = -P$.

$$g^{00} T_{00} = (1)\rho = \rho, \qquad g^{11} T_{11} = \frac{T_{11}}{g_{11}} = -\left(\frac{1 - \mathbf{K}r^2}{a^2}\right) \frac{a^2}{1 - \mathbf{K}r^2} P = -P. \tag{22.8}$$

Let's start with the EFE relating R_{00} and T_{00}. We take the full-form EFE for the relationship (Equation 22.9). Note that k is the Einstein constant, not to be confused with the spatial curvature factor $\mathbf{K}$. It is irritating that physicists didn't label things differently!

For Equation 22.10 we multiply all terms by g^{00} using $g^{00} g_{00} = 1$ for diagonal metrics. We then substitute in the values of $g^{00} R_{00}$ and the Ricci scalar R in terms of b and d using the curvature footprint of Equation 22.2. We also put in the value of $g^{00} T_{00} = \rho$ from Equation 22.8. A simple calculation leads to the expression for $-d$. Substituting in the FRW value for this from Equation 22.5 gives what is called the first Friedmann Equation (labelled *Friedmann-1*). It includes $\mathbf{K}$, so it is affected by whether spatial curvature is positive, flat or negative.

$$R_{00} - \frac{1}{2} g_{00} R - g_{00} \Lambda = k T_{00}, \qquad \text{where}: \quad k = 8\pi G, \qquad (c = 1), \tag{22.9}$$

$$g^{00} R_{00} - \frac{1}{2} R - \Lambda = k g^{00} T_{00}, \tag{22.10}$$

$$3b - \frac{1}{2}(6b + 6d) = k\rho + \Lambda,$$

$$-d = \frac{k\rho}{3} + \frac{\Lambda}{3},$$

$$\frac{\dot{a}^2}{a^2} + \frac{\mathbf{K}}{a^2} = \frac{k\rho}{3} + \frac{\Lambda}{3}, \qquad \text{\textit{Friedmann-1 equation.}} \tag{22.11}$$

The second Friedmann equation comes from the spatial equivalent starting from Equation 22.12 and using $g^{11} T_{11} = -P$ from Equation 22.8. The only twist is that we substitute using the value for d from the first Friedmann equation. This gives an expression for $-b$, which we know for the FRW metric is $\frac{\ddot{a}}{a}$ (see Equation 22.4). The result is the second Friedmann Equation (labelled *Friedmann-2*). It is not affected by the spatial curvature factor $\mathbf{K}$.

$$R_{11} - \frac{1}{2} g_{11} R - g_{11} \Lambda = k T_{11}, \qquad \text{where}: \quad k = 8\pi G, \qquad (c = 1), \tag{22.12}$$

$$g^{11} R_{11} - \frac{1}{2} R - \Lambda = k g^{11} T_{11},$$

$$b + 2d - \frac{1}{2}(6b + 6d) = -kP + \Lambda,$$

$$-2b - d = -kP + \Lambda, \qquad \text{\textit{from Friedmann-1}}: \qquad -d = \frac{k\rho}{3} + \frac{\Lambda}{3},$$

$$-b = -\frac{kP}{2} + \frac{\Lambda}{2} - \frac{k\rho}{6} - \frac{\Lambda}{6} = \frac{\Lambda}{3} - \frac{k}{6}(\rho + 3P),$$

$$\frac{\ddot{a}}{a} = \frac{\Lambda}{3} - \frac{k}{6}(\rho + 3P), \qquad \text{\textit{Friedmann-2 equation.}} \tag{22.13}$$

Box 22.2 The Friedmann equations $c = 1$; $k = 8\pi G$; $\mathbf{K} = +1, 0, -1$

$$\left(\frac{\dot{a}}{a}\right)^2 = \frac{k\rho}{3} + \frac{\Lambda}{3} - \frac{\mathbf{K}}{a^2}, \qquad \text{\textit{Friedmann-1: expansion rate.}}$$

$$\frac{\ddot{a}}{a} = \frac{\Lambda}{3} - \frac{k}{6}(\rho + 3P), \qquad \text{\textit{Friedmann-2: acceleration.}}$$

Below are detailed versions including speed of light factor c. Variables ρ_{mass} and P are volumetric *mass* density in kilograms per cubic metre, and pressure in newtons per square metre. You will not need these for this book. I provide them in case you want to compare with other texts.

$$\left(\frac{\dot{a}}{a}\right)^2 = \frac{8\pi G \rho_{mass}}{3} + \frac{\Lambda c^2}{3} - \frac{\mathbf{K}c^2}{a^2}, \qquad \text{\textit{Friedmann-1: expansion rate.}}$$

$$\frac{\ddot{a}}{a} = \frac{\Lambda c^2}{3} - \frac{4\pi G}{3}\left(\rho_{mass} + \frac{3P}{c^2}\right), \qquad \text{\textit{Friedmann-2: acceleration.}}$$

We will be using equations Friedmann-1 and Friedmann-2 to study the development of the universe, so let me briefly summarise where we are. The FRW metric for spatially expanding spacetime caters for positive, flat and negative spatial curvature. We calculated its Ricci curvature using the symmetries of the cosmological principle to simplify the maths. We then used the EFEs to match up this curvature with the presence of energy-momentum in the universe (based on a perfect fluid), including the possibility of a non-zero cosmological constant.

The result is the two Friedmann equations shown again in Box 22.2. Friedmann-1 quantifies the rate of expansion of the universe. Friedmann-2 quantifies its acceleration rate. I will continue to work in natural units ($c = 1$) but have added versions including the speed of light c factor for any readers who need them.

22.3 The Cosmic Rest Frame

The cosmological principle assumes the universe is *isotropic*, meaning it is the same measured in all directions. A few of you fabulous readers may spot that a universe speckled with galaxies cannot be isotropic in all frames of reference. Suppose I position myself at some point in the universe. I look around and observe the galaxies receding at the same rate in every direction. From this viewpoint, the universe is isotropic. Now, I accelerate up to high speed in one direction. As a result, the galaxies in my direction of motion are moving away less quickly and the galaxies behind are receding more quickly. From this viewpoint, the universe is not isotropic.

Another question concerning reference frames is the expansion factor $a(t)$ in the FRW metric. It is dependent on time, but whose time? Two simultaneous moments in time for one observer aren't simultaneous for another moving at speed. This is a basic fact of special relativity (check back to Section 2.6 on *Leading Clocks Lag* if needed).

A similar argument holds for the energy-momentum tensor of the universe. For an observer stationary relative to a non-rotating mass, only the T_{TT} entry is non-zero (energy-density). However, an observer moving relative to the mass observes the mass to have momentum, so $[T_{\mu\nu}]$ is different. In Section 22.2 I selected the energy-momentum tensor of the universe to have only diagonal terms. In doing this I am choosing a particular frame of reference.

The co-moving expanding isotropic universe indeed is a specific frame of reference called the *cosmic rest frame*. It is the reference frame of an observer moving in tandem with the co-moving frame e.g. on any one of the red dots of Figure 21.1. It is this observer who measures the expansion of the universe (the Hubble parameter) to be the same value in every direction and the energy-momentum tensor $[T_{\mu\nu}]$ of a homogeneous universe to have only diagonal terms.

It may come as a surprise to some readers that the cosmic rest frame is clearly identifiable and measurable. One way is to measure movement using the Doppler shift in the *Cosmic Microwave Background* (CMB). This is residual radiation from the early stages of the universe that bathes the sky all around us (if you are not familiar with the CMB, check out Boxes 22.3 and 22.4). We will be discussing the CMB in detail in coming chapters.

Box 22.3 Cosmic Microwave Background (CMB): a faint whisper of the Big Bang

The early universe was too hot for radiation to travel any significant distance, because of its interaction with electrically charged protons and electrons. Any photons produced would instantly be absorbed by surrounding charged particles, leading to a pattern of constant interaction: absorption–emission–absorption... (this is called Thomson scattering). That is until it cooled enough for the protons and electrons to combine into electrically neutral hydrogen atoms, about 380,000 years after the Big Bang (cosmologists call this *recombination*). This allowed what NASA describes as a *wall of light* to travel out in every direction. The wavelength spectrum of this light lengthened in tandem with the expansion of the universe. The residue is the CMB with a mixture of wavelengths equivalent to the emissions of an object (black body) at only 2–3 degrees above absolute zero. The CMB is fascinating because it allows a glimpse of the universe 13 billion years ago. It provides strong evidence for the Big Bang, because its existence requires the universe to have been denser and hotter in the past. Also, small variations in the CMB's pattern of radiation offer a map of the seed for the creation of the galaxies and large-scale structures that we observe today.

Observatories can pick up slight directional variations in the wavelength of the CMB. This tell us that the peculiar motion (i.e. motion relative to the cosmic rest frame) of the Milky Way galaxy is about 600 km s^{-1}. That sounds a lot, but in terms of light speed is only about 0.002c. The motion is towards the centre of what is called Laniakea, a super-cluster of about 100,000 galaxies, of which more later.

Right now, I can imagine readers protesting: *hold on a moment, Simon, isn't this cosmic rest frame a privileged inertial reference frame? I thought relativity didn't allow such a thing. Isn't this equivalent to the ether that Einstein dismissed?*

The first thing to note is that the cosmic rest frame is *observer-dependent*. It is *not* a single inertial frame. I explained earlier that the cosmic rest frame could be the frame of reference on any observer stationary relative to the co-moving grid illustrated in Figure 21.1. An observer, Jill, sitting on the bottom left red dot views the cosmic rest frame as isotropic relative to her red dot. Another observer, Jack, sitting on the top right red dot sees the cosmic rest frame as isotropic relative to his dot. But the bottom left and top right dots are moving relative to each other. Therefore, Jill's and Jack's cosmic rest frames are moving relative to each other and are different inertial frames.

This being said, the cosmic rest frame has a special status for each observer. Is this a problem for relativity? Discussion of this tends to be more philosophical than quantitative and much comes down to linguistics. Is the measurement of peculiar motion versus the cosmic rest frame a measure of *absolute* motion or *relative* motion? And is that distinction really important in relativity? I'm a simple soul who struggles with semantics. I focus on the basics and take comfort that Einstein's two key tenets (shown again below), which I introduced back in Section 2.1, still hold for all inertial frames. This is very different from some sort of ether that light travels through. In that scenario, neither of Einstein's tenets would be true:

- The speed of light is constant for all observers.
- The laws of physics are the same for all observers.

22.4 Energy-Density and Expansion

Putting the question of curvature **K** to one side, let's examine how the energy-density of the universe changes with its expansion. Look at the Friedmann expansion rate equation in Box 22.2. Energy-density always increases the Hubble parameter ($H = \frac{\dot{a}}{a}$), whether it is matter/radiation energy-density ρ or positive vacuum energy, often called dark energy ($\Lambda > 0$).

Now take a look at the Friedmann acceleration equation (again Box 22.2). The relationship is very different and merits some discussion. The presence of matter/radiation (which can affect both energy-density ρ and pressure P) *decreases* the acceleration parameter $\frac{\ddot{a}}{a}$, as you would expect, because its gravitational attraction has a decelerating effect. However, the presence of positive vacuum energy ($\Lambda > 0$) *increases* the acceleration parameter. Matter, radiation and vacuum energy ($\Lambda > 0$) are all positive forms of energy, so why is the effect of vacuum energy so different? It is because the density of each form of energy changes differently with the expansion of the universe.

I should note that, in theory, the cosmological constant could be negative ($\Lambda < 0$). This would be negative vacuum energy and create a decelerating effect. However, as will be revealed later, this doesn't seem to be the case in our universe, so in this book I will not be addressing the possibility of negative vacuum energy.

Box 22.4 CMB: is it (a) pigeon poop or (b) a Nobel Prize?

In 1964, Robert Wilson and Arno Penzias were using a giant horn antenna in New Jersey, USA to study the cosmos. They were irritated by a strange background hum of microwave radiation that persisted whatever direction they pointed the antenna. Trying to eliminate it, they investigated all sorts of possibilities, including the presence of two pigeons that had made a home in the horn. Wondering if the hum was caused by pigeon droppings, they trapped the birds but the hum remained. In the end, they concluded that there really is a background hum of CMB radiation. Their discovery won them the 1978 Nobel Prize for Physics. So, the answer is (b) a Nobel Prize, not (a) pigeon poop!

22.4.1 An Intuitive Introduction

Let me give an intuitive explanation of the changes in energy-density with expansion. I will start with *matter*, by which I mean atoms, planets, stars, you, me and stuff like that. It is slow-moving (non-relativistic), so the vast majority of the energy is rest mass. Say there is total M mass of matter in a volume of space with sides length $a = 1$. That is Mc^2 of energy, which I will label E_M in a volume-1 box (sides $1 \times 1 \times 1$). Therefore, the initial energy-density of matter ρ_M is E_M. As a grows, the volume of space expands and the energy E_M now is in a volume of a^3 (sides $a \times a \times a$) giving $\rho_M = E_M/a^3$. The energy-density of *matter* ρ_M decreases with a^3 (the cube of the scale factor). This is illustrated in Figure 22.1a. The black dots represent matter such as atoms.

In the same figure, the wavy red lines represent radiation. Imagine in the volume-1 box on the left of the figure there are electromagnetic waves of wavelength λ. The energy of the radiation depends on its wavelength because each photon has energy $\frac{h}{\lambda}$ (h is Planck's constant, $c = 1$). Therefore, the radiation density in the box is $\rho_R \propto \frac{h}{\lambda}$. As a grows, the volume increases to a^3, so the density (in terms of number of photons) decreases with a^3. In addition, the wavelength of the radiation is stretched out to $a\lambda$. The energy of each photon reduces to $\frac{h}{a\lambda}$. Radiation energy-density ρ_R reduces with a^4, because the radiation's wavelength is lengthened by a, in addition to the effect of the a^3 increase in volume.

Those of you who know some quantum mechanics may raise a questioning eyebrow. You know about wave-particle duality. Any particle, such as a proton or electron, has an associated wave function whose wavelength is related to energy. Why doesn't this stretch with the expansion of the universe? It doesn't stretch because the matter we are discussing comes in interacting clumps. As an example, consider an electron in an atom. Its wavelength is governed by the much stronger electromagnetic interaction between it and the nucleus (and the surrounding electrons). This immediately offsets the spatial expansion, holding the electron wavelength at its original limit. The same is true of the gravitational interaction within galaxies. These forces vastly outweigh the slow expansion of the universe, which grows at only 7% per billion years. As a result, the steady expansion of space has no effect on atoms, our solar system or the structure of the Milky Way.

I should mention that neutrinos are more complicated. They have virtually no mass and interact weakly with matter. Many travel at relativistic speeds. In most cases, the energy-density of neutrinos tends to be treated in the same way as radiation, decreasing with a^4, but more accurate models may treat a portion as decreasing with a^3, to reflect the effect of the slower moving ones. Neutrinos will not feature prominently in our calculations, but best you know.

The energy associated with a positive cosmological constant ($\Lambda > 0$) is different. The clue is in the word *constant*. It is a constant level of vacuum energy. If there is ρ_Λ energy-density in the volume-1 box, then there also must be ρ_Λ energy-density in the a^3 box. The energy-density ρ_Λ does not change with the scale factor a. As the universe expands, the total amount of vacuum energy in the universe increases. In summary, we have matter density ($\rho_M \propto 1/a^3$), radiation density ($\rho_R \propto 1/a^4$) and vacuum energy-density (ρ_Λ constant).

22.4.2 A Bit More Rigour (Optional)

For readers who want more detail, I will tie this back to the energy-momentum tensor $[T_{\mu\nu}]$ that we used in calculating the Friedmann equations (check out Box 22.5 for amusing confusion surrounding Friedmann's birthday). The way energy-density changes with volume is encapsulated in the material's *equation of state*. For our

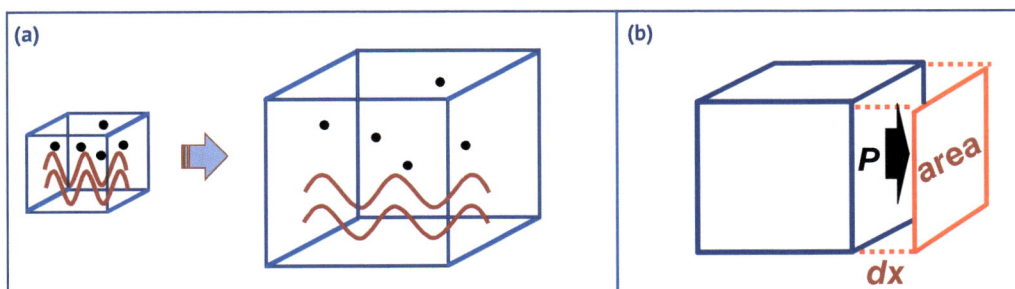

Figure 22.1 (a) Matter and radiation energy-density changes differently with expansion of the universe. (b) Schematic for relationship between pressure and density in expansion.

purposes, we need the relationship between energy-density ρ and pressure P. Typically, physicists label this w, where $P = w\rho$ and w is a constant.

Box 22.5 Friedmann and the expanding birthday

You can find the birthday of Alexander Friedmann listed as June 4th or 16th or 17th or 29th. Four birthdays seem a bit too many for one man. The story starts with the Russian calendar, which differs from the modern one by 12 days. Friedmann was actually born on June 4th according to the Russian calendar, which is June 16th on ours. Friedmann in his curriculum vitae tried to cater for this, but got his sums wrong and arrived at June 17th (a surprising mistake for a mathematician). Biographers assumed this date was the Russian calendar and added *another* 12 days giving the completely wrong but commonly quoted birth date of June 29th!

We link the value of w to how ρ changes with expansion as follows. Consider the change in energy-density of a material such as a gas when we let one side of a containing box expand by dx as shown in Figure 22.1b. The force on the side of the box is pressure multiplied by area ($P \times$ area). As the gas expands, it loses some energy $-dE$ because the speed of the gas molecules drops. The energy lost is equivalent to the work done expanding the box, which is force multiplied by distance ($-dE = F\,dx = P \times$ area $\times\, dx$). The area multiplied by dx is the change in volume dV of the box giving the relationship $-dE = P\,dV$ used in Equation 22.14. Differentiating by parts gives Equation 22.15, which we integrate. With some rearranging, we arrive at Equation 22.16.

$$-P\,dV = dE = d(\rho V), \qquad \text{\textit{E is energy-density} } \rho \times \text{\textit{volume V,}} \qquad (22.14)$$

$$= \rho\,dV + V\,d\rho,$$

$$-w\rho\,dV = \rho\,dV + V\,d\rho, \qquad \text{\textit{by definition of w,}}$$

$$-(1+w)\rho\,dV = V\,d\rho,$$

$$-(1+w)\frac{1}{V}dV = \frac{1}{\rho}d\rho, \qquad \text{\textit{which we integrate...}} \qquad (22.15)$$

$$-(1+w)\ln V + C = \ln\rho, \qquad \text{\textit{C is the constant from integration,}}$$

$$\ln\left(CV^{-(1+w)}\right) = \ln\rho, \qquad \text{\textit{C is a (different) constant,}}$$

$$CV^{-(1+w)} = \rho. \qquad (22.16)$$

Equation 22.16 shows that energy-density ρ decreases with $V^{(1+w)}$. Volume V is a^3, the cube of the scale factor, giving the (well-known) relationship in Equation 22.17 that energy-density ρ decreases with $a^{3(1+w)}$.

$$\rho = \frac{C}{V^{(1+w)}}, \qquad \Longrightarrow \qquad \rho \propto \frac{1}{a^{3(1+w)}}, \qquad \text{\textit{where P = w}}\rho. \qquad (22.17)$$

We can derive the value of w for each energy type from the energy-momentum tensor. Equation 22.18 shows $[T^{\mu\nu}]$ for a perfect fluid in Minkowski spacetime as in Equation 8.14 earlier, including the breakdown of its components, which were explained in Equation 8.10 of Chapter 8.

$$[T^{\mu\nu}] = \rho \begin{bmatrix} u^t u^t & u^t u^x & u^t u^y & u^t u^z \\ u^x u^t & u^x u^x & u^x u^y & u^x u^z \\ u^y u^t & u^y u^x & u^y u^y & u^y u^z \\ u^z u^t & u^z u^x & u^z u^y & u^z u^z \end{bmatrix} = \begin{bmatrix} \rho & 0 & 0 & 0 \\ 0 & P & 0 & 0 \\ 0 & 0 & P & 0 \\ 0 & 0 & 0 & P \end{bmatrix}. \qquad (22.18)$$

For non-relativistic *matter*, the rest mass energy-density is huge compared with any other components. This means that for matter $P \approx 0$. As by definition $P = w\rho$, this gives $w = 0$. Feeding this into the formula in Equation 22.17 gives $\rho_M \propto 1/a^3$ as expected.

For *radiation*, we can calculate w based on radiation moving at light speed, which has a null geodesic (check back to Section 5.5 if needed). This means that the combination of diagonal components in Equation 22.19 from the definition of $[T^{\mu\nu}]$ (the left of Equation 22.18) sum to zero. Therefore, $\rho - P - P - P = 0$, giving $w = \frac{1}{3}$. Feeding this value into Equation 22.17 confirms $\rho_R \propto 1/a^4$.

$$\rho - 3P = \rho \left((u^t u^t) - (u^x u^x) - (u^y u^y) - (u^z u^z) \right), \qquad (c = 1),$$
$$= \frac{\rho}{d\tau^2} \left(dt^2 - dx^2 - dy^2 - dz^2 \right) = 0, \qquad\qquad\qquad (22.19)$$
$$\implies \quad \rho - 3P = 0, \qquad \text{radiation} : P_R = \frac{1}{3}\rho_R.$$

To calculate w for positive vacuum energy ($\Lambda > 0$), we count Λ as an energy-density, shifting it to the right of the EFEs and adding it in to $[T^{\mu\nu}]$ to create combined components that include Λ as shown in Equation 22.20. Note that I have used the upper index version of the EFEs, which is an equally valid form (I showed this earlier in Equation 8.2).

$$R^{\mu\nu} - \frac{1}{2} R g^{\mu\nu} - \Lambda g^{\mu\nu} = k\, T^{\mu\nu}, \qquad\qquad\qquad k = 8\pi G, \quad (c = 1),$$
$$R^{\mu\nu} - \frac{1}{2} R g^{\mu\nu} = k\, T^{\mu\nu} + \Lambda g^{\mu\nu} = k \left(T^{\mu\nu} + \frac{\Lambda}{k} g^{\mu\nu} \right). \qquad\qquad (22.20)$$

We redefine energy-density ρ to include vacuum energy, so that instead of $\rho = T^{00}$, it includes the Λ term. We do the same for pressure P giving the two relations in Equation 22.21. Continuing to use Minkowski spacetime, $g^{00} = 1$ and $g^{11} = -1$. In the case of there being *only* vacuum energy present, this gives the relationships in Equation 22.22. As a result $P_\Lambda = -\rho_\Lambda$, and based on its definition ($P = w\rho$), we know $w = -1$. Substituting this into Equation 22.17 confirms that $\rho_\Lambda \propto a^0$, which is a constant. Therefore, ρ_Λ is independent of a. As expected, ρ_Λ remains constant as the universe expands.

$$\text{Including } \Lambda: \qquad \rho = \left(T^{00} + \frac{\Lambda}{k} g^{00} \right), \qquad P = \left(T^{11} + \frac{\Lambda}{k} g^{11} \right), \qquad (22.21)$$

$$\text{If only } \Lambda: \qquad \rho_\Lambda = \frac{\Lambda}{k}, \qquad\qquad P_\Lambda = -\frac{\Lambda}{k}. \qquad\qquad (22.22)$$

The relationship between the cosmological constant Λ and the vacuum energy-density ρ_Λ is $\Lambda = k\rho_\Lambda$ (see Equation 22.22), where k is the Einstein constant $8\pi G$ (with $c = 1$).

22.5 Dominant Relationships

Including vacuum energy-density ($\Lambda > 0$), the Friedmann expansion equation can be written as Equation 22.23, separating out the energy types.

$$\left(\frac{\dot{a}}{a} \right)^2 = \frac{k}{3} (\rho_M + \rho_R + \rho_\Lambda) - \frac{K}{a^2}, \qquad k = 8\pi G, \qquad (c = 1). \qquad (22.23)$$

If we introduce a label $\overline{\rho}$ for the value of energy-density ρ at $a = 1$, we can write the Friedmann expansion as Equation 22.24. You can insert various values for the energy densities and calculate how H would have changed over time. This then can be compared with observations.

$$H^2 = \left(\frac{\dot{a}}{a} \right)^2 = \frac{k}{3} \left(\frac{\overline{\rho_M}}{a^3} + \frac{\overline{\rho_R}}{a^4} + \overline{\rho_\Lambda} \right) - \frac{K}{a^2}, \qquad k = 8\pi G, \qquad (c = 1). \qquad (22.24)$$

The different relationships with the scale factor a mean that different forms of energy dominated as the universe expanded. In the early life of the universe, when a was very small, radiation energy ρ_R dominated because $\frac{1}{a^4}$ was very large. As the universe grew and a increased, the radiation energy faded in importance (decreasing with a^4), leading to a period of matter energy ρ_M dominance. The scale factor then grew to a level where the matter density diminished in importance (decreasing with a^3) and we are left with vacuum energy ρ_Λ. Modern observational data suggests that we entered this phase of vacuum energy dominance a few billion years ago.

Vacuum energy ($\Lambda > 0$) is more commonly called *dark energy*. Personally, I prefer the term vacuum energy, which emphasises that it is *positive* energy associated with the vacuum. The idea of a vacuum having energy will be familiar to those who have studied quantum theory. The quantum fields have a non-zero ground state energy. Matching the two is a mess because the vacuum energy of general relativity is a great deal *smaller* than predicted by quantum theory. We will discuss this in more detail later.

22.6 The Accelerating Effect of Vacuum Energy

Let's look more closely at the Friedmann acceleration equation. Equation 22.25 shows it, substituting in for matter $P = 0$, meaning $(\rho_M + 3P_M) = \rho_M$. In the case of radiation $P = \frac{\rho}{3}$, meaning $(\rho_R + 3P_R) = 2\rho_R$. For vacuum energy $\Lambda = k\rho_\Lambda$ from Equation 22.22.

As noted earlier, the presence of matter and radiation energy has a negative effect. They *decelerate* the expansion of the universe. But the presence of vacuum energy $(\Lambda > 0)$ is positive and *accelerates* expansion. Why the difference? The proper answer is that this is what the Friedmann equations tell us and the Friedmann equations come directly from the EFEs, but let me try to explain things loosely in intuitive terms.

$$\frac{\ddot{a}}{a} = \frac{\Lambda}{3} - \frac{k}{6}(\rho + 3P) = \frac{k}{3}\left(\rho_\Lambda - \frac{\rho_M}{2} - \rho_R\right), \qquad k = 8\pi G, \quad (c = 1). \qquad (22.25)$$

The decelerating effect of *radiation* energy is twice that of the same amount of *matter* energy. The difference is best understood using Figure 22.1a. The expansion of the universe does not change the total energy of *matter*. It reduces its *density* because the matter is more spread out, but it does not remove any energy from the universe. In contrast, the expansion of the universe actually removes radiation energy from the universe because of the lengthening effect on its wavelength. In layman's language, you might say that the expansion of radiation uses up energy. It takes work and this work slows the expansion compared with that of matter (which causes no such energy reduction).

Vacuum energy $(\Lambda > 0)$ has the opposite effect. As the scale factor increases, new vacuum is introduced into the universe. This means there is a net increase in the total energy in the universe. Rather than energy being removed from the universe in the expansion, energy is added. *P* has the opposite sign and is described as negative pressure. Rather than work being done, energy is released, accelerating the rate of expansion.

22.7 Critical Energy-Density

Observational data reveals that the universe is spatially flat to a very good degree of approximation (more on this in Chapter 23). However, spatial curvature remains important in our analysis because the *lack* of curvature helps us quantify the overall energy-density of the universe. Look back at Equation 22.23. If we assume that the universe is spatially flat, then $\mathbf{K} = 0$ and we have a relationship between the Hubble parameter (left side of the equation) and the total energy-density in the universe (right side). In Equation 22.26, I have rearranged Equation 22.23 using the label H (Hubble parameter) for $\frac{\dot{a}}{a}$. In Equation 22.27, I have multiplied through by $\frac{3}{k}$ and substituted in the value of k. Setting $\mathbf{K} = 0$ gives the energy density of a spatially flat universe, labelled ρ_{crit}. In the calculation I have used the same value of the current Hubble parameter as elsewhere in the book, which is $70\,\text{km}\,\text{s}^{-1}\,\text{Mpc}^{-1}$.

$$\frac{k}{3}(\rho_M + \rho_R + \rho_\Lambda) = H^2 + \frac{\mathbf{K}}{a^2}, \qquad k = 8\pi G, \qquad (c = 1), \qquad (22.26)$$

$$\rho_M + \rho_R + \rho_\Lambda = \frac{3H^2}{8\pi G} + \frac{3}{8\pi G}\frac{\mathbf{K}}{a^2},$$

$$= \rho_{crit} + \frac{3}{8\pi G}\frac{\mathbf{K}}{a^2}, \qquad where\ \rho_{crit} = \frac{3H^2}{8\pi G} \approx 10^{-26}\,\text{kg}\,\text{m}^{-3}. \qquad (22.27)$$

This gives the current value of what is called the *critical density* of the universe as $10^{-26}\,\text{kg}\,\text{m}^{-3}$ (if you want the value in terms of energy-density per cubic metre, then multiply by c^2). The critical density is equivalent to 5–10 hydrogen atoms in a cubic metre of space.

If the universe is spatially flat, this is the energy-density of the universe. If the energy-density in the universe is higher than this, the $\mathbf{K}$ term must be positive. In this case $\mathbf{K} = +1$ and the universe has positive spatial curvature. This is often referred to as *closed* because a universe with positive curvature is finite (if you travel far enough in one direction, you will end up back where you started). On the other hand, if it is lower than the critical density, then $\mathbf{K} = -1$, and the universe has negative spatial curvature. Such a universe is infinite and is described as *open*.

As we have observational evidence that the universe is very nearly spatially flat (I will describe this in Chapter 23), we know $\rho_M + \rho_R + \rho_\Lambda \approx \rho_{crit}$, which gives a good starting point for building a picture of the energy-density of the universe by type (matter, radiation, vacuum energy). This is generally quantified using the symbol Ω, where $\Omega = \rho/\rho_{crit}$. Dividing Equation 22.27 by ρ_{crit} (which is $\frac{3H^2}{8\pi G}$) gives Equation 22.28.

$$\Omega_M + \Omega_R + \Omega_\Lambda = 1 + \frac{\mathbf{K}}{a^2 H^2}, \qquad Example: \quad \Omega_M = \frac{\rho_M}{\rho_{crit}} = \frac{8\pi G}{3H^2}\rho_M. \tag{22.28}$$

22.8 Summary

Our focus in Chapter 23 will be to establish values for Ω_M, Ω_R, Ω_Λ and the $\mathbf{K}$ term by combining theory with observational data. Let me summarise the theoretical work in this chapter before we embark on this. At the same time, I will give some spoilers so your little grey brain cells are prepared for the coming results. Our starting point was the FRW metric, which models homogeneous isotropic spacetime expanding over time. In the metric, $\mathbf{K}$ can toggle in value $+1/0/-1$ for positive/flat/negative spatial curvature.

We worked out the Ricci curvature of the FRW metric. The mathematical calculations were helped by the inherent symmetries of the cosmological principle. We then combined the values for Ricci curvature with the EFEs to give the Friedmann equations. These describe how the scale factor a develops in the presence of energy-density. Friedmann himself realised that the universe could/should be expanding or contracting. At the time, it was a bold assertion (see Box 22.6). This was years before Hubble revealed his results.

The energy-densities of matter, radiation and the vacuum are affected differently by expansion. The total energy of matter doesn't change, so its energy-density decreases with a^3 because the matter is spread through a bigger volume. Radiation energy-density reduces with a^4, because its wavelength is lengthened by a, in addition to the effect of the a^3 increase in volume. The energy-density of the vacuum remains constant in spite of any expansion. This leads to different phases of energy dominance: radiation in the early phase followed by matter and then finally the dominance of vacuum energy.

I offered an optional section giving detail on the relationship ($P = w\rho$) between energy-density ρ and pressure P (the equation of state). For matter $w = 0$, for radiation $w = \frac{1}{3}$ and for positive vacuum energy $w = -1$. I showed how this appears in the energy-momentum tensor of each and how it flows through the EFEs into the Friedmann equations.

Box 22.6 Friedmann: the man who made the universe expand

You might think that Alexander Friedmann built his theory on the back of Hubble's discovery that the universe is expanding. You would be wrong.

It was in 1922 that Friedmann published his paper outlining models of a *non-stationary world* (meaning universe). Einstein reacted negatively, commenting on Friedmann's work: *The results concerning the non-stationary world appear to me suspicious.* Friedmann followed up with a letter to Einstein outlining his arguments and Einstein (to his credit) publicly withdrew: *My criticism... was based on an error in my calculations. I consider that Mr Friedmann's results are correct and shed new light.* Sadly, Friedmann died in 1925 at only 37 years, well before his ideas were vindicated in 1929 by Hubble's observations. To quote his biographers Tropp, Frenkel and Chernin:

His intellect sees what others do not see, and... he rejects the centuries-old tradition that chose, prior to any experience, to consider the universe eternal and eternally immutable. He accomplishes a genuine revolution in science. As Copernicus made the earth go round the sun, so Friedmann made the universe expand.

We discussed why the presence of vacuum energy ($\Lambda > 0$) accelerates the expansion of the universe. The total energy of *radiation* drops during expansion. It entails work $P\,dV$: pressure $\times$ change in volume. This slows the expansion. In contrast, the density of vacuum energy remains constant, so total energy increases during expansion. Rather than work being done, there is what you might call negative pressure and extra energy is released. This accelerates the expansion.

We then returned to the topic of spatial curvature and the value of $\mathbf{K}$. If the total energy-density in the universe matches what is called the critical density ($\rho_{crit} \approx 10^{-26}\ \mathrm{kg\,m^{-3}}$), then $\mathbf{K} = 0$ and the universe is spatially flat. If it is higher than ρ_{crit}, then $\mathbf{K} = +1$ and there is positive spatial curvature. If lower, then $\mathbf{K} = -1$ and there is negative spatial curvature. A recent estimate at time of writing is that the energy-density of the universe is within two parts in a thousand of ρ_{crit}, so if there is any spatial curvature it is very small.

As for the composition of this energy, recent estimates are $\Omega_\Lambda \approx 0.7$, $\Omega_M \approx 0.3$ and Ω_R is negligible, where Ω is ρ/ρ_{crit} as shown in Equation 22.28. This means that vacuum energy is the dominant form of energy in our universe. This came as a big surprise! In Chapter 23, we will discuss all of this and the observational evidence behind it.

Chapter 23

Welcome to the Dark Side

In this chapter, we will look at the observational evidence behind current estimates of radiation, matter and vacuum energy-density Ω_R, Ω_M and Ω_Λ along with curvature $\mathbf{K}$. The big surprises for physicists were dark matter and dark (vacuum) energy. We need dark matter to explain the pattern of gravitational attraction in galaxies (in addition to many other anomalies). Perhaps even more mysterious is the vacuum energy needed to explain the expansion of the universe (see Box 23.1 for a laugh). I prefer the name *vacuum energy*, but *dark energy* is a brand name beloved of sci-fi writers and the press. As I noted earlier, this is not a battle I am going to win!

I will not go through the history of who thought what when. Rather, I will focus on modern evidence to build an up-to-date picture of how things stand at the time of writing this book.

Box 23.1 Just for laughs

The romantic: Love holds the universe together. Hate blows it apart.
The physicist: Wrong. That would be dark matter and dark energy.

23.1   Spatial Curvature: Feeling Flat

We typically measure spatial curvature using $\Omega_\mathbf{K} = -\dfrac{\mathbf{K}}{a^2 H^2}$. This is not an energy-density, but makes it comparable with Ω_R, Ω_M and Ω_Λ, which is helpful. Starting from Equation 22.28 we get Equation 23.1. If $\mathbf{K} = +1$, there is positive spatial curvature (a closed universe). If $\mathbf{K} = -1$, there is negative spatial curvature and if it is zero the universe is spatially flat.

$$\Omega_M + \Omega_R + \Omega_\Lambda - 1 = \frac{\mathbf{K}}{a^2 H^2} = -\Omega_\mathbf{K}, \qquad (c = 1). \qquad (23.1)$$

As discussed in Section 21.5, one test of flatness is to look out and count the galaxies, searching for tell-tale signs of a change in density of their number with distance. With positive curvature, the distance between objects appears larger, so when observing across a given angle you will see fewer galaxies over a particular angular range. Conversely, with negative curvature the distance between objects appears smaller, so more galaxies will fall into the area of observation (check back to Figure 21.3 if needed). These surveys are not as simple as they sound because you have to establish how far away the galaxies are, typically grouping them by their redshift as a measure of distance. Also, you are looking back in time; the further away, the longer ago. As the galaxies formed over time, you need to account in the calculation for their rate of formation (i.e. the earlier the epoch, the fewer the galaxies).

The Baryon Oscillation Spectroscopy Survey (BOSS) is a remarkable database covering 1.2 million galaxies and analysing patterns of clustering. Studies based on the data have seen no sign of curvature and conclude that the magnitude of $\Omega_\mathbf{K}$ if not zero must be less than 0.001.[1]

---

[1] For example, see: *The clustering of galaxies in the completed SDSS-III Baryon Oscillation Spectroscopic Survey: cosmological analysis of the DR12 galaxy sample* 2016.

Untangling General Relativity: The Intuitive Self-Study Guide, First Edition. Simon Sherwood.
© 2026 John Wiley & Sons Ltd. Published 2026 by John Wiley & Sons Ltd.

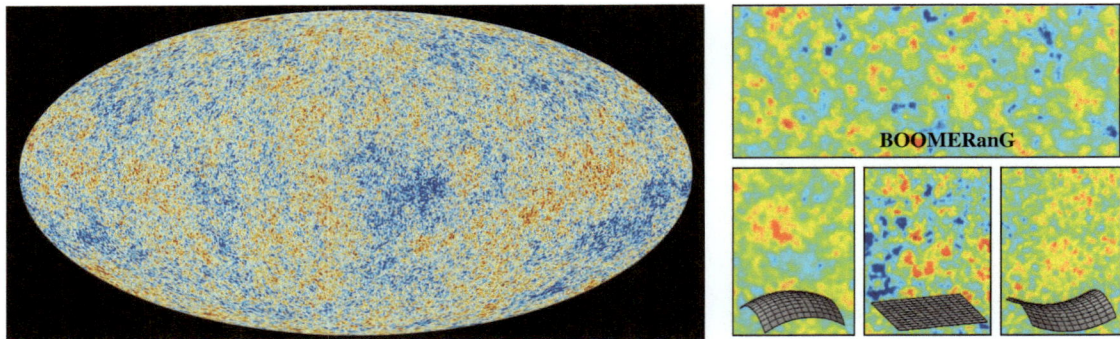

Figure 23.1 Left: Map of the CMB anisotropies. *Source:* ESA and the Planck Collaboration / Wikimedia Commons / CC BY 4.0. Right: Simulation of CMB for positive/flat/negative curvature. *Source:* Courtesy of BOOMERanG.

This is backed up by data from the cosmic microwave background (CMB - refer back to Box 22.3 if needed) that puts the maximum possible magnitude of Ω_K as below 0.003 based on 2013 data from the Wilkinson Microwave Anisotropy Probe (WMAP), then confirmed by the Planck satellite. This built on earlier observations by the BOOMERanG of the CMB that showed a similar pattern. The left of Figure 23.1 shows an image of the CMB scanning across the sky. The different colours are very slight variations in temperature (called anisotropy). The right of Figure 23.1 shows a helpful illustration from BOOMERanG. The presence of spatial curvature would alter the pattern of temperature variation. As you can see in the figure, the CMB pattern (upper image) best matches that of flat space (the middle lower box). There is more detail on CMB measurement of flatness later in this book (optional Section 26.5).

Of course, we cannot say for certain that the universe is *absolutely flat*. Perhaps we have not yet developed a fine enough measure. And don't forget that spatial curvature does exist at least locally; for example, the Schwarzschild metric has spatial curvature. This being said, we have good evidence that the universe is almost spatially flat. A 2-D way of thinking about the universe might be a sheet of bubble wrap that has tiny pockets of spatial curvature, but is flat when viewed on large scales.

23.2 Dark Matter

Well before the BOSS galaxy survey had been thought of or the CMB discovered, physicists suspected there was a large quantity of matter missing in calculations. Galaxies are misbehaving. They refuse to rotate correctly!

Orbital speed varies with distance from the attracting mass. The left of Figure 23.2 shows the orbital speed of the planets compared with their distance from the sun measured along the semi-major axis of orbit (an Astronomical Unit *AU* is equivalent to the distance between earth and sun). The smooth pattern is as expected and follows what is called Kepler's Law. The orbital speed drops away as you move further out from the central attracting mass.

The middle of Figure 23.2 shows the orbital speed of stars around the centre of our Milky Way galaxy. The red line shows actual data. The blue dotted line marked *Kepler* illustrates orbital speed based purely on the mass of stars (any such calculation is highly approximate but the pattern is clear). The deviation indicates that there is a lot of unseen mass orbiting the outer region of the Milky Way. This pattern is repeated in other galaxies. The right of the figure shows another example, Messier 33 (the Triangulum galaxy), a spiral galaxy almost 3 million light years from earth.

The discrepancy is particularly large in the outer regions, so it cannot be caused by a central mass such as a huge undetected black hole at the galactic core. Many have tried to tinker with the equations of general relativity (GR) but without success. Explaining away the galactic orbits without messing up those of the planets is tricky, along with the challenge of matching the many other proven predictions of GR.

Enough evidence has accumulated for most physicists to have confidence in the existence of dark matter. Both the deflection of light in gravitational lensing (Section 15.1) around galaxies and collisions between galaxy clusters reveal its presence. It appears to have played an important role in speeding up the process of galaxy formation. There is also evidence from the CMB that the overall level of matter density in the universe is well in excess of what we can see.

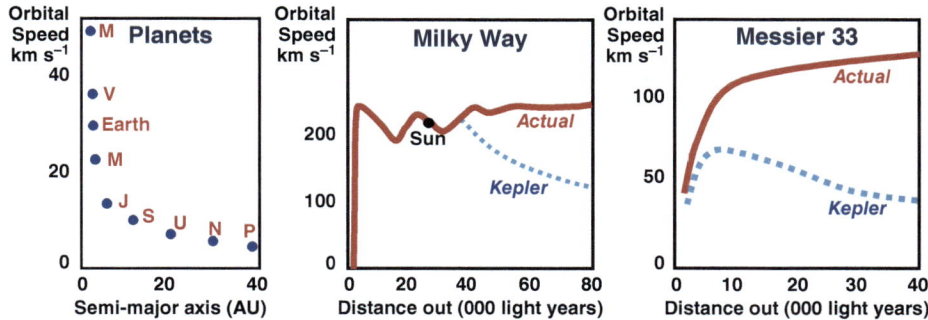

Figure 23.2 The orbits of galaxies are not consistent with the mass of visible stars that they contain. This indicates the presence of dark matter (Messier 33 graphic adapted from work by Mario De Leo, under CC 4.0 International).

There is historical precedent for linking gravitational anomalies with the presence of something new but unseen. Both Neptune and Pluto were discovered based on unexplained perturbations in the orbit of other planets.

What is dark matter? It is believed to be something orbiting in and around the galaxies, helping bind them together. We know it is not electrically charged. We know it does not interact strongly with ordinary matter beyond its gravitational effect. Otherwise, we would have detected it long ago. We believe it represents about 85% of total matter energy. The LHC is searching hard for candidates, but we still don't know what it actually is (see Box 23.2).

Box 23.2 Dark matter: the usual suspects

Let's do a line-up of the usual suspects for you to inspect.

WIMPs: An as yet undiscovered Weakly Interacting Massive Particle could be the culprit and deserves a place in the line-up. This would be like a neutrino in the sense that it carries no electric charge and interacts only via the gravitational and weak forces, but the WIMP must be more massive. Neutrinos move too fast to clump around galaxies and they don't cumulatively represent enough mass to explain dark matter.

SUSY: SUper SYmmetry is a tentative theory of (advanced) particle physics. *Fermions* are the particles that make up matter (e.g. electrons and quarks) while *bosons* can be thought of as particles that carry forces (e.g. photons and gluons). SUSY hypothesises each fermion has a supersymmetric boson partner, and vice versa. To date, no supersymmetric partner has been discovered, but the lightest ones (neutralinos and photinos) could account for dark matter.

Axions: A less massive option is an abundant slow-moving light boson. Such a particle would be related to an as yet unknown scalar quantum field. Such a field might explain some other anomalies in the standard model related to what are called potential charge conjugation parity symmetry (CP) violations in quantum chromodynamics (well beyond the scope of this book). And don't forget that we only recently identified the Higgs scalar boson, so there may well be similar bosons still to be found.

Primordial black holes: Perhaps a less popular proposal is that dark matter is small black holes. These are much too small to form now (check out the limits on black hole formation in Section 13.5), so they must have developed early in the life of the universe.

Modification of Newtonian Dynamics (MOND): Not everybody believes that the orbital discrepancies are due to dark matter. Some argue that at low accelerations, the effective gravitational force falls off more slowly than predicted by GR (i.e. modifying the inverse square law).

Which of the above is correct? We don't know. If you figure out what dark matter is, give me a call... after collecting your Nobel Prize!

23.3 Modelling the Universe

The different ways that matter/radiation/dark (vacuum) energy change with volume can be used to model the expansion of the universe. This is an iterative process. First, estimate the various energy-densities. Then feed them into your model and calculate how the universe would have expanded. Then compare with observational data and repeat as necessary until you find the best fit.

The easiest way to give you a flavour of this is to show how the mix of energy-density affects the expansion of the universe over time. As we know that the universe is close to spatially flat, we can simplify by setting $\Omega_K = 0$. We can then compare the time evolution of a universe containing only matter, only radiation or only vacuum energy.

A flat universe composed solely of matter energy is called an *Einstein-de Sitter universe* because it was a scenario studied by Einstein along with William de Sitter in the 1930s. We start from Equation 22.24 including only matter energy-density in the equation. Don't forget that $\overline{\rho_M}$ is a constant. It is the matter energy-density when $a = 1$. Multiplying both sides by a^2 and taking the square root gives an expression for $\dot{a}$. Inverting this gives $\frac{dt}{da}$, which we integrate as shown (the constant of integration is zero because we set $a = 0$ at $t = 0$). The result shown in Equation 23.2 is the relationship between a and t where C is a constant. We differentiate this to calculate $\dot{a}$ and $\ddot{a}$ in terms of t. Dividing the result for $\dot{a}$ by that of a gives H, which decreases over time.

$$\textit{Matter-only :} \qquad \dot{a} = \frac{da}{dt} = \sqrt{\frac{k\overline{\rho_M}}{3}}\, a^{-\frac{1}{2}}, \qquad \textit{using :} \left(\frac{\dot{a}}{a}\right)^2 = \frac{k\overline{\rho_M}}{3a^3},$$

$$\frac{dt}{da} = \sqrt{\frac{3}{k\overline{\rho_M}}}\, a^{\frac{1}{2}}, \implies t = \frac{2}{3}\sqrt{\frac{3}{k\overline{\rho_M}}}\, a^{\frac{3}{2}}, \tag{23.2}$$

$$a = C t^{\frac{2}{3}}, \qquad \dot{a} = \frac{2}{3} C t^{-\frac{1}{3}}, \qquad \ddot{a} = -\frac{2}{9} C t^{-\frac{4}{3}}, \qquad H = \frac{\dot{a}}{a} = \frac{2}{3t}.$$

Repeating the exercise for a universe with only radiation energy gives the results in Equation 23.3, where C is a different constant (i.e. different from that in the matter-only model). Again, the value of H decreases over time.

$$\textit{Radiation-only :} \qquad \dot{a} = \frac{da}{dt} = \sqrt{\frac{k\overline{\rho_R}}{3a^2}}, \qquad \textit{using :} \left(\frac{\dot{a}}{a}\right)^2 = \frac{k\overline{\rho_R}}{3a^4},$$

$$\frac{dt}{da} = \sqrt{\frac{3}{k\overline{\rho_R}}}\, a, \implies t = \frac{1}{2}\sqrt{\frac{3}{k\overline{\rho_R}}}\, a^2, \tag{23.3}$$

$$a = C t^{\frac{1}{2}}, \qquad \dot{a} = \frac{1}{2} C t^{-\frac{1}{2}}, \qquad \ddot{a} = -\frac{1}{4} C t^{-\frac{3}{2}}, \qquad H = \frac{\dot{a}}{a} = \frac{1}{2t}.$$

The third exercise is to do the same for a universe that contains only vacuum energy. This is called *de Sitter space* when the cosmological constant is positive (expanding effect). It is called *anti-de Sitter space* if the cosmological constant is negative (contracting effect), but Λ is positive in our universe, so I will address only that possibility. Starting with the first Friedmann equation from Box 22.2 (shown again below), we assume that there is only vacuum energy and the universe is spatially flat ($\rho = 0$, $\mathbf{K} = 0$). This gives $\left(\frac{\dot{a}}{a}\right)^2 = \frac{\Lambda}{3}$, which is H^2. As Λ is a constant, we know that H is a constant.

$$\left(\frac{\dot{a}}{a}\right)^2 = \frac{k\rho}{3} + \frac{\Lambda}{3} - \frac{\mathbf{K}}{a^2}, \qquad \textit{Friedmann-1 : expansion rate.}$$

$$\frac{\ddot{a}}{a} = \frac{\Lambda}{3} - \frac{k}{6}(\rho + 3P), \qquad \textit{Friedmann-2 : acceleration.}$$

We can use this result to calculate a, $\dot{a}$ and $\ddot{a}$ in terms of t, as shown in Equation 23.4. The result is that a, $\dot{a}$ and $\ddot{a}$ all grow exponentially over time. If H is constant, then each time the universe scale factor a increases by a factor of two, the speed of growth $\dot{a}$ increases by a factor of two and the acceleration $\ddot{a}$ increases by a factor of four. The expansion of the universe *accelerates*. You can check that this is consistent with the Friedmann acceleration equation by setting Λ positive and the other energy densities and pressure to zero. This also gives the result: $\ddot{a} = \frac{\Lambda}{3}a = H^2 a$.

$$\textit{Vacuum energy-only :} \qquad \dot{a} = \frac{da}{dt} = \sqrt{\frac{\Lambda}{3}}\, a, \qquad \textit{using :} \left(\frac{\dot{a}}{a}\right)^2 = \frac{\Lambda}{3},$$

$$\frac{dt}{da} = \sqrt{\frac{3}{\Lambda}}\, a^{-1}, \implies t = \sqrt{\frac{3}{\Lambda}}\ln a, \implies a = C e^{\sqrt{\frac{\Lambda}{3}}t}, \tag{23.4}$$

$$a = C e^{Ht}, \qquad \dot{a} = HC e^{Ht} = Ha, \qquad \ddot{a} = H^2 C e^{Ht} = H^2 a, \qquad H = \sqrt{\frac{\Lambda}{3}}.$$

I want to take a moment to be clear what we mean when we say that the expansion of the universe is accelerating or decelerating. Suppose our universe were matter-only. If you check back to Equation 23.2, you will see that $\ddot{a}$ is negative and $\dot{a}$ decreases over time. This is a *decelerating* universe. Imagine we look out and measure the Hubble parameter to be 70 km s^{-1} Mpc^{-1}. We observe two galaxies: galaxy A is one megaparsec away and galaxy B is two megaparsecs away. Their separation speeds from us *today* will be 70 and 140 km s^{-1} respectively (obviously). We wait about 20 billion years and measure again when galaxy A is two megaparsecs away. Even though it is twice as far away, its speed of recession would be less than 70 km s^{-1}, because over that time period the value of the Hubble parameter would have dropped by more than a half. This is what we mean by a decelerating universe. The speed of separation of two distant galaxies decreases over time.

Let's now turn to the scenario of a universe containing only vacuum energy. You can see from Equation 23.4 that $\ddot{a}$ is positive and the value of $\dot{a}$ increases over time. This is an *accelerating* universe. In this scenario, the Hubble parameter H is constant. We observe galaxy A today to have a separation speed of 70 km s^{-1} Mpc^{-1}. In about 20 billion years, it would be two megaparsecs away. The Hubble parameter is constant, so it would be moving away at 140 km s^{-1}. This is the acceleration we are referring to. Over time, the speed of separation between two distant galaxies grows.

23.4 Radiation's Trivial Contribution

The contribution of radiation to the energy-density of the universe is trivial. Current estimates put it at $\Omega_R < 10^{-4}$. Some of you may be surprised and be led to wonder why you ever have to wear sunglasses if radiation is so insignificant. The answer is that, in cosmological terms, we are darn close to the sun (this is fortunate because without its energy, life would not have evolved). We are also darn close to the other stars of the Milky Way. If you turned off those stars, all would be black to the naked eye, except for a couple of faint smudges. In cosmological terms, the radiation of galaxies is negligible.

Most significant is the CMB, which is about 60% of total radiation energy. The bulk of the remaining 40% is from neutrinos. These are fermions (matter particles) that we include as radiation because they are low mass and typically move nearly at light speed. Their equation of state is similar to radiation: $P \approx \frac{\rho}{3}$ (see optional Subsection 22.4.2).

At an early stage, radiation energy would have dominated the energy mix. However, it is estimated that radiation domination lasted only about 50,000 years (well before the emission of the CMB) after which the expansion of the universe led to matter taking over the dominant role.

23.5 The Cosmic Age Problem: Globular Clusters

This leaves only two significant sources of energy-density: matter and vacuum energy. Following Hubble's discovery of the expanding universe, everybody initially believed there was no cosmological constant (no vacuum energy). After all, Einstein stuck it in the equation in an attempt to hold the universe static. No static universe... no need for Λ. However, setting $\Lambda = 0$ leads to the matter-only scenario of the Einstein-de Sitter universe. This brings problems of its own and gives me a good excuse to introduce another piece of observational evidence: the dating of globular clusters (stars tightly grouped together in a galaxy).

For a matter-only universe, we calculated in Equation 23.2 that $H = \frac{2}{3t}$, which can also be written as $t = \frac{2}{3H}$. Substituting with the current value for the Hubble parameter, typically labelled H_0, should give us the current age of the universe based on the matter-only expansion profile. For the current value of the Hubble parameter, I am using $H_0 = 70$ km s^{-1} Mpc^{-1}, which is equivalent to about 7% increase (0.07) per billion years. Therefore, quantified per year of time (or per light year of distance) $H_0 \approx 7 \times 10^{-11}$. This puts the age of the universe at approximately 9–10 billion years old, as shown in Equation 23.5. Cosmologists know this is wrong. It is *too young*.

$$t = \frac{2}{3 H_0} \approx \frac{2}{3 \times (7 \times 10^{-11})} \approx 9.5 \; billion \; years, \qquad Matter\text{-}only \; universe. \qquad (23.5)$$

To explain the problem, I must give a brief introduction to the life cycle of stars. I am sure you have noticed a big yellow thing in the sky called the sun. Like other stars, its gravitational pressure fuses hydrogen to helium, releasing energy in the form of light (plus about 1% in the form of neutrinos). In about 5 billion years, it will run

Hertzsprung–Russell Diagrams

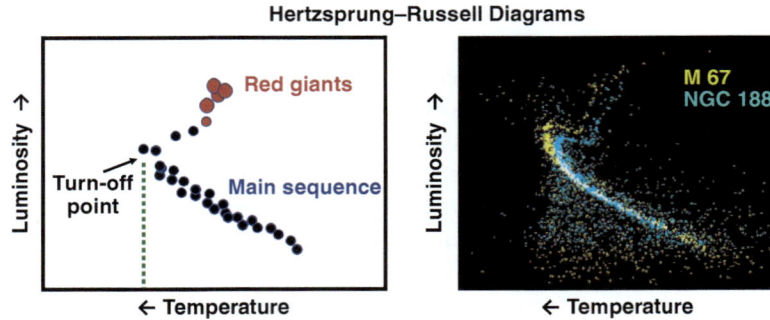

Figure 23.3 Dating globular clusters. Image adapted from work by Worldtraveller.

out of hydrogen. At that stage, its outer layers will collapse inwards, generating enough pressure for helium to fuse into larger atoms such as carbon. The sun then will expand into what is called a red giant.

This pattern of development is typical of most stars. While burning hydrogen, they are said to be on the *main sequence*. When they run out of hydrogen and become red giants, they leave the main sequence. This point of departure is the key to dating globular clusters.

Stars vary in size from about a tenth the mass of the sun up to 200 times its mass. The larger the star, the greater its internal gravitational pressure, and the faster its fusion. More massive stars shine with greater intensity (called *luminosity*) and at a higher temperature (described as bluer/redder for hotter/cooler). Therefore, the temperature and luminosity of a star are closely related. Astronomers map them together on *Hertzsprung–Russell diagrams*. The left of Figure 23.3 illustrates this. The black dots represent stars in the cluster that are still fusing hydrogen. The more massive the star, the higher its luminosity and temperature (note that, by convention, the temperature axis runs from high to low). Choose any star on the main sequence (i.e. hydrogen fusing) and you can calculate its mass either from its temperature or from its luminosity.

It turns out that the more massive the star, the *shorter* its lifespan in terms of hydrogen fusion, because it burns hydrogen so much faster. The relationship is shown as Equation 23.6. The sun ($m = 1$) has a lifespan of about 10 billion years. In contrast, a star 10 times as massive will last less than 50 million years.

$$\text{Hydrogen-fusing lifespan:} \approx \frac{10^{10}}{m^{2.5}} \text{ years}, \qquad m: \text{star mass in solar units.} \tag{23.6}$$

Now for the clever bit. You can lay out the stars in a globular cluster as illustrated on the left of Figure 23.3. As a reminder, the black dots are hydrogen-fusing stars. Once they run out of hydrogen, their temperature drops and they become red giants. This is called the *turn-off point*. You calculate the *mass* of the stars at the turn-off point from their temperature (dotted green line). The mass tells you their *lifespan* (Equation 23.6). As they are at the turn-off point, they are running out of hydrogen and have run through their lifespan, so we know the globular cluster is at least that old.

The right of Figure 23.3 shows two sets of real data. You can see that the cluster labelled in blue (NGC188) is older than that in yellow (M67). The yellow M67 data shows more massive stars still on the main sequence. In the older NGC188 cluster shown in blue, the equivalently massive stars have already burned out. The estimated age of the NGC188 globular cluster is 6.5 billion years compared with 4 billion years for the M67 cluster.

This technique has been used to identify much older globular clusters. For example, M92, which is in the Milky Way 27,000 light years from earth, dates to over 13 billion years old. Therefore, the result in Equation 23.5 of only 9.5 billion years old cannot be correct. You cannot have things in the universe that are older than it is! But the only assumptions we made are that the universe is nearly spatially flat (which we know from observational data) and that there is no cosmological constant. Clearly, we must revisit the assumption that $\Lambda = 0$.

23.6 The Accelerating Universe

As the universe is either spatially flat or very nearly flat and radiation energy-density is negligible ($\Omega_R \approx 0$), a glance back at Equation 22.28 shows that $\Omega_\Lambda + \Omega_M \approx 1$. Let's look at how the presence of Ω_Λ alters the acceleration $\ddot{a}$ of the universe. There is a measure called the *deceleration parameter* typically labelled q, which can be quantified in terms of Ω values (the reason cosmologists used this measure is that they expected the expansion of the universe to be decelerating). Equation 23.7 shows the first Friedmann equation for a flat universe with only matter and vacuum energy (you can refer back to both Friedmann equations in Box 22.2). On the right it shows the formulas for Ω_M and Ω_Λ. Below this is the Friedmann acceleration equation. For Equation 23.8, I have divided both sides by $H^2 = \left(\frac{\dot{a}}{a}\right)^2$ and substituted in the values for Ω_Λ and Ω_M. This gives an expression for $-q$, which I will call the *acceleration parameter*.

$$H^2 = \left(\frac{\dot{a}}{a}\right)^2 = \frac{k\rho_M}{3} + \frac{\Lambda}{3}, \qquad \Longrightarrow 1 = \frac{k\rho_M}{3H^2} + \frac{\Lambda}{3H^2} = \Omega_M + \Omega_\Lambda, \qquad (23.7)$$

$$\frac{\ddot{a}}{a} = \frac{\Lambda}{3} - \frac{k\rho_M}{6}, \qquad Acceleration : matter\ and\ vacuum\ energy\ only,$$

$$-q = \frac{\ddot{a}a}{\dot{a}^2} = \frac{\Lambda}{3H^2} - \frac{k\rho_M}{6H^2} = \Omega_\Lambda - \frac{\Omega_M}{2}, \qquad Acceleration\ parameter : -q. \qquad (23.8)$$

The key point is that if the acceleration parameter $(-q)$ is positive, then the expansion of our universe is accelerating, i.e. $\ddot{a}$ is positive. You can see this in the equation because the scale factor a must be positive, and our universe is expanding, so $\dot{a}$ is positive. You can see from the right of Equation 23.8 that the acceleration parameter turns positive when Ω_Λ is more than half Ω_M. We know $\Omega_\Lambda + \Omega_M \approx 1$. This means the expansion of the universe switches from decelerating to accelerating when $\Omega_\Lambda > \frac{1}{3}$. Let me summarise with two important points:

- According to the Einstein field equations (EFEs) on which the Friedmann equations are based, the expansion of the universe accelerates only in the presence of vacuum energy. Both matter and radiation energy-density always lead to deceleration.
- The expansion of the universe accelerates only if $\Omega_\Lambda > \frac{1}{3}$.

23.6.1 Type 1a Supernovae

It was late in the 1990s that the accelerating expansion of the universe was discovered. Before discussing these results, I want to explain the role of type 1a supernovae in the analysis. To assess the value of the Hubble parameter, we need to know how far away a galaxy is and its speed of recession. The latter is fairly easy using the redshift in its electromagnetic emissions. The former is trickier.

Back in Section 21.2, I told the story of Henrietta Leavitt's discovery of Cepheid variable stars, which have a predictable luminosity allowing them to be used as *standard candles* to judge distance. Hubble used Cepheid variables for his famous discovery that the universe is expanding. Type 1a supernovae also serve as a standard candle but over much greater distances because they are so bright. They all produce standard emissions[2] because they explode with the *same mass*.

The reason for this is illustrated in Figure 23.4. First you need a *white dwarf* . This is the remains of a star that once was like the sun. It has finished fusing available hydrogen, but is too small to form a supernova. Second, it

Figure 23.4 Left: Artist's conception of the process leading up to a type 1a supernova. *Source:* NASA / LAMBDA Archive Team / NASA / Public Domain. Right: Graphic of supernova from Nobel press release.

---

2 Technical point: Supernovae 1a are standardisable because the brighter ones have longer explosions, so you can use a combination of peak brightness and lifetime of explosion to work out their luminosity.

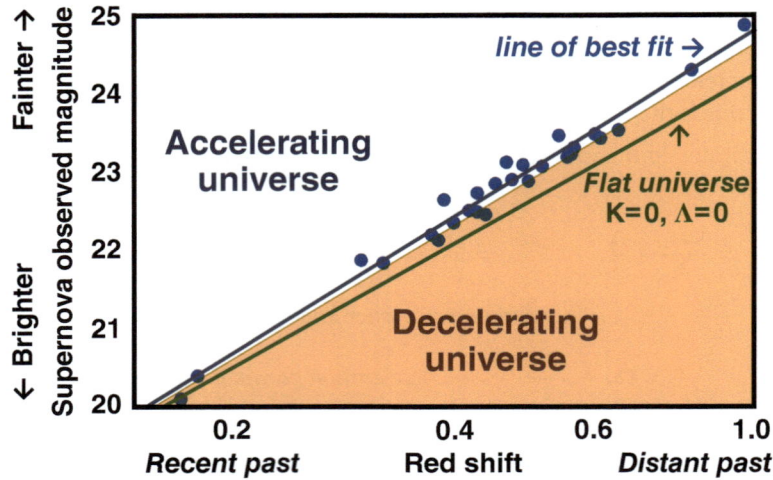

Figure 23.5 *Supernova 1a data adapted from work by Irani Ardeshir, Journal of High Energy Physics, Gravitation and Cosmology, January 2023.*

must be in a binary system or have some form of mass available nearby. This is more common than you might expect. It turns out that over 50% of stars are in binary systems and the larger the star, the more likely it is to be in one.[3]

The gravitational field of the white dwarf can strip off mass from the nearby star as illustrated on the left of the figure. The amusing cartoon on the right shows what happens next. When the white dwarf reaches the Chandrasekhar limit of 1.44 solar masses, it explodes into a supernova (for more details check back to Section 13.5). This is why type 1a supernovae explode with the same mass.

Perhaps 10% of all supernovae are type 1a. There is only about one every 500 years in a galaxy like the Milky Way. However, there is a vast number of galaxies out there, so there are enough to provide the standard candle needed for analysis.

23.6.2 Evidence of Acceleration

Each blue dot in Figure 23.5 is a type 1a supernova. The *x*-axis shows redshift, which is a measure of its speed of recession. The *y*-axis shows observed magnitude. As type 1a supernovae shine with standard brightness, this gives a measure of how far away they are (note that the higher the magnitude the fainter the light, see Box 23.3). The relationship between speed of recession and distance is, of course, the Hubble parameter. The further away the galaxy containing the supernova (higher magnitude, fainter light), the faster its speed of recession (higher redshift).

The green line labelled *flat universe* is the calculated result based on a spatially flat universe with *no* cosmological constant ($K = 0$, $\Lambda = 0$). Quite clearly, it does not fit the data. In the context of the EFEs, the data indicates the presence of vacuum energy in our universe.

The dark blue line is the best fit for the data. As you can see, it is consistent with an accelerating expansion of the universe ($\ddot{a} > 0$). The first evidence of acceleration was published in 1998 by the High-Z Supernova search team, an international collaboration charting type 1a supernovae.

At the time, the findings came as a big surprise for everybody and caused quite a stir. To quote Riess's 1998 paper: *presently, none of these effects appears to reconcile the data with $\Omega_\Lambda = 0$ and $q_0 \geq 0$ (where q_0 is the current value of the deceleration parameter). Everybody assumed the universe would be decelerating, but the results showed the opposite. The expansion of the universe is accelerating. The discovery won Riess, Perlmutter and Schmidt the 2011 Nobel Prize.

---

3 See *Stellar Multiplicity* by Duchene and Kraus: https://arxiv.org/pdf/1303.3028 (link as of February 2025).

> **Box 23.3 Magnitude measure of brightness**
>
> The brightness of stars is often measured in units of *magnitude*. This can be confusing because the higher the magnitude, the dimmer the star. It all started with the observations of Hipparchus circa 150 BC. Claudius Ptolemy (100–170 AD), famous for his geocentric model of the universe, built on Hipparchus' catalogue. He rated stars from magnitude 1 (clearly visible to the naked eye) to magnitude 6 (much dimmer). It is this that grew into the modern magnitude measure.
>
> *Apparent* or *observed* magnitude is as seen from earth. If you are comparing two standard candle stars, such as type 1a supernovae, apparent/observed magnitude provides a measure of their distance away. *Absolute* magnitude is the calculated magnitude measured from a set distance and is used to compare emission levels of different types or sizes of star.

23.7 Summary: The Energy Mix of the Universe

The current estimates of the energy-density mix of the universe are that the majority is dark (vacuum) energy $\Omega_\Lambda \approx 0.7$, with matter energy (including dark matter) at $\Omega_M \approx 0.3$. Radiation energy is negligible at $\Omega_R < 10^{-4}$ but would have been the dominant energy form early in the life of the universe (as discussed in Section 22.5). And so far, across the distances we can observe, the universe looks spatially flat: $\Omega_K \approx 0$. This combination of estimates is called the Lambda-CDM or ΛCDM model (CDM stands for Cold Dark Matter).

Figure 23.6 combines various observational estimates in a graphic that I will use to summarise the story so far. We know that Ω_R is very small. This means that the only significant contributions to total energy-density are matter energy-density Ω_M (shown on the x-axis) and vacuum energy-density Ω_Λ (y-axis). Each axis is a measure of energy-density as a fraction of the critical density. In the case that $\Omega_M + \Omega_\Lambda = 1$, the universe is flat. This is shown as the dashed grey line in the figure. If $\Omega_M + \Omega_\Lambda < 1$, there is positive curvature ($\mathbf{K} = +1$) and the universe is closed. If $\Omega_M + \Omega_\Lambda > 1$, there is negative curvature ($\mathbf{K} = -1$) and the universe is open.

Spatial curvature would affect the apparent size of objects at distance. Positive curvature would make distant objects appear larger, so in a given area of view you would see fewer galaxies. With negative curvature you would see more. Surveys have detected no such distortion. In addition, any spatial curvature would affect the relative size of temperature variations in the CMB. All evidence points to the universe being spatially flat, or close to spatially flat. The results from analysing the CMB are shown in Figure 23.6 in green. As an aside, please don't forget that the universe being *spatially* flat doesn't mean the *spacetime* has no curvature. The spacetime contains energy-momentum, so it definitely has curvature.

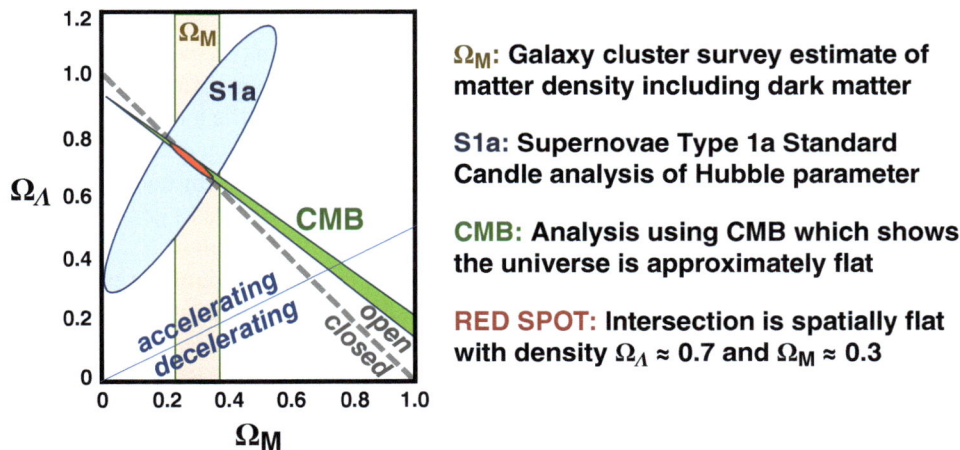

Ω_M: Galaxy cluster survey estimate of matter density including dark matter

S1a: Supernovae Type 1a Standard Candle analysis of Hubble parameter

CMB: Analysis using CMB which shows the universe is approximately flat

RED SPOT: Intersection is spatially flat with density $\Omega_\Lambda \approx 0.7$ and $\Omega_M \approx 0.3$

Figure 23.6 Energy mix graphic adapted from Vallastro.

To calculate the value of matter energy-density Ω_M, we take into account the presence of dark matter. The orbital speeds of galaxies deviate significantly from those expected based on the presence of visible matter. Ω_M can also be measured by studies of gravitational lensing, where the extra mass is assessed by the amount that passing light is deflected. Current surveys combined with data from galaxy clustering put the total Ω_M at somewhere around 0.3, with 85% of this being dark matter. This is shown by the brown band in the figure. The energy-density in a spatially flat universe must match critical density i.e. $\Omega_{total} = 1$. If $\Omega_M \approx 0.3$, then Ω_Λ must be about 0.7.

Another indicator is the cosmic age problem. In Section 23.5, we calculated the age of a spatially flat universe with no vacuum energy. Based on the current value of the Hubble parameter, such a universe would be less than 10 billion years old. I showed you how astronomers date globular clusters using the turn-off point from the main sequence. Older globular clusters such as M92 are over 13 billion years old. For the age of the universe to match this, we must accept the presence of vacuum energy.

The next piece of evidence comes from analysing the expansion of the universe. Astronomers use type 1a supernovae as standard candles to assess the distance to galaxies. This data is mapped against their recession rate based on the redshift of their emissions. Figure 23.5 illustrated such a plot. It is inconsistent with $\Omega_\Lambda = 0$. Furthermore, it shows the expansion of the universe now is accelerating. For this to be the case, $\Omega_\Lambda > \frac{1}{3}$ as discussed at the start of Section 23.6.

Returning to Figure 23.6, the solid blue line distinguishes between assumptions that lead to an accelerating versus decelerating expansion (for acceleration $\Omega_\Lambda > 0.5\Omega_M$). The implication of the type 1a supernovae data is shown in light blue and labelled S1a. Combining the CMB, galaxy survey and S1a data, we arrive at the patch shown in bright red: $\Omega_\Lambda \approx 0.7$, $\Omega_M \approx 0.3$ and $\Omega_K \approx 0$. This is the ΛCDM model, which is the current best estimate of energy-density in the universe. It is summarised for reference in Box 23.4

Box 23.4 ΛCDM model of cosmology	present day estimates
$\Omega_\Lambda \approx 0.7,$ $\quad$ $\Omega_M \approx 0.3,$ $\quad$ $\Omega_R < 10^{-4},$ $\quad$ $\Omega_K \approx 0.$	

23.8 What Is Dark (Vacuum) Energy?

Note carefully in Figure 23.6 that for any model to even vaguely match observations, it must include the presence of dark (vacuum) energy ($\Lambda > 0$). Based on Einstein's theory of GR and his field equations, the universe has to contain it. Throughout this chapter, I have used the term *vacuum* energy, because that is what it is: positive energy of the vacuum.

Written in terms of mass equivalence, the critical density for the universe to be spatially flat is about 10^{-26} kg m^{-3} (refer back to Equation 22.27 if needed). This means that vacuum energy is equivalent to mass of about 7×10^{-27} kg m^{-3}, which is a few hydrogen atoms per cubic metre. This doesn't sound much. For comparison, a cubic metre of hydrogen gas at atmospheric pressure contains about 10^{25} atoms. However, there is so much of what we might call empty space that vacuum energy accounts for about 70% of the total energy of the universe.

Perhaps the most obvious explanation of vacuum energy is that it is the energy of quantum fluctuations in the vacuum. This is an important component of quantum field theory (let me plug again my book *Quantum Untangling* if you want to learn more). The minimum energy of a quantum field, which is called its *ground state* energy, is *not* zero. Therefore, it should be no surprise that the vacuum has energy associated with it.

Problem solved? Not so fast. There is a major stumbling block with this simple-sounding explanation. Based on quantum field theory, there is a ground state energy for every possible quantum oscillation, such as every possible frequency of light. The smaller the wavelength of the oscillation, the higher its ground state energy. As you account for ever shorter wavelengths, the ground state energy becomes ever higher. This leads to the prediction of an infinite amount of vacuum energy. Oops!

You can rationalise eliminating some of these higher energy options. Any oscillation with a wavelength less than the Planck length (10^{-35}m) would put so much energy in such a small space that it would form a black hole. However, even with this adjustment, the calculation predicts about $10^{120}\times$ the observed value of vacuum energy.

To get to the right number, you would need to justify a high-energy cut-off of oscillation wavelengths less than about 1 millimetre. If this were true, we would see new physics at that scale. Yet the LHC has probed down to 10^{-18} m and found no such change.

Try as they might, physicists cannot see a way to make the numbers match, because any calculations based on quantum theory suggest there should be much *more* vacuum energy in the universe. This is known as the *vacuum catastrophe* or the *fine-tuning problem*.

We know something is affecting the rate of expansion of the universe, but currently we cannot explain what this vacuum energy is. Perhaps it is quantum fluctuations of the vacuum, but then why is it so small? Some suggest it might be a new as yet undiscovered quantum field. Others hypothesise that it reveals an error in Einstein's equations.

Another explanation is that there is a multiverse composed of a huge number of universes (we will discuss this later in Section 25.8) with randomly different levels of vacuum energy. Any universe with significantly more vacuum energy than ours (say $5 - 10\times$) would expand so fast that it would be empty. Therefore, for us to exist, we *must* be in one of the universes with a very low level of vacuum energy. This is called the *anthropic principle*. I think it's fair to say that cosmologists look on it as the argument of last resort.

This leaves cosmology in a rather embarrassing position. Something that we call dark matter holds the galaxies together. We can see its effect. It represents about 85% of the matter in the universe. But we don't know what it is. And then there is vacuum energy, which pushes the universe apart, accelerating the expansion. We calculate this represents 70% of the energy in the universe, but we cannot properly explain it. In total, the combination of dark matter and dark (vacuum) energy represents about 95% of the energy in the universe and we don't understand either. Oops! Oops!

Don't forget that the arguments in this chapter are built from observational data (the ΛCDM measure of energy-density) combined with the EFEs and the assumptions inherent in the cosmological principle (Section 21.1). Clearly, there is much we know, but also much we don't know. At the time of writing, this is the best we have.

I end this chapter with Box 23.5, which is an excerpt from the 2011 Nobel Prize press release announcing its award for the discovery of the accelerating universe. It sums things up nicely.

Box 23.5 Accelerating expansion of the universe, 2011 Nobel Prize

This rather wonderful press release, titled *Written in the Stars*, accompanied the award of the 2011 Nobel Prize to Riess, Perlmutter and Schmidt for showing in 1998 that the expansion of the universe is accelerating (i.e. $\ddot{a} > 0$):

Some say the world will end in fire; some say in ice... Robert Frost, *Fire and Ice*, 1920

What is the fate of the universe? Probably it will end in ice if we are to believe this year's Nobel Laureates. They have carefully studied several dozen exploding stars, called supernovae, in faraway galaxies and have concluded that the expansion of the universe is speeding up. The discovery came as a complete surprise even to the Nobel Laureates themselves. What they saw would be like throwing a ball up in the air, and instead of having it come back down, watching as it disappears more and more rapidly into the sky, as if gravity could not manage to reverse the ball's trajectory. Something similar seemed to be happening across the entire universe....

Whatever dark energy is, it seems to be here to stay. It fits very well in the cosmological puzzle that physicists and astronomers have been working on for a long time. According to current consensus, about three quarters of the universe consist of dark energy. The rest is matter. But the regular matter, the stuff that galaxies, stars, humans and flowers are made of, is only 5% of the universe. The remaining matter is called dark matter and is so far hidden from us. The dark matter is yet another mystery in our largely unknown cosmos. Like dark energy, dark matter is invisible. So we know both only by their effects – one is pushing, the other one is pulling. They only have the adjective "dark" in common. Therefore the findings of the 2011 Nobel Laureates in Physics have helped to unveil a universe that is to 95% unknown to science. And everything is possible again.

Chapter 24

After the Big Bang

Box 24.1 The big stretch   **by Max Williams (armchair physicist)**

The term Big Bang is misleading... our best theories of the early universe say that there was a time when our visible universe was incredibly small, hot and dense. That's it. You could extrapolate back from that and say: *Well, looking at the graph of the size of the universe, it crosses the zero line at this time, therefore, it must have come from nothing or a point of zero size, which we call a singularity*, but you would be mistaken in doing so. Lots of people have made this mistake, and the name also helps to create the commonly held misconception that there was a big explosion where previously there was nothing.

I could do the same for your mass, as a foetus. I'd say: *Well, looking at this graph, we can see the mass of the foetus crossing the zero line at this time here, therefore, the foetus came from a singularity!* But you didn't come from nothing. There was a single fertilized cell, which we know about, and then before that, something happened, but we don't know exactly what (and we don't really want to know, thanks, that's between your mom and dad).

Personally, I think we could cure a lot of public misunderstanding if we ditched the phrase *The Big Bang* and replaced it with *The Big Stretch*.

Box 24.1 is a comment that I stumbled across online. I love its mixture of wit and insight (well done Max, whoever you are). In this chapter we will analyse the evolution of the universe building up from the ΛCDM current energy mix (as summarised earlier in Box 23.4). This allows us to calculate the expansion profile over time and create a picture of the early universe. We then will track the development of the universe through to the modern era with its stars, galaxies and the cosmic web.

I should warn in advance that this is not the complete picture. In Chapter 25, we will examine some serious problems that emerge if you use the ΛCDM model as a stand-alone description of the Big Bang, and how an initial period of exponential *inflation* (possibly less than 10^{-30} seconds) offers a solution. According to this, it was at the end of inflation that the universe was flooded with radiation and particles (called *reheating*). But for now we will focus on the development of the universe after all of this.

Back in Section 23.3, *Modelling the Universe*, we analysed how the presence of each type of energy-density (matter/radiation/vacuum) affects the growth of the scale factor. In Equation 23.5, we used this to calculate the age of the universe based on a matter-only model. A similar approach using the ΛCDM model (Box 23.4) correctly weighted for each component in the energy-density mix gives the age of the universe as about 13.8 billion years. The change in scale factor goes hand in hand with changes in energy-density. Radiation energy dominated the early universe, perhaps for about 50,000 years. Matter energy then dominated until the importance of vacuum energy grew, leading to the accelerating expansion of the universe that we see today.

Figure 24.1 gives an overview of the history of the universe, which I will regularly refer to. Let me take the liberty of dividing the history of the universe into two parts (ignoring at this stage any initial inflationary period).

The first is up to the release of the cosmic microwave background (CMB), some 400,000 years after the Big Bang. Observational evidence is limited for this period. Telescopes rely on electromagnetic (EM) radiation such

Untangling General Relativity: The Intuitive Self-Study Guide, First Edition. Simon Sherwood.
© 2026 John Wiley & Sons Ltd. Published 2026 by John Wiley & Sons Ltd.

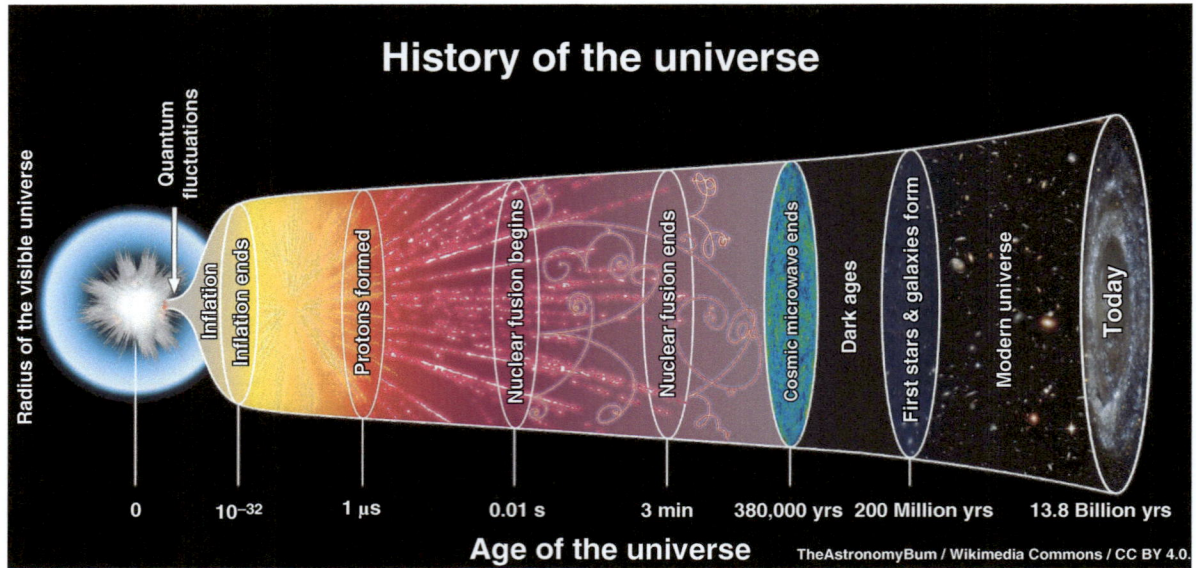

Figure 24.1 Schematic of the evolution of the universe. *Source:* TheAstronomyBum / Wikimedia Commons / CC BY 4.0.

as visible light, infrared and ultraviolet. They can peer no further back in time because the universe was opaque to EM radiation prior to the CMB, as will be explained later. This first early period included baryogenesis, which is the formation of composite particles from the elementary particles created shortly after the Big Bang (such as protons from quarks), followed by their nuclear fusion into atomic nuclei.

The second period is after the CMB, during which the galaxies formed. This is from about 400,000 years up to the present day. At the end of the chapter, I will touch on the long-term future of the universe, which looks a bit depressing, as the expanding universe carries more and more of the galaxies around us out of view, leaving us all alone in a sea of darkness.

24.1 Dating the Early Universe

In discussions of the early universe, you will see timelines, such as that protons formed when the temperature was between 10^{12} and 10^{13} Kelvin, which was about *one microsecond* after the Big Bang. An inquiring mind might question how physicists make the leap from temperature to age, given that none of them were around during the Big Bang to glance at their watches and take notes.

The answer lies in the Friedmann equations. From the temperature, we can calculate energy-density. Assuming the universe is spatially flat, this must be at critical energy-density ρ_{crit} from which we can calculate the Hubble parameter at the time. We then use the relationship between the Hubble parameter and the age of the universe, which we established back in Equation 23.3 for a radiation-dominated universe. The result is the simple approximate rule of thumb shown in Equation 24.1, equating the age of the universe in seconds to the temperature in Kelvin.

$$Age\ of\ early\ universe \approx \frac{10^{20}}{T^2}\ seconds, \qquad T\ :\ temperature\ in\ Kelvin. \tag{24.1}$$

This formula is based on the early radiation-dominated phase, which is the first 50,000 years after the Big Bang. After that, the growing presence of matter in the energy mix alters the relationship, but it remains a useful, if approximate, rule of thumb up to the emission of the CMB. The gory calculation is shown in Box 24.2 for those readers who want it. It also shows how to convert from energy level in electron volts (eV) to temperature in Kelvin.

Box 24.2 Big Bang dating

The title of this box sounds like a dubious website for hooking up physicists. Sadly, I must disappoint readers. Our topic is how to translate from energy/temperature levels to the age of the universe during its early development. The relationship in Equation 24.1 is designed for the early stages, when radiation dominated the energy mix (the first 50,000 years or so after the Big Bang). However, it provides a useful approximate rule of thumb up to the emission of the CMB about 400,000 years after the Big Bang.

You may see energy levels expressed in electron volts (eV). The temperature in Kelvin is *approximately* $10^4 \times$ the measure in eV. Note that one GeV is 10^9 eV. Thus, for example, 10^{16} GeV equates to a temperature of 10^{29} Kelvin, which in turn equates to an age of about 10^{-38} seconds.

The conversion from eV to Kelvin uses *Wien's law* relating the temperature of a hot object to the energy of the peak in its radiation spectrum, giving: 1eV = 11,605 Kelvin.

Assuming that radiation dominates the energy mix, we can approximate the energy-density ρ from temperature using the *Stefan-Boltzmann law* shown on the left of Equation 24.2: α is called the *radiation constant*, k_B is the Boltzmann constant and $\hbar$ is the reduced Planck's constant.

The next step is to calculate the Hubble parameter H using the relationship from Box 22.2. From the value of H, we estimate the age of the universe using the formula $t = \frac{1}{2H}$ for a radiation-dominated universe (Equation 23.3), arriving at the relationship between age (seconds) and temperature (Kelvin) shown on the right of Equation 24.3.

$$\rho = \alpha T^4, \qquad \alpha = \frac{\pi^2 k_B^4}{15c^2\hbar^3} \approx 8 \times 10^{-16} \text{ Jm}^{-3}\text{Kelvin}^{-4}, \tag{24.2}$$

$$H = \sqrt{\frac{8\pi G\rho}{3c^2}} = \sqrt{\frac{8\pi G\alpha \, T^4}{3c^2}} \approx \sqrt{\frac{8(3)(7 \times 10^{-11})(8 \times 10^{-16})}{3(9 \times 10^{16})}} \, T^2,$$

$$H \approx (2 \times 10^{-21}) \, T^2 \text{ (seconds)}^{-1}, \quad \Longrightarrow \quad Age :\approx \frac{1}{2H} \approx \frac{10^{20}}{T^2} \text{ seconds.} \tag{24.3}$$

24.2 Baryogenesis (Protons and Neutrons Form)

We cannot look further back than about 400,000 years after the Big Bang with telescopes because before this, the universe was too hot for neutral atoms to form. It was an opaque plasma of interacting radiation and particles: protons, neutrons, electrons and photons. Perhaps it will be possible in the future for gravity wave detectors such as Laser Interferometer Gravitational-Wave Observatories (LIGO) (see Section 19.4) to delve deeper, but for now, we are a bit stuck for observational evidence.

All is not lost! Much has been learned from theory combined with results at high-energy colliders. The early universe was filled with quarks, electrons, neutrinos and (still-mysterious) dark matter. As the universe expanded, energy-density decreased. This was the era of radiation-dominance. Alongside the radiation, most of the particles were relativistic, moving at close to light speed. Section 22.4.2 explains how this affects the relationship between energy-density ρ and pressure P. As a result, total energy-density decreased at a rate approaching a^4 (see Section 22.5).

Only when the temperature dropped to under 10^{13} Kelvin (energy levels under 1 GeV) were the quarks able to group in threes and form protons and neutrons, collectively called *baryons*. At higher temperatures there would have been a *quark–gluon plasma*, more familiarly called *quark soup*. Although some million times hotter than the centre of the sun, we can reproduce this in high-energy colliders such as the Large Hadron Collider (LHC) at CERN and the Relativistic Heavy Ion Collider (RHIC) at Brookhaven. At the RHIC, gold and lead atoms are ionized, and then the nuclei are accelerated close to light speed and smashed together. What fun!

The transition below 10^{13} Kelvin from a universe of quark soup into one with baryons would have occurred about a microsecond (10^{-6} seconds) after the Big Bang as shown in Figure 24.1. You can calculate this timeline easily using Equation 24.1. The result was a baryon–electron–photon plasma. I should mention that we will discuss later the effect of oscillations through this interactive plasma (Baryon Acoustic Oscillations). Most of the resulting baryons would be protons, because free neutrons are unstable outside of an atomic nucleus (conditions were still too hot for neutral atoms to form).

High-energy collisions in colliders produce particles and antiparticles in equal numbers, so one would expect the quark soup to have been composed of equal numbers of quarks and anti-quarks. As the temperature dropped, logically these would have grouped together into equal numbers of baryons and anti-baryons (e.g. protons and anti-protons). When a baryon meets its anti-baryon partner, they annihilate. Thus, the outcome would be a large number of photons (produced by annihilating baryon/anti-baryon pairs) and a smaller but equal number of residual baryons and anti-baryons.

Instead, the result in our universe is copious baryons, but only a *negligible* number of anti-baryons. What happened? It seems that the baryons and anti-baryons annihilated each other in such a way that baryons were left over. This is called *baryogenesis*. What caused this excess of matter over antimatter? While the Standard Model of particle physics allows for asymmetry (see Box 24.3), it is not enough to explain the discrepancy. In short, we don't know.

Box 24.3 Matter-antimatter asymmetry

Imagine David Beckham kicks a football with his left foot, bending the ball to the right. If you watched this in a mirror, you would see him kicking with his right foot and bending it to the left. The physics is the same. This symmetry is labelled *P* for *parity*. The scientific community was stunned by Chien-Shiung Wu's experiment in 1957, which showed that parity is not conserved in weak force interactions (the force involved in radioactive decay). Wolfgang Pauli's initial reaction was: *That's total nonsense!* Wu showed that certain interactions always produced neutrinos that spin anticlockwise in their direction of motion. Conversely, any anti-neutrinos spin clockwise. Wu's experiment is described in detail in Chapter 24 of my book *Quantum Untangling*.

Although shocked by the violation of parity, there was reassurance that a combined change in parity *P* (mirror-image) plus a change from particle to anti-particle (labelled *C*) restored the symmetry (together labelled *CP* symmetry). However, in 1964, James Cronin and Val Fitch showed in a study of the decay of neutral K-mesons (curious particles composed of a quark and an anti-quark) that even *CP* symmetry is violated by the weak force. They were awarded the 1980 Nobel Prize with the press release commenting: *The new knowledge permits us to make a distinction between matter and antimatter in an absolute and not only relative way.*

24.3 Nuclear Fusion (Light Atomic Nuclei Form)

Let's return to our story. As discussed in the last section, about one microsecond after the Big Bang, the temperature had dropped below about 10^{13} Kelvin and the universe transitioned from a quark soup (quark–gluon plasma) to a baryon–electron–photon plasma. There are three quarks per baryon. Initially, the baryons formed in the plasma would have been protons (electric charge +1), which are composed of two up quarks (each has charge $+\frac{2}{3}$) and one down quark (charge $-\frac{1}{3}$). They are stable and form the atomic nucleus of hydrogen.

Neutrons (electrically neutral) are composed of two down quarks and one up quark but are unstable unless combined with protons in an atomic nucleus. Once the temperature of the early universe had dropped further, then, in addition to the hydrogen nuclei (formed by the protons), some of the protons and neutrons fused into helium nuclei (typically two protons, two neutrons) and a trace of lithium (typically three protons, four neutrons).

For nuclear fusion to occur (called *nucleosynthesis*), the temperature had to be below about 10^{11} Kelvin (0.01 seconds after the Big Bang). This is equivalent to about 10 MeV, which is close to the binding energy of each proton and neutron in helium (7.1 MeV/baryon). Therefore, the temperature had to be lower than this for any helium nuclei to remain stable. Until the temperature was below 10^{11} Kelvin, energetic collisions would have broken apart any newly formed helium nuclei.

On the other hand, the temperature had to be high enough for the positively charged protons to have enough energy to overcome their mutual EM repulsion. A few minutes later, the temperature fell below this level and the nucleosynthesis stopped. It would take another several hundred million years of what is called the *dark ages* before stars formed, and their immense gravitational pressure rekindled the nuclear fusion process such as we see today in our lovely life-giving sun.

Please note carefully that what I have described in this section is the formation of electrically charged *atomic nuclei*. The temperature was still much too hot for electrically neutral atoms. If an atom was forged from electrons and an atomic nucleus, it would be re-ionised instantly. The constant interaction of EM radiation with electrons and atomic nuclei made the universe opaque, a bit like a fluorescent light.

24.4 Cosmic Microwave Background

The next important reference point is the CMB. This is as far as we can see back in time with EM detectors (microwave telescopes in the case of the CMB). CMB is the radiation that was released when the hot early universe cooled down to a low enough temperature for stable atoms to form (check back to Box 22.3 if needed). This bizarrely is called *recombination*. I have no idea why as they had never stably combined before. The photons could then travel largely unhindered around the universe, which is called the *decoupling* of radiation from matter. You may see this process dated variously as 300,000 years after the Big Bang, 380,000 years (as in Figure 24.1), 400,000 years or even the somewhat unnervingly exact estimate of 378,000 years. This is about 2% of the age of the universe stretching from the Big Bang to now.

We estimate the age of the CMB by calculating the temperature of recombination/decoupling. The temperature calculation is not as easy as it sounds. You might be tempted to think that stable atoms would form as soon as the energy of the *majority* of photons fell below the ionisation energy of hydrogen ($13.6\,eV$, equivalent to a temperature of about 10^5 Kelvin). However, there were about 10^9 photons per atomic nucleus, so using the average energy doesn't work. Stable atoms could form only when *less than about one in every billion* photons had the ionisation energy. As a result, recombination/decoupling doesn't occur until a much lower temperature of about 3,000 Kelvin.[1]

For a rough estimate, we can use the formula of Equation 24.1, albeit this is designed for the early radiation-dominated stages. At $T = 3,000$ Kelvin, $T^2 \approx 10^7$ and the age of the universe is about 10^{13} seconds, which is 300,000–400,000 years after the Big Bang.

Another approach is to look at how the CMB radiation has changed between its emission and what we detect today. The radiation was released at the wavelength of visible light (think yellowish-white) but is now microwave. The equivalent temperature of the CMB spectrum today is 2.7 Kelvin, which is about 1,100 times less energetic than at decoupling. Hopefully, you have come across the formula relating the energy of a photon to its frequency: $E = hf$ (where h is Planck's constant). As frequency f is inversely proportional to wavelength λ, the wavelength of the CMB photons has increased by a factor of about 1,100 since decoupling. This means that the scale factor a of the universe has increased roughly 1,100× since decoupling.

We then ask, how long ago was the scale factor a of the universe 1:1,100 of what it is today? As our universe was matter-dominated for most of the period, we can approximate with the maths for a pure matter-only universe. We established back in Equation 23.2 that, in the case of matter-only, $a = Ct^{\frac{2}{3}}$ where C is a constant. This gives the relationship in Equation 24.4 between the current age of the universe t_{now} (13.8 billion years), and the time of the CMB t_{cmb}. This in turn tells us that the CMB was emitted about 380,000 years after the Big Bang (which fits with our first estimate).

$$\frac{a_{now}}{a_{cmb}} = \left(\frac{t_{now}}{t_{cmb}}\right)^{\frac{2}{3}}, \quad \implies \quad t_{cmb} \approx \frac{t_{now}}{1,100^{\frac{3}{2}}} \approx \frac{13.8 \times 10^9}{3.6 \times 10^4} \approx 380,000 \text{ years.} \tag{24.4}$$

In summary, about 380,000 years after the Big Bang, the temperature of the expanding universe dropped to a level that allowed electrons to combine with atomic nuclei and form electrically neutral atoms (without the atoms being re-ionised by high-energy photons). Because the resulting atoms were electrically neutral, they no longer interacted constantly with the photons in what had been a baryon–electron–photon plasma. The photons were released as a burst of EM radiation. The residue of this is the CMB.

---

1 The equation quantifying the equilibrium point between ionised and neutral atoms is called the Saha equation.

The nature of the CMB tells us a lot about conditions at the time of recombination/decoupling. The microwave radiation from the CMB is highly homogeneous. Its radiation temperature observed in all directions varies only by about one in a hundred thousand from the average value. At the time of CMB emission, regions with higher than average energy/matter density would have affected its wavelength due to the redshift of that CMB light escaping a more intense gravitational field. These photons would be slightly less energetic, resulting in the observed temperature of the CMB from the region being slightly cooler.

The highly homogeneous CMB indicates that regional variation in energy-density was very small. Explaining this level of homogeneity is quite challenging and is one of the arguments for an earlier period of inflationary expansion, which we will discuss in detail in the next chapter.

The important point for galaxy formation is that there *is* variation in the CMB, albeit small. This indicates that there were regions of differing energy-density at the time of recombination/decoupling. These regions of higher energy-density are important because they attracted surrounding dark matter, leading to what are called *dark matter haloes* hosting the galaxies we observe today. The next part of the story is to examine the process of star and galaxy formation.

24.5   A Star Is Born

After recombination, it is harder to build an accurate timeline of events because it is no longer possible to tie those events to a particular temperature of the universe. How long were the dark ages? How long did it take for gravity to pull together what was predominantly atomic hydrogen and helium into dense masses with enough internal pressure to re-ignite the fusion process? How long after the Big Bang did stars start to shine? We don't exactly know.

However, post-CMB we have the benefit of modern telescope observations such as those of the James Webb telescope, which began delivering images in 2022. As we gaze into the distance, we are gazing back through time, seeing light emitted long long ago. As I write this chapter (June 2024), the James Webb telescope has just announced the detection of light from galaxy JADES-GS-z14-0, which its redshift reveals was emitted over 13.5 billion years ago.

This recently discovered galaxy is surprisingly well-developed and may lead to an adjustment in the date label of the first stars and galaxies in Figure 24.1. By the time you are reading this, I suspect telescopes will have detected galaxies even further out, but the current consensus is that it would have taken about 200 million years from the Big Bang for stars and galaxies to form, a time sometimes referred to as the *cosmic dawn*.

The fusion of hydrogen to heavier elements such as helium, lithium, carbon and oxygen releases energy. This is because the binding energy per nucleon (the general term for protons and neutrons) *increases*. Figure 24.2 shows the binding energy per nucleon compared with the number of nucleons in the nucleus of the element. The fusion

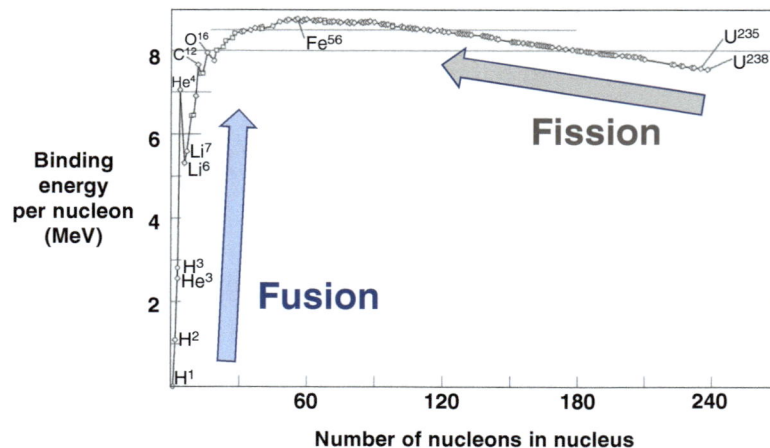

Figure 24.2   Nucleon binding energy. *Source:* Adapted from D. HUMMEL and David NOVOG, 2011.

process is resisted by the EM repulsion between the positively charged protons, but the huge pressure of gravity inside stars overcomes this resistance. Fusion occurs, allowing stars to shine and heavier elements to form.

The binding energy per nucleon continues to grow with the size of the atomic nucleus up to the formation of iron (Fe), which has 26 protons and about 30 neutrons (the exact number of neutrons depends on the isotope). For atoms larger than iron, the binding energy per nucleon *decreases*, so rather than releasing energy, it actually requires energy to create them. You might have thought that no atoms heavier than iron would exist, but they do. Take a look around you. I am sure some of you will be wearing silver (47 protons) or gold (79 protons).

Most elements heavier than iron are likely to have been forged by the energy of cataclysmic events such as supernova explosions (see also Section 13.5) and the merger of neutron stars. As discussed in Section 23.5, the lifetime of a star depends on its mass: the larger the star, the shorter its life. A star about 10× the mass of the sun has a lifespan of about 50 million years, so there have been many generations of stars since their first appearance over 13.5 billion years ago.

Extremely large atoms can be unstable and can have a tendency to break apart into smaller atomic nuclei. Nuclear *fission* reactors take advantage of the energy release from elements such as uranium and plutonium. Sadly, so do atomic bombs.

You might wonder how much of the original hydrogen and helium has been consumed by stellar fusion. The answer is not much! Spectroscopic analysis of the abundance of elements in the sun closely matches results for the wider Milky Way. In both cases, hydrogen and helium still account for 98% by mass, with oxygen (1%) being the only additional significant contributor. As for the production in supernovae of elements heavier than iron, their presence is negligible. The largest is nickel at only 0.008%.

On earth, the story is different. The earth formed from solid particles, with most of the lighter elements being blown away by solar radiation. As a result, the earth is composed predominantly of elements generated in the nuclear fusion of stars. Most abundant are iron (32%), oxygen (30%), silicon (16%) and magnesium (15%). But the sun is over 300,000× more massive than the earth, so the earth's contribution to the overall mix of elements in the solar system is minimal.

As for elements heavier than iron that would have come from supernovae, they are rare even in the case of the earth; the most significant being nickel, weighing in at only 2%. And the gold and silver with which some of you will be adorned only account for 0.0000004% and 0.0000075%, respectively, of the earth's continental crust, which I guess is why they cost so much.

24.6 Our Place in the Cosmic Web

The formation of stars and galaxies is an intuitive consequence of general relativity. The curvature of spacetime makes matter accelerate towards any nearby regions with higher than average energy-density. The energy-density of these regions grows, increasing the accelerating effect on surrounding matter... and so on. It makes sense that stars and galaxies form over time.

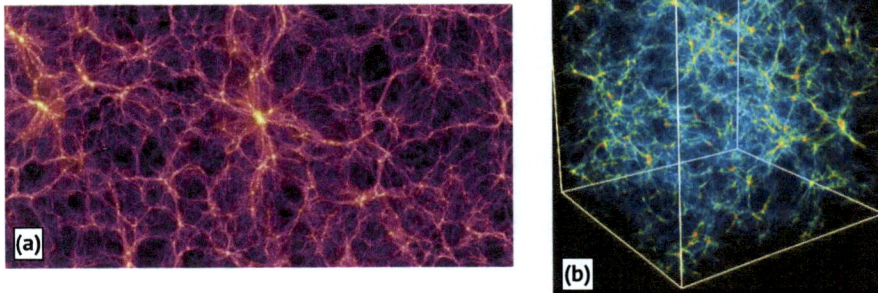

Figure 24.3 The cosmic web. (a) computer simulation of the distribution of dark matter, *Source:* Volker Springel/Wikimedia Commons/CC BY-SA 4.0. (b) Adapted from simulation in 3-D of cosmic web. *Source:* International Gemini Observatory/NOIRLab/NSF/AURA/G. L. Bryan/M. L. Norman/Wikimedia Commons/CC BY 4.0.

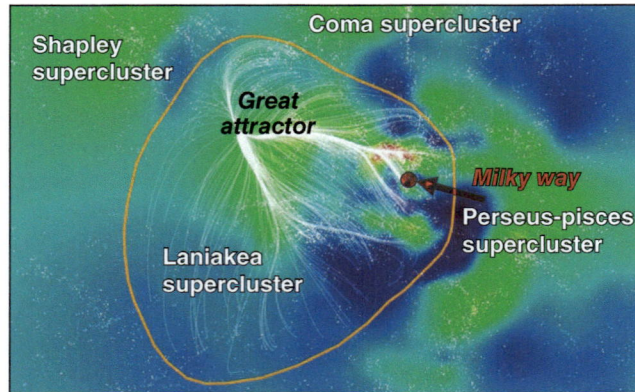

Figure 24.4 The Milky Way (not shown to scale) sits at the edge of the Laniakea supercluster. *Source*: R. Brent Tully (U. Hawaii) et al., SDvision, DP, CEA/Saclay / NASA / Public Domain.

What of the larger-scale structure of the universe? You may have heard that the distribution of dark matter and galaxies in the universe appears to be far from random. At a larger scale, they appear grouped together in a network called the *cosmic web*. Figure 24.3a shows a computer simulation of how the distribution of dark matter develops from a starting point of small variations in energy-density, consistent with those observed in the CMB. The strands in the figure are regions with higher density of dark matter, which draw in visible matter, leading to the formation of galaxies. Figure 24.3b is another simulation represented in 3-D.

The pattern of galaxy formation was affected also by what are called *baryon acoustic oscillations* (BAO). The universe became a baryon–electron–photon plasma about 10^{-6} seconds after the Big Bang as described in Section 24.2. This plasma was attracted towards regions of dark matter over-density. You can think of these regions as gravity-wells pulling in the nearby plasma, which in turn pulls in the plasma surrounding it. The effect was a compression wave that travelled through the plasma at just over half the speed of light (about $0.57c$)[2] until the universe had cooled enough to allow recombination 380,000 years later (the moment the CMB was emitted). After that, the reactive plasma was gone, and the BAO stopped propagating. Effectively it froze, leaving a spherical shell of slightly higher-than-average plasma density surrounding the original region of dark matter over-density.

If you calculate the distance travelled between the formation of the plasma and recombination, adjusting for the expansion of the universe between those events, these BAO spherical shells had a radius of about 500,000 light years at the time of the CMB. If you adjust for the expansion of the universe between the CMB and today, their radius has grown to about 500 million light years. While the original over-dense spots of dark matter were the primary seeds of galaxy formation, the surrounding BAO shells were secondary seeds, and you can detect an excess clustering at separation of 500 million light years in the statistical analysis of galaxy surveys such as BOSS (Baryon Oscillation Spectroscopic Survey) or DESI (The Dark Energy Spectroscopic Instrument).

How do we Earthlings fit into the larger picture? We might like to pretend that we are special because we sit at the centre of the visible universe, but that is untrue of course - the visible universe is defined around us. In truth, we seem to be boringly average. The sun as a star is not unusual in terms of size or structure. The Milky Way is an average galaxy. And we sit in the Virgo supercluster, which is one of the four parts of our larger supercluster Laniakea, a Hawaiian expression which I am told means *immeasurable heaven*. As far as I can tell, it is fairly average too. There are estimated to be somewhere between 200 billion and two trillion galaxies in the observable universe. For illustration, let's say there are a trillion (10^{12}). With each containing about 100 billion stars (10^{11}), this gives a total of about 10^{23} stars, which is 100 billion trillion ($100 \times 10^9 \times 10^{12}$). And that is just the visible universe.

Figure 24.4 illustrates our position at the edge of the Laniakea supercluster. You may wonder what defines where a supercluster starts and ends. It is somewhat subjective, but the trick is to look at the *peculiar velocity* of the galaxies within it. This is the velocity of a galaxy relative to the underlying expansion of the co-moving frame. The universe is expanding, so at a given distance from the earth, the co-moving frame carries galaxies away from it. This underlying movement related to the expansion of the universe is called the *Hubble flow*. The further away from the earth, the faster the Hubble flow.

---

2 For more information on the calculation of the BAO speed, check out Page 14 of *Baryon Acoustic Oscillations* by Bassett and Hlozek, 2009: *https://arxiv.org/pdf/0910.5224* (link as at February 2025).

The peculiar velocity is the difference between the observed velocity of a galaxy and the calculated Hubble flow at that distance from earth. It reveals the gravitational effect of surrounding matter. This is shown in Figure 24.4 by white lines, which indicate that the galaxies in our Laniakea supercluster are pulled towards a central point nicknamed the *Great Attractor*. Don't forget that most of this attraction is due to dark matter. Regions in green and red have a denser presence of galaxies. Those in blue have fewer (sometimes called voids).

This does *not* mean that most of the galaxies in Laniakea are moving towards each other. While the peculiar velocity of a distant galaxy may be towards the Great Attractor, the velocity of the *Hubble flow* tends to be much faster with the net result that the galaxies are growing ever more distant from each other. In the case of the Milky Way, our closest neighbour Andromeda has a net velocity towards us, but it is the exception. At greater distances, the expansion of the universe is overwhelming.

It is hard to get a detailed look at Laniakea or the Great Attractor because they sit in the *zone of avoidance* on the other side of the Milky Way. This is a fancy way of saying that our view is blocked by all the stuff in the Milky Way. Our view will be better when the earth has rotated around to the opposite side. But don't hold your breath—that will be in about 100 million years' time, because the Milky Way takes over 200 million years to fully rotate.

From what we can see, Laniakea contains about 100,000 galaxies and is about 500 million light years across (note that this is similar in size to the BAO shells discussed earlier in this section). The visible universe is about 90 billion light years across (as I will explain in the next section). This means that the volume of space occupied by Laniakea is only about 1:6,000,000 of the total visible universe. So even this so-called immeasurable heaven is microscopic compared with the visible universe. And we have no idea how much might exist beyond the visible universe. In galactic terms we are average and insignificant, which I find humbling. How far we have drifted from the old days when Aristotle believed that the earth sat alone at the centre, with everything revolving around it.

24.7   Horizons and the Fate of the Universe

Let's now turn our attention to the future. Based on the energy-density mix of the ΛCDM model, the fate of the universe is clear. It will end in *heat death of the universe*. In its early age, the energy mix was radiation-dominated. As it expanded, the energy-density of radiation declined more rapidly than that of matter. This led to a long period when matter energy dominated. But over time, the energy-density of matter was diluted by the continued expansion and, several billion years ago, vacuum energy became dominant and the expansion of the universe started to accelerate.

From the moment that Ω_Λ went over $\frac{1}{3}$ and $\ddot{a}$ became positive, there was no turning back (see Section 23.6). The expansion is accelerating. Assuming vacuum energy-density remains constant, it will continue to fuel the expansion. Matter energy-density will drop, so its braking effect on the expansion will decline and fade away. The outcome doesn't depend on spatial curvature. The universe may be finite with slight positive spatial curvature that we haven't spotted yet. Or it might be infinite with slight negative spatial curvature. The end will be the same.

This is illustrated schematically in Figure 24.5. Distance is shown on the *x*-axis, and time since the Big Bang is shown on the *y*-axis. The black horizontal line is *now* the current age of the universe estimated at 13.8 billion years at time of writing. The dotted blue line is the *particle horizon* showing the distance light can have

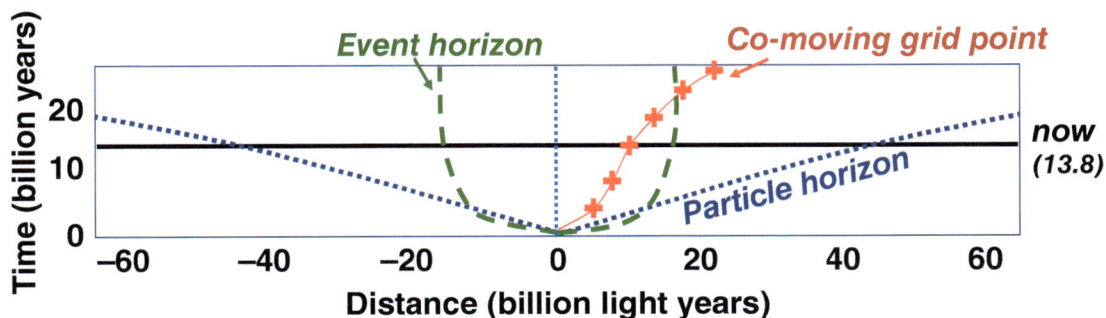

Figure 24.5 Particle and event horizons based on the ΛCDM cosmological model. The red crosses illustrate an example of a co-moving grid position.

travelled, which as of *now* is about 45 billion light years in the 13.8 billion years since the Big Bang. Any observer always sees light travel at speed *c*. What adds the extra distance is the stretching of spacetime while the light is travelling. I will address the mathematics of this in Chapter 25.

As a result, our current *observable universe* is a sphere of radius 45 billion light years (i.e. 90 billion light years across). By this we mean that we are seeing the light from grid points that are *now* up to 45 billion light years away. However, that light was emitted long long ago when they were much closer to us. Gazing into the distance, we are gazing into the past.

The *event horizon* is shown as a dotted green line. It is the distance beyond which there is no possible current or future contact. The observable universe (45 billion light years along the *now* line) is *inside the particle horizon*, so we can see light emissions from the distant past (hence the name *observable universe*). However, much of it is *outside the event horizon*, so any emissions made today or in the future from the more distant locations can never reach us.

An intuitive (albeit slightly inaccurate) way to think of the event horizon is the distance out to where the cumulative stretching of spacetime is such that the separation between us and a distant galaxy is increasing at light speed, so any current light emissions from the galaxy will never reach us. This has the dual benefits of being intuitive and easy to calculate, which is perfect for someone as lazy as me. We know that $H = \frac{\dot{a}}{a}$ for any distance. For recession at light speed, $\dot{a} = c$ giving: $a = \frac{c}{H}$. This is called the *Hubble length*. Written in natural units ($c = 1$), it is $\frac{1}{H}$. Equation 24.5 shows you the current Hubble length. We derived the value of H_0 in units of (years)$^{-1}$ for time or (light years)$^{-1}$ for distance back in Equation 23.5.

$$\text{Current} \qquad\qquad \frac{1}{H_0} = \frac{1}{7 \times 10^{-11}} \approx 14 \ billion \ light \ years, \qquad\qquad (24.5)$$

$$\text{Future} \qquad\qquad \frac{1}{H_\Lambda} = \frac{1}{0.83\,H_0} = \frac{1.2}{7 \times 10^{-11}} \approx 17 \ billion \ light \ years. \qquad\qquad (24.6)$$

Equation 24.6 shows the calculation for the Hubble length in the distant future when the energy-density of matter becomes insignificant compared with vacuum energy. As Ω_Λ is 0.7 of the current energy-density in the universe (Box 23.4), $H_\Lambda^2 = 0.7\,H_0^2$ and therefore $H_\Lambda = 0.83\,H_0$. Based on the presence of vacuum energy alone, the event horizon will settle at about 17 billion light years. If you struggle to reconcile a constant event horizon with an accelerating expansion of the universe, check back to Section 23.3, which explains that the acceleration refers to the increasing speed of separation over time between two galaxies.

To avoid cosmologists foaming at the mouth, I must mention that the event horizon currently is slightly different from the Hubble length. The Hubble parameter is decreasing, and therefore the Hubble length is growing. Imagine light beamed to us from a galaxy exactly the Hubble length away. The light beam travels towards us, effectively offsetting the universe's expansion. The Hubble length grows, and suddenly that light is within the event horizon even though the galaxy emitting it is not. As a result, the event horizon currently is about 16 billion light years, which is slightly bigger than the Hubble length. The discrepancy will disappear in the future as vacuum energy becomes more dominant. If there is only vacuum energy, the Hubble parameter is a constant (check back to Equation 23.4 if needed). From then on, the event horizon will exactly match the Hubble length. Both will remain constant while the universe expands exponentially... for eternity.

As the universe expands, more and more galaxies will recede across the event horizon and disappear from view, as illustrated by the red track in Figure 24.5, which shows an example position that is co-moving with the expanding grid. Note that in the early life of the universe this track curves upwards as the expansion is decelerating. Later and in the future, it curves outwards as the expansion accelerates due to the dominance of vacuum energy in the energy mix.

Only local neighbours such as the Andromeda galaxy are close enough for the gravitational pull to offset the effect of this expansion. In fact, the Andromeda galaxy will crash into the Milky Way in about 5 billion years. And in the words of Joachim Harnois-Deraps who helped edit this module: *at the same time the sun will explode; what a bright future awaits us!*

24.8   Summary

I started this chapter with a warning about what is called the Big Bang. The EFEs and measurements of the energy-mix and expansion of the universe tell us that there was a time when our current visible universe was smaller, hotter and denser. It is a mistake to extrapolate this back to a singularity. The universe may be, and may always have been, infinite. We don't know. In the words of Max Williams in Box 24.1, perhaps we should talk about the Big Stretch rather than the Big Bang.

I shared a rule of thumb that allows us to equate the temperature of the early universe to its age (Box 24.2): $\frac{10^{20}}{T^2}$ seconds after the Big Bang, where T is the temperature in Kelvin. This works for the first 50,000 years after the Big Bang when radiation dominated and is reasonably accurate up to 380,000 years (i.e. up to the time of CMB emission).

After the Big Bang, the universe expanded and its temperature fell. Once it dropped below about 10^{13} Kelvin (10^{-6} seconds), the quark–gluon plasma condensed with quarks joining in threes to form protons and neutrons creating a baryon–electron–photon plasma. This is called *baryogenesis*. One might have expected this to produce equal amounts of particles and antiparticles that would then annihilate with each other. However, baryogenesis resulted in an excess of matter. In the standard model of particle physics there is an underlying asymmetry between matter and antimatter (Box 24.3), but the level of asymmetry in baryogenesis is still unexplained.

Once the temperature dropped below 10^{11} Kelvin (0.01 seconds), nuclear fusion occurred for a few minutes. Some of the hydrogen nuclei fused into helium nuclei and a trace of lithium nuclei. The temperature had to fall to about 10^5 Kelvin (about 380,000 years after the Big Bang) before stable atomic nuclei could form (*recombination*). Electrically neutral atoms are much less disruptive to the passage of EM radiation and the CMB was released (*decoupling*).

From this point on, it is harder to give an accurate timeline because events are no longer closely tied to the temperature of the universe. At time of writing, the consensus is that stars and galaxies formed about 200 million years after the Big Bang. Stellar nuclear fusion can lead to heavier elements because there is a net energy release all the way up to producing iron. The production of elements such as gold and silver, which have a higher atomic number than iron, requires energy. They likely are forged in cataclysmic events such as supernovae and the merger of neutron stars.

There have been many generations of stars, but the overall elemental composition of the universe is largely unchanged since recombination. Hydrogen and helium still represent about 98% of atomic matter by mass (this does not, of course, include dark matter). The earth is an outlier as most of the lighter elements were blown away by the solar wind early in its development.

We discussed the way that galaxies are organised in a network called the *cosmic web* and that it is somewhat humbling to reflect that the earth rotates around an average star, in an average galaxy that is one of a 100,000 in the Laniakea supercluster. And even this supercluster is insignificant compared with the total size of the visible universe.

The final topic was the future fate of the universe. During the billions of years when radiation and then matter heavily dominated the mix of energy-density, the expansion of the universe decelerated. But there came a moment a few billion years ago when the presence of radiation/matter had been sufficiently diluted by the expansion for vacuum energy Ω_Λ to grow to over $\frac{1}{3}$. From then on, the expansion of the universe started to accelerate. Galaxies will grow ever further apart, resulting in what is described as the heat death of the universe. More and more galaxies will cross the Hubble length and disappear from our view. In 100 billion years time, the Milky Way and Andromeda galaxies will be left all alone in a sea of darkness. That sounds depressing, so check out Box 24.4 for a more humorous twist on the end of the universe.

Box 24.4 *The restaurant at the end of the universe*   **by Douglas Adams**

In Douglas Adam's book, Arthur Dent is disappointed when Zaphod Beeblebrox says they need to leave the viewing area just before the universe ends:
But what about the end of the universe? We'll miss the big moment.
I've seen it. It's rubbish, said Zaphod, *nothing but a gnab gib.*
A what?
Opposite of a big bang.

The Friedmann equations and ΛCDM model combine with observational data to give us a reasonable and consistent picture of the development of the universe from creation of the quark–gluon plasma through to the current day. I am not saying that every detail is right and there is still much to discover. For example, we still don't know much about vacuum energy and dark matter. However, although there are missing pieces, the overall story makes sense.

But the ΛCDM model faces serious problems when you track back to the very (very) early development of the universe. Logical inconsistencies appear, leading cosmologists to conclude that there was a period of extremely rapid expansion of the universe in the first few moments of its existence. This theory is called *inflation* and is the focus of Chapter 25.

Chapter 25

Inflation

Box 25.1 Inflation and the standard Big Bang theory

The following are the words of the American physicist Alan Guth, who originally proposed the theory of inflation:

There is a key issue that the standard Big Bang theory does not discuss at all: it does not tell us what banged, why it banged, or what happened before it banged. Despite its name, the Big Bang theory does not describe the bang at all. It is really only the theory of the aftermath of a bang.

So, in particular, the standard Big Bang theory does not address the question of what caused the expansion; rather, the expansion of the universe is incorporated into the equations of the theory as an assumption about the initial state—the state of the universe when the theory begins its description.

Similarly, the standard Big Bang theory says nothing about where the matter in the universe came from. In the standard Big Bang theory, all the matter that we see here, now, was already there, then. The matter was just very compressed and in a form that is somewhat different from its present state. The theory describes how the matter evolved from one form to another as the universe evolved, but the theory does not address the question of how the matter originated.

While inflation does not go so far as to actually describe the ultimate origin of the universe, it does attempt to provide a theory of the bang: a theory of what it was that set the universe into expansion and at the same time supplied essentially all of the matter that we observe in the universe today.

The theory of inflation postulates that there was an early period when the universe underwent a process of enormously rapid expansion. I am going to start this chapter with a discussion on semantics. There is some confusion in the literature about what the term *Big Bang* refers to.

Most cosmologists use the term Big Bang in reference to the beginning of everything and anything. Although we cannot say it started as a singularity, the Big Bang is associated with $a = 0$ in the FRW metric, the very birth of the universe. These cosmologists would say that the Big Bang occurred *before* (or started with) inflation.

Others associate the Big Bang with the moment that the universe was flooded with radiation and particles. In inflationary theory, this follows the inflationary epoch. Therefore, these cosmologists would say that the Big Bang happened *after* inflation.

I don't really care about semantics, but I do want to provide clarity. In this book I use the term *Big Bang* in reference to the birth of the universe (FRW metric $a = 0$) and use the widely accepted term *reheating* to refer to the moment after inflation when the universe was flooded with radiation and particles.

Let's get back to our main topic. Inflation offers a solution to some significant shortcomings of the standard ΛCDM Big Bang theory. In addition, it addresses a number of issues that are outside the scope of that theory (see Alan Guth's quote in Box 25.1).

Untangling General Relativity: The Intuitive Self-Study Guide, First Edition. Simon Sherwood.
© 2026 John Wiley & Sons Ltd. Published 2026 by John Wiley & Sons Ltd.

25.1 Arguments for Inflation

As usual in physics, just when you think you have a solid theory, holes start to appear. There are three topics typically cited when discussing problems with the standard ΛCDM Big Bang theory: the *flatness problem*, the lack of *magnetic monopoles* and the *horizon/homogeneity problem*. My personal opinion (others may disagree) is that the last is the most compelling.

25.1.1 The Flatness Problem

The *flatness problem* asks why the universe is so close to spatially flat. Why is its energy-density so close to critical density?

Current estimates of curvature in terms of the value of Ω_K are that it is zero to a level of about one in a thousand, i.e. $< 10^{-3}$ (see BOSS and WMAP data in Section 23.1). In the ΛCDM model, the absolute value $|\Omega_K|$ *increases* over time. If $\mathbf{K}$ is +1, it becomes more positive. If $\mathbf{K}$ is -1, it becomes more negative. This may surprise you because underlying curvature decreases as the universe (or anything else) expands. However, Ω_K is a measure of curvature's *relative* contribution to the effective energy-density H^2. Equation 25.1 below is a copy of Equation 22.24. The value of $\mathbf{K}$ is +1, 0, or -1 depending on the type of curvature. The $\bar{\rho}$ values (the energy-density when $a = 1$) are constants. Vacuum energy-density Λ doesn't change with scale factor a.

As the universe grows, the value of $\frac{\mathbf{K}}{a^2}$ decreases. However, the values of the matter and radiation terms fall with a^3 and a^4. Therefore, they fall faster and the value of $|\Omega_K|$ increases. You can see what is happening from the definition of Ω_K on the right of Equation 25.2. Over time H falls faster than a increases, so the denominator in the fraction decreases over time. This remains the case until late in the development of the universe, when vacuum energy dominates.

$$H^2 = \left(\frac{\dot{a}}{a}\right)^2 = \frac{k}{3}\left(\frac{\overline{\rho_M}}{a^3} + \frac{\overline{\rho_R}}{a^4} + \Lambda\right) - \frac{\mathbf{K}}{a^2}, \qquad k = 8\pi G, \quad (c=1), \tag{25.1}$$

$$1 = \Omega_M + \Omega_R + \Omega_\Lambda + \Omega_K, \qquad \Omega_K = -\frac{\mathbf{K}}{a^2 H^2}. \tag{25.2}$$

Let's use the value of $|\Omega_K|$ now ($< 10^{-3}$) to calculate its value during what is called the Grand Unified Theory (GUT) epoch. The GUT epoch is the era when physicists believe the electromagnetic, strong and weak forces (but not gravity) were unified as one force. The strength of each force changes with energy level and physicists have a GUT feeling (geddit?) that they converge at about 10^{16} GeV, which, using the rules of thumb in Box 24.2, is a temperature of 10^{29} Kelvin and an age of 10^{-38} seconds after the Big Bang. In comparison, the current age of the universe is about 13.8 billion years, which is well over 10^{17} seconds.

The next step is to work out the value of its magnitude $|\Omega_K|$ at 10^{-38} seconds after the Big Bang. This is shown in Equation 25.3. From the relationship on the right of Equation 25.2, $|\Omega_K|$ varies inversely with $(aH)^2$. We can relate the value of (aH) with time t by multiplying together the values of a and H in the matter-only model from Equation 23.2 (the result is shown to the right of Equation 25.3). We use the matter-only model because the universe was matter-dominated for most of its existence. Substituting in the ages of the universe now (10^{17} seconds) and during the GUT era (10^{-38} seconds) shows that $|\Omega_K|_{GUT}$ must have been $10^{37}\times$ smaller than $|\Omega_K|_{now}$. The value of $|\Omega_K|_{now}$ is less than 10^{-3}, so $|\Omega_K|$ in the GUT epoch must have been less than 10^{-40}.

$$\frac{|\Omega_K|_{now}}{|\Omega_K|_{GUT}} = \frac{(aH)^2_{GUT}}{(aH)^2_{now}} = \left(\frac{t_{GUT}^{-\frac{1}{3}}}{t_{now}^{-\frac{1}{3}}}\right)^2 = \left(\frac{t_{now}}{t_{GUT}}\right)^{\frac{2}{3}}, \qquad note: aH \approx \frac{2C}{3}t^{-\frac{1}{3}}, \tag{25.3}$$

$$\frac{|\Omega_K|_{now}}{|\Omega_K|_{GUT}} \approx \left(\frac{10^{17}}{10^{-38}}\right)^{\frac{2}{3}} \approx 10^{37}, \implies |\Omega_K|_{GUT} < 10^{-40}.$$

Many argue that this is too close to zero to be a coincidence, so there must be a cause for such *fine-tuning*. Personally, I am somewhat sceptical of this reasoning. You equally could argue that there was massive spacetime curvature in the early universe because it was so much hotter and denser. Given the vast amounts of energy present at that time, perhaps it is not surprising that the relative effect of spatial curvature (quantified in terms of equivalent energy-density) was so small as a proportion of the total.

Even if one is sceptical on fine-tuning, the standard ΛCDM Big Bang offers no rationale for the current universe being flat, while inflation gives a ready explanation. An earlier period of rapid expansion would reduce any spatial curvature. As a 2-D example, consider the curvature of the surface of a football. Now blow that up to the size of the earth. The surface curvature is reduced (there even are a few weird flat-earthers who struggle with the idea of it being curved at all). Without inflation, you must conclude that the universe just happens to be almost exactly spatially flat. With inflation, the universe could have started with any level of spatial curvature, which would have diluted away in the inflation process.

25.1.2 Where Are the Magnetic Monopoles?

The issue of *magnetic monopoles* is more technical. I suspect it will be unfamiliar to most readers, so there is some background in Box 25.2. Modern efforts at modelling conditions in the standard ΛCDM Big Bang predict the production of lots of magnetic monopoles. This may require very high energies, which would explain why we have not detected them at the LHC and other colliders. However, it is harder to explain what has happened to the original monopoles that physicists believe must have been produced in the early universe. Conservation of charge (in this case magnetic charge) means that they could only annihilate in pairs or decay into other monopoles. Inflation offers a solution. If (and this is a big unexplained *if*) the inflationary period occurred after the high-energy production of monopoles but before the production of other particles (reheating), then the inflationary exponential expansion would have diluted the monopole presence down to an insignificant level, perhaps not leaving even a single magnetic monopole in our entire observable universe.

Box 25.2 Magnetic monopoles

Physicists have been hunting (unsuccessfully) for magnetic monopoles for many years. The theory is that, just as some particles have net electric charge, there could be particles with net magnetic charge. All magnets known to date have both a north and south pole. Break a magnet into two, and each piece still has both poles. Shown below on the left are Maxwell's equations (written for simplicity in Gaussian units with $c = 1$). E is the electric field, B the magnetic field and ρ_e and J_e the electric charge density and current. The second line ($\nabla \cdot B = 0$) shows the divergence of the magnetic field as zero, which assumes there are no net sources or sinks of magnetic charge (i.e. no magnetic monopoles). The same equations are shown on the right allowing for the existence of magnetic monopoles (ρ_m and J_m are magnetic charge density and current, respectively). You can see this creates an enticing symmetry.

Without monopoles

$$\nabla \cdot E = 4\pi \rho_e$$

$$\nabla \cdot B = 0$$

$$\nabla \times E = -\frac{\partial B}{\partial t}$$

$$\nabla \times B = -\frac{\partial E}{\partial t} + 4\pi J_e$$

With monopoles

$$\nabla \cdot E = 4\pi \rho_e$$

$$\nabla \cdot B = 4\pi \rho_m$$

$$\nabla \times E = -\frac{\partial B}{\partial t} + 4\pi J_m$$

$$\nabla \times B = -\frac{\partial E}{\partial t} + 4\pi J_e$$

The physicist Paul Dirac (1902–1984) concluded that magnetic monopoles are consistent with quantum theory providing electric charge is quantised (i.e. comes in specific lumps), which it is. And magnetic monopoles are predicted in most theories of particle production in the early universe.

In the words of the string theorist Joseph Polchinski, the existence of magnetic monopoles is *one of the safest bets that one can make about physics not yet seen.* Inflation offers a possible reason why we haven't been able to detect any.

25.1.3   The Horizon/Homogeneity Problem

The *horizon/homogeneity problem* is complicated to describe. The microwave radiation from the cosmic microwave background (CMB) is highly homogeneous (see Box 22.3 if you need a reminder about the CMB). The equivalent temperature of its radiation observed in all directions typically varies by about one in a hundred thousand from the average value. At the time of CMB emission, regions with higher than average energy/matter density would have affected its wavelength due to the redshift of that CMB light escaping a more intense gravitational field. How could it be possible to have such a consistent temperature unless at some stage all the regions were causally linked? The obvious explanation is that all the regions indeed were causally linked, but here we run into difficulties with the ΛCDM model.

An alert reader might comment that it is self-evident that everywhere is causally linked because all the regions came from the same Big Bang. But *no*, the stand-alone Big Bang of the ΛCDM model does not allow this. Indeed, it is not the case in any decelerating expansion of the universe. Many of the regions of CMB that we observe are not causally linked. This is counter-intuitive, so grab yourself a nice hot cup of tea and let's look at the problem.

Figure 25.1 is a *highly schematic* illustration that will help our discussion. It is *not* to scale nor is it intended to accurately portray mathematical relationships. The basic layout is similar to that of Figure 24.5, which I used to explain event horizons. The x-axis is physical distance and the y-axis is time. The dotted blue line is the particle horizon, which is the distance light could have travelled out from the starting point, which I have labelled as the Big Bang.

Any observer sees light travel at speed *c*, but we must take into account that spacetime stretches while the light travels. How far might the light travel in terms of the FRW radial coordinate *r* in time *t*? Think of the *r* coordinate as the number of grid points that the light crosses in time *t* (for example, the red dots back in Figure 21.1). If it crosses 1 in time *t* when the scale factor $a = 1$, then it will cross only $\frac{1}{2}$ when the scale factor is 2 and $\frac{1}{3}$ when it is 3. Setting $c = 1$, the relationship is $dr = \frac{dt}{a}$ as is shown in Equation 25.4 using the FRW metric for light ($d\tau = 0$) with motion in the radial direction ($d\theta = d\phi = 0$). Integrating this from time 0 to time *t* gives the total distance *r* travelled by the light in time *t*.

$$dt^2 - a^2 dr^2 = d\tau^2 = 0, \quad \implies \quad dr = \frac{dt}{a}, \quad \implies \quad r = \int_0^t \frac{1}{a}\, dt. \tag{25.4}$$

To get the physical distance that the light might have covered (i.e. at time of arrival, its physical distance away from the original source), we must multiply the distance travelled in terms of *r* by the value of the scale factor at time *t* (i.e. the final value of the scale factor). This calculation is shown below for a matter-only universe and a radiation-only universe starting from the expressions for the relevant scale factors that we derived in Equations 23.2 and 23.3. Note that the capital *C* values are constants and not to be confused with the speed of light! The distance travelled by light is 3*t* in a matter-only universe and 2*t* in a radiation universe. If you want to include the speed of light factor, the answers are 3*ct* and 2*ct*, respectively. The physical distance travelled is increased by the expansion of the universe, as you would expect. It would be *ct* if there were no expansion.

$$\textit{Matter-only:} \qquad a = Ct^{\frac{2}{3}}, \quad a(t)r = a(t)\int_0^t \frac{t^{-\frac{2}{3}}}{C}\, dt = Ct^{\frac{2}{3}}\frac{3t^{\frac{1}{3}}}{C} = 3t. \tag{25.5}$$

$$\textit{Radiation-only:} \qquad a = Ct^{\frac{1}{2}}, \quad a(t)r = a(t)\int_0^t \frac{t^{-\frac{1}{2}}}{C}\, dt = Ct^{\frac{1}{2}}\frac{2t^{\frac{1}{2}}}{C} = 2t. \tag{25.6}$$

The universe has been matter-dominated for most of the time since the Big Bang. Using the result from Equation 25.5, light could in theory have travelled what is now a physical distance of about 45 billion light years between source and point of arrival in the 13.8 billion years since the Big Bang. This is the radius of the current visible universe shown in Figure 25.1. Consider light arriving at our location at the centre of our visible universe from the point labelled *N* at the very edge. That light would have been emitted at the time of the CMB when the *N* grid point was much closer to our central grid point. The light has travelled for 13.4 billion years to reach us (13.8 billion years minus the 0.4 billion years between the Big Bang and CMB). The emitting grid point *N* is *now* a distance of 45 billion light years from us.

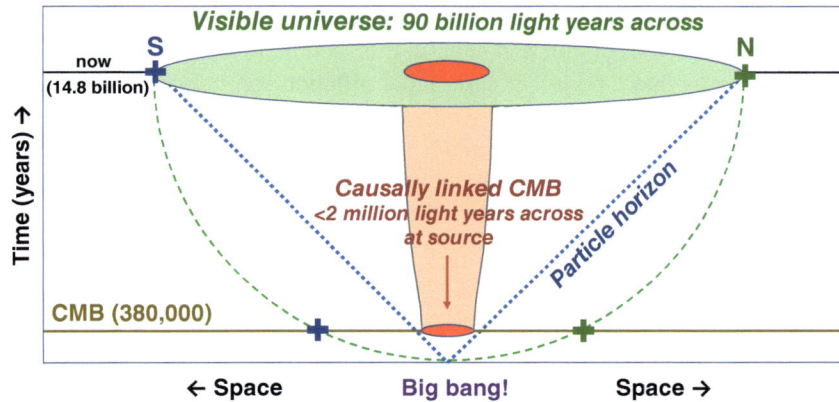

Figure 25.1 Illustrative schematic of the CMB horizon problem (not to scale).

Let's do a similar calculation to assess possible causal linkage between regions of the CMB. Causal linkage could in theory allow the temperatures of different regions to match. How far apart could two such regions be at the time when the CMB was emitted? Obviously, they cannot separate at faster than light speed. Let's calculate how far two light beams could travel if moving freely in opposite directions during the 380,000 years between the Big Bang and CMB.

In the traditional Big Bang model, the important early sources of energy were radiation and matter. Vacuum energy was irrelevant (an assumption we will revisit later). Using the matter-only model ($3t$) light in theory could have travelled a distance of around 1.2 million light years between the Big Bang and the CMB (note that this is a theoretical calculation, because the plasma was opaque, so light wouldn't have been able to travel long distances). The causally linked volume of space cannot have been much more than 2 million light years across. This is shown in red at the bottom of Figure 25.1.

Do you see the problem? Between CMB emission and today, that volume of potentially causally linked CMB space will have grown about 1,000× from 2 million light years across, up to about 2 billion light years across. To make the point in the figure, it is shown as a growing red circle but still covering a much smaller current red zone than the size of the visible universe.

When we look out at the CMB, we are seeing the microwave radiation emitted from what is now a spherical shell with a radius of 45 billion light years. It has a circumference of about 280 billion light years ($2\pi r$). If you pan with a telescope, a shift of about 2 billion light years around the circumference (about 2 degrees panning) takes you from one causally linked region to another. How can the CMB look so homogeneous (i.e. be such a similar temperature) in every direction if the different parts can never have interacted in any way?

Before moving on, I want to address an obvious question. How can there be a horizon problem if everything started in a Big Bang at scale factor $a = 0$? The first part of the answer is that you cannot take the scale factor to zero. The ΛCDM model has its limits. We know that quantum effects must become important, and we have no idea how a singularity behaves (we faced the same issue with black hole singularities in Section 13.10).

This being said, look in Figure 25.1 at the grid points labelled N and S (for north and south) at the far reaches of the observable universe. Track them back to the time of the CMB, as illustrated schematically by the blue and green crosses in the figure. At that time, they were 1,000× closer (the CMB expansion factor), i.e. 90 million light years apart compared to the CMB causal zone, which was only 2 million light years across. Track back further towards the Big Bang. The N and S grid points get closer and closer as shown by the green dotted lines, but they are *always* outside the particle horizon. As the distance between N and S shrinks, the particle horizon shrinks. The mathematics of the ΛCDM model tells us that those N and S grid points were never causally linked in the past... unless...

25.2 Introduction to Inflation

In July 1980, Alan Guth published his paper, *The Inflationary Universe: A Possible Solution to the Horizon and Flatness Problems*. This started as a fringe idea, but it has now, after some adjustment to the original theory, become widely accepted. Guth was a particle physicist and had been working on the problem of magnetic monopoles. The relevance of his knowledge of particle physics, quantum field theory (QFT) and the Higgs field will become clear when we dig into the technical details of inflation later. In this section, I will focus on the broader concept.

Intuitively, it is fairly obvious that a period of early rapid inflation can help solve the flatness and monopole problems. For the former, it smooths out any spatial curvature. For the latter, if we assume magnetic monopoles were created *before* inflation, it dilutes their physical presence. It may also seem intuitively obvious that inflation might ease the horizon/homogeneity problem by creating larger causally linked zones. However, we need to tread carefully and check that the maths works.

The horizon/homogeneity problem occurs because in the classic ΛCDM Big Bang model, the initial expansion of the universe is *decelerating* over time. Check back to Equations 25.5 and 25.6. The scale factor grows with $t^{\frac{2}{3}}$ or $t^{\frac{1}{2}}$ depending on whether you use the matter-only or radiation-only model. In contrast, the particle horizon grows in line with the distance travelled by light shown on the right, which in both cases is proportional to time t. Therefore, the growth of the scale factor decelerates relative to the growth of the particle horizon. As a result, over time more and more causally unlinked regions come into the observable universe.

To produce a larger causally linked region of universe, we need to inject an early period of *accelerating* expansion. And, hey ho, we already know something that drives such an expansion... vacuum energy. The essence of inflation is that there was an early period with extremely high vacuum energy because of the potential energy of a scalar field dubbed the *inflaton*. This vacuum energy drove exponential expansion of the universe. Subsequently, the potential energy of the inflaton dropped, and the vacuum energy in the universe fell to the level we see today. This fall in the potential energy of the vacuum led to what is called *reheating*, which released a huge amount of energy into the universe in the form of radiation and particles.

It is worth taking a moment to check back to the *vacuum energy-only* calculation we did in Equation 23.4. This is called a *de Sitter* space. The scale factor accelerates exponentially ($a = Ce^{Ht}$) and the Hubble parameter H is a constant. The maths is slightly painful, but if you calculate the physical distance travelled by light (i.e. how far the source is away at moment of arrival) in the same way we did earlier in Equations 25.5 and 25.6, you get the result in Equation 25.7. This shows the physical distance travelled by light in time t (vacuum energy only).

$$Vacuum\ energy: \quad a(t)r = a(t)\int_0^t \frac{dt}{a} = Ce^{Ht}\int_0^t \frac{e^{-Ht}}{C}dt = e^{Ht}\left[-\frac{e^{-Ht}}{H}\right]_0^t,$$

$$= e^{Ht}\left(-\frac{e^{-HT}}{H} + \frac{1}{H}\right) = \frac{e^{Ht}}{H} - \frac{1}{H}.$$

(25.7)

Both the scale factor and the distance travelled by light (the particle horizon) increase exponentially with e^{Ht}. A suitable period of this sort of inflationary expansion solves the horizon problem. The universe and particle horizon grow exponentially hand in hand, creating a larger causally linked region from which the ΛCDM expansion can proceed.

In summary, a universe with a large amount of vacuum energy will expand rapidly. With this one theoretical sleight of hand called inflation, we solve the flatness problem, the magnetic monopole problem (assuming the monopoles were created before inflation) and the horizon/homogeneity problem.

25.3 The Inflaton Field

The starting point is to hypothesise the existence of a scalar field named the *inflaton field* that affects the density level of vacuum energy in the universe. Figure 25.2 illustrates the theory. The value of the inflaton field is labelled μ along the x-axis. The y-axis shows the energy-density of positive vacuum energy ($\Lambda > 0$) labelled V. The dark blue line shows the type of relationship theorised between the value of μ and the density of vacuum energy V (it is highly schematic as we don't know the details). The present-day situation is shown by the little green spot at the bottom of what looks like a valley labelled ΛCDM. At this point, V is at a level that matches the current level of vacuum energy.

The theory is that there was a time when the value of μ was different and V was much (much) higher. This is shown as the green spot on the left of the schematic. This enormous level of vacuum energy led to an enormous value of H. This in turn drove an extremely rapid exponential growth in the scale factor of the universe as we

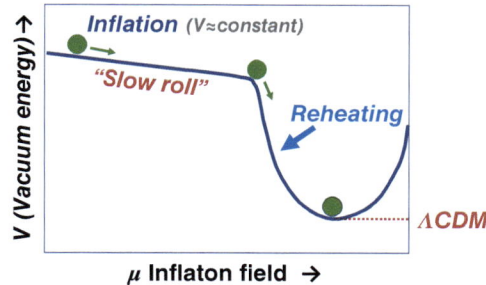

Figure 25.2 Illustrative schematic of the evolution of the inflaton field leading to reheating.

showed in Equation 25.7. It flattened out any spatial curvature in the universe (solving the flatness problem), diluted the presence of any magnetic monopoles (solving the monopole problem) and expanded causally linked regions (solving the horizon/homogeneity problem).

The value of a field tends in the direction of lower potential energy. As a result, the value of μ moved to the right along the schematic x-axis and V dropped over time. Initially, the change in value of μ was what is described as a *slow roll*, using the analogy of a ball rolling down a slight slope. The value of V remained high and was approximately constant. This is the inflationary period.

At some stage, the change in μ reached a point where its relationship with V fell dramatically. The level of vacuum energy dropped to the current level (matching the Λ term in the EFEs). This is the steep slope shown in the schematic. The potential energy stored in the vacuum was released, creating the dark matter, visible matter and radiation present today. This is called *reheating*, presumably because the rapid expansion of inflation would have cooled the universe, and so the subsequent release of matter and radiation would have been a reheating of it.

After reheating, the universe continued to expand. This reduced the density of radiation and matter. During the following 13.8 billion years, the value of H gradually dropped and the expansion slowed to current levels. Ever since reheating, the value of the inflaton field μ has stayed at the point of minimum energy V. As a result, the density of vacuum energy now is a constant.

25.4 How Much Inflation Had to Occur?

Let's look at how much inflation had to occur. At a minimum, it had to be enough for all the currently visible universe to be causally linked some time in the past. Otherwise it wouldn't explain why the CMB is homogeneous in every direction (the horizon/homogeneity problem). We can only calculate a *minimum* inflation. There is no limit above this, because more inflation simply means more dilution of monopoles, more spatial flattening of the universe, and a larger region of causally linked space (beyond the size of the visible universe).

Let's get started. In order to put numbers to the inflation process, you first need to estimate the temperature of the universe at the time of reheating. This must be low enough not to create new magnetic monopoles, but high enough to create oodles of dark matter plus quarks and electrons. Accelerators have reached energies in excess of 10^3 GeV (equivalent to a temperature of over 10^{16} Kelvin) without a hint of dark matter, so the temperature at reheating is likely to have been well in excess of that. Beyond this is a guess, using what Alan Guth amusingly describes as *sample numbers*. If you have a spare hour, I thoroughly recommend his introductory MIT lecture on inflation (for details see resources in Box 27.2).

While we don't know the true number, it turns out that the overall result is similar whatever temperature you choose (at the end of this section there is a table comparing results for different temperatures at reheating). To illustrate the calculations, I'm going to assume the reheating temperature was that estimated for the GUT epoch, about 10^{16} GeV or 10^{29} Kelvin. Before this, it is believed that the electromagnetic, strong and weak forces were unified, as described earlier in Subsection 25.1.1.

The first question is, what was the level of vacuum energy-density V before reheating, i.e. before the green spot in Figure 25.2 rolled down the slope? In reheating, almost all the dark (vacuum) potential energy was released into the universe. The earlier rapid inflationary expansion of the universe will have diluted any existing radiation/matter to negligible levels, so energy-density V before reheating must match the temperature of 10^{29} Kelvin after reheating. As radiation was the dominant energy source in this epoch, Equation 25.8 uses the Stefan–Boltzmann relationship for radiation (Equation 24.2) to give an approximate equivalent energy-density in joules per cubic metre.

$$\rho_V = \alpha T^4 = (8 \times 10^{-16})(10^{29})^4 \approx 10^{101} \text{ Jm}^{-3}, \quad \textit{prior to reheating at } 10^{29} \text{ Kelvin.} \tag{25.8}$$

To put this in context, the current energy-density of vacuum energy Λ is about 10^{-9} Jm^{-3}. This means that energy-density V at the top of the slope was about $10^{110}\times$ larger than it is at the bottom. You can see that Figure 25.2 is schematic and not to scale! The next question is how much inflation occurred, i.e. the minimum amount that the scale factor must have increased from the moment inflation started (which I call the *starting patch*) to the moment inflation ended with reheating (which I call the *ending patch*).

Starting patch: as we are calculating the minimum amount of inflation, we want to begin from the largest possible causally linked starting patch. The maximum radius of a causally linked patch of de Sitter space is the Hubble length $\frac{c}{H}$, as discussed in Section 24.7. This is where the physical separation between the horizon and origin increases at the equivalent of light speed. Any point beyond the Hubble length is expanding away from the central point too fast to be causally linked. Equation 25.9 shows the calculation of the radius of this starting patch. The value of H comes from the expression on the left of Equation 24.3 earlier. The answer is about 10^{-29} metres for an energy-density equivalent to 10^{29} Kelvin. Although small, this is well in excess of the Planck length (10^{-35} metres), so we would still expect our laws of physics to apply.

$$H \approx (2 \times 10^{-21})\, T^2 \approx (2 \times 10^{-21})(10^{29})^2 = 2 \times 10^{37} \text{ seconds}^{-1}$$

$$\textit{Maximum starting patch radius} = \frac{c}{H} \approx \frac{3 \times 10^8 \text{ ms}^{-1}}{2 \times 10^{37} \text{ s}^{-1}} \approx 10^{-29} \text{ metres.} \tag{25.9}$$

Ending patch: this must be large enough to have expanded since reheating to encompass the entire current visible universe, giving a common causal origin that explains the homogeneity of the CMB. Radiation dominated the energy, so the change in scale factor between the time of reheating (a_{reheat}) and today (a_{now}) is approximately inversely proportional to the change in temperature at reheating (assumed as 10^{29} Kelvin) and the current temperature of the CMB (about 3 Kelvin). This gives the relationship in Equation 25.10.

$$\frac{a_{reheat}}{a_{now}} \approx \frac{T_{CMB(now)}}{T_{reheat}} \approx \frac{3}{10^{29}}. \tag{25.10}$$

Let me clarify how this works. As discussed in Section 24.4, in a radiation-dominated scenario the temperature T is proportional to the average energy of the photons ($E = hf$), which is inversely proportional to their wavelength, giving $T \propto \frac{1}{\lambda}$. This wavelength increases with the scale factor giving $T \propto \frac{1}{a}$ and therefore $a \propto \frac{1}{T}$. This relationship holds from reheating up to emission of the CMB because the universe was radiation-dominated for most of that period. Another way to think about the same thing is that radiation energy-density ρ is proportional to T^4 (Equation 24.2) and proportional to a^{-4} (Equation 22.24), leading to $a \propto \frac{1}{T}$.

From the emission of the CMB onwards, the mix of energy-density changes, but we can track the change in temperature of the CMB radiation up to today. This change is also inversely proportional to the scale factor, which means that the relationship $a \propto \frac{1}{T}$ applies from reheating right up to current times, providing we compare the temperature at reheating with the current temperature of the CMB as is the case in Equation 25.10.

For the visible universe to be causally linked, the ending patch after inflation must be big enough to have grown with the scale factor to the radius of the current visible universe (about 45×10^9 light years). As shown in Equation 25.11, the ending patch of universe after reheating must have a radius in excess of one centimetre to have grown to this size (note that one light year is about 9×10^{15} metres).

$$\textit{Ending patch} > \frac{3}{10^{29}} (45 \times 10^9)\,(9 \times 10^{15}) > 10^{-2} \text{ metres.} \tag{25.11}$$

Minimum inflation: we now can compare the maximum radius of the *starting patch* before inflation (Equation 25.9) with the minimum radius of the *ending patch* after inflation (Equation 25.11) to show that during the inflationary period the scale factor of the universe must have increased by a factor of at least 10^{27}. Some physicists prefer to call this a minimum of 62 *e-foldings*. Rather than powers of ten, this uses powers of e, where $e \approx 2.7$.

Table 25.1 Minimum inflation scenarios for different reheating temperatures.

Reheating temperature	10^{19} Kelvin	10^{24} Kelvin	10^{29} Kelvin
Energy equivalent	10^6 GeV	10^{11} GeV	10^{16} GeV
Maximum starting patch (radius)	10^{-9} metres	10^{-19} metres	10^{-29} metres
Minimum ending patch (radius)	10^8 metres	10^3 metres	10^{-2} metres
Minimum inflation in scale factor	$> 10^{17}\times$	$> 10^{22}\times$	$> 10^{27}\times$

Let me briefly summarise the logic. If reheating happened at 10^{29} Kelvin, then the maximum causally linked *starting patch* for inflation is radius 10^{-29} metres, because this is the corresponding Hubble length. We know the *ending patch* must have radius greater than 10^{-2} metres. If it were less than this, it could not have grown big enough to match the current visible universe and explain how the CMB is so homogeneous even at the extremes. Comparing the patches at the start and end of inflation gives an increase in scale factor of over 10^{27}, which in *volume* terms is an increase of over 10^{81}.

For context, if you applied this inflation to a grain of sand ($r \approx 10^{-5}$ metres), it would end up with a radius over a hundred times that of the entire Milky Way ($r \approx 10^{20}$ metres)! We can use the formula in Equation 25.7 to show that this inflationary expansion would have taken only about 10^{-35} seconds (Equation 25.12).

$$a \approx \frac{e^{Ht}}{H}, \quad \Longrightarrow \quad t \approx \frac{\ln(aH)}{H} \text{ seconds.}$$

$$\Delta t = \frac{\ln(a_{end}H)}{H} - \frac{\ln(a_{start}H)}{H} = \frac{1}{H}\ln\left(\frac{a_{end}}{a_{start}}\right), \tag{25.12}$$

$$= \frac{\ln(10^{27})}{10^{37}} \approx 10^{-35} \text{ seconds.}$$

Table 25.1 compares inflation at reheating temperatures varying down to 10^{19} Kelvin, which is equivalent to 10^6 GeV, about 1,000× what we currently achieve in colliders. I show you this to demonstrate that the story of inflation is fairly similar, whatever temperature you choose. For some mathematical fun, take one of the reheating temperatures and run it through the calculations we have just done. Don't forget that this is the *minimum* required inflation. It could be much larger.

You may see inflation described as *super-luminary*, which means faster than light speed. While the pace of inflation was extraordinarily fast (Equation 25.12), bear in mind that any object further out from the earth than the Hubble length (14 billion light years as in Equation 24.5) is currently moving away at faster than light speed. The visible universe is larger than this because we see light emissions from long ago. Thus, calling something super-luminary doesn't afford it a special status when discussing the expansion of the universe. The speed of light remains an absolute speed limit for local measurements through space. It does not apply when comparing to distant galaxies because their recession speed largely is due to the expansion of the universe, so the space in which any distant galaxy sits is moving relative to the space in which you sit.

25.5 The Maths Behind the Inflaton Field (Optional)

Section 25.4 covered all the basic maths that a normal human being needs. However, some readers may want more detail on the inflaton field. The underlying mathematical computations are not too gnarly but require knowledge of general relativity *and* QFT or at least a good grasp of classical field theory. I will do my best to keep the broad storyline accessible and highlight where special knowledge is required.

The inflaton field has never been directly detected. The most prevalent theory assumes it's a *scalar* field, which means that at each point it has a value but no direction. An example of a scalar field is temperature. In contrast, something like wind velocity is a vector field, having magnitude and direction. In the case of the inflaton field, the assumption is that its value changed over time during the inflationary epoch but had (almost) the same value throughout space at any given time.

Technical point for QFT students: the inflaton has similarities with another scalar field, the Higgs field. The famous Higgs Mexican hat potential means that the Higgs field value is also non-zero in its state of minimum energy. Indeed, Alan Guth quips that the inflaton is some sort of cousin of the Higgs field. There is even speculation that the two might have a common origin.

We want to evaluate how the value of the inflaton developed. Classical field theory tells us that the change in value of a field is driven by variation in the potential associated with the field. A useful example of this is the mathematics of the *classical* field theory of gravity. There is a refresher on gravitational potential energy back in Subsection 10.1.1. For ease of comparison, I now use the label V_g for gravitational potential energy (I used the label Φ in the refresher) and work in only one spatial dimension x. If V_g varies with x, there is acceleration in the direction in which V_g decreases. The equation of motion is shown in Equation 25.13 (check back to the right of Equation 10.3 if needed).

$$\frac{\partial^2 x}{\partial t^2} = -\frac{\partial V_g}{\partial x}, \qquad \textit{Classical acceleration in a gravitational field.} \qquad (25.13)$$

In the case of the gravitational field, the potential V_g varies with x. In the case of the inflaton field, the potential V varies with μ, but the basic argument is the same, albeit with a twist. In a moment I will reveal an unusual feature of the inflaton field, but let me start by describing what would happen if its behaviour matched that of the gravitational field. Equation 25.14 shows on this (incorrect) basis how the equation of motion would look for the inflaton field (compare with Equation 25.13). The value of μ would accelerate in the direction of lowering potential V at a rate of $\frac{\partial V}{\partial \mu}$.

This is *not* the full story, but let's look at what it would mean, starting from the upper left green dot in Figure 25.2. At this point, the potential V of vacuum energy is enormous. Even the slightest slope in the relationship would make the value of μ accelerate rapidly to the right. The acceleration would grow further at the steep decline. By the time the value of μ reached the low point of V (the bottom of the valley), it would have a high velocity of change $\frac{\partial \mu}{\partial t}$. The value of μ would flash past the minimum. Once past the minimum, the acceleration would pull it back towards the minimum, gradually reducing $\frac{\partial \mu}{\partial t}$ until finally the value of μ moves back again towards the minimum. The value of $\frac{\partial \mu}{\partial t}$ then would grow but with μ moving in the opposite direction, so that μ would flash past the minimum ... and so forth. The result would be a periodic oscillation in the value of μ around the V low point, rather like the simple harmonic oscillation of a ball rocking to and fro up and down the slope. The value of V would oscillate. The final (current) level of vacuum energy would not be constant.

$$\frac{\partial^2 \mu}{\partial t^2} = -\frac{\partial V}{\partial \mu}, \qquad \textit{Without expansion of universe.} \qquad \textit{NO!} \qquad (25.14)$$

$$\frac{\partial^2 \mu}{\partial t^2} = -\frac{\partial V}{\partial \mu} - 3H\frac{\partial \mu}{\partial t}, \qquad \textit{With expansion of universe.} \qquad \textit{YES!} \qquad (25.15)$$

An oscillating level of vacuum energy doesn't match reality. The problem arises because Equation 25.14 is for a static universe. The *expansion* of the universe adds an additional term to the equation of motion as shown in Equation 25.15. The origin of this term is explained in Box 25.3, but be warned that it requires specialist knowledge (available in my book *Quantum Untangling*). Some of you may recognise the new term as being equivalent to *drag*.

Let's start again from the upper left green dot in Figure 25.2, but this time applying Equation 25.15. The value of μ still accelerates in the direction of lowering potential to the right, but is reduced by the new third term. This term depends on $\frac{\partial \mu}{\partial t}$ multiplied by the Hubble parameter, which is enormous given the high level of vacuum energy. As the value of μ starts to move to the right, this term increases, reducing the accelerating effect, like the drag on a ball rolling through thick honey. The result is what is called a *slow roll* during which the value of V remains high. This is the inflationary period.

Once the value μ of the inflaton reaches the region of steepest slope, the accelerating effect grows, but the heavy drag on motion continues. This damping means that the value of μ settles at the V minimum. The reduction in vacuum energy is released as radiation and matter in reheating. Immediately after reheating, the value of the Hubble parameter remains high. It doesn't reduce until the expansion of the universe dilutes the density of radiation and matter. The value of V does *not* oscillate. The level of vacuum energy in the universe remains constant at the new lower level.

Box 25.3 Inflaton field equation of motion (specialist knowledge required)

This requires background knowledge not covered in this book. The Lagrangian density $\mathcal{L}$ of the inflaton field is shown below using the symbol $\dot{\mu}$ for $\frac{\partial \mu}{\partial t}$ and $\ddot{\mu}$ for $\frac{\partial^2 \mu}{\partial t^2}$. Note that there are no spatial gradient terms because μ is constant throughout space, albeit varying over time (well, *almost* constant throughout space as will be revealed). The Lagrangian density per unit volume of the co-moving grid increases with scale factor a as the unit volume grows in terms of space.

$$\mathcal{L} = a^3 \left(\frac{\dot{\mu}^2}{2} - V \right), \qquad \textit{Lagrangian density per unit volume of co-moving grid.}$$

The potential V varies with μ. We can apply Euler–Lagrange to generate the equation of motion of the inflaton field. An extra term appears because of the time dependence of the scale factor a. Don't forget that by definition $H = \frac{\dot{a}}{a}$.

$$\frac{\partial}{\partial t}(a^3 \dot{\mu}) = -a^3 \frac{\partial V}{\partial \mu}, \qquad \textit{from Euler-Lagrange}: \quad \frac{\partial}{\partial t}\left(\frac{\partial \mathcal{L}}{\partial \dot{\mu}} \right) = \frac{\partial \mathcal{L}}{\partial \mu},$$

$$a^3 \ddot{\mu} + 3\dot{a}a^2 \dot{\mu} = -a^3 \frac{\partial V}{\partial \mu},$$

$$\ddot{\mu} + 3\frac{\dot{a}}{a}\dot{\mu} = -\frac{\partial V}{\partial \mu}, \quad \implies \ddot{\mu} = -\frac{\partial V}{\partial \mu} - 3H\dot{\mu}.$$

(25.16)

The result in Equation 25.16 gives the equation of motion in Equation 25.15. As described in the text, this creates drag, damping any change in the value of μ.

To summarise, the theory of inflation postulates that there was an early epoch when the universe had much higher vacuum energy. This led it to expand rapidly. In the scenario discussed, the inflation occurred in only about 10^{-35} seconds, while increasing the scale factor over $10^{27}\times$, equivalent to a volume increase of over $10^{81}\times$. This rapid expansion would explain how the current visible universe might have a common causally linked origin, and therefore why the CMB is so homogeneous.

Following this inflationary period, the level of vacuum energy dropped to a much lower level, releasing energy into the universe as radiation and particles in what is called reheating.

25.6 Quantum Field Fluctuations

Having explained why the CMB is so homogeneous, we now need to explain its small variations! The CMB is not *perfectly* homogeneous. The variation (called *anisotropy*) is small: only about 1:100,000 away from the average equivalent temperature. Can the theory of inflation explain the origin of this? Bear in mind that this anisotropy is important because it provides the seeds of variation that led to today's galaxies in the cosmic web.

Jumping straight to the answer, the reason for the CMB anisotropy is that the inflaton is a scalar *quantum field*. I told you in Section 25.5 that the value μ of the inflaton field was almost the same everywhere in space (just varying over time). This is true to a high level of approximation, but it is not the whole story. All quantum fields have what are called *ground state fluctuations*. These are tiny excitations even in their lowest energy state.

Those of you familiar with quantum mechanics will know this is a consequence of Heisenberg's uncertainty principle, which says that nothing can be perfectly stationary (momentum $p = 0$) and at the same time have a definite position (e.g. $x = 0$). Some of you also will know that a quantum field must have fluctuations at every possible frequency with minimum ground state energy of $\frac{\hbar\omega}{2}$ where ω is the angular frequency (in terms of frequency: $\omega = 2\pi f$). This is a very small amount of energy but the sudden change in the relationship between μ and V (the steep slope) amplifies these quantum variations.

I will demonstrate this in a moment, but first I want to follow what happens to these quantum fluctuations when the universe rapidly expands in the inflationary epoch (i.e. during the slow roll). A tiny ground state fluctuation in the inflaton field means that the value of μ will be slightly different at different locations. This might strike you as unimportant. After all, during the slow roll a slightly different value of μ doesn't change V significantly.

That is true, but the expansion of the universe adds an extra term to the inflaton equation of motion that dampens out fluctuations in μ (Equation 25.15). The expansion of the universe rapidly increases the wavelength of quantum fluctuations until this damping effect finally stops them at slightly differing fixed values of μ. The maths is beyond our scope, but the fluctuation is fully damped (i.e. stops) when its wavelength is equal to the Hubble length. An intuitive explanation is that each crest loses causal contact with the next, because any distance beyond the Hubble length is moving away faster than light speed. You might think of the Hubble length as a sort of event horizon. Indeed, the process is called *freezing out when crossing the horizon.*

Each frozen fluctuation continues to grow spatially as the universe expands. At the same time, other later quantum fluctuations cross the horizon and freeze. They grow. Others freeze then grow and so forth. The result is small fluctuations in the value of μ at *all scales*. The variations are almost perfectly scale-invariant. The pattern is called *fractal*. At whatever scale you measure, you always measure the same level of variation.

Some readers will be wondering how these almost undetectable fractal variations in μ can be important. Think of the end of the slow roll, when μ reaches the critical value and V falls (the steep slope of Figure 25.2). Any difference in the value of μ will alter the moment when V falls. Reheating happens at (very) slightly different times in different locations!

Let's compare a point in space A with slightly higher value of μ, with a neighbouring space B that has a lower value. Point A reaches the critical value of μ before B does, so the reheating occurs slightly earlier. Vacuum energy is released at A as radiation and particles. The universe still is expanding at breakneck speed, so the energy-density at A quickly falls. Point B reaches the critical value of μ a little later. By the time reheating has happened at B, the energy-density at A has already diluted through expansion. This effectively amplifies the tiny differences from quantum fluctuations to detectable differences in energy-density.

About 380,000 years later, these small differences in the timing of reheating were responsible for variations in energy-density at the time of the CMB. Higher energy-density means a higher redshift for the CMB radiation leaving that point, leading to the anisotropy we detect today.

25.7   Evidence for Inflation in the CMB

In order to analyse the anisotropies of the CMB, we need to quantify them. This is done using the *CMB power spectrum* shown in Figure 25.3. This maps out the variation in the equivalent temperature of the CMB (vertical axis) measured across different angles (horizontal axis). In this section, I will touch on a few highlights. For those who want more, optional Chapter 26 contains a detailed explanation of the CMB power spectrum.

The prominent peaks in Figure 25.3 are the result of the interplay of underlying variations in energy-density at the time of reheating with baryon acoustic oscillations (BAO) that travelled through the baryon-electron-photon plasma in the 380,000 years leading up to the emission of the CMB (I introduced BAO back in Section 24.6).

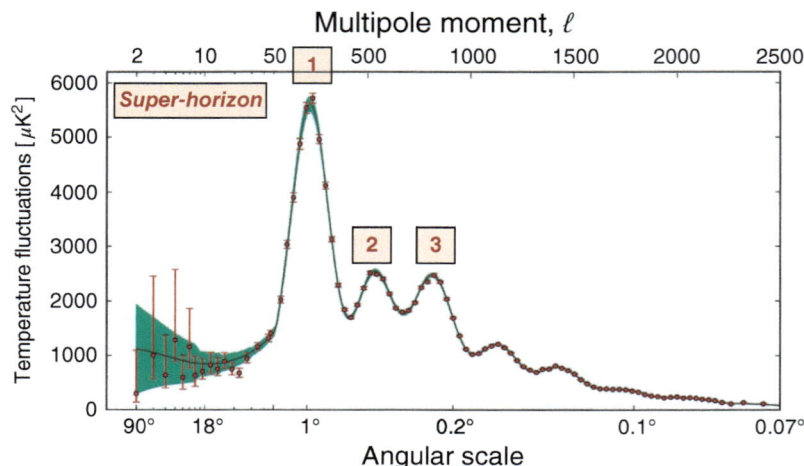

Figure 25.3   Planck mission (2009–2013) CMB power spectrum. Note that angular scale decreases to the right. Note also that µK are micro-Kelvin. *Source:* Courtesy of ESA/NASA.

Let's look at the power spectrum on the left side of the figure labelled super-horizon. Here we are looking at anisotropies of a scale much larger than might have been influenced by the BAO. It reveals the underlying anisotropy immediately after reheating. You can see that the line is fairly flat in this region. A close to flat line means that the anisotropy is almost the same at all scales. This matches the fractal pattern we would expect from quantum fluctuations (see Section 25.6). In fact, there is a slight deviation from a pure fractal pattern. Even that slight deviation matches the predictions of inflation theory.

The patterns of anisotropy at smaller angular scales are more complicated. However, these peaks in the power spectrum accurately match the predictions of inflation. Indeed, the green line through the data is the output of an inflationary ΛCDM model. The model has only six major input parameters. The fit is remarkable.

I think it fair to say that most physicists regard the CMB power spectrum as overwhelming evidence in favour of inflation. For most readers, the level of detail in this description is probably enough. However, those nuts (like me) who always want more can check out the optional Chapter 26, which steps through the CMB power spectrum in more detail.

I should mention *B-mode polarisation of the CMB*. One prediction of inflation is that the CMB will exhibit this special type of polarisation. Many feel it would be a smoking gun and proof of inflation. In spite of an unfortunate incorrect announcement (see Box 25.4), no CMB B-mode polarisation has yet been spotted. The general consensus is that the level of this polarisation is too low for current detectors. With so much evidence in favour of inflation, I wouldn't be surprised if it's detected before you read this. If so, I suspect a Nobel Prize will follow.

Box 25.4 BICEP discomfort

On March 17th, 2014, the team of the Background Imaging of Cosmic Extragalactic Polarization (BICEP) 2 telescope at the South Pole must have been beyond excited. They had just announced the detection of B-mode polarisation in the CMB. This was strong confirmation of gravitational waves and inflation. It also gave hope of a golden gong in Stockholm—a Nobel Prize.

Within a few months these hopes were dashed. In May, a paper was submitted using data from the Planck satellite showing higher levels of polarised dust than BICEP had incorporated into their analysis. The suspicion was that the B-mode polarisation was the effect of intervening dust, not the CMB. Things began to unravel and in January 2015 *Nature* published an article with the crushing title *Gravitational waves discovery now officially dead*. Ouch!

25.8 The Inflationary Multiverse

One intriguing aspect of inflation is that it leads fairly naturally to the idea of a *multiverse*—that we may be living in just one of many universes. In Section 25.2, I showed you that the causally related patch of a universe containing purely vacuum energy cannot be greater in radius than the Hubble length (see Equation 25.7). It cannot be affected by the environment outside its Hubble length, because anything outside is moving away at faster than light speed. The Hubble length acts as a type of event horizon.

In our inflationary model, we arbitrarily select an initial patch that expands exponentially. This is followed by the reheating and, hey presto, we have our universe. But how and when do we select that one-Hubble-length patch? What about the patch that was next to the one we selected? And the one next to that? And the one next to that? And so on. The inflation scenario naturally leads to an ever-growing number of patches inflating into an ever-growing number of what Alan Guth calls *pocket universes*, of which ours is just one. For readers who want to know more about the inflationary multiverse, there is a reference in Box 27.2 to Alan Guth's 2007 paper.

In Section 24.6, I described our place in the cosmic web as average, insignificant and *humbling*. The inflationary multiverse raises the even more humbling possibility that ours is one trifling universe amongst countless others. That is beyond humbling. In fact, I struggle to visualise how trivial it makes our solar system in the greater scheme of things.

25.9 Summary

Let me start with a reminder that I am using the term *Big Bang* for the birth of the universe (FRW metric $a = 0$) and using the term *reheating* for the moment following inflation when the universe was flooded with radiation and particles.

This chapter started with a review of three general arguments for inflation. The first argument is that it explains why the universe appears spatially flat based on current measurements. The value of spatial curvature $|\Omega_K|$ would have grown by a factor of 10^{37} from the time of the GUT epoch up to today, and yet we cannot detect any deviation from flat space. Rapid inflation of the universe would have flattened out any existing curvature. It would explain why the universe was (and is) so close to spatially flat.

The second argument is the absence of magnetic monopoles, which many models predict should have been created in the early universe. If we assume that inflation happened after these monopoles were created, then their presence would have been diluted, perhaps to a level where not a single magnetic monopole is present in our visible universe.

The third argument is the horizon/homogeneity problem. Measurements of the CMB reveal that it is homogeneous to a level of 1:100,000. Yet without inflation, regions over 2° apart in the sky can never have been causally linked. How can the temperature be so homogeneous if there was no possibility of the regions reaching thermal equilibrium with each other? Adding inflation to the story causally links all the CMB regions we see, thus solving this horizon/homogeneity paradox.

I then described the theory. Inflation assumes the existence of a scalar field called the inflaton. Its value, which I labelled μ, affects the density of vacuum energy V in the universe (see graph in Figure 25.2). Early in the development of the universe, there was a period when vacuum energy-density was much higher. Based on 10^{29} Kelvin as a model for the temperature of reheating, it was perhaps $10^{110}\times$ higher than it is today. During this period, the Hubble parameter was enormous and the universe rapidly inflated, increasing the scale factor by at least $10^{27}\times$, which is over 60 e-foldings.

The expansion of the universe adds an extra term to the equation of motion of the inflaton scalar quantum field, creating an effect equivalent to drag. This slows down any change in the value of the inflaton field (μ). This allowed the change in μ during the initial inflationary epoch to be a slow roll during which the value of V was approximately constant and the universe rapidly expanded. Following this, the value of μ reached a point where V declined precipitously. This reduced the vacuum energy-density, leading to a release of energy in the form of matter and radiation: reheating.

The drag term in the equation of motion of the inflaton field meant that the value of μ settled quickly at the minimum level of V (the bottom of the valley). Any oscillations around the point of minimum potential were quickly damped out. Since that time, the values of μ and V have been constant, resulting in the constant level of vacuum energy in the universe today.

The inflaton is a quantum field. As such, there are ground state fluctuations in the value of μ at all scales. These fluctuations created very slight differences in the timing of reheating in different regions. The universe was still rapidly expanding, so the inflaton ground state fluctuations generated matching fluctuations in the energy-density from reheating. Such fluctuations would be almost scale-invariant (fractal).

The CMB power spectrum provides evidence that is consistent with the theory of inflation. At larger angular scales, the anisotropies measured are super-horizon meaning that the energy-density of these zones couldn't be affected by events between reheating and the CMB. They are a direct image of the original primordial anisotropies. The fluctuations are (almost) fractal as predicted by inflation. At smaller angular scales the results also match the predictions of inflation.

The final topic was the inflationary multiverse. The inflaton model creates many patches of inflating space. In our analysis, we have selected and tracked the development of just one. Arguably many patches could inflate into separate universes. Might our universe be just one of many?

Chapter 26, *Interpreting the CMB*, is optional. It provides further detail on the CMB power spectrum, how it was formed and what it tells us. Following that there is a summary of the *Cosmology* module, plus one final chapter touching on some of the attempts to tie general relativity together with quantum theory. Well done. You are almost finished!

Chapter 26

Interpreting the CMB (Optional)

In this chapter, my aim is to explain the shape of the cosmic microwave background (CMB) power spectrum. I will review and build on the material in Section 25.7. I have labelled this optional because there is more detail here than many readers will want. I have included it in the book in part because I find it fascinating and in part because I personally struggled with it when studying the subject.

Any readers jumping into this chapter without having read earlier ones, should note that I use the term *Big Bang* for the theoretical birth of the universe (FRW metric $a = 0$) and the term *reheating* for the moment at the end of inflation when the universe was flooded with radiation and particles. I do this to avoid semantic confusion, although the difference in timing between the two may well be less than 10^{-30} seconds.

Figure 26.1 shows a map of the CMB taken by the Planck satellite. Looking out at the CMB in every direction, we observe around us a spherical surface viewed from the inside. The CMB map displays this surface flat in the same way as our maps of earth do. Looking at the night sky, the very left side of the map in Figure 26.1 is physically next to the far right side. CMB radiation has a mixture of wavelengths that closely resemble the pattern of radiation from a black body, so we can associate a temperature with it. In the map, different colours show the temperature variations (hotter spots red, cooler spots blue), which typically are of the order of only 1:100,000. Before we dig further into the CMB data, let me describe in simple terms the two main causes of the temperature variation (anisotropy).

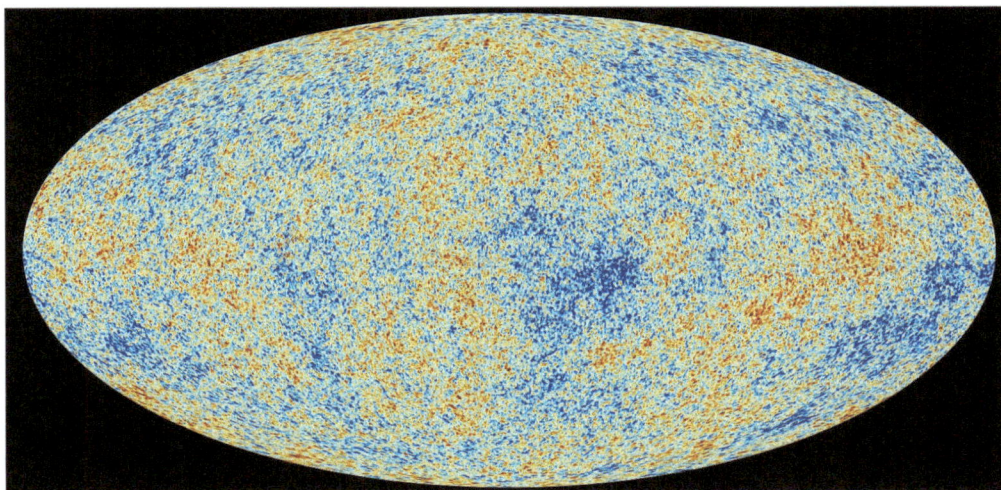

Figure 26.1   Map of the CMB anisotropies. *Source:* ESA and the Planck Collaboration/Wikimedia Commons/CC BY 4.0.

Untangling General Relativity: The Intuitive Self-Study Guide, First Edition. Simon Sherwood.
© 2026 John Wiley & Sons Ltd. Published 2026 by John Wiley & Sons Ltd.

(a)

(b)

BAO schematic

Fourier schematic

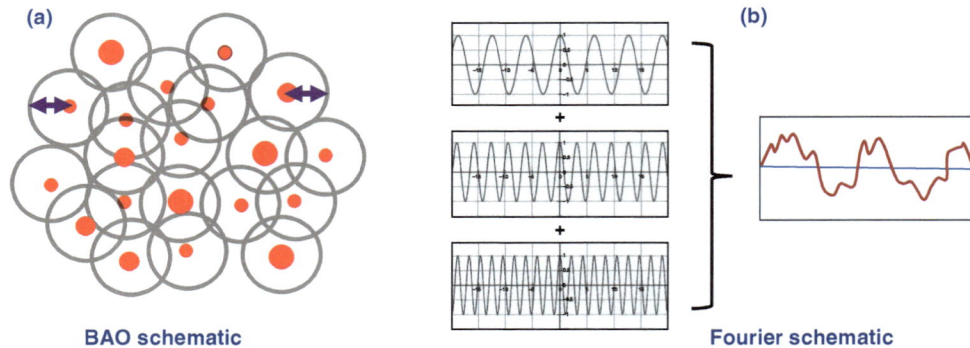

Figure 26.2 (a) Schematic of dark matter hot spots and BAO rings, (b) Fourier schematic.

26.1 Underlying Causes of CMB Temperature Variation

The first major cause of CMB temperature anisotropy was quantum fluctuations in the inflaton field, which meant that the reheating happened at very slightly different times in different locations. This created some variation in the resulting energy-density. Where reheating happened slightly earlier, the expansion of the universe diluted the energy-density compared to regions where reheating happened slightly later (see Section 25.6 for more detail). This led to what you might describe as hot spots with higher dark matter density. Figure 26.2a shows these dark matter hot spots as red spots of different sizes (to represent different levels of dark matter density). The quantum nature of the original fluctuations means that the pattern of variation in the density is fractal, i.e. there is the same level of variation at all scales.

The second major cause of CMB temperature anisotropy is baryon acoustic oscillations (BAO), described in Section 24.6. After reheating, the dark matter hot spots sat in a sea of strongly interacting baryon-electron-photon plasma. They acted like gravity wells drawing in the surrounding plasma, which in turn pulled in the surrounding plasma, and so forth. The result was a BAO compression wave travelling out from the dark matter hot spot at 0.57 of the speed of light. This speed can be computed analytically (see footnote in Section 24.6 for more details).

Effectively, a spherical shell of plasma over-density travelled outwards for 380,000 years until recombination when the CMB was released. If you take into account the expansion of the universe between reheating and CMB, the distance travelled is about 500,000 light years (called the *sound horizon*). Accounting also for the approximate 1,000× expansion of the universe from the CMB to today, the radius of the BAO spherical compression wave is now over 500 million light years. This means that at what is now 500 million light years from the original hot spot, there was an over-density of plasma at the time of the CMB, shown in Figure 26.2a as grey rings.

The net result is a mix of over-dense dark matter hot spots (fractal variation) surrounded by over-dense baryon rings, all of equal radius. At each point of over-density, the photons released in the CMB will have lost a little energy escaping from the higher-than-average gravitational field. The CMB temperature from these regions will be slightly lower than average. That is something we can measure. The challenge, as you can see from the figure, is that the combined picture of over-densities (the red dots and grey circles) is quite a mess!

We know there are some underlying patterns. If we examine the data at the scale of 500 million light years across, which is the present-day radius of what were the BAO rings (represented by the purple arrows in Figure 26.2a), we should pick up all the temperature variations between the over-dense hot spot, the under-density surrounding region and the over-density ring. On the other hand, scales much larger than this (for example, 4 billion light years across) don't distinguish the hot spot from the BAO ring, and any variation will be due to the varying density of the hot spots themselves (not the BAO rings). How can we pull the CMB temperature data apart to reveal these differences?

26.2 The CMB Power Spectrum

The answer is to use Fourier analysis. Fourier showed that any signal can be expressed in terms of sinusoidal waves. Figure 26.2b shows this schematically. The complicated pattern on the right can be expressed as a sum of sinusoidal waves. This sum is called its Fourier transform and is widely used in physics and astronomy for signal processing. We want to do something similar with the CMB so that we can analyse temperature variations across different scales (distances).

In the case of the CMB we are looking at a pattern on the surface of a sphere. This can be analysed as a sum of *spherical harmonics*. Each spherical harmonic is the equivalent of one of the sinusoidal waves in Fourier analysis. In Fourier analysis you enhance the detail of the signal you are replicating by adding higher and higher frequency sinusoids. With spherical harmonics you achieve the same by adding harmonics with higher and higher values of the multipole moment index l. The frequency of a sinusoid reflects the distance between points of symmetry along the axis (e.g. zero nodes). The multipole moment l depends on the number of pieces the spherical surface is divided into when comparing levels of anisotropy. Therefore, it can be expressed roughly as an angular scale. The higher l, the smaller the angular scale evaluated. The relationship between multipole moment l and angular scale is (very) approximately $\theta \approx \frac{180°}{l}$.

This brings us to the CMB power spectrum, shown again for convenience in Figure 26.3. The horizontal axis is the angular scale, which runs from large to small (or it shows the multipole moment running from small to large, if you prefer). The vertical axis shows the *variance* in the temperature fluctuations $(\Delta T)^2$ in (microKelvin)^2. A microKelvin is 10^{-6} Kelvin. We use the square so that both positive and negative fluctuations are counted.

As an example, consider peak-1 in the power spectrum. This tells us that if we break down the CMB temperature anisotropy into spherical harmonics, we find a peak at an angular scale of just under 1°. At smaller angular scales there are other peaks such as peak-2 and peak-3. Does this mean that we could take a telescope, compare the sky at the angular scale of peak-2 and detect the anisotropy? *No!* This is an important point. The power spectrum is not a bunch of individual observations at different angular scales. It is the *statistical analysis* of what combination of scale variations would lead to the CMB we see. The anisotropy of peak-2 would be lost in the huge sum of variations at other scales (don't forget the mess in Figure 26.2a). Physically picking out a single scale of anisotropy would be like spotting one of the sinusoidal waves in the Fourier transform of Figure 26.2b.

Before we dig into the analysis, I must highlight a couple more things. There are $(2l + 1)$ harmonics for every value of multipole moment l. Each is labelled with a different value of index m (this is just an index number and not to be confused with mass). If this looks somewhat familiar, it may be because the spherical harmonics play a central role in the distribution of electrons in atoms, giving us the l and m atomic quantum numbers. Basically the more you divide up the spherical surface (the higher l), the more independent ways there are of comparing. This is why the error bars in the power spectrum grow at low values of the multipole moment. The sample size is smaller, so the average value of anisotropy is less reliable.

Also, the power spectrum in the figure has been adjusted for distortions. One example is the motion of the earth relative to the CMB (Section 22.3) because it increases the temperature of the CMB ahead in the direction of motion (blueshift) and reduces that behind (redshift) creating a distorting *dipolar* anisotropy (dipolar means $l = 1, \theta = 180°$), which is removed from the data set.

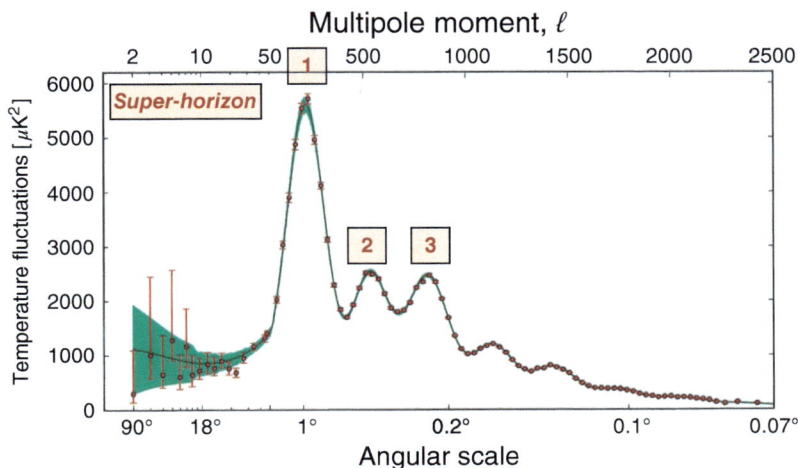

Figure 26.3 Planck mission (2009–2013) CMB power spectrum. *Source:* Courtesy of ESA/NASA.

26.3 Super-Horizon Anisotropy

Angular scales in excess of 2° are labelled *super-horizon* because the patches being compared are larger than the sound horizon. The patches are much bigger than the grey rings of BAO over-density in Figure 26.2a. The BAO can't have had any significant effect over this scale, so we are seeing the pattern of comparative energy-density as it was at the time of reheating.

The almost flat line relationship matches the prediction of inflation at these super-horizon scales. It reflects the approximately fractal (scale-invariant) nature of the quantum fluctuations in the inflaton field (see Section 25.6). I use the word *almost* because inflation models predict a small deviation that depends on factors such as the assumed change in V during the slow roll. Measurements by Planck and other observatories of the super-horizon scale-invariance fit well with the inflationary thesis.

26.4 Effect of Baryon Acoustic Oscillations

If the pattern of energy-density stayed the same from the time of reheating to the time of CMB emission, then the anisotropy in Figure 26.3 would be almost scale-invariant at all angular scales. The power spectrum would be approximately flat. Clearly, this isn't the case. What causes the variations such as at peak-1, peak-2 and peak-3?

Consider the top schematic in Figure 26.4a. The red spot is the dark matter hot spot. The dotted orange circle shows the over-dense region of baryons and photons at the sound horizon. The top of Figure 26.4b shows that the plasma in the BAO wave is compressed at the time of the CMB (indicated by the baryons shown as two orange balls drawing closer together). Calculating the anisotropy at the scale of the sound horizon (the double-headed purple arrow) produces peak-1. This is not surprising. At scales larger than this, you are missing the anisotropy from the BAO circular rings around the dark matter hot spots.

Let's turn our attention to the other peaks. You can think of the dark matter in the hot spot as somewhat passive. It has no electric charge. In contrast, the baryon-electron-photon plasma is a soup of electrically charged particles. The protons, electrons and photons strongly interact, and the plasma behaves somewhat like a fluid. In fact, physicists sometimes refer to it as a *baryon-photon fluid*. When compressed, a counter-pressure gradually grows as the photons resist the compression (check back to Equation 22.19 on pressure and radiation). The result is a bounce back. The plasma undergoes compression, then expansion (called *rarefaction*), then compression and so on.

The middle schematic of Figure 26.4a illustrates the origin of peak-2 in the power spectrum. Consider the baryons at the green dotted circle around the hot spot. What is their density at the time of the CMB? They have had enough time to be drawn into the hot spot and then to be bounced back out by the photon pressure. This is illustrated in green in the middle of Figure 26.4b. After an initial period of compression (heading into the hot spot), they bounce back out and are at maximum rarefaction at the moment of the CMB. Analysis of the data at this smaller scale (the double-headed arrow in the middle schematic of Figure 26.4a) should reveal this anisotropy.

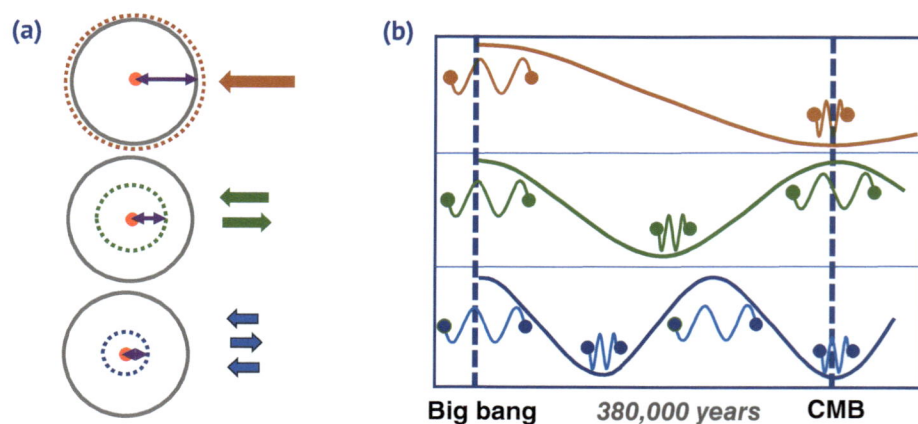

Figure 26.4 Schematic of how BAO affects the CMB power spectrum. In (a) the dotted circles show analysis at different scales relative to the original hot spot and BAO. (b) shows how the BAO compression wave has developed at each of these scales.

The bottom schematics in Figures 26.4a and b (in blue) show the rationale for peak-3. Baryons even closer to the hot spot have time to undergo compression, rarefaction and then recompression. At the time of the CMB, these baryons are at maximum compression again. Analysis at this even smaller scale should reveal this.

You may wonder about the baryons between the dotted green and blue circles of the middle and lower schematics of Figure 26.4a: the ones closer to the hot spot than those related to peak-2, but further out than those related to peak-3. These baryons have time for compression then rarefaction, and at the time of the CMB are in the middle of recompressing. They aren't at maximum compression or rarefaction, so they have less impact on the temperature variation of the CMB. The effect on anisotropy is related to those wave patterns at maximum compression or rarefaction.

Let me rather boringly remind you that this doesn't mean you can detect the anisotropy of, say, peak-3 by observing the CMB at different scales. The CMB is the sum of all the anisotropy of the spherical harmonics. Individual pieces will be smeared out in the overall sum. However, while we cannot see the patterns directly, we can *statistically* deduce their presence in the sum from the power spectrum of the observed data.

26.5 Peak-1 and Measuring Flatness

Let me offer a rough calculation of where approximately peak-1 is in the power spectrum. The rings created by the BAO sound horizon had a radius of about 500,000 light years at the time of the CMB. Since that time the universe has expanded by a factor of 1,100× (elsewhere in this book I have approximated this as a 1,000×, but see the calculation in Section 24.4). The radius of the visible universe is about 45 billion light years. This gives an angular separation between the centre of the dark matter hot spot and the BAO ring of about 0.012 radians: $(0.5 \times 1,100)/45,000$. This is about 0.7°. The observed peak-1 is at just under 1°, so we are in the right ball park. To put this in context, 1° angular separation is about twice the diameter of the full moon viewed from earth. Let me stress that the above calculation is just back-of-the-envelope. It ignores a host of important things such as redshift effects in the early universe. I offer it just to give you a feel for the logic. Please don't use it for any serious physics!

You will be thrilled to hear that peak-1 delivers more nuggets for physicists to savour (cosmological revelations, not crumb-coated chicken ones). In Section 23.1 I told you that the CMB is used to measure how spatially flat the universe is. Let me explain how. You take an *accurate* calculation of the location of peak-1 (*not* my back-of-the-envelope effort) and compare it with the observed position. Positive spatial curvature increases the angular separation of distant objects. Such an increase in angular separation would shift the observed peak-1 to the left (i.e. to a higher angular scale). Negative spatial curvature decreases it. Check back to Figure 21.3 if you need a reminder on why this happens. Looking out at the CMB, we are staring across 45 billion light years of space, but no spatial curvature has been detected to date.

I should mention that the exact location of peak-1 is affected by the density of baryons in the plasma. Higher concentrations of baryons damp the BAO oscillation, which slightly changes the BAO speed through the plasma and therefore the sound horizon. Therefore, an accurate peak-1 calculation reveals the energy-density of baryons, typically labelled Ω_b.

26.6 Comparing Peaks: Another Measure of Baryon Density

The density of baryons in the plasma has a number of effects on the CMB power spectrum. As mentioned, it damps the BAO oscillations, which affects the length of the sound horizon. This also has a general dampening effect on shorter wavelength BAO waves, because they undergo more oscillations between the time of reheating and CMB. As a result, the higher the baryon density, the lower the anisotropy in the tail of the power spectrum (i.e. at small angular scales).

A bit more fun (at least I find it fun) is that the *relative* height of the peaks in the power spectrum gives a measure of baryon density. The level of resistance to compression in the plasma depends on the mix of photons versus baryons. It is the photons that resist compression and drive the bounce from compression to rarefaction. The baryons are happy to crunch down into the gravitational potential well, but reluctant to move back out. The higher the baryon density in the plasma, the more intense the compression phase in the BAO waves and the less intense the rarefaction phase.

If you look back at Figure 26.4b, the BAO waves causing peak-1 and peak-3 (shown in orange and blue, respectively) end in compression, whereas that of peak-2 (shown in green) ends in rarefaction. If the baryon mix in the plasma is higher, the compression grows and the odd-numbered peaks become more pronounced. On the other hand, the rarefaction decreases and the even-numbered peaks diminish. This effect is visible in Figure 26.3. You can see that peak-2 and peak-3 are similar in spite of the general dampening expected at smaller angular scales. This is because the baryons are diminishing peak-2 and enhancing peak-3. The relative size of the odd and even peaks is used by cosmologists to estimate the original baryon mix in the plasma.

It is not only the baryon mix in the plasma that affects the CMB. The presence of neutrinos also has an effect and it is possible to model this. I find it amazing how much information can be gleaned from the CMB, a weak microwave signal initially mistaken for pigeons roosting in a telescope (as described back in Box 22.4).

Chapter 27

Module Summary: Cosmology

27.1 Theory

Much of the study of cosmology relies on applying the *cosmological principle*, which assumes that the universe is homogeneous (the same viewed at any location) and isotropic (the same viewed in any direction). The idea is that if you zoom out to encompass large chunks of the universe, any irregularities become insignificant. On this basis, the universe is like a homogeneous isotropic dust field, with galaxies playing the role of the dust particles.

One might criticise this as an imprecise approximation. However, if we don't take advantage of the symmetries of the cosmological principle, the maths becomes unmanageable without the help of computers.

We want to understand how the universe developed to its current state and what will happen to it in the future. The universe is expanding. We account for this by thinking of galaxies as being positioned on a co-moving grid. Rather than being flung out into space by the Big Bang, they are stationary relative to the grid (ignoring any peculiar motion they have). The galaxies are carried along for the ride as the grid expands.

The distance between grid points increases with time by a scale factor labelled as a. It is time-dependent. If we assume that the universe is spatially flat (which we know is true to good approximation), the metric can be written simply in Cartesian coordinates as shown in Equation 27.1. As time goes by, the scale factor a increases, which means that the distance grows between grid points along the x, y and z axes.

$$\text{FRW metric, flat space } (t,x,y,z;\ c=1): \qquad [g_{\mu\nu}] = \begin{bmatrix} 1 & 0 & 0 & 0 \\ 0 & -a^2 & 0 & 0 \\ 0 & 0 & -a^2 & 0 \\ 0 & 0 & 0 & -a^2 \end{bmatrix}. \tag{27.1}$$

The next step was to adopt the full Friedmann–Robertson–Walker (FRW) metric. This is more complicated because it caters for the possibility of spatial curvature. Let me remind you that the presence of energy-density *always* curves *spacetime*. However, the gravity we experience day to day primarily is due to changes in how *time* passes at different locations.

The easiest way to think about *spatial* curvature is with a 2-D surface. If flat, it is like a sheet. With positive curvature, it is like the surface of a sphere and is called closed, because travelling in a straight line brings you back to where you started. With negative curvature, it is like a saddle and is called open. Spatial curvature affects how distant objects appear. If the universe has positive spatial curvature, distant objects appear larger. With negative spatial curvature they appear smaller (see Figure 21.3).

I stepped through in detail how the FRW metric accounts for spatial curvature. It is a bit of a mathematical nightmare, but the good news is that it can be summarised as shown in Equation 27.2. This uses spherical coordinates. For positive spatial curvature (like the surface of a sphere in 2-D), the value of $\mathbf{K}$ is $+1$. For no spatial curvature (this is a flat surface in 2-D), $\mathbf{K}$ is 0. For negative spatial curvature (a saddle surface in 2-D), $\mathbf{K}$ is -1.

Untangling General Relativity: The Intuitive Self-Study Guide, First Edition. Simon Sherwood.
© 2026 John Wiley & Sons Ltd. Published 2026 by John Wiley & Sons Ltd.

$$FRW\ metric\ (t, r, \theta, \phi;\ c = 1): \quad [g_{\mu\nu}] = \begin{bmatrix} 1 & 0 & 0 & 0 \\ 0 & -\dfrac{a^2}{1 - Kr^2} & 0 & 0 \\ 0 & 0 & -a^2 r^2 & 0 \\ 0 & 0 & 0 & -a^2 r^2 \sin^2\theta \end{bmatrix}. \tag{27.2}$$

In Chapter 22, we used the FRW metric to derive the famous Friedmann equations. The cosmological principle introduces symmetries into the curvature footprint. This helps to simplify the calculation of the Ricci tensor components of the FRW metric.

We substituted these values into the Einstein Field Equations (EFEs) and matched them up with the energy-momentum tensor for a perfect fluid. For example, we matched up the values of R_{00} and R of the FRW metric with T_{00}, which is the energy-density. This gave us the Friedmann-1 equation. We repeated the exercise, substituting in the values of R_{11} and R of the FRW metric and matching them to the value of T_{11}, which is the pressure of the perfect fluid. This gave us the Friedmann-2 equation. The two Friedmann equations are shown again below.

$$\left(\frac{\dot{a}}{a}\right)^2 = \frac{k\rho}{3} + \frac{\Lambda}{3} - \frac{\mathbf{K}}{a^2}, \qquad \textit{Friedmann-1: expansion (c = 1).}$$

$$\frac{\ddot{a}}{a} = \frac{\Lambda}{3} - \frac{k}{6}(\rho + 3P) = \frac{\Lambda}{3} - \frac{k}{3}\left(\frac{\rho_M}{2} + \rho_R\right), \quad \textit{Friedmann-2: acceleration (c = 1).}$$

The Friedmann-1 equation tells us the *expansion rate* of the universe at any given moment in time. On the left of the equation is $\left(\frac{\dot{a}}{a}\right)^2$, which is H^2, where H is the Hubble parameter. The expansion rate depends on the amount of vacuum energy Λ and on the *total* amount of matter and radiation energy-density. It includes a term containing $\mathbf{K}$, so it is affected by any spatial curvature.

The Friedmann-2 equation tells us the *acceleration* in the expansion of the universe, i.e. how the rate of expansion is changing over time. Vacuum energy accelerates the expansion, while matter and radiation energy slow it. The acceleration is affected also by the mix between matter and radiation energy. This shows up as the pressure P term, which is 0 for matter and $\frac{\rho}{3}$ for radiation.

As the universe expands, radiation energy-density falls faster than matter energy-density because radiation energy-density decreases with a^4, of which a^3 comes from the increase in spatial volume, and an additional a comes from the increase in its wavelength. As a result, radiation energy has a greater braking effect on expansion than the equivalent amount of matter energy. To make things clear, I have separated out the energy densities of matter ρ_M and radiation ρ_R on the right of the Friedman-2 equation (check back to Equation 22.25 if you need more detail).

We can put the Friedmann-1 equation in a form that makes it easier to match the energy-densities with observational data. We know that the universe is approximately spatially flat, either because it is actually flat ($\mathbf{K} = 0$) or because the scale factor is very large (a 2-D comparison is that the surface of the earth has a much smaller curvature than that of a football). Either way, the spatial curvature term is negligible. We set this to zero and substitute in the value of the Einstein constant k and the observed value of the Hubble parameter: $H = \frac{\dot{a}}{a}$. This lets us calculate the total energy-density ρ_{crit} that the universe must have for it to be spatially flat and expanding at this rate. The answer is about $10^{-26}\,\mathrm{kg\,m^{-3}}$. The next challenge was to figure out what proportions of this energy-density are matter, radiation and vacuum energy. We label the proportions Ω_M, Ω_R and Ω_Λ.

27.2 Observation

Having worked through the theory, we turned in Chapter 23 to observational data. We used this to estimate Ω_M, Ω_R and Ω_Λ, and build a model of the historical (and future) expansion of the universe. The best current cosmological model is called ΛCDM and is shown again in Box 27.1.

Box 27.1 ΛCDM model of cosmology	**present day estimates**		
$\Omega_\Lambda \approx 0.7,$	$\Omega_M \approx 0.3,$	$\Omega_R < 10^{-4},$	$\Omega_K \approx 0.$

Note that radiation will have been an important source of energy-density in the early universe. However, the energy-density of radiation drops more quickly with expansion than that of matter. The largest remaining element of radiation energy is the CMB, but its spectrum is equivalent to a temperature of about 3 Kelvin. As a result, radiation's contribution to the universe's total energy-density now is negligible.

27.2.1 Flatness

The contribution of spatial curvature is measured as Ω_K. It is estimated to be less than 0.001 by the Baryon Oscillation Spectroscopy Survey (BOSS) and less than 0.003 based on the cosmic microwave background (CMB) (Planck and WMAP). The BOSS survey compares the observed distribution of galaxies with the expected distribution based on the effect of baryon acoustic oscillations (BAO). We also can compare the peaks in the CMB power spectrum with their calculated value based on the distance travelled by BAO waves through the baryon-electron-photon plasma before the emission of the CMB. In both cases, the presence of significant spatial curvature would distort the results because positive curvature makes distant objects appear larger, and negative curvature makes them appear smaller.

Let me remind you that this does *not* mean the universe is perfectly spatially flat at the local level. We know this is not the case. For example, the Schwarzschild metric has some spatial curvature. However, if we zoom out to view the universe at large scale, it appears to be flat (or very nearly flat). I liken this in 2-D to looking at a sheet of bubble wrap. Viewed close-up, you can see the bubbles, but viewed from afar it appears flat.

27.2.2 Dark Matter

The value of Ω_M is heavily influenced by the presence of dark matter, which is estimated to account for about 85% of the total matter in the universe. It reveals itself in the rotation of galaxies. Their orbital speed is higher than would be expected based on the amount of visible matter they contain. This indicates the presence of unseen mass orbiting the outer regions. The presence of dark matter also reveals itself in the way that light bends around galaxies (gravitational lensing).

We know dark matter is not electrically charged and doesn't interact strongly with visible matter apart from its gravitational effect; if it did, we would have identified it long ago. We know that dark matter formed early in the Big Bang, because patterns in the CMB power spectrum show that variations in the density of dark matter led to BAO oscillations in the plasma. Possible candidates for dark matter include a WIMP (weakly interacting massive particle), a particle linked to SUSY (this assumes there are supersymmetric partners for each fermion and boson), an axion (slow-moving light boson) or perhaps the effect of primordial black holes. Some even suspect it might be explained by a modification (MOND) to the EFEs. At the time of writing, we still don't know what dark matter is.

27.2.3 Dark (Vacuum) Energy

The other big surprise for cosmologists is the dominance of dark (vacuum) energy in the mix of energy-density. One early clue to its presence was the cosmic age problem. If you model a matter-dominated universe based on $\Omega_\Lambda = 0$, the age of the universe calculates to about 10 billion years old. However, studies of globular clusters reveal galaxies over 13 billion years old. The numbers work only if you factor in the presence of vacuum energy.

This is supported by data from type 1a supernovae, which form when a white dwarf in a binary system consumes mass from its partner. Once the white dwarf reaches the critical mass, it explodes into a supernova with a standard pattern of luminosity. These supernovae serve as standard candles, revealing how far away a galaxy is. This allows us to map how the Hubble parameter has changed over time by comparing the speed of recession of closer galaxies with distant ones. The results show that the expansion of the universe is *accelerating*. This means the Friedmann-2 equation has a positive value, which can happen only if there is vacuum energy.

Based on all the observational evidence, it appears that vacuum energy accounts for about 70% of the total. In terms of matter, this would be equivalent to a few hydrogen atoms in each cubic metre of vacuum. That doesn't sound much, but there is an awful lot of vacuum, so it adds up to most of the energy-density in the universe today. Yet we don't know what it is. Efforts to explain it as the ground state energy of quantum fluctuations in the vacuum arrive at a number $10^{120}\times$ too large. And there is even speculation that the amount of vacuum energy might be changing over time.

So the somewhat embarrassing truth is that we now believe that we have decent grip on the amount of energy-density in the universe and on the mix of components. However, we don't know what vacuum energy or dark matter are, and they sum to about 95% of the total!

27.3 From the Big Bang to Today

The ΛCDM model gives the age of the universe as about 13.8 billion years from the Big Bang up to the present day. We track the early stages based on energy-density. I gave you a rule of thumb that the age is $\frac{10^{20}}{T^2}$ seconds, where T is the temperature in Kelvin.

Baryogenesis occurred when the temperature dropped to about 10^{13} Kelvin (10^{-6} seconds after the Big Bang), low enough for quarks to form protons and neutrons (collectively called baryons). One would have expected equal amounts of matter and antimatter to be created, resulting in equal numbers of baryons and anti-baryons. However, the actual result was a large number of baryons but a negligible number of anti-baryons. This level of asymmetry between matter and antimatter is higher than would be predicted by the Standard Model of particle physics and is as yet unexplained.

As the universe expanded, the temperature dropped further. Once below 10^{11} Kelvin (0.01 seconds after the Big Bang), there were a few minutes of *nuclear fusion* (nucleosynthesis) creating helium nuclei and a trace of lithium nuclei. At this stage, the hydrogen, helium and lithium were positively charged atomic nuclei, not neutral atoms.

It remained too hot for these nuclei to capture electrons and form electrically neutral atoms until the temperature dropped to about 3,000 Kelvin (about 380,000 years after the Big Bang). Up to this point, the photons in the plasma were constantly interacting with the charged nuclei and electrons. However, once neutral atoms formed (*recombination*), the photons were released (*decoupling*). We can still see the relic of this wave of radiation in the CMB. Its emission spectrum has a temperature equivalent to about 2.7 Kelvin compared with an original emission temperature of about 3,000 Kelvin, indicating that the scale factor of the universe has increased by about 1,100$\times$ between the time of CMB emission and today.

Our best guess is that it took another 200 million years for the first stars and galaxies to form (the *cosmic dawn*). This period is called the *dark ages* for obvious reasons. Nuclear fusion in the stars leads to the production of helium and then heavier elements. Energy is released by fusion right the way up to iron. The creation of even heavier elements than this (such as silver and gold) requires energy. They are most likely forged by cataclysmic events such as supernovae and the merger of neutron stars.

Today, galaxies are clustered together in the *cosmic web*. It is influenced by the effect of BAO that created a predictable pattern of anisotropy in energy-density at the time of the CMB. These areas of higher energy-density pulled in surrounding matter and energy to give us the distribution of galaxies we observe today.

We also can use the ΛCDM model to predict the *future* of the universe. With vacuum energy being the dominant component in the energy mix, the expansion should continue to accelerate, finally ending in the heat death of the universe. As the universe expands, more and more galaxies will cross our event horizon and disappear from view, except for our Milky Way and the Andromeda galaxy (which will crash into the Milky Way in about 5 billion years' time).

27.4 Some Bits That Might Not Fit

I don't want to create the impression that everything stacks up perfectly. In addition to the big questions about antimatter, dark matter and vacuum energy, there are some niggling discrepancies in the ΛCDM model that may (or may not) prove important.

Perhaps best known is the *Hubble tension*. There is a significant discrepancy in the value of the Hubble parameter depending on how it is measured. In this book, I have used for H the value $70 \text{ km s}^{-1} \text{Mpc}^{-1}$, where Mpc is a megaparsec. The problem is the following. If you calculate H with data from the CMB, you arrive at $68 \text{ km s}^{-1} \text{Mpc}^{-1}$. However, if you calculate H from the recession of galaxies using standard candles to measure distance (such as type 1a supernovae), you get $73 \text{ km s}^{-1} \text{Mpc}^{-1}$. This may not seem a big difference, but a lot of effort has gone into assessing the accuracy of each approach. The conclusion is that this is over *four sigma* mismatch and there is less than a 1:10,000 probability of it occurring by chance. Oops!

Clearly something is wrong. One possibility is a systematic measurement error that is unaccounted for. For example, there might be an error in the detailed calibration of the standard candles or perhaps there is an as yet unknown factor affecting the CMB power spectrum. However, there has been a lot of checking and re-checking. Could all this work still have missed a major error of this type, or is there a more serious theoretical problem?

There is also what is called the *S8 tension*. It is a discrepancy in the amount of matter clustering. Different measurement techniques lead to different results. Again we start with the CMB. Analysing its anisotropy and modelling how this changes over time gives a prediction of the pattern of matter clustering we should see today in the cosmic web. However, observational data suggests that there is less clustering or clumpiness than would be expected.

This could be a measurement problem, but different observational techniques consistently show a discrepancy in the same direction (e.g. simply counting galaxies versus using galaxy lensing to measure the presence of matter). The space-based Euclid telescope launched in July 2023 should help resolve any measurement errors. Some preliminary analysis of its observational data should be available in 2025, but we will need to wait until 2026 for the full results.

27.5 Inflation

Many students wonder if inflation really is science. How can we possibly know what happened back then? But the theory of inflation is more than idle speculation. There is a scientific logic for it and it produces testable predictions. So far, it has passed the tests well. I think it is fair to say that the vast majority of cosmologists believe there was an early inflationary epoch during which the scale factor of the universe increased exponentially.

27.5.1 The Rationale for Inflation

The theory of inflation was inspired by three problems with the traditional ΛCDM Big Bang model of the universe. The first was the *flatness problem*. A simple calculation shows that if $|\Omega_K| < 0.003$ now (in line with observational data), then at a time early in the life of the universe, such as the GUT epoch, $|\Omega_K| < 10^{-40}$. How do we explain this when we know that spatial curvature is a feature of general relativity at the local level (for example, the Schwarzschild metric has spatial curvature)? Inflation offers a solution. A rapid period of early inflation before the standard ΛCDM expansion would increase the scale factor and reduce the effect of any pre-existing spatial curvature.

The second problem was the absence of *magnetic monopoles*, which many theories predict should exist, albeit at high energy. If the inflationary epoch occurred after the production of magnetic monopoles, then the presence of the monopoles would be diluted by the expansion. If the inflation was intense enough, this might leave not a single magnetic monopole present in our visible universe.

The third problem (which I find the most compelling) was the *horizon/homogeneity problem*. If we look out at the CMB in any direction, its equivalent temperature is the same within 1:100,000. Yet, based on the traditional ΛCDM Big Bang model, regions of the CMB more than about 2° apart cannot possibly have had any causal connection. There would not have been time before the CMB for light (or anything else) to travel from one patch to the next. How then can their temperatures be so homogeneous? A period of exponential expansion causally links these regions. It explains the homogeneity.

27.5.2 The Mechanism of Inflation

The underlying concept is that there is a scalar quantum field called the *inflaton* that bears some resemblance to the Higgs field. In the earliest epoch, there was a much higher amount of vacuum energy, perhaps over $10^{100}\times$ the level today. As a result, the Hubble parameter was huge and the universe expanded extraordinarily quickly, doubling perhaps every 10^{-37} seconds.

The value of the inflaton field would have moved in the direction of lower potential energy (as all fields do). However, the expansion of the universe adds an extra term to the equation of motion of a scalar field, which resembles a drag-like term. This slows the rate of change like a ball rolling through honey, leading to what is described as a *slow roll* towards the lower potential. During this inflationary epoch, the level of vacuum energy-density remained approximately constant.

At some point, the slowly changing value of the inflaton reached a transition point and the level of vacuum energy fell precipitously. As a result, at the end of the inflationary epoch, an enormous amount of vacuum energy was released into the universe in reheating. This was the genesis of the matter and radiation present in the universe today.

Typically, you might expect the value of the inflaton field to oscillate around the point of lowest potential (rolling to and fro up and down the bottom of the valley in Figure 25.2). However, the drag-like term damped any oscillations. Since then, the value of the inflaton field has remained constant, as has the level of vacuum energy-density in the universe.

We can estimate how much inflation is the *minimum* required to explain the homogeneity of the CMB. The numbers vary depending on the temperature of reheating, but most assumptions indicate an expansion in the scale factor of over $10^{20}\times$ or more than about 60 e-foldings.

27.5.3 Evidence for Inflation from the CMB

The inflaton field would have had quantum fluctuations, as is true of any quantum field. These would create regions with slightly different values of the inflaton in the slow roll. As a result, they would arrive at the transition point (the steep fall) at slightly different times, creating small spatial variations in the timing of reheating. This means that the pattern of quantum fluctuations would be reflected as areas of different energy-density after reheating.

Quantum fluctuations of this nature have a distinct *scale-invariant* pattern (often described as *fractal*). We can detect this pattern in the power spectrum of the CMB. If we analyse at scales larger than the sound horizon of the BAO, the patterns haven't been influenced by events between the time of reheating and the CMB, so we are seeing the original primordial pattern... and that pattern matches the prediction of inflation.

In addition, when we model the CMB power spectrum at smaller scales using assumptions from the inflationary theory, and calculate how it would have affected the plasma created by reheating, the results exquisitely match the observed CMB power spectrum (there is more detail on this in optional Chapter 26).

27.6 Cosmology: Watch the News

The ΛCDM/inflation model fits well with observational data. For example, based on only six major parameters, it gives an almost perfect match with the CMB power spectrum. However, there remain missing pieces of the puzzle. Some are obvious, such as the composition of dark matter and vacuum energy. Others, such as the Hubble and S8 tensions, may require a major rebuild of the theory, or prove to be simple measurement errors.

Who knows what will have been discovered by the time you read this book? When you have completed this module, I urge you to read up on the latest news. Cosmology is a rapidly developing field of study. Finally, for a laugh at the end of all the work you have done in this module, I offer you the cosmologist's periodic table of elements in Figure 27.1.

27.7 Module Memory Jogger

Below is a memory jogger of some of the key topics in this module, and Box 27.2 suggests some further resources.

- *The cosmological principle assumes the universe is homogeneous and isotropic.*

- *FRW metric $(t, r, \theta, \phi;\ c = 1)$:*
$$[g_{\mu\nu}] = \begin{bmatrix} 1 & 0 & 0 & 0 \\ 0 & -\dfrac{a^2}{1 - \mathbf{K}r^2} & 0 & 0 \\ 0 & 0 & -a^2\,r^2 & 0 \\ 0 & 0 & 0 & -a^2\,r^2\,\sin^2\theta \end{bmatrix}.$$

- *For positive spatial curvature $\mathbf{K}=+1$ (like surface of a sphere in 2-D); for zero curvature $\mathbf{K}=0$ (flat); for negative curvature $\mathbf{K}=-1$ (like surface of a saddle in 2-D).*

- *Friedmann-1 $(c = 1)$:* $\left(\dfrac{\dot{a}}{a}\right)^2 = \dfrac{k\rho}{3} + \dfrac{\Lambda}{3} - \dfrac{\mathbf{K}}{a^2}.$

- *Friedmann-2 $(c = 1)$:* $\dfrac{\ddot{a}}{a} = \dfrac{\Lambda}{3} - \dfrac{k}{6}(\rho + 3P) = \dfrac{\Lambda}{3} - \dfrac{k}{3}\left(\dfrac{\rho_M}{2} + \rho_R\right).$

- *Critical energy-density for flat universe:* $\rho_{crit} = \dfrac{3H^2}{8\pi G} \approx 10^{-26}\ kg\,m^{-3}.$

- *ΛCDM model (present day estimates):* $\Omega_\Lambda \approx 0.7,$ $\Omega_M \approx 0.3,$ $\Omega_R < 10^{-4},$ $\Omega_\mathbf{K} \approx 0.$

- *Dark matter: about 85% of matter or 25% of total energy-density. Composition unknown.*

- *Dark (vacuum) energy: about 70% of total energy-density. Composition unknown.*

- *History of the universe since the Big Bang (current estimates):*
  - *Baryogenesis: 10^{-6} seconds.*
  - *Nuclear fusion: 0.01 seconds.*
  - *CMB, recombination and decoupling: 380,000 years.*
  - *Cosmic dawn, first stars (fusion restarts): 2 million years.*
  - *Current age of the universe: 13.8 billion years.*

- *Nuclear fusion releases energy up to production of iron (Fe).*

- *Hydrogen and helium still account for over 98% of visible matter.*

- *Our place in the cosmic web: the Milky Way sits on the edge of the Laniakea supercluster.*

- *Inflation.*
  - *Inflaton: a scalar quantum field somewhat similar to the Higgs field.*
  - *Slow-roll inflationary epoch, vacuum energy-density about $10^{100}\times$ current level.*
  - *Exponential expansion: scale factor increased over $10^{20}\times$ (60 e-foldings).*
  - *Reheating: transition to lower level of vacuum energy released as matter and radiation.*

- *CMB Power Spectrum (Optional).*
  - *Super-horizon anisotropy: scale-invariant pattern matches inflationary theory.*
  - *Comparison of peak-1 with BAO sound horizon as a measure of spatial flatness.*
  - *Relative size of peaks provides a measure of baryons in reheating plasma.*
  - *BAO sound horizon can be used a standard ruler for galaxy distribution.*

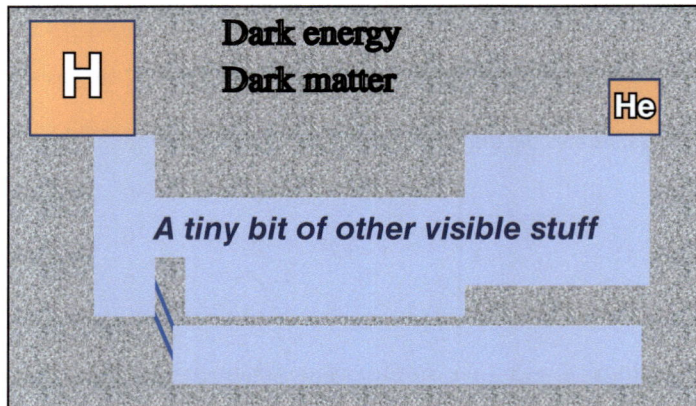

Figure 27.1 Just for laughs: a cosmologist's periodic table of elements.

Box 27.2 Module 3: further resources   (web links as of February 2025)

Leonard Susskind's lecture series on cosmology has less mathematical detail than this book, but is accessible and provides good intuitive insights:
https://theoreticalminimum.com/courses/cosmology/2013/winter

Check out the excellent lecture by Alan Guth, the father of inflation theory:
https://ocw.mit.edu/courses/8-286-the-early-universe-fall-2013/resources/lecture-23-inflation/

Alan Guth's paper on the inflationary multiverse: *Eternal Inflation and its Implications, Journal of Physics A: Mathematical and Theoretical*, 22 February 2007.

Chapter 28

The Big Challenge

I'm not going to burden readers with a long review of the contents of this book. If you want a trip down memory lane, Chapters 11, 20 and 27 provide summaries and module memory joggers. And I urge all of you to flick back to the *Overview* in Chapter 1. Read through it again. I hope it will make sense now.

In the last chapter, I discussed several questions concerning the ΛCDM model, such as:

- *Baryogenesis: Where has all the antimatter gone?*
- *What is dark matter?*
- *What is dark (vacuum) energy?*

These are important questions. Perhaps one or more will have been answered by the time you read this book. The answers may require making some changes to the ΛCDM model. However, there is a much bigger challenge facing physicists: *How do we reconcile general relativity (GR) with quantum mechanics (QM)?* This is *not* going to be achieved with a few tweaks here and there. Without a complete rethink, the two theories are incompatible.

Much of what is to come will be familiar to readers of *Quantum Untangling*; at least to readers who made it through to the last chapter, which discusses the same challenge.

28.1 General Relativity Versus Quantum Mechanics

GR and QM have different, arguably irreconcilable, structures. As there may be readers with no knowledge of QM, I am going to give a brief tour of some of its weirder features, in order to demonstrate how hard it is to integrate with GR, or indeed any classical field theory.

Quanta of energy: At the most basic level, QM is, well, *quantum*. In QM, everything comes in lumps that are discrete. What we call particles actually are the excitation of quantum fields. The quantum fields have resonant frequencies, a bit like musical instruments. Energy is absorbed or released in indivisible lumps. For the electromagnetic field, we call these lumps photons. For the electron field, we call them electrons. In contrast, GR explains gravity as the curving of spacetime. It is smooth and continuous. Where are the quantum lumps and bumps?

Superposition: A key feature of QM is that a particle can be in a *superposition* of states. For example, in what is called the Double-slit experiment, there is a screen with two slits between an emitting source of particles (photons, electrons and larger) and a detecting screen. As a result, there are two pathways that the particles can travel to the screen. The result is affected by both pathways, even if you send particles through one at a time. Each particle seems to sample both available routes. It turns out that particles can be in a superposition of locations, states, energy levels and so forth. For instance, the location of electrons around the nucleus of an atom is a fuzzy probability density (not a specific position) and may be a superposition of energy levels. In contrast, in GR, the presence of energy-momentum curves spacetime. In GR, the spacetime metric is well-defined and has a single value at any given location in spacetime. GR does not cater for superpositions of different energy-momentum levels across multiple locations.

Uncertainty: Another striking difference is the *uncertainty principle* of QM. Heisenberg proved that there is an inherent uncertainty between the position and momentum of a particle. There is also uncertainty in its energy.

Untangling General Relativity: The Intuitive Self-Study Guide, First Edition. Simon Sherwood.
© 2026 John Wiley & Sons Ltd. Published 2026 by John Wiley & Sons Ltd.

In fact, an essential feature of particle interactions is that they can suddenly find some extra energy from nowhere, albeit often for only a short period of time. If you apply this to GR, it means there is uncertainty in the value of the energy-momentum tensor on the right of the Einstein Field Equations (EFEs)... uncertainty in the amount of energy, the amount of momentum and the position. On the basis of the EFEs, this feeds through to uncertainty in the structure of spacetime. Ouch!

Probability: This weirdness of QM means that there are multiple possible outcomes for every interaction. The chance that *A* leads to outcome *B* is quantified as a probability. GR is the opposite. In the equations of GR, matter precisely and exactly curves spacetime. This curvature precisely and exactly affects how the matter behaves. $A = B$. There is nothing in the GR equations that speaks of superpositions, uncertainty, or probability.

Mechanism of force: QM and GR describe forces with completely different mechanisms. In QM, the electromagnetic, strong and weak forces are transmitted by the exchange of mediating particles. The energy moves through the quantum field. It is affected by the resonance of the field, so it reflects many, but not all, of the characteristics of the particle associated with the field (hence the name *virtual particle*). QM would describe the gravitational force as the exchange of virtual gravitons, which are massless bosons. GR describes the same thing as a change to the curvature of spacetime.

28.2 The Challenge

This is the conundrum. GR has been shown again and again to describe accurately the gravitational force. QM has been shown again and again to describe accurately the electromagnetic force, the weak force and the strong force. These two magnificent theories collide majestically at the theoretical level. This is the big challenge that physics faces. We really need a unified theory that includes all four forces to describe conditions in the early universe or at the centre of a black hole. How can we reconcile QM and GR?

The obvious solution is to find a quantum theory of gravity that fits with QM but, at a large scale, gives GR. This may sound simple. It is not. Physicists have been grappling with this for decades with little progress. In the following sections, I provide a short introduction to two of the best-known efforts, *string theory* and *loop quantum gravity (LQG)*, although both are struggling to make progress. Following this, I will touch on a couple of more recent directions of research.

One thing GR and QM have in common is a mutual enemy: the problem of singularities. In the case of GR, singularities cause problems at the heart of a black hole (and at time $t = 0$ in the FRW metric). In the case of QM, the maths becomes unmanageable in scenarios where you get infinitely close to, say, an electron. The force becomes infinite, its charge becomes infinite and so on. Both string theory and LQG address this problem, but in different ways. In the case of string theory, particles are 1-D strings of finite length whose interactions avoid singularities. In the case of LQG, spacetime itself is quantised, meaning there is a limit to how small a volume of space can be, eliminating the possibility of singularities.

Both string theory and LQG build on the maths of QM. Readers of this book have not been exposed to this, so I must offer truncated descriptions of them. For a more complete description (albeit still only a taste), please refer to the last chapter of *Quantum Untangling*.

28.3 String Theory

The idea behind string theory is disarmingly simple. The excitations of quantum fields that we call particles are actually tiny vibrating one-dimensional strings. Different vibrations in these strings are associated with different elementary particles. There is an obvious attraction to such a reductionist theory that seeks to describe all the particles and forces as having a common underlying structure.

One of the immediate benefits is that the strings have a length, albeit very small (perhaps 10^{-33} metres compared to the 10^{-10} metre diameter of an atom). This avoids many mathematical problems associated with point particles. In particular, efforts to describe gravity as a quantum theory struggle because the self-interaction in the gravitational field leads to unmanageable infinities as you explore closer and closer to a point particle. String theory neatly sidesteps this difficulty because the strings have a finite size.

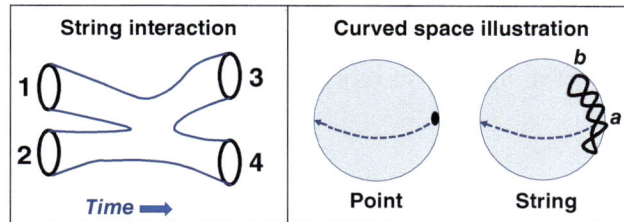

Figure 28.1 String theory illustrations.

Strings can interact by joining together at the ends or splitting apart. Consider the left side of Figure 28.1. It could represent any interaction, such as an electron and positron annihilating to form a photon that subsequently decays back into an electron and positron. At the start, there are two particles labelled 1 and 2. These are *closed loop* strings. Note that *open loop* strings also play a role in many string theories. From left to right, these closed loops each trace out a two-dimensional surface called a *world sheet*. Their world sheets merge. The loops join to form one larger loop before splitting into loops 3 and 4. The interaction effectively is smeared out over a region of spacetime. At no stage is there a singularity.

28.3.1 Gravity in String Theory

In QM, gravity is expected to be transmitted by the graviton, which is spin-2. Each particle in QM reflects the characteristic of its quantum field. You will find this hard to understand unless you are familiar with QM. The best I can do is give examples. A scalar field varies only in magnitude, so the associated particle is spin-0 (e.g. the Higgs field and the Higgs boson). A vector field has magnitude and a direction, so its particle is spin-1 (e.g. the electromagnetic field and the photon). Gravity is linked to the metric. This is a tensor field whose magnitude is linked with two directions, hence, the graviton should be spin-2.

One early attraction of string theory is that it readily accommodates the graviton as a closed loop string. A closed loop can carry a vibration in either of two directions (think clockwise and anticlockwise). Due to symmetry considerations, the energy moving in each direction must be the same (for highly technical reasons called *level matching*). As a result, excitation comes in units of two, matching a spin-2 graviton.

Even more compelling is that GR falls straight out of string theory. The maths is way beyond the scope of this book, but let me give you a flavour of how this happens. Every string can be described as an infinite number of quantum harmonic oscillators. This is similar in concept to the way you describe any signal as the sum of sinusoidal waves (its Fourier transform). The volume of space that might be occupied by the points between the ends of the string depends on the number of oscillators. If you consider a string with, say, 5 oscillators, it will occupy a certain volume, but if you add in higher frequencies up to, say, 10 oscillators, that volume will be larger. Add more and it is larger still. The growth never stops, although it slows dramatically. This means that the amount of space covered by a 1-D quantum string cannot be quantified. It is undefined.

In flat space this has no effect. A string moving through flat space has momentum mv where v is the velocity of its centre of mass. The position of various parts of the string does not change this. However, the picture is very different in *curved* space. Let's consider what happens if space is curved in the shape of the surface of a sphere as shown on the right side of Figure 28.1. A *point* particle moves on the surface of the sphere in a geodesic (great circle). In addition to its mv of linear momentum, it has mvr of angular momentum because in this curved space it is rotating around the vertical axis of the sphere.

In the case of a *string*, the space it occupies is undefined. Consider it spread out as shown on the far right in Figure 28.1. The part of the string at point *a* has angular momentum mvr, but the part of the string at point *b* has less angular momentum because it is closer to the axis of rotation. If the volume of space occupied by the string is undefined, its angular momentum is undefined. As a result, you cannot have a consistent string theory in the spherically bent space of Figure 28.1, but there is a solution to this paradox...

The trick is to limit string theory to those spacetimes that have just the right curvature not to create the paradox. The curvature of spacetime is quantified by the metric $[g_{\mu\nu}]$. String theorists calculated what limits would need to be placed on $[g_{\mu\nu}]$ to resolve the string paradox. Can you guess what popped out? The result is Einstein's EFEs in a vacuum. I find this amazing. In order for string theory to be consistent, spacetime must curve in exactly the way that Einstein showed it does! Furthermore, this relationship is driven by a symmetry in string theory called *conformal symmetry* or *Weyl gauge invariance*. This symmetry is very particular to the world sheets of strings. There is a neat mathematical fit. This got string theorists very excited.

28.3.2 Difficulties with String Theory

String theory is appealing and seductive because it resolves many of the problems related to the singularities that accompany point-like particles. It postulates that all the particles and forces are made up of the same thing... strings. Not only do gravitons fit neatly into the model, but Einstein's EFEs for a vacuum are required by it. Sadly, we must now discuss the problems with string theory and why some physicists describe it as a dead end.

The biggest theoretical change is that string theory needs more than just the EFEs of GR to be a viable model in curved space. It requires the universe to have at least 10 dimensions: one of time, the three of space (x, y, z) that we know, plus *six additional* spatial dimensions. Obviously, we cannot see these six extra dimensions. The idea is that they are curled up to a tiny size (perhaps about 10^{-30} metres) that we cannot sense, but the tiny strings can move in. This is described as *compactification* with the dimensions intertwined in what are called *Calabi–Yau manifolds*. To date, no experiments have given a hint of the existence of additional dimensions. Indeed, it is far from clear what kind of experiment might reveal them.

The second challenge and perhaps the biggest experimental disappointment is the lack of evidence for super-symmetry (SUSY). This postulates the existence of a bunch of new particles. I mentioned in Box 23.2 that one of them might be a candidate for dark matter. Modern string theory relies on SUSY. Without it, many particles (the fermions) don't really fit, and the number of required dimensions jumps from 10 to 26! There were high hopes that the Large Hadron Collider might reveal one or more SUSY particles, but none have been found. String theory can be tweaked to explain the lack of these particles at lower energies but, as some critics say, each tweak makes it that bit uglier.

A third significant problem is that string theory can be defined in so many different ways. Over time, five different geometries of string theory developed.[1] However, in 1995, Edward Witten tied together the five versions of string theory into *M-theory*, albeit at the cost of adding another dimension, increasing the count to 11.[2] M-theory includes *D-branes* of various dimensions that act as anchor points for strings. While this link sparked new excitement, there is still a vast array of theoretical approaches, especially in the geometries of the hidden dimensions and how the strings behave in them. The hope must be to explain quantities such as the mass of the elementary particles and the coupling constants of the forces, but after 50 years of study, there are still an estimated 10^{500} different possible versions of string theory. How do you narrow down the options?

String theory still has many fans and believers. I am told that its mathematical beauty is enticing, but there is no hard evidence and the LHC's failure to find SUSY particles has been a serious blow. Moreover, with so many possible geometries for the compactified dimensions, some go as far as to call it untestable. I think it is fair to say that, in recent years, enthusiasm for the subject has slowly declined (hence the joke in Box 28.1).

Box 28.1 Just for laughs

Is there hard evidence that string theory is right?　　*Frayed knot!*

28.4 Loop Quantum Gravity

The approach of LQG is radically different. In LQG spacetime itself is *quantised*. Singularities cannot exist because there is a limit to how small a chunk of spacetime can be. In layman's language, spacetime is *granular*.

LQG theorists argue that we know from QM that the energy-momentum tensor $[T_{\mu\nu}]$ is quantised. In order for the left side of the EFEs to equal the right side, this effect must flow across into the spacetime curvature and the metric. Spacetime must come in some sort of quanta.

---

1 Type 1, Type IIA, Type IIB, SO(32) heterotic and E8 × E8 heterotic.
2 Technical point: the extra dimension of M-theory actually was received positively because it aligned string theory with some 11-D theories of supergravity.

Another way to think about this is the following. According to QM, there are quantum fields associated with each force. For example, the electromagnetic force is associated with the electromagnetic quantum field. LQG argues that the quantum field associated with the gravitational force *is* spacetime. Take a moment to think about this. LQG says that spacetime is nothing more than the gravitational field. And if spacetime is a quantum field, then it must be quantised.

28.4.1 LQG Space as a Quantum Entity

In this book, I have not introduced the QM tools required to understand the maths. Sorry! But this is the last chapter, so I am not going to play by the rules. I am going to offer a simplistic summary of the structure of LQG. I apologise in advance if some of this is a bit much for QM novices.

LQG postulates that uncertainty exists within the very geometry of space itself. The model is best described by comparing with the QM uncertainty relationship of angular momentum. The uncertainty between position and momentum (Heisenberg's famous principle) leaks into angular momentum because the angular momentum of an object depends on its position (distance from the axis of rotation) *and* on its momentum. If you run through the maths, it turns out that you cannot know the angular momentum (typically labelled L) around all three axes at the same time. However, you *can* know the sum of the squares of the three ($L^2 = L_X^2 + L_Y^2 + L_Z^2$) and this is quantised. L^2 can have only certain values.

LQG theorises that there is a similar uncertainty relationship between areas of space. Figure 28.2 shows a comparison of the uncertainty in angular momentum (on the left) with that of spatial area (on the right). Don't worry if you cannot understand the details. The consequence is that there is a minimum area (about 10^{-70}m^2) and a minimum volume (about 10^{-103}m^3). For context, there are almost as many quanta of this size in the volume of one atom as there are atoms in the visible universe. Small it may be, but it is important because if it is true, it eliminates the possibility of spacetime singularities such as at the centre of a black hole.

28.4.2 Difficulties with LQG

Before entertaining any criticism of LQG, I want to reflect on its strengths, which are numerous. It represents a fundamental shift in thinking by proposing that spacetime *is* the gravitational field. This is a logical simple message, albeit the mathematical consequences are complicated. And the theory is built on the base of things we know: GR, QM and the Standard Model of particle physics. It does not postulate new concepts like strings, extra dimensions or SUSY particles. As some quip, it is a theory with *no strings attached*. This being said, LQG experts such as Carlo Rovelli are the first to admit that it is a *tentative theory* (his words). There is no direct evidence because the quanta of spacetime are much too small to detect.

While it is easy to shoot down a new theory, there are a number of legitimate worries about LQG (see Box 28.2). If there really is a minimum length of space, will that not appear different to different observers (length contraction)? If so, might that allow an observer to know if he or she is moving and drive a stake through the heart of relativity? Another complaint is that LQG is really a theory of space with time added in, and does not address combined spacetime effectively. There are also some questions as to how well LQG leads to GR in the classical limit.

The fundamental challenge of LQG is the same as that of string theory. How do you test it? There was speculation that in LQG different frequencies of light would travel at different speeds because interaction with the quanta of space would make high-energy photons travel very slightly more slowly. Results from the Atmospheric MAGIC

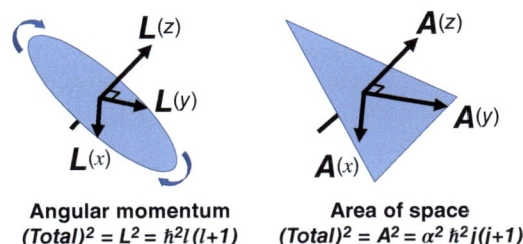

Angular momentum
(Total)^2 = L^2 = $\hbar^2 l\,(l+1)$

Area of space
(Total)^2 = A^2 = $\alpha^2\,\hbar^2 j(j+1)$

Figure 28.2 LQG: spatial uncertainty resembles that of angular momentum in QM.

telescope gave a hint of this and caused much excitement. It might have provided a shot in the arm for LQG in the same way that the discovery of SUSY particles would bolster string theory... but further analysis revealed no detectable difference.

Box 28.2 Are you stringy or loopy?

Like oil and water, the two do not seem to mix:

The reason why I don't work on LQG is the issue with special relativity. If your approach does not respect the symmetries of special relativity from the outset, then you basically need a miracle to happen at one of your intermediate steps... String theorist on LQG.

String theory seems to me to have failed to deliver what it offered in the '80s... I do not really understand how people can still have hope in it... LQG theorist on string theory.

Some worry that string theory and LQG are splitting into separate camps. One researcher in quantum gravity reflected sadly that:

Conferences have segregated. Loopy people go to loopy conferences. Stringy people go to stringy conferences. They don't even go to 'physics' conferences anymore. I think it's unfortunate that it developed this way.

One theoretical prediction of LQG is that black holes could bounce back as white holes. This would be instantaneous inside the hole, but for an observer outside, might take billions of years because of gravitational time dilation. Might we be able to detect white hole explosions? Researchers toil on in the hope that, through something of this sort, nature might show her hand and give us an indication of the right path forwards. Short of that, it is hard to see how we will know if LQG is reality or just a very pretty idea (and so one final joke in Box 28.3).

Box 28.3 Just for laughs

Did you hear about the LQG researcher who retired?   *He's out of the loop!*

28.5 Spacetime Is Doomed

So where does this leave us? Having spent decades trying to come up with a quantum version of GR and spacetime, many physicists have concluded that the whole concept of spacetime is *doomed*; that spacetime is an illusion or what might be described as an *emergent phenomenon* coming from some other underlying reality. This is not born completely out of frustration. It is backed by some curious and unexpected mathematical connections that encourage speculation.

Perhaps best known is *AdS/CFT correspondence*, which was discovered by Juan Maldacena in 1997 and has led to the theory of the *holographic universe*. The idea is that our universe is the manifestation of an underlying reality based on its boundary surface. I mean... what? I will try to explain, but this is a complicated topic.

Let me start by discussing the Schwarzschild black hole. A falling observer continues on in, but from the perspective of a distant observer, nothing ever crosses the surface of its event horizon (Section 13.6). It is as if everything happening inside the black hole remains imprinted on its event horizon. Furthermore, calculations indicate that the entropy of the black hole (see Box 28.4) is proportional to the surface area of the event horizon. This led to speculation that there must be some correspondence between conditions inside the black hole (spatially 3-D) and conditions on its event horizon surface (spatially 2-D).

> **Box 28.4 Entropy and the second law of thermodynamics**
>
> The entropy of a black hole is the number of ways everything inside it can be organised. You might describe entropy as a measure of randomness. Imagine putting 100 balls in the bottom of a box: 50 red ones on the left and 50 blue ones to the right. Shake the box and what happens? In theory, the result could still be 50 red balls on one side and 50 blue on the other. In practice, you will get a mix of balls, simply because there are more random ways for the coloured balls to be mixed than separate. There will be billions of patterns with mixed balls, but only one with them separate.
>
> If you spend energy in any form to put back the red balls on one side and the blue balls on the other, the energy spent will always be associated with a release of heat, no matter what means you use to split the balls again; that heat inevitably increases the entropy. This natural statistical tendency towards greater randomness is described by the second law of thermodynamics: *the total entropy of a system must always increase.*

Maldacena's discovery hints that something like this might help explain what is happening in our universe as a whole. He studied an Anti-de Sitter (AdS) universe. This has negative curvature and is open, so it is infinitely large (AdS space is discussed briefly in Section 23.3). You may be surprised to learn that it is possible mathematically to define the boundary around an infinite AdS universe. The trick is to use conformal mapping to create a finite map of an infinite universe, similar in concept to the Penrose-Carter diagrams explained in Section 16.8. Maldacena found an exact correspondence between what happens inside the AdS universe (called the *bulk*) and what happens on its boundary surface. Even more bizarrely, what looks like QM combined with gravity in the bulk looks like only QM on the boundary surface (the CFT in AdS/CFT stands for conformal field theory). Gravity appears with the change in perspective from surface to bulk.

If this were true for us, then what we see as our 4-D universe (3× space, 1× time, plus gravity) might be better understood as a sort of holographic projection of a simpler 3-D reality (2× space, 1× time, no gravity) on the mathematically defined surface of our universe infinitely far away. We might be able to develop an understanding of quantum gravity by studying what is happening on the simpler boundary surface and during the projection process. Dear reader... is your head spinning? If you have understood any of this, well done!

Now for the caveats. First, Maldacena worked on a static AdS universe, which ours definitely is not. Second, he showed this correspondence only for a 5-D interior bulk (4× space, 1× time) and a 4-D surface (3× space, 1× time). To apply to our universe, it needs to be proved for a 4-D interior bulk and 3-D surface. Third, these theories tend to incorporate SUSY. Unless there is a discovery soon at the Large Hadron Collider (LHC), this is looking hard to justify (see Subsection 28.3.2). Maldacena's paper is the most cited in the history of high energy physics. This illustrates how excited the scientific community is about the idea, but is also evidence of the huge amount of work already done on the subject.

28.6 Entropic Gravity

In 2010, the Dutch theoretical physicist Erik Verlinde built on the model of a holographic universe to propose his theory of *entropic gravity*. He argues that gravity is not a force in the usual sense but happens because of an associated increase in entropy on the holographic boundary surface.

In his paper, Verlinde illustrates the underlying concept with the example of a long molecule tethered at one end, as shown on the left of Figure 28.3. If we release the other end of the molecule, then it is likely to end up in some folded state nearer the tethered end, simply because there are many more ways for the atoms to be folded up than in a straight line. We might mistake this for an attractive force pulling the atoms together, whereas in reality,

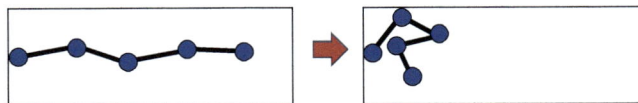

Figure 28.3 Illustration of how entropy can lead to behaviour resembling a force of attraction.

the attraction is the emergent result of the molecule moving towards a state of greater randomness (increasing entropy). Verlinde argues that we are making the same error when we describe gravity as an attractive force.

His reasoning goes as follows. As discussed in the last section, the entropy of the contents of a black hole is proportional to the surface area of the event horizon. The entropy of the bulk (the centre) is coded on the boundary (the event horizon). Let's scale up to the holographic universe. If, as Maldacena's discovery hints, there is a correspondence between the bulk and the infinitely distant boundary surface, then some part of that surface must encode the entropy in the bulk. Now consider an object falling into the black hole. This increases the total entropy of the black hole and therefore must increase the entropy of the part of the holographic boundary surface corresponding to the event horizon.

You might describe this as gravitational force causing the mass to fall towards the black hole, which in turn causes an increase in entropy on the boundary surface. Verlinde turns this on its head. He argues that an increase in entropy on the boundary surface is what you would expect. Rather than the gravitational force leading to an increase in entropy, it is the increase in entropy that creates the gravitational effect.

Verlinde argues more broadly that every shell around a mass will be encoded somewhere on the boundary surface of the universe. As one object approaches another, the entropy within more and more of these shells increases. The same increase in entropy is occurring somewhere on the boundary surface. Gravity is an emergent phenomenon driven by an inevitable growth in the total entropy of the holographic boundary surface.

Using some mathematical approximations about the change in entropy, Verlinde arrives at something similar to Newton's law of gravitation. Adjusting his equations for the expansion of the universe (i.e. the presence of dark energy), he finds a small deviation from Einstein's equations, specifically that in extremely weak gravitational fields the force falls with $\frac{1}{r}$, not $\frac{1}{r^2}$. Might this create measurable differences? Might it explain the orbital rotation of galaxies (Section 23.2) without the need to invoke the presence of dark matter?

To date there is no firm evidence for his hypothesis and we must remain sceptical, given that we don't yet even have a working model of a holographic universe (Maldacena's work is for a very different universe). However, the search for an emergent form of gravity continues.

28.7 Postquantum Gravity

The last topic I am going to discuss in this book is something new and completely different: Jonathan Oppenheim's theory of *postquantum gravity*. He starts with the premise that gravity is *classical* and then looks at how it can be meshed with the quantum nature of particles and the other forces. This is radical, but perhaps the difficulty in finding a quantum solution indicates that gravity really is different.

Let me first lay out the challenge. A key part of quantum theory is that a particle can be in a superposition of, say, two locations (or more). While in the superposition, there can be no objective way to determine that the particle is at one location or the other. Now suppose that gravity is classical. In classical field theory, we can measure things to any arbitrary degree of accuracy. There is no uncertainty principle to hinder us. Therefore, we could measure the gravitational effect of any particle at any moment and know exactly where it is. But if we can know even in theory where the particle is, it cannot be in a superposition. Do you see the problem? Using the classical field as a measure (even just in theory) destroys any possibility of superposition—and puff—quantum theory collapses!

Oppenheim's cool idea is that the gravitational field is classical but has *random* variation built in. This allows it to be consistent with the presence of quantum fields. In layman's language, the random variation creates uncertainty in measurements of the gravitational field. This uncertainty makes it impossible to distinguish between the quantum superpositions—and hurrah—quantum theory is resuscitated!

Let's take an extreme scenario. Imagine we somehow get the earth into a superposition of two locations (A and B), set well apart (ridiculous, but bear with me). We release an apple out in space. Without Oppenheim's postquantum gravity, we have a big problem. If gravity is a classical field, it can't be in a superposition, so it can accelerate the apple in only one direction at any time. If the apple moves steadily towards A or B, we immediately know where the earth is and there is no superposition. If the apple heads to the average location between A and B, the earth is not there. If later the superposition collapses say to A, the apple has been heading somewhere completely different. That doesn't work.

With postquantum gravity, the random fluctuations in the gravitational force/acceleration are affected by the superposition. The apple can move towards A or towards B. Its path is a sort of drunken walk, taking a step one way or the other (Oppenheim uses Brownian motion as a model). Each step is random, but the probability depends on the superposition of the earth's location. Over time by chance, let's say the apple gets slightly closer to B. The apple's presence then has a subtly stronger influence on the earth's superposition at B. They get closer and closer until the apple ends up at B. In the language of quantum physics, the approach of the apple is a measurement. The apple and earth now are at B, but might just as well have ended up at A.

Importantly, the presence of the gravitational field cannot be used to distinguish between A and B, except in tandem with the measurement progressively being made by the motion of the apple. As the apple moves, the relative probability of the earth being at A or B changes. Gradually, the superposition favours one option over the other. Throughout the process, the superposition does not lose its quantum nature. There is no sudden moment of change. There is no discontinuity.

One feature of postquantum gravity is that it naturally leads to gravitational collapse of the wave function. One of the puzzles of quantum physics is what causes a particle in a superposition suddenly to collapse into a final state. We talk about *measurement* being the cause, but what is a measurement? What determines how long superpositions last? Roger Penrose suggested the collapse might be due to gravity. While the mechanism in postquantum gravity is different, the concept is the same. This gives a natural explanation for the process and a definitive collapse.

There is a possible test of postquantum gravity. It turns out that the random fluctuations in the classical gravitational field have to be much larger than the quantum fluctuations. They might be detectable. Another super-cool possibility emerges. Oppenheim and his team believe the random fluctuations (the drunken walk) slightly alter the acceleration equations of GR. This might (and I emphasise might) account for some of the effects that we currently attribute to the presence of dark (vacuum) energy as well as dark matter. It is early days, but what an exciting idea. And even if postquantum gravity is wrong, it's good to see new ways of thinking.

Box 28.5 Chapter 28: further resources   (web links as of February 2025)

I should preface this with a warning. The mathematics behind both string theory and LQG is very advanced. The sources below will give you more on each topic but will still leave you with many unanswered questions. There is a limit to how much even great teachers such as Susskind and Rovelli can explain in simple terms.

For string theory, Leonard Susskind, one of the founding fathers, has a series of lectures on string theory and M theory. These are Stanford University videos available on YouTube:
https://www.youtube.com/watch?v=25haxRuZQUk

For LQG, Carlo Rovelli, who was and is one of the driving forces in the subject has produced a number of lectures. The first listed below is an introduction. The second is a more complete (and mathematically complicated) lecture series:
https://www.youtube.com/watch?v=IkqHu1BqzDg
https://www.youtube.com/playlist?list=PLwLvxaPjGHxR6zr421tXXlaDGbq8S36Un

PBS has an entertaining video on the holographic universe:
https://www.youtube.com/watch?v=klpDHn8viX8

PBS has a video on entropic gravity:
https://www.youtube.com/watch?v=qYSKEbd956M

Verlinde's paper is called *On the Origin of Gravity and the Laws of Newton*:
https://arxiv.org/abs/1001.0785

Oppenheim's paper on postquantum gravity is well written, although the maths will be too much for some:
https://journals.aps.org/prx/pdf/10.1103/PhysRevX.13.041040

28.8 Toodle-Pip!

Congratulations! You have reached the end of the book. I wish I had a better ending for you. I wish I could offer a way to unify GR with QM. But I cannot. If I could, I would be in Stockholm picking up a nice shiny gold trophy. Perhaps this book will inspire some of you to take up the challenge. Perhaps one of you will find the way forwards. If you do, don't forget to mention me in your Nobel acceptance speech!

Good luck in your studies. Some additional resources for the topics in this chapter are listed in Box 28.5. If you enjoyed the book and learned a few interesting things, please remember to leave a review in order to encourage others and, yes, positive reviews also mean a lot to me, your lowly author.

Toodle-pip! (an old English way of saying goodbye).

Index

Untangling General Relativity: The Intuitive Self-Study Guide, First Edition. Simon Sherwood.
© 2026 John Wiley & Sons Ltd. Published 2026 by John Wiley & Sons Ltd.